ADVANCES IN RESTING-STATE FUNCTIONAL MRI

Neuroimaging Methods and Applications Series

Series Editor

Peter Bandettini, PhD

Chief, Section on Functional Imaging Methods, Laboratory of Brain and Cognition
Director, Functional MRI Core Facility
National Institute of Mental Health, National Institutes of Health, Bethesda, MD, United States

About the Series:

Neuroimaging Methods and Applications is a book series that embodies the collective expertise and experience of the neuroimaging community. The books in this series cover neuroimaging approaches with an emphasis on fMRI. They include the latest insights and practical information about instrumentation, acquisition methods, processing, multimodal integration, and both clinical and neuroscience research applications. They also include such topics as neuromodulation and computational modeling. The series is intended to provide useful information, insight, and perspective to neuroimaging researchers and clinicians at all levels—from undergraduates to full professors. The topics and content are intended to complement a wide range of reader experience and expertise, including neuroscience, psychology, psychiatry, neurology, engineering, physics, statistics, mathematics, computer science, and physiology.

Titles published:

Advances in Resting-State Functional MRI—Edited by Jean Chen and Catie Chang
Connectome Analysis: Characterization, Methods, and Applications—Edited by Markus D. Schirmer, Tomoki Arichi, and Ai Wern Chung

Visit the Series webpage at https://www.elsevier.com/catalog/all/all/all/advances-in-magneticresonance-technology-and-applications

ADVANCES IN RESTING-STATE FUNCTIONAL MRI

Methods, Interpretation, and Applications

Edited by

JEAN CHEN
Rotman Research Institute, Baycrest, Toronto, ON, Canada; Medical Biophysics, University of Toronto, Toronto, ON, Canada

CATIE CHANG
Vanderbilt University, Nashville, TN, United States

ELSEVIER

ACADEMIC PRESS
An imprint of Elsevier

Academic Press is an imprint of Elsevier
125 London Wall, London EC2Y 5AS, United Kingdom
525 B Street, Suite 1650, San Diego, CA 92101, United States
50 Hampshire Street, 5th Floor, Cambridge, MA 02139, United States
The Boulevard, Langford Lane, Kidlington, Oxford OX5 1GB, United Kingdom

Notices
Knowledge and best practice in this field are constantly changing. As new research and experience broaden our understanding, changes in research methods, professional practices, or medical treatment may become necessary.

Practitioners and researchers must always rely on their own experience and knowledge in evaluating and using any information, methods, compounds, or experiments described herein. In using such information or methods they should be mindful of their own safety and the safety of others, including parties for whom they have a professional responsibility.

To the fullest extent of the law, neither the Publisher nor the authors, contributors, or editors, assume any liability for any injury and/or damage to persons or property as a matter of products liability, negligence or otherwise, or from any use or operation of any methods, products, instructions, or ideas contained in the material herein.

ISBN 978-0-323-91688-2

For information on all Academic Press publications
visit our website at https://www.elsevier.com/books-and-journals

Publisher: Mara Conner
Acquisitions Editor: Tim Pitts
Editorial Project Manager: Zsereena Rose Mampusti
Production Project Manager: Prasanna Kalyanaraman
Cover Designer: Greg Harris

Typeset by STRAIVE, India

Working together
to grow libraries in
developing countries

www.elsevier.com • www.bookaid.org

Contents

Contributors

Rui Duarte Armindo Department of Neuroradiology, Hospital Beatriz Ângelo, Loures—Lisbon, Portugal

Aliza Ayaz Department of Psychiatry, Weill Cornell Medicine, New York, NY, United States

Ana Cecilia Saavedra Bazan Department of Biomedical Engineering, University of Michigan, Ann Arbor, MI, United States

Eyal Bergmann Department of Neuroscience, Mortimer B. Zuckerman Mind Brain Behavior Institute, Columbia University, New York, NY, United States

Rasmus Birn Department of Psychiatry, University of Wisconsin - Madison, Madison, WI, United States

Bharat B. Biswal Department of Biomedical Engineering, New Jersey Institute of Technology, Newark, NJ, United States

Molly Bright Department of Physical Therapy and Human Movement Sciences, Feinberg School of Medicine, Northwestern University, Chicago; Department of Biomedical Engineering, McCormick School of Engineering, Northwestern University, Evanston, IL, United States

Jiayue Cao Department of Biomedical Engineering, University of Michigan, Ann Arbor, MI, United States

Donna Y. Chen Department of Biomedical Engineering, New Jersey Institute of Technology, Newark, NJ, United States

Mark Chiew Sunnybrook Research Institute; Department of Medical Biophysics, University of Toronto, Toronto, ON, Canada

Sarah Genon Institute of Neuroscience and Medicine, Brain and Behaviour (INM-7), Research Center Jülich, Jülich; Institute for Systems Neuroscience, Medical Faculty, Heinrich-Heine University Düsseldorf, Düsseldorf, Germany

Alessandro Gozzi Functional Neuroimaging Laboratory, Center for Neuroscience and Cognitive systems, Istituto Italiano di Tecnologia, Rovereto, Italy

Daniel Gutierrez-Barragan Functional Neuroimaging Laboratory, Center for Neuroscience and Cognitive systems, Istituto Italiano di Tecnologia, Rovereto, Italy

Elizabeth De Guzman Functional Neuroimaging Laboratory, Center for Neuroscience and Cognitive systems, Istituto Italiano di Tecnologia, Rovereto, Italy

Lia M. Hocke McLean Hospital, Harvard Medical School, Belmont, MA, United States

Yuanyuan Jiang Athinoula A. Martinos Center for Biomedical Imaging, Massachusetts General Hospital and Harvard Medical School, Charlestown, MA, United States

Itamar Kahn Department of Neuroscience, Mortimer B. Zuckerman Mind Brain Behavior Institute, Columbia University, New York, NY, United States

Shella Keilholz Biomedical Engineering, Emory University/Georgia Tech, Atlanta, GA, United States

Vesa Kiviniemi Oulu Functional NeuroImaging, Health Science and Technology (HST), Oulu University; Diagnostic Imaging, Medical Research Center (MRC), Oulu University Hospital, Oulu, Finland

Hsin-Ju Lee Sunnybrook Research Institute; Department of Medical Biophysics, University of Toronto, Toronto, ON, Canada

Jingwei Li Institute of Neuroscience and Medicine, Brain and Behaviour (INM-7), Research Center Jülich, Jülich, Germany

Fa-Hsuan Lin Sunnybrook Research Institute; Department of Medical Biophysics, University of Toronto, Toronto, ON, Canada

Peiying Liu Department of Diagnostic Radiology & Nuclear Medicine, University of Maryland School of Medicine, Baltimore, MD, United States

Thomas T. Liu Center for Functional MRI; Department of Radiology; Department of Psychiatry; Department of Bioengineering, University of California San Diego, La Jolla, CA, United States

Zhongming Liu Department of Biomedical Engineering; Department of Electrical Engineering and Computer Science, University of Michigan, Ann Arbor, MI, United States

Weitao Man Athinoula A. Martinos Center for Biomedical Imaging, Massachusetts General Hospital and Harvard Medical School, Charlestown, MA, United States

Marco Pagani Functional Neuroimaging Laboratory, Center for Neuroscience and Cognitive systems, Istituto Italiano di Tecnologia, Rovereto, Italy; Autism Center, Child Mind Institute, New York, NY, United States

Jonathan D. Power Department of Psychiatry, Weill Cornell Medicine, New York, NY, United States

Yunjie Tong Weldon School of Biomedical Engineering, Purdue University, West Lafayette, IN, United States

Xiaokai Wang Department of Biomedical Engineering, University of Michigan, Ann Arbor, MI, United States

Chao-Gan Yan CAS Key Laboratory of Behavioral Science, Institute of Psychology, Beijing, China

Xin Yu Athinoula A. Martinos Center for Biomedical Imaging, Massachusetts General Hospital and Harvard Medical School, Charlestown, MA, United States

Greg Zaharchuk Department of Radiology, Stanford University, Stanford, CA, United States

Xiaoqing Alice Zhou Athinoula A. Martinos Center for Biomedical Imaging, Massachusetts General Hospital and Harvard Medical School, Charlestown, MA, United States

Shasha Zhu Department of Psychiatry, Weill Cornell Medicine, New York, NY, United States

Introduction to resting-state fMRI

Donna Y. Chen and Bharat B. Biswal

Department of Biomedical Engineering, New Jersey Institute of Technology, Newark, NJ, United States

Introduction

Resting-state functional magnetic resonance imaging (fMRI) has gained much attention in the fMRI community since its first use. Since 1995, the number of publications with the keywords "resting-state fMRI" has increased exponentially (Biswal, 2012). It is now used in many different fields of neuropsychiatric research, allowing us to study the brain in an unconstrained state. Initial recordings of fMRI data during the "resting-state" were discarded as noise; however, this "noise" appeared to be correlated. This would later be known as *resting-state functional connectivity* (RSFC), in which distinct areas of the brain are correlated across time. A brief introduction to resting-state fMRI is given here, and various aspects of resting-state fMRI are provided in this chapter, including experimental design, analysis methods, limitations, and future directions. These aspects will be covered in detail in the subsequent chapters of this book.

First case of resting-state fMRI

When functional magnetic resonance imaging (fMRI) was first used in the early 1990s with blood-oxygenation-level-dependent (BOLD) contrast, many researchers used a task design in which the participants being scanned were asked to perform a specific task pertinent to the research question. In between periods of the task, i.e., a finger-tapping motor task or a checkerboard visual task, there would be a period of rest. During these periods of rest, the participant would be instructed to not perform any task and typically keep their gaze fixed at a crosshair on a blank screen. These blocks of rest were often used as a "baseline" condition in block and event-related task-based studies; however, it was not entirely clear what occurred during the baseline state. The periods of rest were also regarded as "noise" and the signals were often not analyzed further or simply discarded. However, in 1995, Biswal and colleagues found that this "noise" contained meaningful

Advances in Resting-State Functional MRI. https://doi.org/10.1016/B978-0-323-91688-2.00011-4

information which was temporally correlated between spatially distinct brain regions (Biswal et al., 1995).

At the time, it was known that task-induced neuronal activity was correlated with changes in both regional cerebral blood flow and blood oxygenation; however, it was not clear what occurred during the resting-state, when one was not engaged in any particular task. Biswal and colleagues performed seed-based correlation analysis on data from 11 healthy adults who performed a resting-state task and a bilateral finger-tapping task (Biswal et al., 1995). The seed was a region in the sensorimotor cortex and activity in this region was correlated with the activity in all the other brain regions. The spatial correlation pattern from the resting-state data was found to be very similar to the activation response from the task-based data. Regions of the sensorimotor cortex showed significant functional connectivity in the resting state that were not due to "imagined" motor tasks, since some of the participants were not told beforehand of any motor task.

The sensorimotor network was one of the first resting-state networks (RSNs) discovered; RSNs are thought to be composed of distinct brain regions that share a common function. Subsequent studies confirmed similar correlation patterns between different brain regions' low-frequency fluctuations in the resting state and would show resting-state networks in the auditory, visual, and higher-order cognitive areas (Lowe et al., 1998; Biswal et al., 1997a; Xiong et al., 1999; Stein et al., 2000; Cordes et al., 2001; Hampson et al., 2002; Greicius et al., 2003). These RSNs consist of regions that are spatially distinct from each other yet are temporally correlated in the low-frequency range from 0.01 to 0.1 Hz. One particular resting-state network that is widely studied is the default-mode network (DMN), which has shown decreased activity during goal-directed tasks. The DMN is active during mind-wandering, self-referential thinking, or daydreaming (Raichle et al., 2001; Raichle, 2015). Resting-state signals reveal information about the brain's intrinsic activity, which is never quite at rest.

Experimental design of resting-state fMRI

The experimental design of resting-state fMRI is relatively simple to implement. Participants are typically asked to be in the eyes-fixed condition, in which they keep their eyes open, and gaze fixed at a cross-hair on a blank computer screen. However, studies have also used paradigms in which the participants have their eyes closed. Having one's eyes closed may increase the chances of falling asleep during the experiment, and the state of alertness is difficult to ascertain. Yan and colleagues showed higher connectivity in the DMN during the eyes-open resting-state condition compared to the eyes-closed and eyes-fixed conditions (Yan et al., 2009). This suggests greater mind-wandering during the eyes-open condition with non-specific visual information gathering. Patriat and colleagues reported that in the eyes-closed condition, there was higher functional connectivity in the auditory network compared to the eyes-open or eyes-fixed conditions (Patriat et al., 2013). They also found greater reliability in the eyes-fixed conditions for the default-mode, attention, and auditory networks, while the eyes-open condition showed greater reliability in the visual networks (Patriat et al., 2013). Therefore, based on the research question or clinical population, one may choose one resting-state paradigm over the other. It is important to describe the type of resting-state condition implemented, since there are differences in the resting-state functional connectivity patterns between the eyes-open, eyes-open with object fixation, and eyes-closed states (Yan et al., 2009; Patriat et al., 2013; Van Dijk et al., 2010; Liu et al., 2013; Agcaoglu et al., 2019; Wei et al., 2018).

The typical resting-state fMRI scan ranges from 5 to 10 min. Birn and colleagues have shown that increasing the resting-state scan lengths from 5 min to up to 13 min can improve the reliability of resting-state functional connectivity, especially for scans taken during the same session; however, the reliability plateaued at around 12–16 min between different sessions (Birn et al., 2013). Increasing the duration of the resting-state scan may increase the possibility that the participant will become fatigued, bored, or fall asleep. Therefore, it is important that the scan duration is not too long. Overall, implementation of the resting-state paradigm is fairly simple compared to task-based fMRI studies.

Origin of resting-state fMRI signals

It is yet not clear the extent to which spontaneous low-frequency resting-state signals are contributed by neuronal or nonneuronal hemodynamic activity, which makes the interpretation of RSFC difficult. Different hypotheses of the origin of the resting-state signal may help explain RSFC, such as the *biophysical-origin hypothesis* and the *cognitive-origin hypothesis* (Chen et al., 2020). The biophysical-origin hypothesis suggests that the resting-state fMRI signal is affected primarily by neurovascular physiology. To study the relationship between regional oxygen metabolism and blood flow, Cooper and colleagues used gold electrodes and microthermistors that were implanted in patients to measure the local regional oxygen availability (O_2a) and relative measures of local cerebral blood flow (CBF) (Cooper et al., 1966). They found fluctuating O_2a at a frequency of about 6 waves per minute in all electrodes, which is similar to the frequency of resting-state signals observed in resting-state fMRI (Biswal et al., 1995).

The fluctuations were also found to be different in different brain regions, which is consistent with resting-state fMRI findings of frequency-specific resting-state networks (Wu et al., 2008). Winder and colleagues found that the spontaneous fluctuations persisted even when local neural spiking and glutamatergic input were blocked, which supports the nonneuronal origin of spontaneous resting-state signals (Winder et al., 2017). The vascular origins of rs-fMRI are further discussed in Chapters 6 and 7. Furthermore, the low-frequency correlations have shown to be a general feature of neural systems, which are also present in the spinal cord and white matter in nonhuman primates (Chen et al., 2017a).

The cognitive-origin hypothesis suggests that the resting-state signals are more neuronal in nature, in contrast to being influenced primarily by vasculature. Biswal and colleagues have shown that the BOLD resting-state functional connectivity maps coincide more with task-activation maps rather than the blood-flow signal maps, supporting the cognitive-origin hypothesis (Biswal et al., 1997a). Furthermore, the low-frequency resting-state signals have been shown to be diminished reversibly through a hypercapnic-induced stimulus, which suggests that these signals are similar to and coupled with neuronal activity (Biswal et al., 1997b). Hypercapnia, which induces reduced neural activity, results in vascular dilation due to increased CO_2, and thus reduces the agents responsible for coupling neuronal activity to local cerebral hemodynamics. The study found reduced low-frequency fluctuations, showing a potential coupling of the resting-state signal with neuronal activity. Lu and colleagues have referred to this as the "neurocentric" model, which describes the ongoing neuronal processes in the spontaneous resting-state fluctuations and supports the neuronal origin of the resting-state signals (Lu et al., 2019). The neuronal constituents of rs-fMRI are further discussed in Chapters 3, 8, and 9. However, these mechanisms may not apply to all brain regions. Jaime et al. (2019) and Winder et al. (2017) found weak local field potential (LFP)-BOLD correlation, which suggests that the "neurocentric" model may have to be updated, or that "resting-state" is more complex than previously defined, including other variables such as behavior, and both neural and nonneural activities (Winder et al., 2017; Jaime et al., 2019).

Applications of resting-state fMRI

Resting-state fMRI (rs-fMRI) has shown us that in order to obtain meaningful brain activity information, tasks are not necessarily needed. This has opened up many doors in fMRI research since certain clinical populations may have difficulty performing specific tasks. For example, for people with Alzheimer's disease, carrying out a task involving memory and attention may be difficult due to the decline in cognitive capacity and the difficulty in understanding the task instructions. Studies have found altered RSFC in individuals with Alzheimer's disease, such as reduced connectivity in the DMN (Sorg et al., 2007; Binnewijzend et al., 2012), increased connectivity in the frontal-attention network as a compensatory mechanism (Agosta et al., 2012), and functional dysconnectivity between anterior-posterior regions with greater within-lobe functional connectivity (Wang et al., 2007).

Resting-state fMRI is also helpful for studying pediatric populations since task comprehension is lacking particularly for infants or toddlers. This allows researchers to study the pediatric population with greater ease and has been used to detect certain disorders such as attention-deficit hyperactivity disorder (ADHD) (Zang et al., 2007; Tian et al., 2008) and autism spectrum disorder (ASD) in children (Hull et al., 2017; Weng et al., 2010). Resting-state fMRI has also helped to contribute to the development of brain growth charts to evaluate the typical trajectories of children's neuroimaging data and predict whether one would be at risk of developing a neuropsychiatric disorder (Kessler et al., 2016). In addition to charting normal brain growth, rs-fMRI has been used to predict individual brain maturity with support vector-machine-based multivariate pattern analysis (Dosenbach et al., 2010). Resting-state fMRI has also been used in the clinic for presurgical planning for patients with epilepsy (Boerwinkle et al., 2020; Kollndorfer et al., 2013), as will be discussed in Chapter 14. However, studying pediatric populations also poses limitations due to the large amount of head movement that is typically found in children's fMRI scans. Power and colleagues have shown that head motion artifacts could yield spurious results in which shorter range brain correlations are increased while long-distance correlations are decreased (Power et al., 2012, 2015).

Resting-state fMRI provides many advantages over conventional task-based fMRI due to the ease of experimental design and its application to a wide range of clinical populations. Notably, rs-fMRI allows researchers to study the whole-brain, unconstrained by any particular task. This allows researchers to investigate the interactions between all the brain regions comprehensively in contrast to focusing on pre-defined regions of interest.

Preprocessing of resting-state fMRI data

After collecting rs-fMRI data, one has to preprocess the data to check for motion artifacts and prepare the data for statistical analyses (see also Chapters 6 and 10). The preprocessing steps for rs-fMRI data are similar to that of task-based fMRI studies, except a bandpass filter (passband from 0.01 to 0.1 Hz) is commonly used in rs-fMRI data, which are primarily composed of low-frequency signals (Biswal et al., 1995). However, there is still ongoing research in this area, particularly in characterizing higher-frequency resting-state networks (Wu et al., 2008; Boubela et al., 2013; Chen and Glover, 2015; Chen et al., 2017b; Gohel and Biswal, 2015). Lee and colleagues used magnetic resonance encephalography (MREG), which can resolve signals at higher frequencies, to observe resting-state signal fluctuations in the range from 0.5 to 0.8 Hz and found stable visual and motor networks (Lee et al., 2013) (also see Chapter 6). The bandpass filtering step also helps to remove unwanted respiratory or cardiac signals; however, it is important to check that the cardiac cycle does not encroach into the low-frequency signals through aliasing. One way to do this is by collecting independent measures of cardiac and respiratory signals using a pulse oximeter and pneumatic belt, respectively. Global signal regression is another method used for resting-state data, which regresses out the average whole-brain global signal from each voxel's data since the global signal may include physiological confounds; however, its usefulness has been debated and depends on the research study or quality of data (Murphy and Fox, 2017; Saad et al., 2012). Common preprocessing strategies include realignment, co-registration, segmentation, normalization, bandpass filtering, and smoothing (Esteban et al., 2019). Some freely accessible rs-fMRI data sets also have minimally preprocessed versions, such as the Human Connectome Project (HCP), using preprocessing pipelines consisting of realignment, nuisance regression, registration using FreeSurfer, and concatenation of all transforms for each registration and distortion process into a single transformation (Glasser et al., 2013). Various programs and software exist to preprocess rs-fMRI data, such as analysis of functional neuroimages (AFNI) (Cox, 1996), statistical parametric mapping 12 (SPM12) (Penny et al., 2011), and FMRIB Software Library (FSL) (Jenkinson et al., 2012).

Resting-state fMRI analysis

There are various methods used to analyze rs-fMRI data following preprocessing. These can be categorized into regional (or local) resting-state metrics and global (whole-brain) RSFC metrics. Regional resting-state metrics include amplitude of low-frequency fluctuations (ALFF), fractional ALFF (fALFF), and regional homogeneity (ReHo). Whole-brain RSFC metrics include seed-based correlation, data-driven methods such as independent component analysis (ICA) or principal component analysis (PCA), and graph theory analysis methods. Additionally, recent advances in machine learning and deep learning algorithms have found applications for rs-fMRI, as large data sets become more accessible to the brain imaging community. These analysis methods are not limited to resting-state data, as some methods initially used for resting-state data have been applied to data sets containing tasks and naturalistic conditions. Originally, Pearson's correlation analysis was borrowed from task-based fMRI but many methods from rs-fMRI are now in turn being applied to task and naturalistic data, such as graph-theory and dynamic functional-connectivity methods.

Regional fMRI metrics

ALFF is calculated by taking the preprocessed, bandpass-filtered rs-fMRI data and using a fast Fourier transform on the data to transform it into the frequency domain to obtain the power spectrum (Zang et al., 2007). The square root of power for each given frequency is then calculated and averaged to obtain the ALFF value. Since this is a regional rs-fMRI metric, an ALFF value is calculated for each voxel in the brain regions of interest. In contrast to ALFF, fALFF is taken as the ratio of the power spectrum in the low-frequency range (0.01–0.08 Hz) to the power spectrum in the entire frequency range of the data (Zou et al., 2008). fALFF is an improvement over ALFF since it can suppress the nonspecific signal components in the rs-fMRI data and reduce physiological noise in cisterns, ventricles, and areas near large blood vessels. This allows for improved sensitivity and specificity, particularly in brain areas within the DMN that are known to have high levels of spontaneous fluctuations (Greicius et al., 2003; Zou et al., 2008). Another regional resting-state metric is ReHo, which calculates the similarity of a voxel's activity to that of neighboring voxels by using Kendall's coefficient of concordance (Zang et al., 2004). This is repeated for all the voxels of interest and a cluster size for the number of neighboring voxels needs to be specified to calculate the ReHo value.

In contrast to voxel-wise resting-state metrics, whole-brain resting-state metrics give us information from multiple voxels and can yield information about the relationship between different brain regions. One global rs-fMRI method is seed-based correlation, in which a "seed" or brain region is chosen. The rs-fMRI signal from the seed region is then correlated with those of all the other regions of the brain to see which brain areas are significantly correlated with the seed region of interest (Biswal et al., 1995). This yields a functional connectivity matrix for each subject. However, dynamic functional connectivity methods (further described in Chapter 13) also exist in contrast to the static functional connectivity methods mentioned above (Preti et al., 2017). In sliding-window correlation analysis, one would get multiple functional connectivity matrices rather than just one since the time-series data is broken into multiple windows of time prior to correlating them (Hutchison et al., 2013; Chang and Glover, 2010; Sakoglu et al., 2010).

For this method, one would need to choose the time for each window length for correlation and the time-step to take when "sliding a window" through the time-series data. Other dynamic functional connectivity methods include wavelet transform coherence (Chang and Glover, 2010), tapered sliding-window (Allen et al., 2014), dynamic conditional correlation (Lindquist et al., 2014), single-volume co-activation patterns (Liu and Duyn, 2013), dynamic phase synchronization (Glerean et al., 2012; Ponce-Alvarez et al., 2015), and state-space models (Eavani et al., 2013; Taghia et al., 2017; Yaesoubi et al., 2018). Dynamic functional connectivity can also show different "brain states" of activity since the brain is not static and changes over time. These transient states have been studied in neuropsychiatric research such as in schizophrenia (Sakoglu et al., 2010; Damaraju et al., 2014), major depressive disorder (Demirtas et al., 2016; Zhi et al., 2018; Wu et al., 2019), and attention-deficit/hyperactivity disorder (Sun et al., 2021; Agoalikum et al., 2021).

ICA and PCA

Data-driven methods used in rs-fMRI include independent component analysis (ICA) and principal component analysis (PCA), with ICA more commonly used than PCA when obtaining RSNs. These methods are dimension-reduction techniques that help to describe the rs-fMRI data by a small number of components. ICA decomposes the data into a set of independent time-courses and spatial maps by assuming non-Gaussian distribution (Beckmann et al., 2005; McKeown et al., 1998; Biswal and Ulmer, 1999). PCA on the other hand captures variability in the signal, commonly by using singular value decomposition. The principal components are ranked based on how much variability of the data they explain. Using PCA on rs-fMRI data can reveal intrinsic structures in the data or "eigenconnectivities" that have been interpreted as the building blocks of dynamic functional connectivity (Leonardi et al., 2013). For PCA, the components are orthogonal to each other, while in ICA, the components are independent of each other. Since orthogonality does not necessarily imply independence, one might choose ICA over PCA and vice versa depending on one's research question. For example, if the resting-state or task fMRI changes are only a small part of the total signal variance, then the principal components may reveal little information about brain activation, since most of the data are represented in the first few principal components. In contrast, ICA may reveal more information about brain activation since the data can be represented by multiple independent components (McKeown et al., 1998, 2003).

Graph theory

Graph theory is also commonly applied to rs-fMRI data such that the brain regions are defined as nodes, while the edges represent the connectivity between the nodes (Bullmore and Sporns, 2009; Bullmore and Bassett, 2011). In this manner, researchers can evaluate measures of information flow and efficiency in the brain. Different measures of graph theory used in rs-fMRI include the path length, degree, clustering coefficient, and modularity. The path length is the number of edges crossed from one node to another node, while the degree refers to the number of edges connecting one node to the rest of the graph network. Small-world characteristics of the brain denote efficiency in brain organization in which the brain contains "small-world" features such as short paths across the network with a large number of clusters of neighboring nodes, lying in between a regular and completely random network (Watts and Strogatz, 1998). These graph theoretical measures of brain network activity have shown differences between clinical populations, allowing us to better understand the brain in various disease states (Bassett and Bullmore, 2009). For example, in the case of schizophrenia, small-world properties were found to be disrupted compared to healthy controls, and the clustering and small-world characteristics were found to be inversely correlated with illness duration (Liu et al., 2008). Salvador and colleagues found large-scale systems organization with small-world topology in low-frequency rs-fMRI data (Salvador et al., 2005).

There are potential challenges in interpreting certain graph metrics for RSNs. In some cases, graph theory metrics may yield contradictory findings between functional and structural networks, despite the close relationship between the two (Farahani et al., 2019). For example, using functional RSNs, van den Heuvel and colleagues found reduced local efficiency and segregation in individuals with schizophrenia; however, with structural networks, increased segregation and reduced global efficiency were found (van den Heuvel and Fornito, 2014). The structural studies suggest prominent reduction of long-distance association pathways in schizophrenia while functional studies tend to show an excess of long-distance functional coupling between brain regions (van den Heuvel et al., 2012, 2013; Alexander-Bloch et al., 2013). Some possible reasons for these contradictory results include the usage of different preprocessing steps, methods to characterize functional connectivity, methods for node definition, dealing with negative correlation values and thresholding procedures (van den Heuvel and Fornito, 2014).

Machine learning and deep learning

Advances in big-data rs-fMRI have allowed the expanded application of machine-learning and deep-learning algorithms to extract meaningful information and make predictions regarding brain and behavioral characteristics. Many such studies have been able to accurately distinguish between those with neuropsychiatric disorders and those without (Khosla et al., 2019). Khazaee and colleagues combined graph theory and a support vector machine to automatically identify healthy controls, mild cognitive impairment, and Alzheimer's disease groups, and were able to achieve an accuracy of 88.4% (Khazaee et al., 2016). For a more in-depth review of machine learning using rs-fMRI data, see Khosla et al. (2019). It is important to note that as larger rs-fMRI data sets are used, the prediction accuracy of the machine learning models may decrease due to the tendency to over-fit smaller data sets and the greater heterogeneity in larger data samples. Therefore, it is important to consider larger rs-fMRI data sets in the application of machine learning models to prevent overfitting of the data and yield more robust models.

Challenges and limitations of rs-fMRI

Despite the advantages of rs-fMRI, some challenges and limitations still exist. Head motion and other artifacts are challenging since there is no model of expected temporal effects, i.e., a reference vector. It is understood that motion artifacts alter the BOLD time series data, but the exact effects are not clear and is likely dynamic (Satterthwaite et al., 2013). In particular, Frew and colleagues note that due to the large head-to-body ratio of children, there is already a degree of neck flexion when children are in the scanner, thus a nodding motion is commonly seen in children as a motion artifact (Frew et al., 2022). Head motion artifacts are also present across different age groups and clinical populations (Power et al., 2014; Goto et al., 2016). These artifacts may preclude an accurate representation of brain activity during the resting state. Increased head motion or physiological artifacts may also reduce reliability of the data (Parkes et al., 2018; Yan et al., 2013). To address these artifacts, head-motion correction methods have been applied as well as more stringent and careful rs-fMRI methodological protocols instructing the participants to stay as still as possible. Combinations of different head-motion correction algorithms have also been implemented, which may be more powerful than using only single motion artifact correction algorithms (Maknojia et al., 2019).

The interpretation of rs-fMRI data still poses a challenge due to the lack of research studies on the neurobiological explanation of the rs-fMRI signal. Furthermore, it may be difficult to keep a participant awake during the whole duration of the resting-state scan. It is difficult to know what the participant is thinking about during a resting-state scan. Alternative methods such as movie-watching may be more engaging due to the visual stimuli presented to the viewer. Movie-watching has been shown to yield lower levels of head-motion compared to resting-state paradigms in children, particularly during dialogue-heavy scenes with close camera angles (Frew et al., 2022; Vanderwal et al., 2015, 2019). In contrast to task-based fMRI studies, movie-watching allows some flexibility in terms of mind-wandering or intrinsic processes; compared to rs-fMRI, it allows one to be more alert, engaged, and less drowsy. However, it is not truly unconstrained as rs-fMRI is. Depending on the research question, one may want something in between the resting-state and task-state paradigm in terms of mind-wandering. Since there have been a few studies showing that movie watching results in less head motion compared to the resting state (Frew et al., 2022; Vanderwal et al., 2015, 2019; Finn and Bandettini, 2021), it may be preferred in scenarios where children are involved, or the research focus involves eye-tracking. Functional connectome *"fingerprinting"* studies (relying on the finding that each individual's functional connectome is unique yet stable within individuals) have also shown that movie-watching performs better than resting state in terms of predicting individuals' behavioral traits in cognition and emotion (Finn and Bandettini, 2021).

It is also difficult to study a specific brain region or function in depth with rs-fMRI. For example, if a researcher is focused on the auditory network, then a task that focuses on an auditory stimulus may provide greater information about auditory function in comparison to a resting-state paradigm. In this case, task-based fMRI may be more suitable than rs-fMRI, for its specificity of brain function. Additionally, the neural interaction of RSFC with behavior is not as well understood as the interaction between task-based brain activity and behavior. This poses a challenge for studies that attempt to predict behavior from brain activity, since the neural underpinning is not well understood.

Future of rs-fMRI

With the increase in rs-fMRI studies, the number of open-source data sets has also increased. Resting-state fMRI allows for greater ease of the sharing of open-source data sets, since the experimental design is very similar across research groups. Participants typically have their eyes open and stare at a cross-hair on a computer screen. Due to this ease of paradigm, collecting data from multiple fMRI centers is greatly facilitated. In 2010, Biswal and colleagues showed that rs-fMRI could be applied to discovery science and created the 1000 Functional Connectomes Project, analogous to the 1000 Genomes Project (Biswal et al., 2010). Today, open-source rs-fMRI data sets can be found in Neuroimaging Tools and Resources Collaboratory (NITRC), Open fMRI, Open Neuro, etc. Researchers in the field of rs-fMRI are encouraged to upload their data onto open-source online repositories for access by other researchers and this contributes to big-data rs-fMRI.

Some future directions for rs-fMRI include single-subject analysis, developing resting-state biomarkers, and using multimodal imaging techniques in combination with fMRI such as electroencephalography (EEG), positron-emission tomography (PET), and transcranial magnetic stimulation (TMS). These areas will greatly facilitate the use of rs-fMRI in clinical settings and allow us to better understand the resting-state signal. Single-subject analysis would allow for rs-fMRI to have greater clinical applicability (Stephan et al., 2017; Arbabshirani et al., 2017), since clinical care is delivered and assessed individually. In contrast, most research studies depend on large sample sizes of different groups to achieve reliable conclusions. Therefore, using normative data to compare an individual's resting-state data to that of a group's would allow clinicians to better understand RSFC deviations in a shorter amount of time. For clinical usage, further development in rs-fMRI toolboxes for single-subject analyses is needed (O'Connor and Zeffiro, 2019). Additionally, it is not yet clear which rs-fMRI metric is best to use for different patient populations. The development of resting-state biomarkers is an ongoing effort that would allow the testing of different treatments for efficacy in specific patient populations. This can also further be enhanced by multimodal imaging techniques such as EEG-fMRI, PET-fMRI, and TMS-fMRI.

Conclusion

Throughout the years, rs-fMRI has evolved from being considered "noise" to now being used to study a wide range of clinical populations. Despite some limitations, rs-fMRI has been used across different fields of neuroscience research due to its ease of implementation and versatility in understanding the brain in its intrinsic state. Advancements in the field of fMRI methodology and equipment will further advance the field of rs-fMRI. Higher resolution fMRI images also allow for greater specificity and better interpretation of resting-state data. The increase of open-source resting-state data will also help in machine learning and deep learning fields for the classification of different neuropsychiatric disorders.

References

Agcaoglu, O., Wilson, T.W., Wang, Y.P., Stephen, J., Calhoun, V.D., 2019. Resting state connectivity differences in eyes open versus eyes closed conditions. Hum. Brain Mapp. 40 (8), 2488–2498. Epub 20190205 https://doi.org/10.1002/hbm.24539. 30720907. PMCID: PMC6865559.

Agoalikum, E., Klugah-Brown, B., Yang, H., Wang, P., Varshney, S., Niu, B., et al., 2021. Differences in disrupted dynamic functional network connectivity among children, adolescents, and adults with attention deficit/hyperactivity disorder: a resting-state fMRI study. Front. Hum. Neurosci. 15, 697696. Epub 20211005 https://doi.org/10.3389/fnhum.2021.697696. 34675790. PMCID: PMC8523792.

Agosta, F., Pievani, M., Geroldi, C., Copetti, M., Frisoni, G.B., Filippi, M., 2012. Resting state fMRI in Alzheimer's disease: beyond the default mode network. Neurobiol. Aging 33 (8), 1564–1578. Epub 20110803 https://doi.org/10.1016/j.neurobiolaging.2011.06.007. 21813210.

Alexander-Bloch, A., Giedd, J.N., Bullmore, E., 2013. Imaging structural co-variance between human brain regions. Nat. Rev. Neurosci. 14 (5), 322–336. Epub 20130327 https://doi.org/10.1038/nrn3465. 23531697. PMCID: PMC4043276.

Allen, E.A., Damaraju, E., Plis, S.M., Erhardt, E.B., Eichele, T., Calhoun, V.D., 2014. Tracking whole-brain connectivity dynamics in the resting state. Cereb. Cortex 24 (3), 663–676. Epub 20121111 https://doi.org/10.1093/cercor/bhs352. 23146964. PMCID: PMC3920766.

Arbabshirani, M.R., Plis, S., Sui, J., Calhoun, V.D., 2017. Single subject prediction of brain disorders in neuroimaging: promises and pitfalls. NeuroImage 145 (Pt B), 137–165. Epub 20160321 https://doi.org/10.1016/j.neuroimage.2016.02.079. 27012503. PMCID: PMC5031516.

Bassett, D.S., Bullmore, E.T., 2009. Human brain networks in health and disease. Curr. Opin. Neurol. 22 (4), 340–347. https://doi.org/10.1097/WCO.0b013e32832d93dd. 19494774. PMCID: PMC2902726.

Beckmann, C.F., DeLuca, M., Devlin, J.T., Smith, S.M., 2005. Investigations into resting-state connectivity using independent component analysis. Philos. Trans. R. Soc. Lond. Ser. B Biol. Sci. 360 (1457), 1001–1013. https://doi.org/10.1098/rstb.2005.1634. 16087444. PMCID: PMC1854918.

Binnewijzend, M.A., Schoonheim, M.M., Sanz-Arigita, E., Wink, A.M., van der Flier, W.M., Tolboom, N., et al., 2012. Resting-state fMRI changes in Alzheimer's disease and mild cognitive impairment. Neurobiol. Aging 33 (9), 2018–2028. Epub 20110820 https://doi.org/10.1016/j.neurobiolaging.2011.07.003. 21862179.

Birn, R.M., Molloy, E.K., Patriat, R., Parker, T., Meier, T.B., Kirk, G.R., et al., 2013. The effect of scan length on the reliability of resting-state fMRI connectivity estimates. NeuroImage 83, 550–558. Epub 2013/06/12 https://doi.org/10.1016/j.neuroimage.2013.05.099. 23747458. PMCID: PMC4104183.

Biswal, B.B., 2012. Resting state fMRI: a personal history. NeuroImage 62 (2), 938–944. Epub 2012/02/14 https://doi.org/10.1016/j.neuroimage.2012.01.090. 22326802.

Biswal, B.B., Ulmer, J.L., 1999. Blind source separation of multiple signal sources of fMRI data sets using independent component analysis. J. Comput. Assist. Tomogr. 23 (2), 265–271. https://doi.org/10.1097/00004728-199903000-00016. 10096335.

Biswal, B., Yetkin, F.Z., Haughton, V.M., Hyde, J.S., 1995. Functional connectivity in the motor cortex of resting human brain using echo-planar MRI. Magn. Reson. Med. 34 (4), 537–541. Epub 1995/10/01 https://doi.org/10.1002/mrm.1910340409. 8524021.

Biswal, B.B., Van Kylen, J., Hyde, J.S., 1997a. Simultaneous assessment of flow and BOLD signals in resting-state functional connectivity maps. NMR Biomed. 10 (4-5), 165–170. Epub 1997/06/01 https://doi.org/10.1002/(sici)1099-1492(199706/08)10:4/5<165::aid-nbm454>3.0.co;2-7. 9430343.

Biswal, B., Hudetz, A.G., Yetkin, F.Z., Haughton, V.M., Hyde, J.S., 1997b. Hypercapnia reversibly suppresses low-frequency fluctuations in the human motor cortex during rest using echo-planar MRI. J. Cereb. Blood Flow Metab. 17 (3), 301–308. https://doi.org/10.1097/00004647-199703000-00007. 9119903.

Biswal, B.B., Mennes, M., Zuo, X.N., Gohel, S., Kelly, C., Smith, S.M., et al., 2010. Toward discovery science of human brain function. Proc. Natl. Acad. Sci. U. S. A. 107 (10), 4734–4739. Epub 2010/02/24 https://doi.org/10.1073/pnas.0911855107. 20176931. PMCID: PMC2842060.

Boerwinkle, V.L., Mirea, L., Gaillard, W.D., Sussman, B.L., Larocque, D., Bonnell, A., et al., 2020. Resting-state functional MRI connectivity impact on epilepsy surgery plan and surgical candidacy: prospective clinical work. J. Neurosurg. Pediatr., 1–8. Epub 20200320 https://doi.org/10.3171/2020.1.PEDS19695. 32197251.

Boubela, R.N., Kalcher, K., Huf, W., Kronnerwetter, C., Filzmoser, P., Moser, E., 2013. Beyond noise: using temporal ICA to extract meaningful information from high-frequency fMRI signal fluctuations during rest. Front. Hum. Neurosci. 7, 168. Epub 20130501 https://doi.org/10.3389/fnhum.2013.00168. 23641208. PMCID: PMC3640215.

Bullmore, E.T., Bassett, D.S., 2011. Brain graphs: graphical models of the human brain connectome. Annu. Rev. Clin. Psychol. 7, 113–140. Epub 2010/12/07 https://doi.org/10.1146/annurev-clinpsy-040510-143934. 21128784.

Bullmore, E., Sporns, O., 2009. Complex brain networks: graph theoretical analysis of structural and functional systems. Nat. Rev. Neurosci. 10 (3), 186–198. Epub 20090204 https://doi.org/10.1038/nrn2575. 19190637.

Chang, C., Glover, G.H., 2010. Time-frequency dynamics of resting-state brain connectivity measured with fMRI. NeuroImage 50 (1), 81–98. Epub 20091216 https://doi.org/10.1016/j.neuroimage.2009.12.011. 20006716. PMCID: PMC2827259.

Chen, J.E., Glover, G.H., 2015. BOLD fractional contribution to resting-state functional connectivity above 0.1 Hz. NeuroImage 107, 207–218. Epub 20141212 https://doi.org/10.1016/j.neuroimage.2014.12.012. 25497686. PMCID: PMC4318656.

Chen, L.M., Yang, P.F., Wang, F., Mishra, A., Shi, Z., Wu, R., et al., 2017a. Biophysical and neural basis of resting state functional connectivity: evidence from non-human primates. Magn. Reson. Imaging. 39, 71–81. Epub 20170202 https://doi.org/10.1016/j.mri.2017.01.020. 28161319. PMCID: PMC5410395.

Chen, J.E., Jahanian, H., Glover, G.H., 2017b. Nuisance regression of high-frequency functional magnetic resonance imaging data: denoising can be noisy. Brain Connect. 7 (1), 13–24. Epub 20170105 https://doi.org/10.1089/brain.2016.0441. 27875902. PMCID: PMC5312601.

Chen, K., Azeez, A., Chen, D.Y., Biswal, B.B., 2020. Resting-state functional connectivity: signal origins and analytic methods. Neuroimaging Clin. N. Am. 30 (1), 15–23. Epub 2019/11/25 https://doi.org/10.1016/j.nic.2019.09.012. 31759568.

Cooper, R., Crow, H.J., Walter, W.G., Winter, A.L., 1966. Regional control of cerebral vascular reactivity and oxygen supply in man. Brain Res. 3 (2), 174–191. https://doi.org/10.1016/0006-8993(66)90075-8. 5971521.

Cordes, D., Haughton, V.M., Arfanakis, K., Carew, J.D., Turski, P.A., Moritz, C.H., et al., 2001. Frequencies contributing to functional connectivity in the cerebral cortex in "resting-state" data. AJNR Am. J. Neuroradiol. 22 (7), 1326–1333. Epub 2001/08/11 11498421.

Cox, R.W., 1996. AFNI: software for analysis and visualization of functional magnetic resonance neuroimages. Comput. Biomed. Res. 29 (3), 162–173. https://doi.org/10.1006/cbmr.1996.0014. 8812068.

Damaraju, E., Allen, E.A., Belger, A., Ford, J.M., McEwen, S., Mathalon, D.H., et al., 2014. Dynamic functional connectivity analysis reveals transient states of dysconnectivity in schizophrenia. Neuroimage Clin. 5, 298–308. Epub 20140724 https://doi.org/10.1016/j.nicl.2014.07.003. 25161896. PMCID: PMC4141977.

Demirtas, M., Tornador, C., Falcon, C., Lopez-Sola, M., Hernandez-Ribas, R., Pujol, J., et al., 2016. Dynamic functional connectivity reveals altered variability in functional connectivity among patients with major depressive disorder. Hum. Brain Mapp. 37 (8), 2918–2930. Epub 20160428 https://doi.org/10.1002/hbm.23215. 27120982. PMCID: PMC5074271.

Dosenbach, N.U., Nardos, B., Cohen, A.L., Fair, D.A., Power, J.D., Church, J.A., et al., 2010. Prediction of individual brain maturity using fMRI. Science 329 (5997), 1358–1361. Epub 2010/09/11 https://doi.org/10.1126/science.1194144. 20829489. PMCID: PMC3135376.

Eavani, H., Satterthwaite, T.D., Gur, R.E., Gur, R.C., Davatzikos, C., 2013. Unsupervised learning of functional network dynamics in resting state fMRI. Inf. Process. Med. Imaging 23, 426–437. https://doi.org/10.1007/978-3-642-38868-2_36. 24683988. PMCID: PMC3974209.

Esteban, O., Markiewicz, C.J., Blair, R.W., Moodie, C.A., Isik, A.I., Erramuzpe, A., et al., 2019. fMRIPrep: a robust preprocessing pipeline for functional MRI. Nat. Methods 16 (1), 111–116. Epub 2018/12/12 https://doi.org/10.1038/s41592-018-0235-4. 30532080. PMCID: PMC6319393.

Farahani, F.V., Karwowski, W., Lighthall, N.R., 2019. Application of graph theory for identifying connectivity patterns in human brain networks: a systematic review. Front. Neurosci. 13, 585. Epub 20190606 https://doi.org/10.3389/fnins.2019.00585. 31249501. PMCID: PMC6582769.

Finn, E.S., Bandettini, P.A., 2021. Movie-watching outperforms rest for functional connectivity-based prediction of behavior. NeuroImage 235, 117963. Epub 20210402 https://doi.org/10.1016/j.neuroimage.2021.117963. 33813007. PMCID: PMC8204673.

Frew, S., Samara, A., Shearer, H., Eilbott, J., Vanderwal, T., 2022. Getting the nod: pediatric head motion in a transdiagnostic sample during movie- and resting-state fMRI. PLoS One 17 (4), e0265112. Epub 20220414 https://doi.org/10.1371/journal.pone.0265112. 35421115. PMCID: PMC9009630.

Glasser, M.F., Sotiropoulos, S.N., Wilson, J.A., Coalson, T.S., Fischl, B., Andersson, J.L., et al., 2013. The minimal preprocessing pipelines for the human connectome project. NeuroImage 80, 105–124. Epub 20130511 https://doi.org/10.1016/j.neuroimage.2013.04.127. 23668970. PMCID: PMC3720813.

Glerean, E., Salmi, J., Lahnakoski, J.M., Jaaskelainen, I.P., Sams, M., 2012. Functional magnetic resonance imaging phase synchronization as a measure of dynamic functional connectivity. Brain Connect. 2 (2), 91–101. Epub 20120611 https://doi.org/10.1089/brain.2011.0068. 22559794. PMCID: PMC3624768.

Gohel, S.R., Biswal, B.B., 2015. Functional integration between brain regions at rest occurs in multiple-frequency bands. Brain Connect. 5 (1), 23–34. Epub 2014/04/08 https://doi.org/10.1089/brain.2013.0210. 24702246. PMCID: PMC4313418.

Goto, M., Abe, O., Miyati, T., Yamasue, H., Gomi, T., Takeda, T., 2016. Head motion and correction methods in resting-state functional MRI. Magn. Reson. Med. Sci. 15 (2), 178–186. Epub 20151222 https://doi.org/10.2463/mrms.rev.2015-0060. 26701695. PMCID: PMC5600054.

Greicius, M.D., Krasnow, B., Reiss, A.L., Menon, V., 2003. Functional connectivity in the resting brain: a network analysis of the default mode hypothesis. Proc. Natl. Acad. Sci. U. S. A. 100 (1), 253–258. Epub 2002/12/31 https://doi.org/10.1073/pnas.0135058100. 12506194. PMCID: PMC140943.

Hampson, M., Peterson, B.S., Skudlarski, P., Gatenby, J.C., Gore, J.C., 2002. Detection of functional connectivity using temporal correlations in MR images. Hum. Brain Mapp. 15 (4), 247–262. https://doi.org/10.1002/hbm.10022. 11835612. PMCID: PMC6872035.

Hull, L., Mandy, W., Petrides, K.V., 2017. Behavioural and cognitive sex/gender differences in autism spectrum condition and typically developing males and females. Autism 21 (6), 706–727. Epub 2017/07/28 https://doi.org/10.1177/1362361316669087. 28749232.

Hutchison, R.M., Womelsdorf, T., Allen, E.A., Bandettini, P.A., Calhoun, V.D., Corbetta, M., et al., 2013. Dynamic functional connectivity: promise, issues, and interpretations. NeuroImage 80, 360–378. Epub 20130524 https://doi.org/10.1016/j.neuroimage.2013.05.079. 23707587. PMCID: PMC3807588.

Jaime, S., Gu, H., Sadacca, B.F., Stein, E.A., Cavazos, J.E., Yang, Y., et al., 2019. Delta rhythm orchestrates the neural activity underlying the resting state BOLD signal via phase-amplitude coupling. Cereb. Cortex 29 (1), 119–133. https://doi.org/10.1093/cercor/bhx310. 29161352. PMCID: PMC6490973.

Jenkinson, M., Beckmann, C.F., Behrens, T.E., Woolrich, M.W., Smith, S.M., 2012. FSL. NeuroImage 62 (2), 782–790. Epub 20110916 https://doi.org/10.1016/j.neuroimage.2011.09.015. 21979382.

Kessler, D., Angstadt, M., Sripada, C., 2016. Growth charting of brain connectivity networks and the identification of attention impairment in youth. JAMA Psychiatry 73 (5), 481–489. https://doi.org/10.1001/jamapsychiatry.2016.0088. 27076193. PMCID: PMC5507181.

Khazaee, A., Ebrahimzadeh, A., Babajani-Feremi, A., 2016. Application of advanced machine learning methods on resting-state fMRI network for identification of mild cognitive impairment and Alzheimer's disease. Brain Imaging Behav. 10 (3), 799–817. https://doi.org/10.1007/s11682-015-9448-7. 26363784.

Khosla, M., Jamison, K., Ngo, G.H., Kuceyeski, A., Sabuncu, M.R., 2019. Machine learning in resting-state fMRI analysis. Magn. Reson. Imaging 64, 101–121. Epub 20190605 https://doi.org/10.1016/j.mri.2019.05.031. 31173849. PMCID: PMC6875692.

Kollndorfer, K., Fischmeister, F.P., Kasprian, G., Prayer, D., Schopf, V., 2013. A systematic investigation of the invariance of resting-state network patterns: is resting-state fMRI ready for pre-surgical planning? Front. Hum. Neurosci. 7, 95. Epub 20130326 https://doi.org/10.3389/fnhum.2013.00095. 23532457. PMCID: PMC3607808.

Lee, H.L., Zahneisen, B., Hugger, T., LeVan, P., Hennig, J., 2013. Tracking dynamic resting-state networks at higher frequencies using MR-encephalography. NeuroImage 65, 216–222. Epub 20121013 https://doi.org/10.1016/j.neuroimage.2012.10.015. 23069810.

Leonardi, N., Richiardi, J., Gschwind, M., Simioni, S., Annoni, J.M., Schluep, M., et al., 2013. Principal components of functional connectivity: a new approach to study dynamic brain connectivity during rest. NeuroImage 83, 937–950. Epub 20130718 https://doi.org/10.1016/j.neuroimage.2013.07.019. 23872496.

Lindquist, M.A., Xu, Y., Nebel, M.B., Caffo, B.S., 2014. Evaluating dynamic bivariate correlations in resting-state fMRI: a comparison study and a new approach. NeuroImage 101, 531–546. Epub 20140630 https://doi.org/10.1016/j.neuroimage.2014.06.052. 24993894. PMCID: PMC4165690.

Liu, X., Duyn, J.H., 2013. Time-varying functional network information extracted from brief instances of spontaneous brain activity. Proc. Natl. Acad. Sci. U. S. A. 110 (11), 4392–4397. Epub 20130225 https://doi.org/10.1073/pnas.1216856110. 23440216. PMCID: PMC3600481.

Liu, Y., Liang, M., Zhou, Y., He, Y., Hao, Y., Song, M., et al., 2008. Disrupted small-world networks in schizophrenia. Brain 131 (Pt 4), 945–961. Epub 2008/02/27 https://doi.org/10.1093/brain/awn018. 18299296.

Liu, D., Dong, Z., Zuo, X., Wang, J., Zang, Y., 2013. Eyes-open/eyes-closed dataset sharing for reproducibility evaluation of resting state fMRI data analysis methods. Neuroinformatics 11 (4), 469–476. https://doi.org/10.1007/s12021-013-9187-0. 23836389.

Lowe, M.J., Mock, B.J., Sorenson, J.A., 1998. Functional connectivity in single and multislice echoplanar imaging using resting-state fluctuations. NeuroImage 7 (2), 119–132. Epub 1998/04/29 https://doi.org/10.1006/nimg.1997.0315. 9558644.

Lu, H., Jaime, S., Yang, Y., 2019. Origins of the resting-state functional MRI signal: potential limitations of the "neurocentric" model. Front. Neurosci. 13, 1136. Epub 20191023 https://doi.org/10.3389/fnins.2019.01136. 31708731. PMCID: PMC6819315.

Maknojia, S., Churchill, N.W., Schweizer, T.A., Graham, S.J., 2019. Resting State fMRI: going through the motions. Front. Neurosci. 13, 825. Epub 2019/08/29 https://doi.org/10.3389/fnins.2019.00825. 31456656. PMCID: PMC6700228.

McKeown, M.J., Makeig, S., Brown, G.G., Jung, T.P., Kindermann, S.S., Bell, A.J., et al., 1998. Analysis of fMRI data by blind separation into independent spatial components. Hum. Brain Mapp. 6 (3), 160–188. Epub 1998/07/23 9673671.

McKeown, M.J., Hansen, L.K., Sejnowsk, T.J., 2003. Independent component analysis of functional MRI: what is signal and what is noise? Curr. Opin. Neurobiol. 13 (5), 620–629. Epub 2003/11/25 https://doi.org/10.1016/j.conb.2003.09.012. 14630228. PMCID: PMC2925426.

Murphy, K., Fox, M.D., 2017. Towards a consensus regarding global signal regression for resting state functional connectivity MRI. NeuroImage 154, 169–173. Epub 2016/11/27 https://doi.org/10.1016/j.neuroimage.2016.11.052. 27888059. PMCID: PMC5489207.

O'Connor, E.E., Zeffiro, T.A., 2019. Why is clinical fMRI in a resting state? Front. Neurol. 10, 420. Epub 20190424 https://doi.org/10.3389/fneur.2019.00420. 31068901. PMCID: PMC6491723.

Parkes, L., Fulcher, B., Yucel, M., Fornito, A., 2018. An evaluation of the efficacy, reliability, and sensitivity of motion correction strategies for resting-state functional MRI. NeuroImage 171, 415–436. Epub 2017/12/27 https://doi.org/10.1016/j.neuroimage.2017.12.073. 29278773.

Patriat, R., Molloy, E.K., Meier, T.B., Kirk, G.R., Nair, V.A., Meyerand, M.E., et al., 2013. The effect of resting condition on resting-state fMRI reliability and consistency: a comparison between resting with eyes open, closed, and fixated. NeuroImage 78, 463–473. Epub 2013/04/20 https://doi.org/10.1016/j.neuroimage.2013.04.013. 23597935. PMCID: PMC4003890.

Penny, W.D., Friston, K.J., Ashburner, J.T., Kiebel, S.J., Nichols, T.E., 2011. Statistical Parametric Mapping: The Analysis of Functional Brain Images. Elsevier.

Ponce-Alvarez, A., Deco, G., Hagmann, P., Romani, G.L., Mantini, D., Corbetta, M., 2015. Resting-state temporal synchronization networks emerge from connectivity topology and heterogeneity. PLoS Comput. Biol. 11 (2), e1004100. Epub 20150210 https://doi.org/10.1371/journal.pcbi.1004100. 25692996. PMCID: PMC4333573.

Power, J.D., Barnes, K.A., Snyder, A.Z., Schlaggar, B.L., Petersen, S.E., 2012. Spurious but systematic correlations in functional connectivity MRI networks arise from subject motion. NeuroImage 59 (3), 2142–2154. Epub 20111014 https://doi.org/10.1016/j.neuroimage.2011.10.018. 22019881. PMCID: PMC3254728.

Power, J.D., Mitra, A., Laumann, T.O., Snyder, A.Z., Schlaggar, B.L., Petersen, S.E., 2014. Methods to detect, characterize, and remove motion artifact in resting state fMRI. NeuroImage 84, 320–341. Epub 20130829 https://doi.org/10.1016/j.neuroimage.2013.08.048. 23994314. PMCID: PMC3849338.

Power, J.D., Schlaggar, B.L., Petersen, S.E., 2015. Recent progress and outstanding issues in motion correction in resting state fMRI. NeuroImage 105, 536–551. Epub 20141024 https://doi.org/10.1016/j.neuroimage.2014.10.044. 25462692. PMCID: PMC4262543.

Preti, M.G., Bolton, T.A., Van De Ville, D., 2017. The dynamic functional connectome: state-of-the-art and perspectives. NeuroImage 160, 41–54. Epub 20161226 https://doi.org/10.1016/j.neuroimage.2016.12.061. 28034766.

Raichle, M.E., 2015. The brain's default mode network. Annu. Rev. Neurosci. 38, 433–447. Epub 2015/05/06 https://doi.org/10.1146/annurev-neuro-071013-014030. 25938726.

Raichle, M.E., MacLeod, A.M., Snyder, A.Z., Powers, W.J., Gusnard, D.A., Shulman, G.L., 2001. A default mode of brain function. Proc. Natl. Acad. Sci. U. S. A. 98 (2), 676–682. Epub 2001/02/24 https://doi.org/10.1073/pnas.98.2.676. 11209064. PMCID: PMC14647.

Saad, Z.S., Gotts, S.J., Murphy, K., Chen, G., Jo, H.J., Martin, A., et al., 2012. Trouble at rest: how correlation patterns and group differences become distorted after global signal regression. Brain Connect. 2 (1), 25–32. https://doi.org/10.1089/brain.2012.0080. 22432927. PMCID: PMC3484684.

Sakoglu, U., Pearlson, G.D., Kiehl, K.A., Wang, Y.M., Michael, A.M., Calhoun, V.D., 2010. A method for evaluating dynamic functional network connectivity and task-modulation: application to schizophrenia. MAGMA 23 (5-6), 351–366. Epub 2010/02/18 https://doi.org/10.1007/s10334-010-0197-8. 20162320. PMCID: PMC2891285.

Salvador, R., Suckling, J., Coleman, M.R., Pickard, J.D., Menon, D., Bullmore, E., 2005. Neurophysiological architecture of functional magnetic resonance images of human brain. Cereb. Cortex 15 (9), 1332–1342. Epub 20050105 https://doi.org/10.1093/cercor/bhi016. 15635061.

Satterthwaite, T.D., Elliott, M.A., Gerraty, R.T., Ruparel, K., Loughead, J., Calkins, M.E., et al., 2013. An improved framework for confound regression and filtering for control of motion artifact in the preprocessing of resting-state functional connectivity data. NeuroImage 64, 240–256. Epub 20120825 https://doi.org/10.1016/j.neuroimage.2012.08.052. 22926292. PMCID: PMC3811142.

Sorg, C., Riedl, V., Muhlau, M., Calhoun, V.D., Eichele, T., Laer, L., et al., 2007. Selective changes of resting-state networks in individuals at risk for Alzheimer's disease. Proc. Natl. Acad. Sci. U. S. A. 104 (47), 18760–18765. Epub 20071114 https://doi.org/10.1073/pnas.0708803104. 18003904. PMCID: PMC2141850.

Stein, T., Moritz, C., Quigley, M., Cordes, D., Haughton, V., Meyerand, E., 2000. Functional connectivity in the thalamus and hippocampus studied with functional MR imaging. AJNR Am. J. Neuroradiol. 21 (8), 1397–1401. 11003270. PMCID:PMC7974059.

Stephan, K.E., Schlagenhauf, F., Huys, Q.J.M., Raman, S., Aponte, E.A., Brodersen, K.H., et al., 2017. Computational neuroimaging strategies for single patient predictions. NeuroImage 145 (Pt B), 180–199. Epub 20160622 https://doi.org/10.1016/j.neuroimage.2016.06.038. 27346545.

Sun, Y., Lan, Z., Xue, S.W., Zhao, L., Xiao, Y., Kuai, C., et al., 2021. Brain state-dependent dynamic functional connectivity patterns in attention-deficit/hyperactivity disorder. J. Psychiatr. Res. 138, 569–575. Epub 20210508 https://doi.org/10.1016/j.jpsychires.2021.05.010. 33991995.

Taghia, J., Ryali, S., Chen, T., Supekar, K., Cai, W., Menon, V., 2017. Bayesian switching factor analysis for estimating time-varying functional connectivity in fMRI. NeuroImage 155, 271–290. Epub 20170304 https://doi.org/10.1016/j.neuroimage.2017.02.083. 28267626. PMCID: PMC5536190.

Tian, L., Jiang, T., Liang, M., Zang, Y., He, Y., Sui, M., et al., 2008. Enhanced resting-state brain activities in ADHD patients: a fMRI study. Brain Dev. 30 (5), 342–348. Epub 2007/12/07 https://doi.org/10.1016/j.braindev.2007.10.005. 18060712.

van den Heuvel, M.P., Fornito, A., 2014. Brain networks in schizophrenia. Neuropsychol. Rev. 24 (1), 32–48. Epub 20140206 https://doi.org/10.1007/s11065-014-9248-7. 24500505.

van den Heuvel, M.P., Kahn, R.S., Goni, J., Sporns, O., 2012. High-cost, high-capacity backbone for global brain communication. Proc. Natl. Acad. Sci. U. S. A. 109 (28), 11372–11377. Epub 20120618 https://doi.org/10.1073/pnas.1203593109. 22711833. PMCID: PMC3396547.

van den Heuvel, M.P., Sporns, O., Collin, G., Scheewe, T., Mandl, R.C., Cahn, W., et al., 2013. Abnormal rich club organization and functional brain dynamics in schizophrenia. JAMA Psychiatry 70 (8), 783–792. https://doi.org/10.1001/jamapsychiatry.2013.1328. 23739835.

Van Dijk, K.R., Hedden, T., Venkataraman, A., Evans, K.C., Lazar, S.W., Buckner, R.L., 2010. Intrinsic functional connectivity as a tool for human connectomics: theory, properties, and optimization. J. Neurophysiol. 103 (1), 297–321. Epub 20091104 https://doi.org/10.1152/jn.00783.2009. 19889349. PMCID: PMC2807224.

Vanderwal, T., Kelly, C., Eilbott, J., Mayes, L.C., Castellanos, F.X., 2015. Inscapes: a movie paradigm to improve compliance in functional magnetic resonance imaging. NeuroImage 122, 222–232. Epub 20150801 https://doi.org/10.1016/j.neuroimage.2015.07.069. 26241683. PMCID: PMC4618190.

Vanderwal, T., Eilbott, J., Castellanos, F.X., 2019. Movies in the magnet: naturalistic paradigms in developmental functional neuroimaging. Dev. Cogn. Neurosci. 36, 100600. Epub 2018/12/16 https://doi.org/10.1016/j.dcn.2018.10.004. 30551970. PMCID: PMC6969259.

Wang, K., Liang, M., Wang, L., Tian, L., Zhang, X., Li, K., et al., 2007. Altered functional connectivity in early Alzheimer's disease: a resting-state fMRI study. Hum. Brain Mapp. 28 (10), 967–978. https://doi.org/10.1002/hbm.20324. 17133390. PMCID: PMC6871392.

Watts, D.J., Strogatz, S.H., 1998. Collective dynamics of 'small-world' networks. Nature 393 (6684), 440–442. https://doi.org/10.1038/30918. 9623998.

Wei, J., Chen, T., Li, C., Liu, G., Qiu, J., Wei, D., 2018. Eyes-open and eyes-closed resting states with opposite brain activity in sensorimotor and occipital regions: multidimensional evidences from machine learning perspective. Front. Hum. Neurosci. 12, 422. Epub 20181018 https://doi.org/10.3389/fnhum.2018.00422. 30405376. PMCID: PMC6200849.

Weng, S.J., Wiggins, J.L., Peltier, S.J., Carrasco, M., Risi, S., Lord, C., et al., 2010. Alterations of resting state functional connectivity in the default network in adolescents with autism spectrum disorders. Brain Res. 1313, 202–214. Epub 20091211 https://doi.org/10.1016/j.brainres.2009.11.057. 20004180. PMCID: PMC2818723.

Winder, A.T., Echagarruga, C., Zhang, Q., Drew, P.J., 2017. Weak correlations between hemodynamic signals and ongoing neural activity during the resting state. Nat. Neurosci. 20 (12), 1761–1769. Epub 20171106 https://doi.org/10.1038/s41593-017-0007-y. 29184204. PMCID: PMC5816345.

Wu, C.W., Gu, H., Lu, H., Stein, E.A., Chen, J.H., Yang, Y., 2008. Frequency specificity of functional connectivity in brain networks. NeuroImage 42 (3), 1047–1055. Epub 20080715 https://doi.org/10.1016/j.neuroimage.2008.05.035. 18632288. PMCID: PMC2612530.

Wu, X., He, H., Shi, L., Xia, Y., Zuang, K., Feng, Q., et al., 2019. Personality traits are related with dynamic functional connectivity in major depression disorder: a resting-state analysis. J. Affect. Disord. 245, 1032–1042. Epub 20181102 https://doi.org/10.1016/j.jad.2018.11.002. 30699845.

Xiong, J., Parsons, L.M., Gao, J.H., Fox, P.T., 1999. Interregional connectivity to primary motor cortex revealed using MRI resting state images. Hum. Brain Mapp. 8 (2-3), 151–156. Epub 1999/10/19 https://doi.org/10.1002/(sici)1097-0193(1999)8:2/3<151::aid-hbm13>3.0.co;2-5. 10524607.

Yaesoubi, M., Adali, T., Calhoun, V.D., 2018. A window-less approach for capturing time-varying connectivity in fMRI data reveals the presence of states with variable rates of change. Hum. Brain Mapp. 39 (4), 1626–1636. Epub 20180109 https://doi.org/10.1002/hbm.23939. 29315982. PMCID: PMC5847478.

Yan, C., Liu, D., He, Y., Zou, Q., Zhu, C., Zuo, X., et al., 2009. Spontaneous brain activity in the default mode network is sensitive to different resting-state conditions with limited cognitive load. PLoS One 4 (5), e5743. Epub 20090529 https://doi.org/10.1371/journal.pone.0005743. 19492040. PMCID: PMC2683943.

Yan, C.G., Cheung, B., Kelly, C., Colcombe, S., Craddock, R.C., Di Martino, A., et al., 2013. A comprehensive assessment of regional variation in the impact of head micromovements on functional connectomics. NeuroImage 76, 183–201. Epub 20130315 https://doi.org/10.1016/j.neuroimage.2013.03.004. 23499792. PMCID: PMC3896129.

Zang, Y., Jiang, T., Lu, Y., He, Y., Tian, L., 2004. Regional homogeneity approach to fMRI data analysis. NeuroImage 22 (1), 394–400. https://doi.org/10.1016/j.neuroimage.2003.12.030. 15110032.

Zang, Y.F., He, Y., Zhu, C.Z., Cao, Q.J., Sui, M.Q., Liang, M., et al., 2007. Altered baseline brain activity in children with ADHD revealed by resting-state functional MRI. Brain Dev. 29 (2), 83–91. Epub 2006/08/22 https://doi.org/10.1016/j.braindev.2006.07.002. 16919409.

Zhi, D., Calhoun, V.D., Lv, L., Ma, X., Ke, Q., Fu, Z., et al., 2018. Aberrant dynamic functional network connectivity and graph properties in major depressive disorder. Front. Psychiatry 9, 339. Epub 20180731 https://doi.org/10.3389/fpsyt.2018.00339. 30108526. PMCID: PMC6080590.

Zou, Q.H., Zhu, C.Z., Yang, Y., Zuo, X.N., Long, X.Y., Cao, Q.J., et al., 2008. An improved approach to detection of amplitude of low-frequency fluctuation (ALFF) for resting-state fMRI: fractional ALFF. J. Neurosci. Methods 172 (1), 137–141. Epub 2008/05/27 https://doi.org/10.1016/j.jneumeth.2008.04.012. 18501969. PMCID: PMC3902859.

2

Resting state fMRI connectivity mapping across species: Challenges and opportunities

Marco Pagani[a,b], Daniel Gutierrez-Barragan[a], Elizabeth De Guzman[a], and Alessandro Gozzi[a]

[a]*Functional Neuroimaging Laboratory, Center for Neuroscience and Cognitive systems, Istituto Italiano di Tecnologia, Rovereto, Italy,* [b]*Autism Center, Child Mind Institute, New York, NY, United States*

The elusive origin of resting-state "fMRI connectivity"

Since it was first observed that the activity of the right and left motor cortex is highly synchronized during rest (Biswal et al., 1995), resting-state fMRI (rs-fMRI) has been extensively used to describe the intrinsic functional architecture of the human brain. By using statistical dependencies between spontaneous fluctuations in BOLD signals as an index of "functional connectivity" (Fox and Raichle, 2007), rs-fMRI is one of the few neuroimaging methods allowing the identification of whole-brain patterns of interareal communication with high spatial resolution (Beckmann et al., 2005). The ensuing rsfMRI signal dependencies display robust and reproducible spatiotemporal structure, defining a well-characterized set of resting-state networks (RSNs) exhibiting a strong correspondence with known functional systems of the human brain (Smith et al., 2009; Power et al., 2014b). These features, together with the ease of implementation of rsfMRI and the fast-paced development of advanced methods to map and analyze rsfMRI network dynamics (Mitra et al., 2015; Liu and Duyn, 2013; Thomas Yeo et al., 2011), have dramatically improved our ability to map spontaneous network activity and its intrinsic architectural organization in healthy and pathological states.

Most applications of rsfMRI connectivity mapping to date have investigated the human brain organization at the macroscale, with the goal of understanding how spontaneous network activity affects and biases cognitive function (Pezzulo et al., 2021; Power et al., 2014b).

Advances in Resting-State Functional MRI. https://doi.org/10.1016/B978-0-323-91688-2.00009-6

rsfMRI has also been extensively used to map functional network organization in brain disorders, providing robust evidence of atypical or disrupted functional connectivity in most psychiatric and neurological conditions (Di Martino et al., 2014; Whitfield-Gabrieli and Ford, 2012; Fornito and Bullmore, 2014). However, the purely correlative nature of rsfMRI and our current inability to physiologically decode rsfMRI coupling strongly limit the impact and interpretability of this readout in cognitive and experimental neuroscience. As a result, several fundamental questions related to the origin and significance of rsfMRI connectivity remain unaddressed. For one, what are the neurophysiological underpinnings and neural rhythms underlying brain-wide functional connectivity? How are they influenced by arousal and brain states? To what extent does structural anatomy constrain and bias rsfMRI connectivity? What are the determinants and significance of functional dysconnectivity observed in brain disorders? And how does rsfMRI network activity dynamically reconfigure in response to intrinsic (e.g., neuromodulatory) and extrinsic (i.e., TMS, tACS, etc.) perturbations?

Advocates of the use of rsfMRI as an imaging endophenotype or proxy for interregional communication may contend that these limitations have not prevented its proficient use in cognitive neuroscience. Nonetheless, a deeper understanding of the physiological basis of rsfMRI may equally benefit scientists interested in using this approach to investigate the functional organization of the healthy brain, as well as those interested in applying this approach to investigate dysfunctional brain activity in mental illness. A telling example of how a deeper understanding of brain physiology is warranted is the ongoing debate about the meaning of the so-called fMRI "global signal" (GS) (Raut et al., 2021; Gutierrez-Barragan et al., 2019; Billings and Keilholz, 2018; Power et al., 2017; Van De Ville, 2019). In an attempt to control for head motion and spurious physiological contributions, many researchers apply fMRI global signal regression during time-series preprocessing. While this procedure improves network specificity (Li et al., 2019; Fox et al., 2009), it also discards globally distributed neural information putatively central to interareal communication and introduces negative correlations between certain brain regions (Caballero-Gaudes and Reynolds, 2017). Importantly, recent investigations have convincingly linked global fMRI signals to arousal states and global neural dynamics (Turchi et al., 2018; Liu et al., 2018), hence suggesting that this often disregarded signal (when devoid of patent motion-related contaminations) does actually serve as a key contributor to the emergence and dynamics of rsfMRI network activity (Gutierrez-Barragan et al., 2019). Nonetheless, controversy over the interpretation and benefits of global signal regression lingers (Uddin, 2017; Murphy and Fox, 2017). As our understanding of the physiological underpinnings and significance of fMRI global signal will improve, a more rational approach toward the use (or removal) of this contribution can be envisaged, hopefully helping settle the controversy over its use in the field of human rsfMRI mapping. This simple example highlights how a deeper understanding of the neural basis of brain-wide rsfMRI coupling can be of great benefit to both cognitive and translational neuroscience. In the following section, we discuss current and future avenues toward this goal.

Bridging the explanatory gap with cross-species rsfMRI

Mechanistic studies in humans

A key goal of contemporary neuroscience is to uncover the elusive physiological underpinnings of rsfMRI network activity. This effort is part of a broader set of investigations aimed at bridging the major explanatory gap that exists between molecular and biophysical modeling of microscale neural activity, and network-level descriptions of brain function (Fig. 1) (Liska and Gozzi, 2016). Investigations in humans have sought to fill this knowledge gap by relating rsfMRI signals to specific neuroanatomical features, or to intrinsic or exogenous neural events recorded in healthy and pathological conditions. As part of

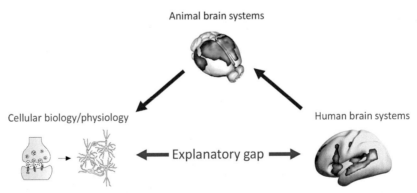

Fig. 1 Bridging the explanatory gap. Human fMRI connectivity mapping allows for the investigation of brain systems in health and disease. Complementary model organisms (such as rodents) can be used to uncover the molecular and cellular principles underlying brain function at different investigational scales. Until recently, we have been unable to cross the "explanatory gap" between human neuroimaging findings and microscale processes discovered in animals due to our inability to translate models of brain function across levels of inquiry. Implementing network level mapping (such as rsfMRI) in model organisms can crucially help bridge this gap by causally linking neural events at the microscale with corresponding systems level changes.

this research, a number of influential human studies have established significant correlation between the spatial organization of rsfMRI and the architecture of white matter connectivity as inferred with diffusion MRI, supporting the notion that rsfMRI activity is strongly constrained by the underlying brain anatomy (reviewed by Suárez et al., 2020).

Other prevalent research approaches in humans have linked genetic variation to specific rsfMRI network attributes via large-scale imaging initiatives in healthy populations (Bycroft et al., 2018; Volkow et al., 2018) or in carriers of genetic and chromosomal alterations associated with brain disorders (Moreau et al., 2020; Zhao et al., 2022; Trachtenberg et al., 2012; Schreiner et al., 2014). Attempts to go beyond the purely correlative nature of human rsfMRI research have entailed rsfMRI mapping at different sleep stages (Tagliazucchi et al., 2013; Mitra et al., 2017; Altmann et al., 2016), or the use of anesthetics (Amico et al., 2014; Scheidegger et al., 2012; Monti et al., 2013; Heine et al., 2012; Akeju et al., 2014) to assess how network activity reconfigures as a function of the level of consciousness (Tagliazucchi and Van Someren, 2017; Demertzi et al., 2019; Akeju et al., 2014; Uehara et al., 2013).

By relating patterns of rsfMRI connectivity with neurological lesions in stroke patients, pseudo-causal associations (Siddiqi et al., 2022) between lesion-induced changes in rsfMRI connectivity and neuro-cognitive deficits have also been described (Corbetta, 2012; Grefkes and Fink, 2014). These studies have revealed that focal lesions can compromise network activity over extended brain areas and that clinical recovery is associated with restored connectivity within the affected networks (Ramsey et al., 2016; Grefkes and Fink, 2014). Further attempts to establish causality between focal patterns of neural activity and rsfMRI network dynamics in the human brain have also been carried out with the use of invasive (i.e., deep brain stimulation—DBS) and noninvasive neuromodulation such as transcranial magnetic stimulation (TMS), transcranial alternate current stimulation (tACS), and related neurostimulation methods (Siddiqi et al., 2022; Horn et al., 2022). Investigations of these tools for treating neurological and psychiatric diseases have revealed that treatment sites with high therapeutic efficacy are more likely to be situated within the same functional network (Fox et al., 2014; Siddiqi et al., 2021). This suggests that brain network activity mapping could guide the development of improved neurostimulation therapy via the identification of sites optimally configured for network-level manipulations.

Finally, a number of studies have employed electroencephalography (EEG)-fMRI or perioperative intracranial recordings to link spatio-temporal rsfMRI dynamics to specific electrophysiological rhythms (Musso et al., 2010; Das et al., 2022; He and Raichle, 2009; He et al., 2008; Nir et al., 2008). These investigations have revealed significant coupling between multiple EEG frequency bands and rsfMRI activity

(Drew et al., 2020; Fox and Raichle, 2007; He and Raichle, 2009; Wang et al., 2012). However, the relative contribution of high- and slow-frequency (including infra- and ultra-slow) oscillations in coordinating brain-wide rsfMRI coupling remains elusive (Rocchi et al., 2022).

Mechanistic studies in nonhuman primates

Human investigations have (and will continue to) significantly advanced our understanding of the functional organization and significance of spontaneous network activity as assessed with fMRI. However, a full physiological disambiguation of a complex, correlative phenomenon like rsfMRI can only be attained by interventional methods that establish causality between neural, molecular, and cellular events occurring at different scales (Fig. 1). Toward this goal, the implementation of rsfMRI in physiologically accessible model organisms like nonhuman primates (NHP) and rodents offers the unprecedented opportunity to bridge scales of inquiry via investigational tools off-limits to human research.

The mechanistic power of rsfMRI mapping in animals is epitomized by foundational work in NHP, where the combined use of fMRI and intracortical electrophysiological recordings has established the groundwork for identifying the neural basis of local hemodynamic activity (Logothetis, 2015; Logothetis et al., 2010; Logothetis et al., 2012; Logothetis et al., 2001; Magri et al., 2012; Shmuel and Leopold, 2008). Further electrophysiological investigations in NHP have been extended to relate rsfMRI activity to resting local field potentials (Schölvinck et al., 2010), interareal neural coherence (Wang et al., 2012), and electrophysiological signatures of arousal (Chang et al., 2016). The implementation of rsfMRI in NHP has also helped to examine the correspondence between rsfMRI functional connectivity and structural anatomy beyond the coarse anatomical resolution attainable with diffusion MRI. Specifically, seminal NHP work has revealed robust consistency (albeit not complete correspondence) between rsfMRI connectivity and experimental tracer studies (Vincent et al., 2007; Margulies et al., 2009; Adachi et al., 2011; Hori et al., 2020). Interventional studies in NHP have also made it possible to assess the causal effect of structural lesions on rsfMRI connectivity, such as in the case of surgical callostomy (O'Reilly et al., 2013) or ischemic insult (Adam et al., 2020), revealing extensive reorganization of rsfMRI connectivity upon interruption of anatomical connectivity. These investigations were recently complemented by studies in which reversible inactivation of regional brain activity was induced via local infusion of drugs or using chemogenetics in animals transfected with designer receptors activated by designer drugs DREADDs (Roth, 2016). The use of these methods in macaque monkeys revealed that inactivation of the amygdala produces patterns of rsfMRI connectivity disruption recapitulating the anatomical connectivity profile of this region (Grayson et al., 2016), hence corroborating the validity of structurally based models of functional connectivity. Lesion studies also showed that cholinergic nuclei crucially contribute to rsfMRI global signal, hence linking this phenomenon to arousal (Turchi et al., 2018). Interventional studies of the rsfMRI connectome in NHP have also included the use of neurostimulation methods to understand how regional activity causally affects global patterns of rsfMRI activity. This goal can be achieved by coupling rsfMRI mapping with electrical microstimulations, focal ultrasound modulation of cortical or subcortical regions (Folloni et al., 2019; Khalighinejad et al., 2020; Verhagen et al., 2019), or infrared stimulations of cortical sites (Xu et al., 2019a).

A few influential studies have investigated how the functional connectome in NHPs reconfigures as a function of brain state. To this aim, rs-fMRI activity has been mapped and compared in awake and anesthetized conditions. This research has shown that rs-fMRI network dynamics is profoundly affected by different brain (and consciousness) states, in the face of an otherwise substantially preserved time-averaged "stationary" rs-fMRI connectivity (Barttfeld et al., 2015; Vincent et al., 2007; Uhrig et al., 2018). Initial attempts to map rs-fMRI connectivity in transgenic primate models of human disorders have also been recently published, offering opportunities for the investigation of the neural basis of functional dysconnectivity in NHP models of developmental and genetic disorders (Cai et al., 2020).

Taken together, these examples reveal a key contribution of NHP research toward a better understanding of the significance and physiological underpinnings of rs-fMRI activity, a goal attainable via the combination of interventional approaches that are off-limits in humans. The translational and mechanistic impact of these studies is greatly strengthened by the close phylogenetic relationship between NHPs and humans, and by the unique opportunity to physiologically decode in these species complex cognitive/behavioral functions via the use of task-based fMRI (Klink et al., 2021; Milham et al., 2022; Milham et al., 2020). However, despite the undisputed impact and value of NHP research, the implementation of neuroimaging in primates remains a complicated endeavor with substantial logistical and procedural demands, the need of highly specialized equipment and personnel, and high investment and running costs. To circumvent these issues, recent years have seen the increased implementation of rs-fMRI in small animals such as rodents. In the next section, we summarize recent advances in the use of rodent rs-fMRI as a tool to elucidate the underpinnings of rs-fMRI connectivity in health and pathology.

Mechanistic studies in rodents

Multiple investigations over the past 10 years have shown the possibility of extending rs-fMRI mapping to rodent species like rats and mice (Gozzi and Schwarz, 2016; Mandino et al., 2020; Sforazzini et al., 2014). While technically challenging and compounded by the need to achieve tight physiological control in a noncompliant species (Ferrari et al., 2012), rodent rs-fMRI offers exciting opportunities to complement and expand experimental manipulations available to human and NHP neuroimagers. Most notably, the implementation of rs-fMRI in rodent models of brain pathology can help shed light on the elusive significance and neural underpinnings of rs-fMRI dysconnectivity observed in clinical populations.

Macroscale functional connectivity is a multifactorial phenomenon, and the determinants of atypical or disrupted functional connectivity observed in human brain disorders remain unclear. Human studies have attempted to tackle this question in a series of increasingly large multisite case-control studies, in which rs-fMRI connectivity is mapped in patients diagnosed with specific conditions such as autism (Di Martino et al., 2017), schizophrenia (Wang et al., 2016), or Alzheimer's disease (Mueller et al., 2005). Using a genetics-first approach, the same paradigm has been extended to map rs-fMRI connectivity in individuals harboring disease-associated genetic alterations (Bertero et al., 2018; Simons VIP Consortium, 2012; Schleifer et al., 2017; Moreau et al., 2020; Savitz and Drevets, 2009; Liu et al., 2019a,b).

These investigations have strongly influenced our conceptualization of brain pathology, supporting the notion that brain disorders are (at least in part) the manifestation of altered network connectivity and brain communication (Vasa et al., 2016; Fornito et al., 2015; Stam, 2014). However, human investigations of brain functional connectopathy suffer from two main problems. The first issue is the identification of appropriate "healthy" or "typically-developing" control cohorts for between-group investigations of fMRI dysconnectivity in brain disorders. Properly accounting for the large interindividual variability in rs-fMRI network activity across control populations (an effect reflecting genetic heterogeneity and/or environmental factors) remains an unresolved issue (Airan et al., 2016; Hacker et al., 2013; Gao et al., 2014; Seghier and Price, 2018). As a result, case control studies, especially in heterogeneous conditions such as schizophrenia and autism spectrum disorders (ASD), are anecdotally dependent on the control population available and on the approach used to analyze and control for site-specific differences (Marquand et al., 2016). This quandary is compounded by the ongoing debate on the impact of different data preprocessing strategies (Nichols et al., 2017; Botvinik-Nezer et al., 2020; Power et al., 2014a) and by difficulties in controlling physiological and motion artifacts in patient populations (Hong et al., 2020; Poldrack et al., 2002; Simhal et al., 2021). While the use of rs-fMRI as a diagnostic biomarker via machine learning classifications shows promise (Abraham et al., 2017; Sarraf and Tofighi, 2016; Saccà et al., 2019; Yang et al., 2010), features extracted by these approaches are often hard to interpret and, in most cases, have only limited mechanistic relevance. These limitations complicate the interpretation of human imaging findings and have fueled controversy about the mechanistic significance of rs-fMRI dysconnectivity in human brain pathology (King et al., 2019).

Within this framework, rs-fMRI mapping in rodent models of pathology can strategically complement human research by offering the opportunity to test (or generate) mechanistically relevant hypotheses under controlled experimental conditions unattainable in clinical investigations. Specifically, rs-fMRI in genetic models of brain disorders allows for tight control of (a) physiological and motion artifacts via light sedation or head-fixation (Ferrari et al., 2012; Grandjean et al., 2020; Gutierrez-Barragan et al., 2022); (b) genetic variability, using inbred transgenic lines; and (c) environmental variability, by testing experimental and control littermate animals bred under identical conditions. This latter point is of special importance as it underscores the fact that, unlike human research, rodent rs-fMRI studies typically leverage a clearly identified, well-controlled reference "population" (i.e., control or wild-type littermate animals), against which effects can be assessed and quantified with great precision.

Research into the underpinnings of functional dysconnectivity associated with autism has benefitted the most from this approach. Autism is a highly heritable disease, with hundreds of risk genes identified to date (Satterstrom et al., 2020). rs-fMRI connectivity mapping in patients with autism has revealed the presence of highly heterogeneous findings, with evidence of decreased or increased rs-fMRI connectivity across brain systems and patient cohorts (Hull et al., 2017). Leveraging the availability of multiple genetic models of autism in the mouse, rs-fMRI has been successfully employed to link genetic or etiological contributions to corresponding macroscale rs-fMRI dysconnectivity signatures, helping to make sense of these heterogeneous findings. Specifically, alterations in rs-fMRI connectivity have been identified and linked to major genetic alterations associated with ASD, such as knockout models for autism-risk genes Cntnap2 (Liska et al., 2018; Zerbi et al., 2018; Balasco et al., 2022), Shank3 (Pagani et al., 2019; Balasco et al., 2021), Tsc2 (Pagani et al., 2021a), Fmr1 (Haberl et al., 2015), Nf1 (Shofty et al., 2019), IQSec2 (Lichtman et al., 2021), microdeletion of 16p11.2 chromosomal region (Bertero et al., 2018), 22q11.2 deletion (Alvino et al., 2021), and agenesis of the corpus callosum (Sforazzini et al., 2016). In some of these cases, regional patterns of rs-fMRI connectivity have been linked to atypicalities in mesoscale axonal wiring (Bertero et al., 2018; Liska et al., 2018; Haberl et al., 2015; Pagani et al., 2019), synaptic abnormalities (Pagani et al., 2021a), or electrophysiological alterations (Bertero et al., 2018). Notable attempts to translate some of these rodent findings to corresponding human populations have also been described, revealing very encouraging correspondences across species (Bertero et al., 2018; Shofty et al., 2019; Pagani et al., 2021a,b).

Importantly, rs-fMRI studies in rodent models have also provided important clues as to the significance of heterogeneous connectivity alterations in autism. Specifically, a long-standing question in this field is whether the autism spectrum is associated with a signature of network dysfunction that is specific (in diagnostic or phenotypic terms) to this condition, or if the etiological heterogeneity that characterizes the spectrum manifests instead as the sum of distinct and different network connectivity alterations, arguing against a common circuital substrate for this disorder. Clinical investigators have long assumed the former to be the case and, in search for reproducible signatures of connectivity dysfunction, have resorted to imaging increasingly large patient cohorts (Holiga et al., 2018; Di Martino et al., 2017; Di Martino et al., 2014; Supekar et al., 2013; Müller et al., 2011). However, these large studies have produced inconsistent findings, prompting some authors to put into question (under the assumption that autism should converge to a single signature of altered connectivity) the reliability and utility of rs-fMRI to assess brain connectopathy in developmental disorders (He et al., 2020).

By pooling and comparing rs-fMRI connectivity alterations in 16 mouse models of autism across two research sites, Zerbi et al. (Zerbi et al., 2021) first showed that under controlled laboratory conditions, rs-fMRI connectivity maps of individual mouse mutants could be clustered into their original genetic group with great confidence. This finding suggests that under controlled experimental conditions, rodent fMRI is sensitive to developmental disruption and can be reliably used to map connectivity changes even at the subject level. Notably, the authors also found that different autism-associated etiologies resulted in a broad spectrum of connectional abnormalities in which diverse and even diverging connectivity signatures were recognizable. These results are translationally relevant as they indicate that etiological variability (e.g., genetics) is a crucial determinant of connectivity heterogeneity in autism, hence reconciling conflicting findings in clinical populations, and exonerating rs-fMRI as the primary cause for the variably and irreconcilable clinical results. Importantly, the identified connectivity alterations could be classified into four subtypes characterized by discrete signatures of network dysfunction, suggesting that the heterogeneity of the spectrum can be parsed into common network subtypes and thus charting future research endeavors toward this goal (Hong et al., 2020). This recent study epitomizes the power of rodent rs-fMRI and its use to understand brain connectopathy.

Other notable applications of rodent rs-fMRI include fine-grained investigations of the structural and functional organization of the mammalian brain connectome, leveraging the "ground-truth" mesoscale axonal connectome produced by the Allen Brain Institute via hundreds of viral injection studies (Oh et al., 2014). These studies have revealed a tight relationship between brain-wide fMRI networks and corresponding patterns of axonal connectivity (Coletta et al., 2020), showing that the organization of the cortical connectome shapes rs-fMRI dynamics (Sethi et al., 2017). A similar study revealed that interhemispheric and cortico-striatal functional connectivity in the mouse emerges primarily via monosynaptic structural connections (Grandjean et al., 2017b). These studies confirm and expand on previous human investigations, corroborating the notion that the structural connectome critically shapes and constrains whole-brain network topography (Suárez et al., 2020). Additional research has extended these investigations to probe the regional and layer-specific axonal organization of subnetworks of the rodent brain, such as the mouse default mode network (DMN) (Whitesell et al., 2021) or salience network (SN) (Tsai et al., 2020).

Recent rodent investigations have also employed cell-type-specific neural manipulations to physiologically decode rs-fMRI signals in pathology, predict how brain manipulations affect the organization and dynamics of brain connectivity, or causally test hypotheses on neuronal drivers of rs-fMRI connectivity. For example, Trakoshis et al. (2020) used cell-type-specific chemogenetic manipulations to increase cortical excitability in the mouse and identify rs-fMRI time-series parameters that could be decoded in clinical populations with autism. Using computational modeling, the authors showed that a parameter termed Hurst index could be used to decode excitatory and inhibitory imbalance in human rs-fMRI time series obtained in patients with autism (Trakoshis et al., 2020).

Chemogenetic manipulations have also been used to deconstruct rs-fMRI connectivity and understand how regional activity affects brain-wide connectivity patterns (Tu et al., 2020; Peeters et al., 2020a). Of note, a recent study from Rocchi and colleagues showed that chronic and acute inactivation of the mouse prefrontal cortex do not necessarily disrupt functional connectivity between the inactivated region and its terminals but can unexpectedly result in patterns of increased rs-fMRI coupling between these areas. Electrophysiological investigations showed that this effect is driven by increased delta and infraslow coupling between manipulated areas. This result challenges prevailing interpretations of rs-fMRI connectivity as an index of direct interareal communication and reveals a key contribution of slow oscillatory processes to the establishment of rs-fMRI connectivity (Rocchi et al., 2022).

Optogenetics has also been used to locally manipulate neuronal activity to determine how single neural or regional elements affect large-scale network coupling as assessed with fMRI (Lee et al., 2010). Notable examples include the photostimulation of thalamocortical networks (Weitz et al., 2019) and dorsal hippocampus (Chan et al., 2017), both of which revealed frequency-dependent network engagement at the whole-brain level. While these studies have charted task-like optogenetically elicited fMRI-activity (assessed as % BOLD response with respect to baseline), future applications of optogenetics can also be conceivably used to disambiguate the rhythms driving large-scale rs-fMRI connectivity by investigating how different optogenetically generated rhythmic frequencies affect brain-wide resting state fMRI coupling.

An additional set of rodent studies have employed chemogenetic or optogenetic manipulations to probe the contribution of subcortical neuromodulatory systems (e.g., noradrenergic, cholinergic, serotonergic, or dopaminergic neurotransmission) to brain-wide patterns of connectivity (Zerbi et al., 2019; Oyarzabal et al., 2022; Giorgi et al., 2017; Peeters et al., 2020b; Ferenczi et al., 2016). These investigations may help disambiguate the contribution of specific neurotransmitter systems to brain-wide network dynamics with a precision unattainable in pharmacological studies or with other methods (i.e., electrical stimulation or lesion) in NHPs.

A last set of rodent investigations we will mention here entail multimodal combinations of signal recordings. These include studies aimed to investigate the basis of neural coupling in rs-fMRI signals via EEG-fMRI mapping (Paasonen et al., 2020), or by the combined use of rs-fMRI and calcium signals detected via fiber photometry (Ma et al., 2022). A promising extension of this latter approach has been recently demonstrated via the combination of cortex-wide calcium imaging and rs-fMRI in the mouse (Lake et al., 2020). This study revealed that calcium signal predicts BOLD signal within a slow frequency range. Combined with the rapidly expanding repertoire of genetically encoded calcium indicators, this novel research approach is expected to provide novel mechanistic insight into the role of neuromodulation and neurotransmitters at different scales.

Convergence and divergence in resting-state network organization across species

Evolutionarily relevant resting-state networks in rodents and monkeys

The examples we have described above illustrate the critical need to mechanistically decode rs-fMRI (dys)connectivity via the use of animal investigations. However, the ultimate impact of this growing field of research is critically dependent on our ability to reliably translate rs-fMRI findings to and from species. In the following section, we discuss advances and challenges toward this goal.

Most of our ability to back-translate and cross-compare functional connectivity findings across species relies on the assumption that functional network structure follows phylogenetically conserved principles. As some of the higher mental faculties embodied in the functional organization of rs-fMRI networks are thought to be uniquely human, questions were initially raised on whether the organization of functional networks could somehow reflect a distinctive organizational mode of the human brain. The identification of distributed resting-state networks in anesthetized macaque monkeys (Vincent et al., 2007) has firmly disproven this hypothesis, revealing that the coherent organization of spontaneous BOLD fluctuations in resting-state networks is not a prerogative of the human brain. Measures of rs-fMRI connectivity in other mammals, such as marmosets (Ghahremani et al., 2017), rats (Liang et al., 2011) and mice (Sforazzini et al., 2014), as well as in birds (De Groof et al., 2013) have revealed that connectivity networks can reliably be identified across species. This suggests that the organization of spontaneous fMRI signals into recurring spatiotemporal modes is a foundational and evolutionarily conserved characteristic of spontaneous brain activity.

These initial observations have prompted interest in the comparative organization of RSN across species. The ultimate goal of this area of research is the use of rs-fMRI as a cross-species bridge to relate region- or network-specific alterations (for example, in models of disease with respect to corresponding human populations). Comparative investigations of the organization of RSNs in rodents and primates have revealed very interesting convergences, corroborating the translational potential of this approach. For example, rs-fMRI studies in rats (Lu et al., 2012; Gozzi and Schwarz, 2016) and mice (Sforazzini et al., 2014; Whitesell et al., 2021) have shown that rodents have a plausible precursor of the human DMN and SN (Gozzi and Schwarz, 2016; Sforazzini et al., 2014; Tsai et al., 2020). Similar findings have been reported in NHPs (Mantini et al., 2011; Tsai et al., 2020). Further investigations of the rodent and monkey cortex have also revealed the presence of additional phylogenetically-relevant systems, including highly synchronous interhemispheric visual and somatomotor networks, the latter being anticorrelated to the DMN like in humans (Gozzi and Schwarz, 2016) and referred to as the latero-cortical network in rodents (Gutierrez-Barragan et al., 2022; Liska et al., 2015; Xu et al., 2019b; Sforazzini et al., 2014; Mandino et al., 2021) (Fig. 2A).

As expected, all the resting state networks so far reliably identified in rodents and NHP encompass phylogenetically conserved regions,

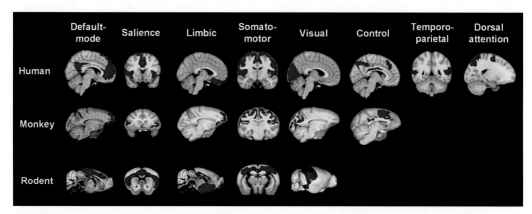

Fig. 2 Resting-state networks across the phylogenetic tree. Spatial extension of representative RSNs in humans (Thomas Yeo et al., 2011), monkeys (Xu et al., 2019b), and mice (Gutierrez-Barragan et al., 2022). Plausible homologues of the human default mode, salience, limbic, somatomotor, and visual networks have been identified in monkeys (Xu et al., 2019b) and mice (Gutierrez-Barragan et al., 2022). A possible homologue of the executive control network has been identified in primates. The dorsal attention and temporoparietal networks (Thomas Yeo et al., 2011) have so far been described in humans only.

such as limbic and somatomotor cortical areas or evolutionarily ancient precursors of the medial prefrontal cortex (Vogt and Paxinos, 2012), and as such only partly recapitulate the broader and richer topography of corresponding systems in humans (Garin et al., 2022; Buckner et al., 2008; Lu et al., 2012). Whether plausible evolutionary precursors of human higher-order cognitive networks, such as the dorsal-attention, temporoparietal network, or control networks (Thomas Yeo et al., 2011), are present and can unambiguously be identified in NHP is still a matter of debate, with the possible exception of a monkey substrate for executive control functions (Corbetta and Shulman, 2002; Stoet and Snyder, 2009; Mantini et al., 2013) (Fig. 2C). On the other hand, there is general consensus that evolutionarily relevant network precursors of these higher-order systems are unlikely to be unambiguously represented in the lissencephalic rodent cortex (Lyamzin and Benucci, 2019; Gozzi and Schwarz, 2016).

Importantly, seed-based mapping or independent component analyses in rodents have decomposed rs-fMRI network activity into more finely partitioned functional systems, crucially revealing subcortical networks of translational relevance, such as hippocampal, basal ganglia, thalamic, and basal-forebrain networks (Schwarz et al., 2013; Gutierrez-Barragan et al., 2022; Liang et al., 2013). Analogous subcortical systems have been identified in NHP (Balsters et al., 2020; Yacoub et al., 2020; Hutchison et al., 2011) and humans (Alves et al., 2019; Zhang and Li, 2017; Kim et al., 2013; Hwang et al., 2017). These subcortical RSNs can be more directly related across species, albeit with the possible caveat that their topographic organization may present with increased complexity along the phylogenetic tree (Balsters et al., 2020).

General organizational principles of rs-fMRI connectivity across species

Notwithstanding some expected differences in the organization of higher-order functional networks across species, the observation of evolutionarily-relevant RSNs in rodents, NHPs, and humans suggests that resting network activity in the mammalian brain follows evolution-invariant general organizational principles. As discussed in the previous section, one of these principles is the organization of cortical activity into spatially segregable associative (i.e., DMN, salience, etc.) and primary somatomotor (i.e., visual, lateral cortical, etc.) networks exhibiting exquisitely bilateral, homotopic topographies. Additional investigations have highlighted further important organizational tenets for rs-fMRI connectivity across species. These encompass the topographic organization of so-called "functional connectivity hubs," i.e., those brain regions exhibiting exceedingly high functional connectivity with the rest of the brain. Cross-species studies have indeed shown that high connectivity density rs-fMRI hubs are similarly anchored to midline components of the DMN in rodents (Liska et al., 2015), NHP (Belcher et al., 2016), and humans (Buckner et al., 2009; Cole et al., 2010), although a shift from rostral (medial prefrontal) to caudal (posterior cingulate) integration is apparent across the evolutionary timeline (Liska et al., 2015; Gozzi and Schwarz, 2016).

Investigations of structure-function coupling in the mammalian axonal connectome have also highlighted notable cross-species correspondences. Specifically, the use of high-resolution connectomes based on tracer injections in rodents and NHP has allowed rs-fMRI network architecture to be related to quantitative and directional measures of connectivity at the mesoscale (Oh et al., 2014; Hori et al., 2020; Knox et al., 2018), hence transcending the limitations of coarse-scale human connectome mapping via diffusion-based MRI. These investigations, recently reviewed by Suárez et al. (2020), have revealed conserved rules of cortical connectivity across species (Goulas et al., 2019), including cell-class-specific projection patterns (Whitesell et al., 2021; Harris and Shepherd, 2015). More importantly, these studies have also documented that the topographic organization of rs-fMRI networks is critically shaped and constrained by the underlying axonal connectome structure in all investigated species, independent of the spatial scale investigated (Coletta et al., 2020; Hori et al., 2020). Collectively, this research supports a tight relationship between the structural and functional organization of the mammalian brain activity at the macroscale, providing robust empirical support to so-called "structurally based" models of rs-fMRI connectivity (Suárez et al., 2020; Alstott et al., 2009), according to which, under resting conditions, fMRI network organization is critically constrained by the anatomical organization of the underlying axonal connectome.

Gradients of fMRI connectivity

The observation of a close relationship between functional and structural organization of the brain has prompted more refined investigations aimed at understanding why the topography of rs-fMRI connectivity exhibits its particular spatial organization. Most of these studies have focused on cortical areas in an attempt to address a fundamental question of high evolutionary relevance: why is each (cortical) brain region located where it is, and how does its microarchitecture influence its role in interareal communication?

Decades of research before the neuroimaging era have dealt with this question by defining a hierarchical ordering of brain regions relating structural constraints to their functional role (See Hilgetag and Goulas, 2020 for review). Specifically, classical neuroanatomical studies demonstrated that cortical areas could be grouped based on their distinct microstructural properties and that variation in these properties exhibits smooth transitions following a gradation principle according to which regions having similar features occupy the same position along this axis (Sanides, 1962). Subsequent investigations showed that a whole host of biological properties, ranging from patterns of gene expression to profiles of axonal connectivity, share a conserved primary axis in spatial variation ranging from sensory to transmodal cortical regions (Huntenburg et al., 2018; Wang, 2020; Mesulam, 1998; Burt et al., 2018; Fulcher et al., 2019; Hilgetag and Goulas, 2020). Of relevance for this review, it has been recently postulated that this axis is representative of the course of cortical evolution, with evolutionary "newer" areas emerging at particular points along the axis (Goulas et al., 2019).

Extending this framework to rs-fMRI-based mapping, Margulies et al. (2016) employed diffusion embedding to show that the topological organization of cortical rs-fMRI connectivity is hierarchically organized along a dominant spatial gradient spanning unimodal-polymodal cortical regions both in humans and in NHPs (Fig. 3A). In the same work, the authors also described a secondary axis of differentiation across primary sensory cortices, with visual and somatomotor areas at the two opposite ends. Furthermore, the dominant (unimodal-polymodal) rs-fMRI connectivity gradient was also found to be spatially aligned to T1-T2-weighted inferences of intracortical myelin content. This configuration spatially reconstitutes the classically identified sensory-transmodal hierarchical organization of cortico-cortical layer connectivity (Mesulam, 1998; Huntenburg

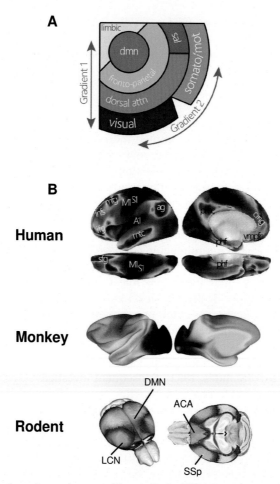

Fig. 3 Functional gradients in the rodent, primate, and human brain. (A) Schematic representation of the spatial relation between rs-fMRI networks and functional gradients of the human cortex. *dmn*, default-mode network; *dorsal attn*, dorsal attention network; *sal*, salience network; *somato/mot*, somatosensory/motor network. (B) Brain topographies of the principal gradient in the human (Margulies et al., 2016), monkey (Xu et al., 2020), and mouse (Coletta et al., 2020) cortex. Human abbreviations: *A1*, primary auditory; *ag*, angular gyrus; *cing*, anterior cingulate cortex; *ifg*, inferior frontal gyrus; *infs*, intermediate frontal sulcus; *L*, limbic; *M1*, primary motor; *mfg*, middle frontal gyrus; *mtc*, middle temporal cortex; *P*, parietal; *Pf*, prefrontal; *phf*, para-hippocampal formation; *pmc*, posteromedial cortex; *ps*, principal sulcus; *S1*, primary somatosensory; *sfg*, superior frontal gyrus; *V1*, primary visual; *vmpfc*, ventromedial prefrontal cortex. Mouse abbreviations: *ACA*, anterior cingulate area; *DMN*, default mode network; *LCN*, latero-cortical network; *SSp*, primary somatosensory area. Panel (A): Adapted with permission from Margulies, D. S., Ghosh, S. S., Goulas, A., Falkiewicz, M., Huntenburg, J. M., Langs, G., Bezgin, G., Eickhoff, S. B., Castellanos, F. X., Petrides, M., 2016. Situating the default-mode network along a principal gradient of macroscale cortical organization. Proc. Natl. Acad. Sci. 113, 12574-12579, Panel B: Adapted with permission from Margulies, D. S., Ghosh, S. S., Goulas, A., Falkiewicz, M., Huntenburg, J. M., Langs, G., Bezgin, G., Eickhoff, S. B., Castellanos, F. X., Petrides, M., 2016. Situating the default-mode network along a principal gradient of macroscale cortical organization. Proc. Natl. Acad. Sci. 113, 12574-12579, Xu, T., Nenning, K.-H., Schwartz, E., Hong, S.-J., Vogelstein, J. T., Goulas, A., Fair, D. A., Schroeder, C. E., Margulies, D. S., Smallwood, J., 2020. Cross-species functional alignment reveals evolutionary hierarchy within the connectome. NeuroImage, 223, 117346, and Coletta, L., Pagani, M., Whitesell, J. D., Harris, J. A., Bernhardt, B., Gozzi, A., 2020. Network structure of the mouse brain connectome with voxel resolution. Sci. Adv. 6, eabb7187.

et al., 2018), crucially relating macroscale RSN organization to its underlying microstructural foundations (Huntenburg et al., 2017).

Corroborating the evolutionary relevance of these architectural features, analogous rs-fMRI-based connectivity gradients have been recently identified in macaques (Xu et al., 2020) and rodents (Coletta et al., 2020), with evidence of a dominant transmodal-unimodal cortical axis of organization and a second modality-specific gradient in both species (Fig. 3B). Using a finely-partitioned model of the mouse axonal connectome, Coletta et al. (2020) extended these findings by showing that axonal (i.e., anatomical) and rs-fMRI connectivity gradients exhibit highly coincident spatial topographies at the voxel level. The authors further showed that gradient organization reflects laminar hierarchy in the rodent cortex, and that their spatial organization critically shapes the dynamic structure of rs-fMRI signal. As observed in humans, the principal gradient of functional connectivity in rodents has also been associated with relevant transcriptomic and cytoarchitectonic features (Huntenburg et al., 2021). These findings suggest that a spatial arrangement of cortical rs-fMRI connectivity along preordered evolutionarily relevant gradients is a fundamental organizational principle of mammalian RSNs.

fMRI coactivation dynamics

Most of the RSN attributes described so far apply to "static" (or "time-averaged") rs-fMRI measurements of functional connectivity, under the assumption that statistical dependencies in rs-fMRI signal are a time-invariant phenomenon. However, the past decade has provided accumulating evidence that, even in resting conditions, spontaneous activity is highly dynamic, with interregional functional coupling undergoing recurring configurations (Hutchison et al., 2013a). A broad spectrum of approaches, termed dynamic functional connectivity analyses (dFC), have been brought forward to map and assess time-varying changes in interregional rs-fMRI coupling (reviewed by Lurie et al., 2020). dFC methods are typically designed to find recurrent "functional connectivity" modes by mapping interareal correlation in the parcellated brain using a sliding time window (Allen et al., 2012). Using this approach, it has been shown that RSNs in rodents and NHPs (Barttfeld et al., 2015) undergo robust reconfiguration over a time scale of seconds as previously described in humans (Grandjean et al., 2017a; Hutchison et al., 2013b).

A set of alternative methods that allow the mapping of rs-fMRI dynamics with voxel-resolution are especially relevant to this review, as they have enabled the detection of brain-wide patterns of BOLD cofluctuation of high evolutionarily relevance (Fig. 4A). These approaches entail the identification of recurrent patterns of BOLD coactivation (termed coactivation patterns, or CAPs) via spatial clustering

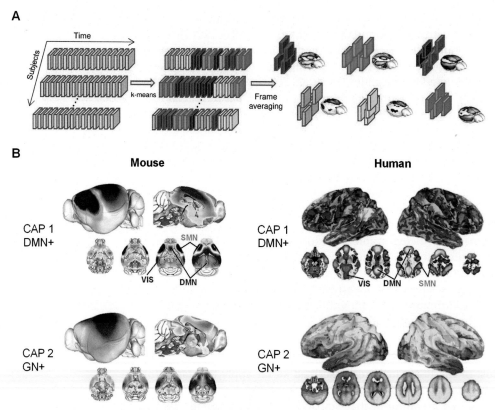

Fig. 4 Mapping fMRI brain dynamics with coactivation patterns. (A) Dynamic network reconfiguration as assessed with fMRI can be mapped with coactivation patterns (CAPs). CAPs describe at voxel-resolution recurrent states of whole-brain BOLD co-activation, which can be identified via k-means clustering of individual fMRI volumes. (B) The topography of dominant CAPs is evolutionarily conserved in mice and human. Panel (A): Adapted with permission from Gutierrez-Barragan, D., Basson, M. A., Panzeri, S., Gozzi, A., 2019. Infraslow state fluctuations govern spontaneous fMRI network dynamics. Curr. Biol. 29, 2295-2306.e5. Panel (B): Adapted with permission from Huang, Z., Zhang, J., Wu, J., Mashour, G. A., Hudetz, A. G., 2020. Temporal circuit of macroscale dynamic brain activity supports human consciousness. Sci. Adv. 6, eaaz0087.

of rs-fMRI frames (Liu et al., 2013; Karahanoğlu and Van De Ville, 2015), or by alternatively mapping time converging spatiotemporal patterns of BOLD activity (Quasi-Periodic Patterns—QPPs) (Yousefi et al., 2018). One key advantage of these methods is that they permit the mapping and description of rs-fMRI dynamics without the spatial constraint of regional parcellations, hence allowing for a more direct cross-comparison of findings across species.

The use of whole-brain CAPs, a concept first described in humans (Liu et al., 2013) and subsequently expanded in a mouse study (Gutierrez-Barragan et al., 2019), is of particular interest as it has generated key insight supporting the notion that dFC is a "bursty" phenomenon (Esfahlani et al., 2020), with most of variance (>60%) being explained by a limited set of recurring CAPs. Importantly, CAPs exhibit a specific set of evolutionarily relevant dynamic properties (Gutierrez-Barragan et al., 2019), namely, (a) dominant CAPs are characterized by opposing peaks of BOLD activity in somatomotor and DMN regions (a finding that explains the anticorrelation between these systems reported in humans and rodents (Fox et al., 2005, Sforazzini et al., 2014)); (b) functional configurations captured by CAPs are characterized by inverse BOLD polarity (i.e., each CAP has an "anti-CAP" exhibiting a reversed pattern of BOLD co-activation) suggesting that CAPs describe spontaneous infraslow oscillatory transitions; (c) accordingly, CAPs show infraslow dynamics peaking between 0.01 and 0.03 Hz, and preferred occurrence within global fMRI signal cycles. These features, initially identified in anesthetized mice, have been recently generalized to awake mice and extended to a cross-species comparison in awake NHP and humans in which the appearance of CAPs exhibiting BOLD cofluctuations encompassing higher order RSNs was also apparent (Gutierrez-Barragan et al., 2021). Similar findings were reported in anesthetized mice and macaques using a region-of-interest approach (Mandino et al., 2021), or by mapping QPPs in anesthetized rats, mice, and awake humans (Yousefi et al., 2018; Majeed et al., 2011; Belloy et al., 2018). Taken together, these findings define an evolutionarily conserved set of dynamic rules which govern rs-fMRI network dynamics (Gutierrez-Barragan et al., 2019) and offer opportunities to map and directly cross-compare evolutionarily relevant patterns of spontaneous BOLD activity across species.

Lost in translation: Challenges in comparing resting-state fMRI connectivity across species

In previous sections, we highlighted significant similarities in the organization and dynamics of RSN across mammalian species. Together with the exquisitely translational nature of rs-fMRI, these observations support cross-species implementations of this approach to investigate the underpinnings of rs-fMRI coupling and generate testable hypotheses about the neural basis of dysconnectivity in brain disorders. Recent advances in the implementation of rs-fMRI in rodents represent a key exciting addition to NHP research, enriching the methodological repertoire available to preclinical researchers. However, while evolutionary convergence exists and supports the use of animal rs-fMRI to bridge the "explanatory" gap, the extrapolation of animal findings to (and from) human populations comes with some challenges, which we discuss in detail in the next sections.

Relating network findings across the phylogenetic tree

The biggest hurdle in comparing changes across species is our limited understanding of the evolutionary trajectory of mammalian cortical functions across the phylogenetic tree. This factor strongly complicates anatomically driven comparisons of connectivity findings across species, a notion epitomized by the classic controversy about whether rodents do (or do not) have a "prefrontal cortex" (Preuss, 1995; Carlén, 2017). While recent cytoarchitectural studies (Vogt and Paxinos, 2012) support the notion that rodents may indeed possess an evolutionarily relevant medial prefrontal cortex (a finding consistent with the organization of the rodent DMN (Gozzi and Schwarz, 2016)), this view is still debated by some in the field (Laubach et al., 2018). This quandary exemplifies the need to exercise caution when using anatomical locations to extrapolate connectivity findings across species. The use of a translational framework based on a few well-characterized and phylogenetically relevant RSNs (e.g. DMN, salience, somatomotor, etc.) may help overcome this hurdle by enabling a proper contextualization of rs-fMRI findings across species. The power of this approach has been recently demonstrated by a series of cross-species investigations in the field of autism research, where network-based inferences have made it possible to reliably relate connectivity changes mapped in rodent models to corresponding rs-fMRI aberrancies in patient populations (Pagani et al., 2021a; Bertero et al., 2018).

Along this line, qualitative approaches that are minimally dependent on brain anatomy can greatly help compare RSN organization across species. The initial application of these methods has produced promising results. For example, cross-species matching based on cortical connectivity gradients has revealed that monkey and human cortical systems can be projected in a common space reflecting broad correspondences in the hierarchical organization of these species, thus improving alignment and interpretation of imaging datasets across modalities and species (Xu et al., 2020). Comparing cortical connectivity architectures identified by gradients comes also with benefit of not requiring the use of discrete brain parcellations, hence avoiding potential bias related to pre-imposed spatial constraints. This approach can also be crucially exploited as a dimensionality reduction method rooted in the functional hierarchy of the cortex. This strategy has been used on human data to identify imbalances in network hierarchy in autism (Hong et al., 2019) and describe gradients of connectivity variation during development (Larivière et al., 2020; Baum et al., 2020).

Independent of the method employed, it should be noted that evolutionary constraints prevent a fully bidirectional translation of findings across the phylogenetic tree, since higher order cortical functions in higher mammalian species are subserved by specialized cortical territories that are less developed or absent in NHPs and rodents (Buckner and Krienen, 2013; Garin et al., 2022). This issue may be somewhat less critical for subcortical network systems, given that the circuit structure of these evolutionarily ancient areas is more conserved across evolution. However, nonnegligible differences in the organization of basal ganglia RSNs where observed when the connectivity profile of these regions was probed in rodents and primates (Balsters et al., 2020). This suggests that investigations of subcortical connectivity systems also need to be interpreted with caution and properly conceptualized within the evolutionarily context of the species probed.

Anesthesia vs. awake imaging

A second important factor to be considered when extrapolating connectivity changes across species is the widespread use of anesthesia in rodent and NHP imaging. The use of anesthetics protocols, typically light sedation, is a convenient choice in rs-fMRI mapping to minimize restraint stress and head motion artifacts (Mandino et al., 2020; Ferrari et al., 2012). A large body of evidence has shown that commonly employed anesthetic regimens broadly preserve the static functional organization of RSN with respect to awake states in rodents, NHPs, and humans (Coletta et al., 2020; Gutierrez-Barragan et al., 2022; Liang et al., 2015; Akeju et al., 2014; Vincent et al., 2007). These observations support the possibility of carefully extrapolating altered RSN connectivity topography across species, independent of the brain state produced by anesthesia (Pagani et al., 2021a; Bertero et al., 2018; Mandino et al., 2021; Tsai et al., 2020; Lu et al., 2012; Gozzi and Schwarz, 2016). The use of anesthesia is thus a practice that still has value in the preclinical rs-fMRI community (Grandjean et al., 2022) as it combines readiness of use with negligible motion artifacts and the possibility of obtaining high-quality time-series. This effect is compounded by the enhanced and more focal distribution of rs-fMRI connectivity observed in anesthetized rodents, an aspect that greatly facilitates network detection in these species (Gutierrez-Barragan et al., 2022).

However, the use of anesthesia also comes with some notable methodological limitations. For example, anesthetic agents can alter hemodynamic and neurovascular coupling (Rungta et al., 2017) or may generate unwanted pharmacological interactions (Gozzi et al., 2008) that can confound the mechanistic interpretation of rs-fMRI signals. Thus, it is of paramount importance that mechanistic studies consider this possible limitation by showing the generalizability of findings across anesthetic mechanisms (Gutierrez-Barragan et al., 2022) and, whenever possible, their extension to awake conditions. The fact that widely used anesthesia protocols in rodents and NHP often entail a cocktail of drugs (Grandjean et al., 2022; Rocchi et al., 2022) also poses interpretational and technical constraints related to the control and measurement of neurophysiological parameters in multimodal studies. While not patently detectable when probing steady-state RSN configuration, this procedure is often accompanied by robust physiological drifts in neural rhythms and brain states reflecting the pharmacokinetic and pharmacological properties of the employed agents. In this respect, the use of single-anesthetic preparations (Ferrari et al., 2012) might be advantageous with respect to drug cocktails.

Lastly, while sedation does not appear to affect the general topography of RSNs, focal changes in rs-fMRI network structure following the use of this procedure are apparent, some of which may be of nonnegligible importance. These include more extensive involvement of arousal-related basal forebrain nuclei in awake states, and a heightened presence of regional anticorrelation, resulting in a breakdown of cortical coupling (Gutierrez-Barragan et al., 2022; Demertzi et al., 2019; Barttfeld et al., 2015). More importantly, brain dynamics appear to be robustly affected by brain state, with a key contribution of arousal-related nuclei (Gutierrez-Barragan et al., 2022), and stereotypic changes in the topography and temporal trajectory of rs-fMRI dynamics that are predictive of consciousness in rodents, NHPs and humans (Gutierrez-Barragan et al., 2022, Demertzi et al., 2019, Barttfeld et al., 2015). These aspects are of key importance for investigations of rs-fMRI network dynamics in animals and humans.

Recent progress in the implementation of awake rs-fMRI in rodents and NHPs offers the opportunity to overcome some of these challenges, hence narrowing the translational gap with humans (Gutierrez-Barragan et al., 2022; Klink et al., 2021). Different approaches to awake imaging in rodents have been proposed, the most prominent ones entailing the use of a head post and extensive behavioral habituation to minimize motion artifacts and reduce restraint-induced stress (Gutierrez-Barragan et al., 2022; Han et al., 2019; Milham et al., 2020; Yoshida et al., 2016), although surgery-free restraint-based approaches have also been employed in rats (Stenroos et al., 2018; Liu et al., 2020). These approaches have shown the possibility of reliably mapping networks with minimal contamination by motion, paving the way for a broader use of awake rs-fMRI mapping in the preclinical field (Liu et al., 2020).

While highly promising, awake animal imaging also comes with technical limitations, such as the use of lengthy procedures to habituate animals to restraint, the need for invasive surgeries for headpost implantation (when required), the possible influence of residual motion in awake animals, and the unclear role of residual stress induced by noise and restraint associated with the scanning procedure. While corticosterone measurements in rodents suggest stress can be well tolerated upon extensive habituation (Gutierrez-Barragan et al., 2022; Stenroos et al., 2018), whether the ensuing brain state is truly comparable to the quiet wakefulness achieved in human scanning remains to be determined. The possibility that awake imaging in rodents and NHPs encompasses higher arousal conditions than human imaging should thus be given serious consideration. This problem could be especially noxious in brain stimulation studies, where states of arousal might result in stimulus-locked involuntary motion in highly aroused animals, potentially contaminating subsequent image analyses. Future implementations of silent fMRI methods can help mitigate this problem (Paasonen et al., 2020). Notwithstanding these technical problems, the field of awake rs-fMRI imaging in animals has been steadily growing, and it is poised to become a dominant approach in the investigations of rs-fMRI network activity.

Conclusions

rs-fMRI applied to nonhuman mammals offers a privileged angle of investigation to explore the significance and organization of spontaneous brain activity at different levels of inquiry. Recent progress in the implementation of functional connectivity mapping in rodents strategically complements the more mature field of NHP neuroimaging by significantly expanding the methodological repertoire available to researchers, and by enabling the investigation of the neural bases of rs-fMRI dysconnectivity in rodent models of human disease. Leveraging emerging correspondences in the architectural organization of networks across the phylogenetic tree, the impact of this versatile research platform toward a better understanding of human brain function is substantial, and it is expected to rapidly grow in the coming years.

Acknowledgments

This work has received funding from the European Research Council (ERC) under the European Union's Horizon 2020 research and innovation program (#DISCONN; no 802371 to A.Go.), the Brain and Behavior Foundation (NARSAD; Independent Investigator Grant; no. 25861), Simons Foundation (SFARI 400101), the NIH (1R21MH116473-01A1), and the Fondazione Telethon (GGP19177).

References

Abraham, A., Milham, M.P., Di Martino, A., Craddock, R.C., Samaras, D., Thirion, B., Varoquaux, G., 2017. Deriving reproducible biomarkers from multi-site resting-state data: an autism-based example. NeuroImage 147, 736–745.

Adachi, Y., Osada, T., Sporns, O., Watanabe, T., Matsui, T., Miyamoto, K., Miyashita, Y., 2011. Functional connectivity between anatomically unconnected areas is shaped by collective network-level effects in the macaque cortex. Cereb. Cortex 22, 1586–1592.

Adam, R., Johnston, K., Menon, R.S., Everling, S., 2020. Functional reorganization during the recovery of contralesional target selection deficits after prefrontal cortex lesions in macaque monkeys. NeuroImage 207, 116339.

Airan, R.D., Vogelstein, J.T., Pillai, J.J., Caffo, B., Pekar, J.J., Sair, H.I., 2016. Factors affecting characterization and localization of interindividual differences in functional connectivity using MRI. Hum. Brain Mapp. 37, 1986–1997.

Akeju, O., Loggia, M.L., Catana, C., Pavone, K.J., Vazquez, R., Rhee, J., Contreras Ramirez, V., Chonde, D.B., Izquierdo-Garcia, D., Arabasz, G., Hsu, S., Habeeb, K., Hooker, J.M., Napadow, V., Brown, E.N., Purdon, P.L., 2014. Disruption of thalamic functional connectivity is a neural correlate of dexmedetomidine-induced unconsciousness. elife 3, e04499.

Allen, E.A., Damaraju, E., Plis, S.M., Erhardt, E.B., Eichele, T., Calhoun, V.D., 2012. Tracking whole-brain connectivity dynamics in the resting state. Cereb. Cortex 24, 663–676.

Alstott, J., Breakspear, M., Hagmann, P., Cammoun, L., Sporns, O., 2009. Modeling the impact of lesions in the human brain. PLoS Comput. Biol. 5, e1000408.

Altmann, A., Schröter, M.S., Spoormaker, V.I., Kiem, S., Jordan, D., Ilg, R., Bullmore, E.T., Greicius, M.D., Czisch, M., Sämann, P.G., 2016. Validation of non-REM sleep stage decoding from resting state fMRI using linear support vector machines. NeuroImage 125, 544–555.

Alves, P.N., Foulon, C., Karolis, V., Bzdok, D., Margulies, D.S., Volle, E., Thiebaut De Schotten, M., 2019. An improved neuroanatomical model of the default-mode network reconciles previous neuroimaging and neuropathological findings. Commun. Biol. 2, 370.

Alvino, F., Galbusera, A., Sastre, D., Rocchi, F., Pagani, M., Montani, C., Papaleo, F., Pasqualetti, M., Bearden, C.E., Gozzi, A., 2021. Tracking the developmental trajectory of 22q11. 2 deletion syndrome in a mouse model. Neuropsychopharmacology, 380. SPRINGERNATURE CAMPUS, 4 CRINAN ST, LONDON, N1 9XW, ENGLAND.

Amico, E., Gomez, F., Di Perri, C., Vanhaudenhuyse, A., Lesenfants, D., Boveroux, P., Bonhomme, V., Brichant, J.-F., Marinazzo, D., Laureys, S., 2014. Posterior cingulate cortex-related co-activation patterns: a resting state fMRI study in Propofol-induced loss of consciousness. PLoS One 9, e100012.

Balasco, L., Pagani, M., Pangrazzi, L., Chelini, G., Ciancone Chama, A.G., Shlosman, E., Mattioni, L., Galbusera, A., Iurilli, G., Provenzano, G., 2021. Abnormal whisker-dependent behaviors and altered cortico-hippocampal connectivity in Shank3b−/− mice. Cereb. Cortex 32, 3042–3056.

Balasco, L., Pagani, M., Pangrazzi, L., Chelini, G., Viscido, F., Chama, A.G.C., Galbusera, A., Provenzano, G., Gozzi, A., Bozzi, Y., 2022. Somatosensory cortex hyperconnectivity and impaired whisker-dependent responses in Cntnap2−/− mice. Neurobiol. Dis. 169, 105742.

Balsters, J.H., Zerbi, V., Sallet, J., Wenderoth, N., Mars, R.B., 2020. Primate homologs of mouse cortico-striatal circuits. elife 9, e53680.

Barttfeld, P., Uhrig, L., Sitt, J.D., Sigman, M., Jarraya, B., Dehaene, S., 2015. Signature of consciousness in the dynamics of resting-state brain activity. Proc. Natl. Acad. Sci. 112, 887–892.

Baum, G.L., Cui, Z., Roalf, D.R., Ciric, R., Betzel, R.F., Larsen, B., Cieslak, M., Cook, P.A., Xia, C.H., Moore, T.M., Ruparel, K., Oathes, D.J., Alexander-Bloch, A.F., Shinohara, R.T., Raznahan, A., Gur, R.E., Gur, R.C., Bassett, D.S., Satterthwaite, T.D., 2020. Development of structure-function coupling in human brain networks during youth. Proc. Natl. Acad. Sci. U. S. A. 117, 771–778.

Beckmann, C.F., Deluca, M., Devlin, J.T., Smith, S.M., 2005. Investigations into resting-state connectivity using independent component analysis. Philos. Trans. R. Soc. Lond. Ser. B Biol. Sci. 360, 1001–1013.

Belcher, A.M., Yen, C.C.-C., Notardonato, L., Ross, T.J., Volkow, N.D., Yang, Y., Stein, E.A., Silva, A.C., Tomasi, D., 2016. Functional connectivity hubs and networks in the awake marmoset brain. Front. Integr. Neurosci. 10.

Belloy, M.E., Naeyaert, M., Abbas, A., Shah, D., Vanreusel, V., Van Audekerke, J., Keilholz, S.D., Keliris, G.A., Van Der Linden, A., Verhoye, M., 2018. Dynamic resting state fMRI analysis in mice reveals a set of quasi-periodic patterns and illustrates their relationship with the global signal. NeuroImage 180, 463–484.

Bertero, A., Liska, A., Pagani, M., Parolisi, R., Masferrer, M.E., Gritti, M., Pedrazzoli, M., Galbusera, A., Sarica, A., Cerasa, A., Buffelli, M., Tonini, R., Buffo, A., Gross, C., Pasqualetti, M., Gozzi, A., 2018. Autism-associated 16p11.2 microdeletion impairs prefrontal functional connectivity in mouse and human. Brain 141, 2055–2065.

Billings, J.C.W., Keilholz, S.D., 2018. The not-so-global BOLD signal. Brain Connect. 8, 121–128.

Biswal, B., Zerrin Yetkin, F., Haughton, V.M., Hyde, J.S., 1995. Functional connectivity in the motor cortex of resting human brain using echo-planar MRI. Magn. Reson. Med. 34, 537–541.

Botvinik-Nezer, R., Holzmeister, F., Camerer, C.F., Dreber, A., Huber, J., Johannesson, M., Kirchler, M., Iwanir, R., Mumford, J.A., Adcock, R.A., 2020. Variability in the analysis of a single neuroimaging dataset by many teams. Nature 582, 84–88.

Buckner, R.L., Andrews-Hanna, J.R., Schacter, D.L., 2008. The Brain's default network. Ann. N. Y. Acad. Sci. 1124, 1–38.

Buckner, R.L., Krienen, F.M., 2013. The evolution of distributed association networks in the human brain. Trends Cogn. Sci. 17, 648–665.

Buckner, R.L., Sepulcre, J., Talukdar, T., Krienen, F.M., Liu, H., Hedden, T., Andrews-Hanna, J.R., Sperling, R.A., Johnson, K.A., 2009. Cortical hubs revealed by intrinsic functional connectivity: mapping, assessment of stability, and relation to Alzheimer's disease. J. Neurosci. 29, 1860–1873.

Burt, J.B., Demirtaş, M., Eckner, W.J., Navejar, N.M., Ji, J.L., Martin, W.J., Bernacchia, A., Anticevic, A., Murray, J.D., 2018. Hierarchy of transcriptomic specialization across human cortex captured by structural neuroimaging topography. Nat. Neurosci. 21.

Bycroft, C., Freeman, C., Petkova, D., Band, G., Elliott, L.T., Sharp, K., Motyer, A., Vukcevic, D., Delaneau, O., O'connell, J., 2018. The UK biobank resource with deep phenotyping and genomic data. Nature 562, 203–209.

Caballero-Gaudes, C., Reynolds, R.C., 2017. Methods for cleaning the BOLD fMRI signal. NeuroImage 154, 128–149.

Cai, D.-C., Wang, Z., Bo, T., Yan, S., Liu, Y., Liu, Z., Zeljic, K., Chen, X., Zhan, Y., Xu, X., Du, Y., Wang, Y., Cang, J., Wang, G.-Z., Zhang, J., Sun, Q., Qiu, Z., Ge, S., Ye, Z., Wang, Z., 2020. MECP2 duplication causes aberrant GABA pathways, circuits and behaviors in transgenic monkeys: neural mappings to patients with autism. J. Neurosci. 40, 3799–3814.

Carlén, M., 2017. What constitutes the prefrontal cortex? Science 358, 478–482.

Chan, R.W., Leong, A.T.L., Ho, L.C., Gao, P.P., Wong, E.C., Dong, C.M., Wang, X., He, J., Chan, Y.-S., Lim, L.W., 2017. Low-frequency hippocampal–cortical activity drives brain-wide resting-state functional MRI connectivity. Biol. Sci. 114 (33), E6972–E6981. https://doi.org/10.1073/pnas.1703309114.

Chang, C., Leopold, D.A., Schölvinck, M.L., Mandelkow, H., Picchioni, D., Liu, X., Ye, F.Q., Turchi, J.N., Duyn, J.H., 2016. Tracking brain arousal fluctuations with fMRI. Proc. Natl. Acad. Sci. 113, 4518.

Cole, M.W., Pathak, S., Schneider, W., 2010. Identifying the brain's most globally connected regions. NeuroImage 49, 3132–3148.

Coletta, L., Pagani, M., Whitesell, J.D., Harris, J.A., Bernhardt, B., Gozzi, A., 2020. Network structure of the mouse brain connectome with voxel resolution. Sci. Adv. 6, eabb7187.

Corbetta, M., 2012. Functional connectivity and neurological recovery. Dev. Psychobiol. 54, 239–253.

Corbetta, M., Shulman, G.L., 2002. Control of goal-directed and stimulus-driven attention in the brain. Nat. Rev. Neurosci. 3, 201–215.

Das, A., De Los Angeles, C., Menon, V., 2022. Electrophysiological foundations of the human default-mode network revealed by intracranial-EEG recordings during resting-state and cognition. NeuroImage 250, 118927.

De Groof, G., Jonckers, E., Güntürkün, O., Denolf, P., Van Auderkerke, J., Van Der Linden, A., 2013. Functional MRI and functional connectivity of the visual system of awake pigeons. Behav. Brain Res. 239, 43–50.

Demertzi, A., Tagliazucchi, E., Dehaene, S., Deco, G., Barttfeld, P., Raimondo, F., Martial, C., Fernández-Espejo, D., Rohaut, B., Voss, H.U., Schiff, N.D., Owen, A.M., Laureys, S., Naccache, L., Sitt, J.D., 2019. Human consciousness is supported by dynamic complex patterns of brain signal coordination. Sci. Adv. 5, eaat7603.

Di Martino, A., O'connor, D., Chen, B., Alaerts, K., Anderson, J.S., Assaf, M., Balsters, J.H., Baxter, L., Beggiato, A., Bernaerts, S., Blanken, L.M., Bookheimer, S.Y., Braden, B.B., Byrge, L., Castellanos, F.X., Dapretto, M., Delorme, R., Fair, D.A., Fishman, I., Fitzgerald, J., Gallagher, L., Keehn, R.J., Kennedy, D.P., Lainhart, J.E., Luna, B., Mostofsky, S.H., Muller, R.A., Nebel, M.B., Nigg, J.T., O'hearn, K., Solomon, M., Toro, R., Vaidya, C.J., Wenderoth, N., White, T., Craddock, R.C., Lord, C., Leventhal, B., Milham, M.P., 2017. Enhancing studies of the connectome in autism using the autism brain imaging data exchange II. Sci. Data 4, 170010.

Di Martino, A., Yan, C.G., Li, Q., Denio, E., Castellanos, F.X., Alaerts, K., Anderson, J.S., Assaf, M., Bookheimer, S.Y., Dapretto, M., Deen, B., Delmonte, S., Dinstein, I., Ertl-Wagner, B., Fair, D.A., Gallagher, L., Kennedy, D.P., Keown, C.L., Keysers, C., Lainhart, J.E., Lord, C., Luna, B., Menon, V., Minshew, N.J., Monk, C.S., Mueller, S., Muller, R.A., Nebel, M.B., Nigg, J.T., O'hearn, K., Pelphrey, K.A., Peltier, S.J., Rudie, J.D., Sunaert, S., Thioux, M., Tyszka, J.M., Uddin, L.Q., Verhoeven, J.S., Wenderoth, N., Wiggins, J.L., Mostofsky, S.H., Milham, M.P., 2014. The autism brain imaging data exchange: towards a large-scale evaluation of the intrinsic brain architecture in autism. Mol. Psychiatry 19, 659–667.

Drew, P.J., Mateo, C., Turner, K.L., Yu, X., Kleinfeld, D., 2020. Ultra-slow oscillations in fMRI and resting-state connectivity: neuronal and vascular contributions and technical confounds. Neuron 107, 782–804.

Esfahlani, F.Z., Jo, Y., Faskowitz, J., Byrge, L., Kennedy, D.P., Sporns, O., Betzel, R.F., 2020. High-amplitude cofluctuations in cortical activity drive functional connectivity. Proc. Natl. Acad. Sci. 117, 28393–28401.

Ferenczi, E.A., Zalocusky, K.A., Liston, C., Grosenick, L., Warden, M.R., Amatya, D., Katovich, K., Mehta, H., Patenaude, B., Ramakrishnan, C., 2016. Prefrontal cortical regulation of brainwide circuit dynamics and reward-related behavior. Science 351, aac9698.

Ferrari, L., Turrini, G., Crestan, V., Bertani, S., Cristofori, P., Bifone, A., Gozzi, A., 2012. A robust experimental protocol for pharmacological fMRI in rats and mice. J. Neurosci. Methods 204, 9–18.

Folloni, D., Verhagen, L., Mars, R.B., Fouragnan, E., Constans, C., Aubry, J.-F., Rushworth, M.F.S., Sallet, J., 2019. Manipulation of subcortical and deep cortical activity in the primate brain using transcranial focused ultrasound stimulation. Neuron 101, 1109–1116.e5.

Fornito, A., Bullmore, E.T., 2014. Connectomics: a new paradigm for understanding brain disease. Eur. Neuropsychopharmacol. 25, 733–748.

Fornito, A., Zalesky, A., Breakspear, M., 2015. The connectomics of brain disorders. Nat. Rev. Neurosci. 16, 159–172.

Fox, M.D., Buckner, R.L., Liu, H., Chakravarty, M.M., Lozano, A.M., Pascual-Leone, A., 2014. Resting-state networks link invasive and noninvasive brain stimulation across diverse psychiatric and neurological diseases. Proc. Natl. Acad. Sci. 111, E4367–E4375.

Fox, M.D., Raichle, M.E., 2007. Spontaneous fluctuations in brain activity observed with functional magnetic resonance imaging. Nat. Rev. Neurosci. 8, 700–711.

Fox, M.D., Snyder, A.Z., Vincent, J.L., Corbetta, M., Van Essen, D.C., Raichle, M.E., 2005. The human brain is intrinsically organized into dynamic, anticorrelated functional networks. Proc. Natl. Acad. Sci. U. S. A. 102, 9673–9678.

Fox, M.D., Zhang, D., Snyder, A.Z., Raichle, M.E., 2009. The global signal and observed anticorrelated resting state brain networks. J. Neurophysiol. 101, 3270–3283.

Fulcher, B.D., Murray, J.D., Zerbi, V., Wang, X.J., 2019. Multimodal gradients across mouse cortex. Proc. Natl. Acad. Sci. U. S. A. 116.

Gao, W., Elton, A., Zhu, H., Alcauter, S., Smith, J.K., Gilmore, J.H., Lin, W., 2014. Intersubject variability of and genetic effects on the brain's functional connectivity during infancy. J. Neurosci. 34, 11288–11296.

Garin, C.M., Hori, Y., Everling, S., Whitlow, C.T., Calabro, F.J., Luna, B., Froesel, M., Gacoin, M., Ben Hamed, S., Dhenain, M., Constantinidis, C., 2022. An evolutionary gap in primate default mode network organization. Cell Rep. 39.

Ghahremani, M., Hutchison, R.M., Menon, R.S., Everling, S., 2017. Frontoparietal functional connectivity in the common marmoset. Cereb. Cortex 27, 3890–3905.

Giorgi, A., Migliarini, S., Galbusera, A., Maddaloni, G., Mereu, M., Margiani, G., Gritti, M., Landi, S., Trovato, F., Bertozzi, S.M., Armirotti, A., Ratto, G.M., De Luca, M.A., Tonini, R., Gozzi, A., Pasqualetti, M., 2017. Brain-wide mapping of endogenous serotonergic transmission via chemogenetic fMRI. Cell Rep. 21, 910–918.

Goulas, A., Margulies, D.S., Bezgin, G., Hilgetag, C.C., 2019. The architecture of mammalian cortical connectomes in light of the theory of the dual origin of the cerebral cortex. Cortex 118, 244–261.

Gozzi, A., Schwarz, A.J., 2016. Large-scale functional connectivity networks in the rodent brain. NeuroImage 127, 496–509.

Gozzi, A., Schwarz, A.J., Reese, T., Crestan, V., Bifone, A., 2008. Drug-anaesthetic interaction in phMRI: the case of the pyschotomimetic agent phencyclidine. Magn. Reson. Imaging 26, 999–1006.

Grandjean, J., Canella, C., Anckaerts, C., Ayrancı, G., Bougacha, S., Bienert, T., Buehlmann, D., Coletta, L., Gallino, D., Gass, N., 2020. Common functional networks in the mouse brain revealed by multi-centre resting-state fMRI analysis. NeuroImage 205, 116278.

Grandjean, J., Desrosiers-Gregoire, G., Anckaerts, C., Angeles-Valdez, D., Ayad, F., Barrière, D.A., Blockx, I., Bortel, A.B., Broadwater, M., Cardoso, B.M., Célestine, M., Chavez-Negrete, J.E., Choi, S., Christiaen, E., Clavijo, P., Colon-Perez, L., Cramer, S., Daniele, T., Dempsey, E., Diao, Y., Doelemeyer, A., Dopfel, D., Dvořáková, L., Falfán-Melgoza, C., Fernandes, F.F., Fowler, C.F., Fuentes-Ibañez, A., Garin, C., Gelderman, E., Golden, C.E., Guo, C.C., Henckens, M.J., Hennessy, L.A., Herman, P., Hofwijks, N., Horien, C., Ionescu, T.M., Jones, J., Kaesser, J., Kim, E., Lambers, H., Lazari, A., Lee, S.-H., Lillywhite, A., Liu, Y., Liu, Y.Y., López-Castro, A., López-Gil, X., Ma, Z., Macnicol, E., Madularu, D., Mandino, F., Marciano, S., Mcauslan, M.J., Mccunn, P., McIntosh, A., Meng, X., Meyer-Baese, L., Missault, S., Moro, F., Naessens, D., Nava-Gomez, L.J., Nonaka, H., Ortiz, J.J., Paasonen, J., Peeters, L.M., Pereira, M., Perez, P.D., Pompilus, M., Prior, M., Rakhmatullin, R., Reimann, H.M., Reinwald, J., De Rio, R.T., Rivera-Olvera, A., Ruiz-Pérez, D., Russo, G., Rutten, T.J., Ryoke, R., Sack, M., Salvan, P., Sanganahalli, B.G., Schroeter, A., Seewoo, B.J., Selingue, E., Seuwen, A., Shi, B., Sirmpilatze, N., Smith, J.A., Smith, C., Sobczak, F., Stenroos, P.J., Straathof, M., Strobelt, S., Sumiyoshi, A., Takahashi, K., Torres-García, M.E., Tudela, R., Van Den Berg, M., Van Der Marel, K., et al., 2022. StandardRat: a multi-center consensus protocol to enhance functional connectivity specificity in the rat brain. bioRxiv. 2022.04.27.489658.

Grandjean, J., Preti, M.G., Bolton, T.A.W., Buerge, M., Seifritz, E., Pryce, C.R., Van De Ville, D., Rudin, M., 2017a. Dynamic reorganization of intrinsic functional networks in the mouse brain. NeuroImage 152, 497–508.

Grandjean, J., Zerbi, V., Balsters, J.H., Wenderoth, N., Rudin, M., 2017b. Structural basis of large-scale functional connectivity in the mouse. J. Neurosci. 37, 8092–8101.

Grayson, D.S., Bliss-Moreau, E., Machado, C.J., Bennett, J., Shen, K., Grant, K.A., Fair, D.A., Amaral, D.G., 2016. The Rhesus monkey connectome predicts disrupted functional networks resulting from pharmacogenetic inactivation of the amygdala. Neuron 91, 453–466.

Grefkes, C., Fink, G.R., 2014. Connectivity-based approaches in stroke and recovery of function. Lancet Neurol. 13, 206–216.

Gutierrez-Barragan, D., Basson, M.A., Panzeri, S., Gozzi, A., 2019. Infraslow state fluctuations govern spontaneous fMRI network dynamics. Curr. Biol. 29, 2295–2306.e5.

Gutierrez-Barragan, D., Panzeri, S., Xu, T., Gozzi, A., 2021. Evolutionarily conserved fMRI network dynamics in the human, macaque and mouse brain. ISMRM. abstract 0574.

Gutierrez-Barragan, D., Singh, N.A., Alvino, F.G., Coletta, L., Rocchi, F., De Guzman, E., Galbusera, A., Uboldi, M., Panzeri, S., Gozzi, A., 2022. Unique spatiotemporal fMRI dynamics in the awake mouse brain. Curr. Biol. 32, 631–644.e6.

Haberl, M.G., Zerbi, V., Veltien, A., Ginger, M., Heerschap, A., Frick, A., 2015. Structural-functional connectivity deficits of neocortical circuits in the Fmr1Gê\Æ/y mouse model of autism. Sci. Adv. 1, e1500775.

Hacker, C.D., Laumann, T.O., Szrama, N.P., Baldassarre, A., Snyder, A.Z., Leuthardt, E.C., Corbetta, M., 2013. Resting state network estimation in individual subjects. NeuroImage 82, 616–633.

Han, Z., Chen, W., Chen, X., Zhang, K., Tong, C., Zhang, X., Li, C.T., Liang, Z., 2019. Awake and behaving mouse fMRI during Go/No-Go task. NeuroImage 188, 733–742.

Harris, K.D., Shepherd, G.M., 2015. The neocortical circuit: themes and variations. Nat. Neurosci. 18, 170–181.

He, Y., Byrge, L., Kennedy, D.P., 2020. Nonreplication of functional connectivity differences in autism spectrum disorder across multiple sites and denoising strategies. Hum. Brain Mapp. 41, 1334–1350.

He, B.J., Raichle, M.E., 2009. The fMRI signal, slow cortical potential and consciousness. Trends Cogn. Sci. 13, 302–309.

He, B.J., Snyder, A.Z., Zempel, J.M., Smyth, M.D., Raichle, M.E., 2008. Electrophysiological correlates of the brain's intrinsic large-scale functional architecture. Proc. Natl. Acad. Sci. U. S. A. 105, 16039–16044.

Heine, L., Soddu, A., Gómez, F., Vanhaudenhuyse, A., Tshibanda, L., Thonnard, M., Charland-Verville, V., Kirsch, M., Laureys, S., Demertzi, A., 2012. Resting state networks and consciousness: alterations of multiple resting state network connectivity in physiological, pharmacological, and pathological consciousness states. Front. Psychol. 3, 295.

Hilgetag, C.C., Goulas, A., 2020. 'Hierarchy' in the organization of brain networks. Philos. Trans. R. Soc. Lond. Ser. B Biol. Sci. 375, 20190319.

Holiga, S., Hipp, J.F., Chatham, C.H., Garces, P., Spooren, W., Logier, D., Ardhuy, X., Bertolino, A., Bouquet, C., Buitelaar, J.K., Bours, C., Rausch, A., Oldehinkel, M., Bouvard, M., Amestoy, A., Caralp, M., Gueguen, S., Ly-Le Moal, M., Houenou, J., Beckmann, C.F., Loth, E., Murphy, D., Charman, T., Tillmann, J., Laidi, C., Delorme, R., Beggiato, A., Gaman, A., Scheid, I., Leboyer, M., Albis, M.-A., Sevigny, J., Czech, C., Bolognani, F., Honey, G.D., Dukart, J., 2018. Reproducible functional connectivity alterations are associated with autism spectrum disorder. bioRxiv.

Hong, S.-J., Vogelstein, J.T., Gozzi, A., Bernhardt, B.C., Yeo, B.T.T., Milham, M.P., Di Martino, A., 2020. Towards neurosubtypes in autism. Biol. Psychiatry 88, 111–128.

Hong, S.-J., Vos De Wael, R., Bethlehem, R.A.I., Lariviere, S., Paquola, C., Valk, S.L., Milham, M.P., Di Martino, A., Margulies, D.S., Smallwood, J., Bernhardt, B.C., 2019. Atypical functional connectome hierarchy in autism. Nat. Commun. 10.

Hori, Y., Schaeffer, D.J., Gilbert, K.M., Hayrynen, L.K., Cléry, J.C., Gati, J.S., Menon, R.S., Everling, S., 2020. Comparison of resting-state functional connectivity in marmosets with tracer-based cellular connectivity. NeuroImage 204, 116241.

Horn, A., Al-Fatly, B., Neumann, W.-J., Neudorfer, C., 2022. Chapter 1—connectomic DBS: an introduction. In: HORN, A. (Ed.), Connectomic Deep Brain Stimulation. Academic Press.

Hull, J.V., Jacokes, Z.J., Torgerson, C.M., Irimia, A., Van Horn, J.D., 2017. Resting-state functional connectivity in Autism spectrum disorders: a review. Front. Psychiatry 7.

Huntenburg, J.M., Bazin, P.L., Goulas, A., Tardif, C.L., Villringer, A., Margulies, D.S., 2017. A systematic relationship between functional connectivity and intracortical myelin in the human cerebral cortex. Cereb. Cortex 27, 981–997.

Huntenburg, J.M., Bazin, P.L., Margulies, D.S., 2018. Large-scale gradients in human cortical organization. Trends Cogn. Sci. 22, 21–31.

Huntenburg, J.M., Yeow, L.Y., Mandino, F., Grandjean, J., 2021. Gradients of functional connectivity in the mouse cortex reflect neocortical evolution. NeuroImage 225, 117528.

Hutchison, R.M., Leung, L.S., Mirsattari, S.M., Gati, J.S., Menon, R.S., Everling, S., 2011. Resting-state networks in the macaque at 7T. NeuroImage 56, 1546–1555.

Hutchison, R.M., Womelsdorf, T., Allen, E.A., Bandettini, P.A., Calhoun, V.D., Corbetta, M., Della Penna, S., Duyn, J.H., Glover, G.H., Gonzalez-Castillo, J., Handwerker, D.A., Keilholz, S., Kiviniemi, V., Leopold, D.A., De Pasquale, F., Sporns, O., Walter, M., Chang, C., 2013a. Dynamic functional connectivity: promise, issues, and interpretations. NeuroImage 80, 360–378.

Hutchison, R.M., Womelsdorf, T., Gati, J.S., Everling, S., Menon, R.S., 2013b. Resting-state networks show dynamic functional connectivity in awake humans and anesthetized macaques. Hum. Brain Mapp. 34, 2154–2177.

Hwang, K., Bertolero, M.A., Liu, W.B., Esposito, M., 2017. The human thalamus is an integrative hub for functional brain networks. J. Neurosci. 37, 5594.

Karahanoğlu, F.I., Van De Ville, D., 2015. Transient brain activity disentangles fMRI resting-state dynamics in terms of spatially and temporally overlapping networks. Nat. Commun. 6.

Khalighinejad, N., Bongioanni, A., Verhagen, L., Folloni, D., Attali, D., Aubry, J.-F., Sallet, J., Rushworth, M.F., 2020. A basal forebrain-cingulate circuit in macaques decides it is time to act. Neuron 105, 370–384. e8.

Kim, D.J., Park, B., Park, H.J., 2013. Functional connectivity-based identification of subdivisions of the basal ganglia and thalamus using multilevel independent component analysis of resting state fMRI. Hum. Brain Mapp. 34, 1371–1385.

King, J.B., Prigge, M.B.D., King, C.K., Morgan, J., Weathersby, F., Fox, J.C., Dean, D.C., Freeman, A., Villaruz, J.A.M., Kane, K.L., Bigler, E.D., Alexander, A.L., Lange, N., Zielinski, B., Lainhart, J.E., Anderson, J.S., 2019. Generalizability and reproducibility of functional connectivity in autism. Mol. Autism 10, 27.

Klink, P.C., Aubry, J.F., Ferrera, V.P., Fox, A.S., Froudist-Walsh, S., Jarraya, B., Konofagou, E.E., Krauzlis, R.J., Messinger, A., Mitchell, A.S., Ortiz-Rios, M., Oya, H., Roberts, A.C., Roe, A.W., Rushworth, M.F.S., Sallet, J., Schmid, M.C., Schroeder, C.E., Tasserie, J., Tsao, D.Y., Uhrig, L., Vanduffel, W., Wilke, M., Kagan, I., Petkov, C.I., 2021. Combining brain perturbation and neuroimaging in non-human primates. NeuroImage 235, 118017.

Knox, J.E., Harris, K.D., Graddis, N., Whitesell, J.D., Zeng, H., Harris, J.A., Shea-Brown, E., Mihalas, S., 2018. High Resolution Data-Driven Model of the Mouse Connectome. Cold Spring Harbor Laboratory.

Lake, E.M., Ge, X., Shen, X., Herman, P., Hyder, F., Cardin, J.A., Higley, M.J., Scheinost, D., Papademetris, X., Crair, M.C., 2020. Simultaneous cortex-wide fluorescence Ca2+ imaging and whole-brain fMRI. Nat. Methods 17, 1262–1271.

Larivière, S., Vos De Wael, R., Hong, S.J., Paquola, C., Tavakol, S., Lowe, A.J., Schrader, D.V., Bernhardt, B.C., 2020. Multiscale structure-function gradients in the neonatal connectome. Cereb. Cortex 30.

Laubach, M., Amarante, L.M., Swanson, K., White, S.R., 2018. What, if anything, is rodent prefrontal cortex? eNeuro 5. ENEURO.0315-18.2018.

Lee, H., Durand, R., Gradinaru, V., Zhang, F., Goshen, I., Kim, D.-S., Fenno, L.E., Ramakrishnan, C., Deisseroth, K., 2010. Global and local fMRI signals driven by neurons defined optogenetically by type and wiring. Nature 465, 788–792.

Li, J., Kong, R., Liégeois, R., Orban, C., Tan, Y., Sun, N., Holmes, A.J., Sabuncu, M.R., Ge, T., Yeo, B.T.T., 2019. Global signal regression strengthens association between resting-state functional connectivity and behavior. NeuroImage 196, 126–141.

Liang, Z., King, J., Zhang, N., 2011. Uncovering intrinsic connectional architecture of functional networks in awake rat brain. J. Neurosci. 31, 3776–3783.

Liang, Z., Li, T., King, J., Zhang, N., 2013. Mapping thalamocortical networks in rat brain using resting-state functional connectivity. NeuroImage 83, 237–244.

Liang, Z., Liu, X., Zhang, N., 2015. Dynamic resting state functional connectivity in awake and anesthetized rodents. NeuroImage 104, 89–99.

Lichtman, D., Bergmann, E., Kavushansky, A., Cohen, N., Levy, N.S., Levy, A.P., Kahn, I., 2021. Structural and functional brain-wide alterations in A350V Iqsec2 mutant mice displaying autistic-like behavior. Transl. Psychiatry 11, 181.

Liska, A., Bertero, A., Gomolka, R., Sabbioni, M., Galbusera, A., Barsotti, N., Panzeri, S., Scattoni, M.L., Pasqualetti, M., Gozzi, A., 2018. Homozygous loss of autism-risk gene CNTNAP2 results in reduced local and long-range prefrontal functional connectivity. Cereb. Cortex 10, 1–13.

Liska, A., Galbusera, A., Schwarz, A.J., Gozzi, A., 2015. Functional connectivity hubs of the mouse brain. NeuroImage 115, 281–291.

Liska, A., Gozzi, A., 2016. Can mouse imaging studies bring order to autism connectivity Chaos? Front. Neurosci. 10, 484.

Liu, X., Chang, C., Duyn, J.H., 2013. Decomposition of spontaneous brain activity into distinct fmri co-activation patterns. Front. Syst. Neurosci. 7.

Liu, X., De Zwart, J.A., Schölvinck, M.L., Chang, C., Ye, F.Q., Leopold, D.A., Duyn, J.H., 2018. Subcortical evidence for a contribution of arousal to fMRI studies of brain activity. Nat. Commun. 9, 395.

Liu, X., Duyn, J.H., 2013. Time-varying functional network information extracted from brief instances of spontaneous brain activity. Proc. Natl. Acad. Sci. 110, 4392–4397.

Liu, F., Gong, X., Yao, X., Cui, L., Yin, Z., Li, C., Tang, Y., Wang, F., 2019a. Variation in the CACNB2 gene is associated with functional connectivity of the Hippocampus in bipolar disorder. BMC Psychiatry 19, 1–7.

Liu, W., Hu, X., An, D., Zhou, D., Gong, Q., 2019b. Resting-state functional connectivity alterations in periventricular nodular heterotopia related epilepsy. Sci. Rep. 9, 1–9.

Liu, Y., Perez, P.D., Ma, Z., Ma, Z., Dopfel, D., Cramer, S., Tu, W., Zhang, N., 2020. An open database of resting-state fMRI in awake rats. NeuroImage 220, 117094.

Logothetis, N.K., 2015. Neural-event-triggered fMRI of large-scale neural networks. Curr. Opin. Neurobiol. 31, 214–222.

Logothetis, N.K., Augath, M., Murayama, Y., Rauch, A., Sultan, F., Goense, J., Oeltermann, A., Merkle, H., 2010. The effects of electrical microstimulation on cortical signal propagation. Nat. Neurosci. 13, 1283–1291.

Logothetis, N.K., Eschenko, O., Murayama, Y., Augath, M., Steudel, T., Evrard, H.C., Besserve, M., Oeltermann, A., 2012. Hippocampal-cortical interaction during periods of subcortical silence. Nature 491, 547–553.

Logothetis, N.K., Pauls, J., Augath, M., Trinath, T., Oeltermann, A., 2001. Neurophysiological investigation of the basis of the fMRI signal. Nature 412, 150–157.

Lu, H., Zou, Q., Gu, H., Raichle, M.E., Stein, E., Yang, Y., 2012. Rat brains also have a default mode network. Proc. Natl. Acad. Sci. U. S. A. 109, 3979–3984.

Lurie, D.J., Kessler, D., Bassett, D.S., Betzel, R.F., Breakspear, M., Kheilholz, S., Kucyi, A., Liégeois, R., Lindquist, M.A., McIntosh, A.R., Poldrack, R.A., Shine, J.M., Thompson, W.H., Bielczyk, N.Z., Douw, L., Kraft, D., Miller, R.L., Muthuraman, M., Pasquini, L., Razi, A., Vidaurre, D., Xie, H., Calhoun, V.D., 2020. Questions and controversies in the study of time-varying functional connectivity in resting fMRI. Netw. Neurosci. 4, 30–69.

Lyamzin, D., Benucci, A., 2019. The mouse posterior parietal cortex: anatomy and functions. Neurosci. Res. 140, 14–22.

Ma, Z., Zhang, Q., Tu, W., Zhang, N., 2022. Gaining insight into the neural basis of resting-state fMRI signal. NeuroImage 250, 118960.

Magri, C., Schridde, U., Murayama, Y., Panzeri, S., Logothetis, N.K., 2012. The amplitude and timing of the BOLD signal reflects the relationship between local field potential power at different frequencies. J. Neurosci. 32, 1395–1407.

Majeed, W., Magnuson, M., Hasenkamp, W., Schwarb, H., Schumacher, E.H., Barsalou, L., Keilholz, S.D., 2011. Spatiotemporal dynamics of low frequency BOLD fluctuations in rats and humans. NeuroImage 54, 1140–1150.

Mandino, F., Cerri, D.H., Garin, C.M., Straathof, M., Van Tilborg, G.A., Chakravarty, M.M., Dhenain, M., Dijkhuizen, R.M., Gozzi, A., Hess, A., 2020. Animal functional magnetic resonance imaging: trends and path toward standardization. Front. Neuroinform. 13, 78.

Mandino, F., Vrooman, R.M., Foo, H.E., Yeow, L.Y., Bolton, T.A.W., Salvan, P., Teoh, C.L., Lee, C.Y., Beauchamp, A., Luo, S., Bi, R., Zhang, J., Lim, G.H.T., Low, N., Sallet, J., Gigg, J., Lerch, J.P., Mars, R.B., Olivo, M., Fu, Y., Grandjean, J., 2021. A triple-network organization for the mouse brain. Mol. Psychiatry 27, 865–872.

Mantini, D., Corbetta, M., Romani, G.L., Orban, G.A., Vanduffel, W., 2013. Evolutionarily novel functional networks in the human brain? J. Neurosci. 33, 3259–3275.

Mantini, D., Gerits, A., Nelissen, K., Durand, J.B., Joly, O., Simone, L., Sawamura, H., Wardak, C., Orban, G.A., Buckner, R.L., Vanduffel, W., 2011. Default mode of brain function in monkeys. J. Neurosci. 31, 12954–12962.

Margulies, D.S., Ghosh, S.S., Goulas, A., Falkiewicz, M., Huntenburg, J.M., Langs, G., Bezgin, G., Eickhoff, S.B., Castellanos, F.X., Petrides, M., 2016. Situating the default-mode network along a principal gradient of macroscale cortical organization. Proc. Natl. Acad. Sci. 113, 12574–12579.

Margulies, D.S., Vincent, J.L., Kelly, C., Lohmann, G., Uddin, L.Q., Biswal, B.B., Villringer, A., Castellanos, F.X., Milham, M.P., Petrides, M., 2009. Precuneus shares intrinsic functional architecture in humans and monkeys. Proc. Natl. Acad. Sci. 106, 20069–20074.

Marquand, A.F., Rezek, I., Buitelaar, J., Beckmann, C.F., 2016. Understanding heterogeneity in clinical cohorts using normative models: beyond case-control studies. Biol. Psychiatry 80, 552–561.

Mesulam, M.M., 1998. From sensation to cognition. Brain 121 (Pt 6), 1013–1052.

Milham, M., Petkov, C., Belin, P., Hamed, S.B., Evrard, H., Fair, D., Fox, A., Froudist-Walsh, S., Hayashi, T., Kastner, S., 2022. Toward next-generation primate neuroscience: a collaboration-based strategic plan for integrative neuroimaging. Neuron 110, 16–20.

Milham, M., Petkov, C.I., Margulies, D.S., Schroeder, C.E., Basso, M.A., Belin, P., Fair, D.A., Fox, A., Kastner, S., Mars, R.B., Messinger, A., Poirier, C., Vanduffel, W., Van Essen, D.C., Alvand, A., Becker, Y., Ben Hamed, S., Benn, A., Bodin, C., Boretius, S., Cagna, B., Coulon, O., El-Gohary, S.H., Evrard, H., Forkel, S.J., Friedrich, P., Froudist-Walsh, S., Garza-Villarreal, E.A., Gao, Y., Gozzi, A., Grigis, A., Hartig, R., Hayashi, T., Heuer, K., Howells, H., Ardesch, D.J., Jarraya, B., Jarrett, W., Jedema, H.P., Kagan, I., Kelly, C., Kennedy, H., Klink, P.C., Kwok, S.C., Leech, R., Liu, X., Madan, C., Madushanka, W., Majka, P., Mallon, A.-M., Marche, K., Meguerditchian, A., Menon, R.S., Merchant, H., Mitchell, A., Nenning, K.-H., Nikolaidis, A., Ortiz-Rios, M., Pagani, M., Pareek, V., Prescott, M., Procyk, E., Rajimehr, R., Rautu, I.-S., Raz, A., Roe, A.W., Rossi-Pool, R., Roumazeilles, L., Sakai, T., Sallet, J., García-Saldivar, P., Sato, C., Sawiak, S., Schiffer, M., Schwiedrzik, C.M., Seidlitz, J., Sein, J., Shen, Z.-M., Shmuel, A., Silva, A.C., Simone, L., Sirmpilatze, N., Sliwa, J., Smallwood, J., Tasserie, J., Thiebaut De Schotten, M., Toro, R., Trapeau, R., Uhrig, L., Vezoli, J., Wang, Z., Wells, S., Williams, B., Xu, T., Xu, A.G., Yacoub, E., Zhan, M., Ai, L., Amiez, C., Balezeau, F., et al., 2020. Accelerating the evolution of nonhuman primate neuroimaging. Neuron 105, 600–603.

Mitra, A., Snyder, A.Z., Tagliazucchi, E., Laufs, H., Elison, J., Emerson, R.W., Shen, M.D., Wolff, J.J., Botteron, K.N., Dager, S., 2017. Resting-state fMRI in sleeping infants more closely resembles adult sleep than adult wakefulness. PLoS One 12, e0188122.

Mitra, A., Snyder, A.Z., Tagliazucchi, E., Laufs, H., Raichle, M.E., 2015. Propagated infra-slow intrinsic brain activity reorganizes across wake and slow wave sleep. elife 4, e10781.

Monti, M.M., Lutkenhoff, E.S., Rubinov, M., Boveroux, P., Vanhaudenhuyse, A., Gosseries, O., Bruno, M.-A., Noirhomme, Q., Boly, M., Laureys, S., 2013. Dynamic change of global and local information processing in propofol-induced loss and recovery of consciousness. PLoS Comput. Biol. 9, e1003271.

Moreau, C.A., Urchs, S.G.W., Kuldeep, K., Orban, P., Schramm, C., Dumas, G., Labbe, A., Huguet, G., Douard, E., Quirion, P.-O., Lin, A., Kushan, L., Grot, S., Luck, D., Mendrek, A., Potvin, S., Stip, E., Bourgeron, T., Evans, A.C., Bearden, C.E., Bellec, P., Jacquemont, S., 2020. Mutations associated with neuropsychiatric conditions delineate functional brain connectivity dimensions contributing to autism and schizophrenia. Nat. Commun. 11, 5272.

Mueller, S.G., Weiner, M.W., Thal, L.J., Petersen, R.C., Jack, C.R., Jagust, W., Trojanowski, J.Q., Toga, A.W., Beckett, L., 2005. Ways toward an early diagnosis in Alzheimer's disease: the Alzheimer's Disease Neuroimaging Initiative (ADNI). Alzheimers Dement. 1, 55–66.

Müller, R., Shih, P., Keehn, B., 2011. Underconnected, but how? A survey of functional connectivity MRI studies in autism spectrum disorders. Cereb. Cortex 21.

Murphy, K., Fox, M.D., 2017. Towards a consensus regarding global signal regression for resting state functional connectivity MRI. NeuroImage 154, 169–173.

Musso, F., Brinkmeyer, J., Mobascher, A., Warbrick, T., Winterer, G., 2010. Spontaneous brain activity and EEG microstates. A novel EEG/fMRI analysis approach to explore resting-state networks. NeuroImage 52, 1149–1161.

Nichols, T.E., Das, S., Eickhoff, S.B., Evans, A.C., Glatard, T., Hanke, M., Kriegeskorte, N., Milham, M.P., Poldrack, R.A., Poline, J.-B., Proal, E., Thirion, B., Van Essen, D.C., White, T., Yeo, B.T.T., 2017. Best practices in data analysis and sharing in neuroimaging using MRI. Nat. Neurosci. 20, 299–303.

Nir, Y., Mukamel, R., Dinstein, I., Privman, E., Harel, M., Fisch, L., Gelbard-Sagiv, H., Kipervasser, S., Andelman, F., Neufeld, M.Y., Kramer, U., Arieli, A., Fried, I., Malach, R., 2008. Interhemispheric correlations of slow spontaneous neuronal fluctuations revealed in human sensory cortex. Nat. Neurosci. 11, 1100–1108.

Oh, S.W., Harris, J.A., Ng, L., Winslow, B., Cain, N., Mihalas, S., Wang, Q., Lau, C., Kuan, L., Henry, A.M., Mortrud, M.T., Ouellette, B., Nguyen, T.N., Sorensen, S.A., Slaughterbeck, C.R., Wakeman, W., Li, Y., Feng, D., Ho, A., Nicholas, E., Hirokawa, K.E., Bohn, P., Joines, K.M., Peng, H., Hawrylycz, M.J., Phillips, J.W., Hohmann, J.G., Wohnoutka, P., Gerfen, C.R., Koch, C., Bernard, A., Dang, C., Jones, A.R., Zeng, H., 2014. A mesoscale connectome of the mouse brain. Nature 508, 207–214.

O'Reilly, J.X., Croxson, P.L., Jbabdi, S., Sallet, J., Noonan, M.P., Mars, R.B., Browning, P.G.F., Wilson, C.R.E., Mitchell, A.S., Miller, K.L., Rushworth, M.F.S., Baxter, M.G., 2013. Causal effect of disconnection lesions on interhemispheric functional connectivity in rhesus monkeys. Proc. Natl. Acad. Sci. 110, 13982–13987.

Oyarzabal, E.A., Hsu, L.-M., Das, M., Chao, T.-H.H., Zhou, J., Song, S., Zhang, W., Smith, K.G., Sciolino, N.R., Evsyukova, I.Y., Yuan, H., Lee, S.-H., Cui, G., Jensen, P., Shih, Y.-Y.I., 2022. Chemogenetic stimulation of tonic locus coeruleus activity strengthens the default mode network. Sci. Adv. 8, eabm9898.

Paasonen, J., Laakso, H., Pirttimäki, T., Stenroos, P., Salo, R.A., Zhurakovskaya, E., Lehto, L.J., Tanila, H., Garwood, M., Michaeli, S., Idiyatullin, D., Mangia, S., Gröhn, O., 2020. Multi-band SWIFT enables quiet and artefact-free EEG-fMRI and awake fMRI studies in rat. NeuroImage 206, 116338.

Pagani, M., Barsotti, N., Bertero, A., Trakoshis, S., Ulysse, L., Locarno, A., Miseviciute, I., De Felice, A., Canella, C., Supekar, K., 2021a. mTOR-related synaptic pathology causes autism spectrum disorder-associated functional hyperconnectivity. Nat. Commun. 12, 1-15.

Pagani, M., Bertero, A., Liska, A., Galbusera, A., Sabbioni, M., Barsotti, N., Colenbier, N., Marinazzo, D., Scattoni, M.L., Pasqualetti, M., Gozzi, A., 2019. Deletion of autism risk gene Shank3 disrupts prefrontal connectivity. J. Neurosci., 2529. 18.

Pagani, M., Zerbi, V., Galbusera, A., Xu, T., Wenderoth, N., Milham, M., Di Martino, A., Gozzi, A., 2021b. Mapping the neuroconnectional landscape in autism via cross-species fMRI. Neuropsychopharmacology, 141-142. SPRINGERNATURE CAMPUS, 4 CRINAN ST, LONDON, N1 9XW, ENGLAND.

Peeters, L.M., Hinz, R., Detrez, J.R., Missault, S., De Vos, W.H., Verhoye, M., Van Der Linden, A., Keliris, G.A., 2020a. Chemogenetic silencing of neurons in the mouse anterior cingulate area modulates neuronal activity and functional connectivity. NeuroImage 220, 117088.

Peeters, L.M., Van Den Berg, M., Hinz, R., Majumdar, G., Pintelon, I., Keliris, G.A., 2020b. Cholinergic modulation of the default mode like network in rats. iScience 23.

Pezzulo, G., Zorzi, M., Corbetta, M., 2021. The secret life of predictive brains: what's spontaneous activity for? Trends Cogn. Sci. 25, 730-743.

Poldrack, R.A., Paré-Blagoev, E.J., Grant, P.E., 2002. Pediatric functional magnetic resonance imaging: progress and challenges. Top. Magn. Reson. Imaging 13, 61-70.

Power, J.D., Laumann, T.O., Plitt, M., Martin, A., Petersen, S.E., 2017. On global fMRI signals and simulations. Trends Cogn. Sci. 21, 911-913.

Power, J.D., Mitra, A., Laumann, T.O., Snyder, A.Z., Schlaggar, B.L., Petersen, S.E., 2014a. Methods to detect, characterize, and remove motion artifact in resting state fMRI. NeuroImage 84, 320-341.

Power, J., Schlaggar, B., Petersen, S., 2014b. Studying brain organization via spontaneous fMRI signal. Neuron 84, 681-696.

Preuss, T.M., 1995. Do rats have prefrontal cortex? The rose-Woolsey-akert program reconsidered. J. Cogn. Neurosci. 7, 1-24.

Ramsey, L.E., Siegel, J.S., Baldassarre, A., Metcalf, N.V., Zinn, K., Shulman, G.L., Corbetta, M., 2016. Normalization of network connectivity in hemispatial neglect recovery. Ann. Neurol. 80, 127-141.

Raut, R.V., Snyder, A.Z., Mitra, A., Yellin, D., Fujii, N., Malach, R., Raichle, M.E., 2021. Global waves synchronize the brain's functional systems with fluctuating arousal. Sci. Adv. 7, eabf2709.

Rocchi, F., Canella, C., Noei, S., Gutierrez-Barragan, D., Coletta, L., Galbusera, A., Stuefer, A., Vassanelli, S., Pasqualetti, M., Iurilli, G., 2022. Increased fMRI connectivity upon chemogenetic inhibition of the mouse prefrontal cortex. Nat. Commun. 13, 1-15.

Roth, B.L., 2016. DREADDs for neuroscientists. Neuron 89, 683-694.

Rungta, R.L., Osmanski, B.-F., Boido, D., Tanter, M., Charpak, S., 2017. Light controls cerebral blood flow in naive animals. Nat. Commun. 8, 14191.

Saccà, V., Sarica, A., Novellino, F., Barone, S., Tallarico, T., Filippelli, E., Granata, A., Chiriaco, C., Bruno Bossio, R., Valentino, P., 2019. Evaluation of machine learning algorithms performance for the prediction of early multiple sclerosis from resting-state FMRI connectivity data. Brain Imaging Behav. 13, 1103-1114.

Sanides, F., 1962. Die Architektonik des menschlichen Stirnhirns. Springer, Berlin.

Sarraf, S., Tofighi, G., 2016. Classification of Alzheimer's disease using FMRI data and deep learning convolutional neural networks. arXiv. preprint arXiv:1603.08631.

Satterstrom, F.K., Kosmicki, J.A., Wang, J., Breen, M.S., De Rubeis, S., An, J.-Y., Peng, M., Collins, R., Grove, J., Klei, L., Stevens, C., Reichert, J., Mulhern, M.S., Artomov, M., Gerges, S., Sheppard, B., Xu, X., Bhaduri, A., Norman, U., Brand, H., Schwartz, G., Nguyen, R., Guerrero, E.E., Dias, C., Aleksic, B., Anney, R., Barbosa, M., Bishop, S., Brusco, A., Bybjerg-Grauholm, J., Carracedo, A., Chan, M.C.Y., Chiocchetti, A.G.,

Chung, B.H.Y., Coon, H., Cuccaro, M.L., Curró, A., Dalla Bernardina, B., Doan, R., Domenici, E., Dong, S., Fallerini, C., Fernández-Prieto, M., Ferrero, G.B., Freitag, C.M., Fromer, M., Gargus, J.J., Geschwind, D., Giorgio, E., González-Peñas, J., Guter, S., Halpern, D., Hansen-Kiss, E., He, X., Herman, G.E., Hertz-Picciotto, I., Hougaard, D.M., Hultman, C.M., Ionita-Laza, I., Jacob, S., Jamison, J., Jugessur, A., Kaartinen, M., Knudsen, G.P., Kolevzon, A., Kushima, I., Lee, S.L., Lehtimäki, T., Lim, E.T., Lintas, C., Lipkin, W.I., Lopergolo, D., Lopes, F., Ludena, Y., Maciel, P., Magnus, P., Mahjani, B., Maltman, N., Manoach, D.S., Meiri, G., Menashe, I., Miller, J., Minshew, N., Montenegro, E.M.S., Moreira, D., Morrow, E.M., Mors, O., Mortensen, P.B., Mosconi, M., Muglia, P., Neale, B.M., Nordentoft, M., Ozaki, N., Palotie, A., Parellada, M., Passos-Bueno, M.R., Pericak-Vance, M., Persico, A.M., Pessah, I., Puura, K., et al., 2020. Large-scale exome sequencing study implicates both developmental and functional changes in the neurobiology of autism. Cell 180, 568–584.e23.

Savitz, J.B., Drevets, W.C., 2009. Imaging phenotypes of major depressive disorder: genetic correlates. Neuroscience 164, 300–330.

Scheidegger, M., Walter, M., Lehmann, M., Metzger, C., Grimm, S., Boeker, H., Boesiger, P., Henning, A., Seifritz, E., 2012. Ketamine Decreases Resting State Functional Network Connectivity in Healthy Subjects: Implications for Antidepressant Drug Action.

Schleifer, C.H., Lin, A., Kushan, L., Ji, J.L., Yang, G., Bearden, C.E., Anticevic, A., 2017. Reciprocal disruptions in thalamic and hippocampal resting-state functional connectivity in youth with 22q11.2 deletions. bioRxiv.

Schölvinck, M.L., Maier, A., Frank, Q.Y., Duyn, J.H., Leopold, D.A., 2010. Neural basis of global resting-state fMRI activity. Proc. Natl. Acad. Sci. 107, 10238–10243.

Schreiner, M.J., Karlsgodt, K.H., Uddin, L.Q., Chow, C., Congdon, E., Jalbrzikowski, M., Bearden, C.E., 2014. Default mode network connectivity and reciprocal social behavior in 22q11. 2 deletion syndrome. Soc. Cogn. Affect. Neurosci. 9, 1261–1267.

Schwarz, A.J., Gass, N., Sartorius, A., Zheng, L., Spedding, M., Schenker, E., Risterucci, C., Meyer-Lindenberg, A., Weber-Fahr, W., 2013. The low-frequency blood oxygenation level-dependent functional connectivity signature of the hippocampal-prefrontal network in the rat brain. Neuroscience 228, 243–258.

Seghier, M.L., Price, C.J., 2018. Interpreting and utilising intersubject variability in brain function. Trends Cogn. Sci. 22, 517–530.

Sethi, S.S., Zerbi, V., Wenderoth, N., Fornito, A., Fulcher, B.D., 2017. Structural connectome topology relates to regional BOLD signal dynamics in the mouse brain. Chaos 27, 047405.

Sforazzini, F., Bertero, A., Dodero, L., David, G., Galbusera, A., Scattoni, M.L., Pasqualetti, M., Gozzi, A., 2016. Altered functional connectivity networks in acallosal and socially impaired BTBR mice. Brain Struct. Funct. 221, 941–954.

Sforazzini, F., Schwarz, A.J., Galbusera, A., Bifone, A., Gozzi, A., 2014. Distributed BOLD and CBV-weighted resting-state networks in the mouse brain. NeuroImage 87, 403–415.

Shmuel, A., Leopold, D.A., 2008. Neuronal correlates of spontaneous fluctuations in fMRI signals in monkey visual cortex: implications for functional connectivity at rest. Hum. Brain Mapp. 29, 751–761.

Shofty, B., Bergmann, E., Zur, G., Asleh, J., Bosak, N., Kavushansky, A., Castellanos, F.X., Ben-Sira, L., Packer, R.J., Vezina, G.L., Constantini, S., Acosta, M.T., Kahn, I., 2019. Autism-associated Nf1 deficiency disrupts corticocortical and corticostriatal functional connectivity in human and mouse. Neurobiol. Dis. 130, 104479.

Siddiqi, S.H., Kording, K.P., Parvizi, J., Fox, M.D., 2022. Causal mapping of human brain function. Nat. Rev. Neurosci. 23, 361–375.

Siddiqi, S.H., Schaper, F.L.W.V.J., Horn, A., Hsu, J., Padmanabhan, J.L., Brodtmann, A., Cash, R.F.H., Corbetta, M., Choi, K.S., Dougherty, D.D., Egorova, N., Fitzgerald, P.B., George, M.S., Gozzi, S.A., Irmen, F., Kuhn, A.A., Johnson, K.A., Naidech, A.M., Pascual-Leone, A., Phan, T.G., Rouhl, R.P.W., Taylor, S.F., Voss, J.L., Zalesky, A.,

Grafman, J.H., Mayberg, H.S., Fox, M.D., 2021. Brain stimulation and brain lesions converge on common causal circuits in neuropsychiatric disease. Nat. Hum. Behav. 5, 1707–1716.

Simhal, A.K., José Filho, O., Segura, P., Cloud, J., Petkova, E., Gallagher, R., Castellanos, F.X., Colcombe, S., Milham, M.P., Di Martino, A., 2021. Predicting multiscan MRI outcomes in children with neurodevelopmental conditions following MRI simulator training. Dev. Cogn. Neurosci. 52, 101009.

Simons VIP Consortium, 2012. Simons variation in individuals project (Simons VIP): a genetics-first approach to studying autism spectrum and related neurodevelopmental disorders. Neuron 73, 1063–1067.

Smith, S.M., Fox, P.T., Miller, K.L., Glahn, D.C., Fox, P.M., Mackay, C.E., Filippini, N., Watkins, K.E., Toro, R., Laird, A.R., Beckmann, C.F., 2009. Correspondence of the brain's functional architecture during activation and rest. Proc. Natl. Acad. Sci. 106, 13040–13045.

Stam, C.J., 2014. Modern network science of neurological disorders. Nat. Rev. Neurosci. 15, 683–695.

Stenroos, P., Paasonen, J., Salo, R.A., Jokivarsi, K., Shatillo, A., Tanila, H., Gröhn, O., 2018. Awake rat brain functional magnetic resonance imaging using standard radio frequency coils and a 3D printed restraint kit. Front. Neurosci. 12.

Stoet, G., Snyder, L.H., 2009. Neural correlates of executive control functions in the monkey. Trends Cogn. Sci. 13, 228–234.

Suárez, L.E., Markello, R.D., Betzel, R.F., Misic, B., 2020. Linking structure and function in macroscale brain networks. Trends Cogn. Sci. 24, 302–315.

Supekar, K., Uddin, L.-Á., Khouzam, A., Phillips, J., Gaillard, W.-Á., Kenworthy, L.-Á., Yerys, B.-Á., Vaidya, C.-Á., Menon, V., 2013. Brain hyperconnectivity in children with autism and its links to social deficits. Cell Rep. 5, 738–747.

Tagliazucchi, E., Van Someren, E.J.W., 2017. The large-scale functional connectivity correlates of consciousness and arousal during the healthy and pathological human sleep cycle. NeuroImage 160, 55–72.

Tagliazucchi, E., Von Wegner, F., Morzelewski, A., Brodbeck, V., Jahnke, K., Laufs, H., 2013. Breakdown of long-range temporal dependence in default mode and attention networks during deep sleep. Proc. Natl. Acad. Sci. 110, 15419–15424.

Thomas Yeo, B., Krienen, F.M., Sepulcre, J., Sabuncu, M.R., Lashkari, D., Hollinshead, M., Roffman, J.L., Smoller, J.W., Zöllei, L., Polimeni, J.R., 2011. The organization of the human cerebral cortex estimated by intrinsic functional connectivity. J. Neurophysiol. 106, 1125–1165.

Trachtenberg, A.J., Filippini, N., Ebmeier, K.P., Smith, S.M., Karpe, F., Mackay, C.E., 2012. The effects of APOE on the functional architecture of the resting brain. NeuroImage 59, 565–572.

Trakoshis, S., Rocchi, F., Canella, C., You, W., Chakrabarti, B., Ruigrok, A.N., Bullmore, E.T., Suckling, J., Markicevic, M., Zerbi, V., 2020. Intrinsic excitation-inhibition imbalance affects medial prefrontal cortex differently in autistic men versus women. elife 9, e55684.

Tsai, P.-J., Keeley, R.J., Carmack, S.A., Vendruscolo, J.C.M., Lu, H., Gu, H., Vendruscolo, L.F., Koob, G.F., Lin, C.-P., Stein, E.A., Yang, Y., 2020. Converging structural and functional evidence for a rat salience network. Biol. Psychiatry 88, 867–878.

Tu, W., Ma, Z., Ma, Y., Dopfel, D., Zhang, N., 2020. Suppressing anterior cingulate cortex modulates default mode network and behavior in awake rats. Cereb. Cortex 31, 312–323.

Turchi, J., Chang, C., Ye, F.Q., Russ, B.E., Yu, D.K., Cortes, C.R., Monosov, I.E., Duyn, J.H., Leopold, D.A., 2018. The basal forebrain regulates global resting-state fMRI fluctuations. Neuron 97, 940–952.e4.

Uddin, L.Q., 2017. Mixed signals: on separating brain signal from noise. Trends Cogn. Sci. 21, 405–406.

Uehara, T., Yamasaki, T., Okamoto, T., Koike, T., Kan, S., Miyauchi, S., Kira, J.-I., Tobimatsu, S., 2013. Efficiency of a "small-world" brain network depends on consciousness level: a resting-state fMRI study. Cereb. Cortex 24, 1529–1539.

Uhrig, L., Sitt, J.D., Jacob, A., Tasserie, J., Barttfeld, P., DuPont, M., Dehaene, S., Jarraya, B., 2018. Resting-state dynamics as a cortical signature of anesthesia in monkeys. Anesthesiology 129, 942–958.

Van De Ville, D., 2019. Brain dynamics: global pulse and brain state switching. Curr. Biol. 29, R690–R692.

Vasa, R.A., Mostofsky, S.H., Ewen, J.B., 2016. The disrupted connectivity hypothesis of autism spectrum disorders: time for the next phase in research. Biol. Psychiatry Cogn. Neurosci. Neuroimag. 1, 245–252.

Verhagen, L., Gallea, C., Folloni, D., Constans, C., Jensen, D.E., Ahnine, H., Roumazeilles, L., Santin, M., Ahmed, B., Lehericy, S., 2019. Offline impact of transcranial focused ultrasound on cortical activation in primates. elife 8, e40541.

Vincent, J.L., Patel, G.H., Fox, M.D., Snyder, A.Z., Baker, J.T., Van Essen, D.C., Zempel, J.M., Snyder, L.H., Corbetta, M., Raichle, M.E., 2007. Intrinsic functional architecture in the anaesthetized monkey brain. Nature 447, 83–86.

Vogt, B.A., Paxinos, G., 2012. Cytoarchitecture of mouse and rat cingulate cortex with human homologies. Brain Struct. Funct. 219, 185–192.

Volkow, N.D., Koob, G.F., Croyle, R.T., Bianchi, D.W., Gordon, J.A., Koroshetz, W.J., Pérez-Stable, E.J., Riley, W.T., Bloch, M.H., Conway, K., 2018. The conception of the ABCD study: from substance use to a broad NIH collaboration. Dev. Cogn. Neurosci. 32, 4–7.

Wang, X.J., 2020. Macroscopic gradients of synaptic excitation and inhibition in the neocortex. Nat. Rev. Neurosci. 21, 169–178.

Wang, L., Alpert, K.I., Calhoun, V.D., Cobia, D.J., Keator, D.B., King, M.D., Kogan, A., Landis, D., Tallis, M., Turner, M.D., 2016. SchizConnect: mediating neuroimaging databases on schizophrenia and related disorders for large-scale integration. NeuroImage 124, 1155–1167.

Wang, L., Saalmann, Y.B., Pinsk, M.A., Arcaro, M.J., Kastner, S., 2012. Electrophysiological Low-frequency coherence and cross-frequency coupling contribute to BOLD connectivity. Neuron 76, 1010–1020.

Weitz, A.J., Lee, H.J., Choy, M., 2019. Input to orbitofrontal cortex drives brain-wide, frequency-dependent inhibition mediated by GABA and zona incerta. Neuron 104 (6), 1153–1167.e4. https://doi.org/10.1016/j.neuron.2019.09.023.

Whitesell, J.D., Liska, A., Coletta, L., Hirokawa, K.E., Bohn, P., Williford, A., Groblewski, P.A., Graddis, N., Kuan, L., Knox, J.E., 2021. Regional, layer, and cell-type-specific connectivity of the mouse default mode network. Neuron 109, 545–559. e8.

Whitfield-Gabrieli, S., Ford, J.M., 2012. Default mode network activity and connectivity in psychopathology. 359, Annu. Rev. Clin. Psychol. 8, 49–76.

Xu, T., Nenning, K.-H., Schwartz, E., Hong, S.-J., Vogelstein, J.T., Goulas, A., Fair, D.A., Schroeder, C.E., Margulies, D.S., Smallwood, J., 2020. Cross-species functional alignment reveals evolutionary hierarchy within the connectome. NeuroImage 223, 117346.

Xu, A.G., Qian, M., Tian, F., Xu, B., Friedman, R.M., Wang, J., Song, X., Sun, Y., Chernov, M.M., Cayce, J.M., 2019a. Focal infrared neural stimulation with high-field functional MRI: a rapid way to map mesoscale brain connectomes. Sci. Adv. 5, eaau7046.

Xu, T., Sturgeon, D., Ramirez, J.S., Froudist-Walsh, S., Margulies, D.S., Schroeder, C.E., Fair, D.A., Milham, M.P., 2019b. Interindividual variability of functional connectivity in awake and anesthetized rhesus macaque monkeys. Biol. Psychiatry Cogn. Neurosci. Neuroimag. 4, 543–553.

Yacoub, E., Grier, M.D., Auerbach, E.J., Lagore, R.L., Harel, N., Adriany, G., Zilverstand, A., Hayden, B.Y., Heilbronner, S.R., Uğurbil, K., Zimmermann, J., 2020. Ultra-high field (10.5 T) resting state fMRI in the macaque. NeuroImage 223, 117349.

Yang, H., Liu, J., Sui, J., Pearlson, G., Calhoun, V.D., 2010. A hybrid machine learning method for fusing fMRI and genetic data: combining both improves classification of schizophrenia. Front. Hum. Neurosci. 4, 192.

Yoshida, K., Mimura, Y., Ishihara, R., Nishida, H., Komaki, Y., Minakuchi, T., Tsurugizawa, T., Mimura, M., Okano, H., Tanaka, K.F., Takata, N., 2016. Physiological effects of a habituation procedure for functional MRI in awake mice using a cryogenic radiofrequency probe. J. Neurosci. Methods 274, 38–48.

Yousefi, B., Shin, J., Schumacher, E.H., Keilholz, S.D., 2018. Quasi-periodic patterns of intrinsic brain activity in individuals and their relationship to global signal. NeuroImage 167, 297–308.

Zerbi, V., Floriou-Servou, A., Markicevic, M., Vermeiren, Y., Sturman, O., Privitera, M., Von Ziegler, L., Ferrari, K.D., Weber, B., De Deyn, P.P., 2019. Rapid reconfiguration of the functional connectome after chemogenetic locus coeruleus activation. Neuron 103, 702–718.e5.

Zerbi, V., Ielacqua, G.D., Markicevic, M., Haberl, M.G., Ellisman, M.H., A-Bhaskaran, A., Frick, A., Rudin, M., Wenderoth, N., 2018. Dysfunctional autism risk genes cause circuit-specific connectivity deficits with distinct developmental trajectories. Cereb. Cortex 28, 2495–2506.

Zerbi, V., Pagani, M., Markicevic, M., Matteoli, M., Pozzi, D., Fagiolini, M., Bozzi, Y., Galbusera, A., Scattoni, M.L., Provenzano, G., 2021. Brain mapping across 16 autism mouse models reveals a spectrum of functional connectivity subtypes. Mol. Psychiatry, 1–11.

Zhang, S., Li, C.-S.R., 2017. Functional connectivity parcellation of the human thalamus by independent component analysis. Brain Connect. 7, 602–616.

Zhao, B., Li, T., Smith, S.M., Xiong, D., Wang, X., Yang, Y., Luo, T., Zhu, Z., Shan, Y., Matoba, N., 2022. Common variants contribute to intrinsic human brain functional networks. Nat. Genet., 1–10.

3

Brain networks atlases

Sarah Genon[a,b] and Jingwei Li[a]

[a]*Institute of Neuroscience and Medicine, Brain and Behaviour (INM-7), Research Center Jülich, Jülich, Germany,* [b]*Institute for Systems Neuroscience, Medical Faculty, Heinrich-Heine University Düsseldorf, Düsseldorf, Germany*

Introduction

One challenge in studying the human brain comes from its multiscale organization and its complexity as a system. One of the most important sources of our knowledge on brain organization stems from ex vivo histological examinations, such as cytoarchitecture mapping (Amunts et al., 2013). By investigating changes in microstructural features across the brain, different brain territories and areas have been delineated. However, this "local" approach to studying brain organization is particularly time- and resource-consuming, and more importantly, it does not take into account functional aspects of brain organization emerging from interaction between regions. The advent of neuroimaging scanners has offered the possibility to study brain organization in vivo, by looking at interactions between brain regions across relatively large numbers of individuals. Accordingly, the last decade has seen a burst of parcellation studies that delineate brain organization based on connectivity features, in particular based on functional connectivity (Eickhoff et al., 2018). Since rs-fMRI could be relatively readily performed, resting-state functional connectivity (RSFC), also sometimes called "intrinsic connectivity," became popular for brain parcellation. By applying matrix factorization approaches to rs-fMRI data, many brain networks and brain regions (representing different scales of brain organization) could be delineated. In the next section, we describe the main matrix factorization techniques that have been used to derive brain atlases. We then present the major milestones of scientific developments in the field. Following this overview of the last decade, we discuss the main limitations, challenges, and opportunities in brain network partitions. Finally, in the last part of this chapter, we review the main applications of function brain atlases, with a particular focus on applications for machine learning, as these approaches represent an important avenue of future development in the neuroimaging field.

Advances in Resting-State Functional MRI. https://doi.org/10.1016/B978-0-323-91688-2.00001-1

Methods to delineate brain networks

The first evidence that features of brain macroscale organization can be observed at rest, with observations of rs-fMRI correlations within the sensorimotor network (Biswal et al., 1995), led to the development of different approaches for delineating brain networks or functional systems from rs-fMRI. As a highly interdisciplinary research area, delineation of functional brain networks borrows methods from image segmentation, machine learning and information theory. In this section, we briefly introduce several widely used methods for building functional atlases, including independent component analysis (ICA), K-means, and graph theory. We also discuss the advantages and limitations of these methods, as well as the challenges in this field.

Independent component analysis (ICA)

The application of ICA to fMRI analysis can be traced back to almost 25 years ago (Mckeown et al., 1998). Technically, ICA is an approach that detects statistically independent components in a given two-dimensional matrix, as also described in Chapter 1. In the application to fMRI data, which can be represented as a (*number of time points*) × (*number of voxels*) matrix per participant, ICA decomposes it into two matrices. The first matrix contains the spatial map of each independent component. The second matrix, named *mixing matrix,* corresponds to the temporal dynamics of each component. In rs-fMRI data, the spatial maps can be further interpreted as different functional brain networks and artifacts (Smith et al., 2009).

Although the mathematical representations of ICA might look similar to principal component analysis (PCA) and general linear model (GLM), readers should be aware of the fundamental difference between these models (Calhoun et al., 2009). PCA aims to detect the underlying orthogonal directions which capture the most data variance. However, ICA identifies independent directions in data, meaning that the probabilistic distribution along one direction does not affect the others. It should be noted that statistical independence does not guarantee orthogonality, and vice versa. Although GLM also models the original data matrix as a product of two matrices, the regressor matrix is usually predesigned by researchers, e.g., task-design matrix in the analyses of task-based fMRI data. In contrast, the mixing matrix in ICA is estimated by optimizing the spatial independence among components. Furthermore, GLM is usually not used for delineating brain networks in rs-fMRI data.

The advantages of ICA are obvious. It is fully data-driven and does not require a training procedure. One can easily apply it independently to a new data set. The techniques of ICA are relatively mature in the field of fMRI studies. For instance, the software package FSL

(FMRIB Software Library) offers a function called MELODIC to perform ICA on fMRI volumes (Beckmann and Smith, 2004) (https://fsl.fmrib.ox.ac.uk/fsl/fslwiki/MELODIC). Compared to most other brain network clustering or parcellation methods, ICA can provide a "soft" delineation by allowing a voxel to be assigned to multiple brain networks (Eickhoff et al., 2015).

However, a persistent debate in the application of ICA to fMRI data concerns the choice of spatial or temporal independency (Calhoun et al., 2001c), since ICA can be used in both ways to identify either spatially or temporally independent components (Biswal and Ulmer, 1999). An important consequence of temporal ICA is the generation of spatially overlapping and dependent components, since temporal ICA has no assumption on spatial independence. This characteristic of temporal ICA is also useful to detect certain types of physiological noise (Boubela et al., 2013), especially global artifacts in rs-fMRI (Glasser et al., 2018) (but see debates from Power, 2019), which are not readily detected by spatial ICA (Beall and Lowe, 2010). However, it should be noted that temporal ICA is more computational demanding when applied to fMRI data because usually the number of voxels is much larger than the length of time series. Care needs to be taken when choosing between spatial and temporal ICA, and this choice should depend on the assumption of the specific study designs and on the data.

K-means

Intuitively, the K-means method aims to classify all data points into K clusters so that the interclass distance is maximized and the within-class distance is minimized. Within-class distance captures the average distance between each cluster center and every single data point in this cluster, then averaged across all clusters, whereas interclass distance captures the mean distance between each cluster center and the center of cluster centers. Different distance measures can be used, for instance, the distance between any two data points can be measured by the Mahalanobis distance (a special case is Euclidean distance), the Manhattan distance, or the Chebyshev distance.

When the K-means algorithm is used for identifying functional brain networks, the time series of each brain voxel is treated as a T-dimensional data point, where T indicates the number of time points (Golland et al., 2008; Goutte et al., 1999). Voxels classified into the same cluster are then considered as sharing similar temporal fluctuations and hence are interpreted as a single functional network. Note that, besides time courses, the features of each brain voxel which are used to measure distances can also consist of patterns in the frequency domain (Mezer et al., 2009), or the functional connectivity of the current voxel to other brain regions (Zhang and Li, 2012), depending on the purpose of the study.

However, the K-means algorithm is circumscribed by a few limitations. First, the number of clusters needs to be set a priori, enforcing the researcher to make explicit assumptions on the true data structure. Second, the choice of distance metric may have an important impact on clustering results (Goutte et al., 1999). Last but not the least, K-means is an iterative algorithm. Cluster centers need to be initialized to start the iteration. Unfortunately, repeating the same algorithm with different initialization often generates diverging clustering results. Some of these limitations can be addressed by hierarchical clustering method, introduced in the next subsection.

Hierarchical clustering

Hierarchical clustering is a family of iterative methods to build a hierarchy of clusters. The most common algorithm starts from individual data points, and in each iteration, a new cluster is formed by merging a pair of clusters (from the previous iteration) between which the distance is minimal among all pairs of clusters. The whole procedure is frequently represented by a *dendrogram* (Fig. 1B). The height in a dendrogram indicates the extent to which the connected clusters can be separated. To put it differently, we can decide the number of clusters by selecting a cut-off distance in the dendrogram. For example, by thresholding the distance at 4, we can obtain 3 clusters from these 7 total data points: {D, E}, {A, B, C}, and {F, G}. This procedure therefore bypasses two problems of K-means: initialization of cluster centers and prespecification of cluster numbers.

An important reason to use hierarchical clustering for neuroimaging data analysis pertains to theories positing a hierarchical structural and functional organization of the brain (Arslan and Rueckert, 2015; Park and Friston, 2013). More specifically, whole-brain functional brain architectures, most commonly derived from rs-fMRI data, can be examined at a hierarchy of resolutions, ranging from hundreds to even a thousand locally integrated areas/parcels (Gordon et al., 2016; Schaefer et al., 2018) to around 5–20 spatially distributed networks (Power et al., 2011; Yeo et al., 2011). It is also worth mentioning that hierarchical clustering can not only be used independently (Cordes et al., 2002; Moreno-Dominguez et al., 2014) but can also be integrated with multiple methods including ICA, K-means, and graph theoretical methods for brain parcellation. For example, hierarchical clustering was used to estimate brain modules and systems by classifying independent components derived from individual fMRI (Doucet et al., 2011). Alternatively, hierarchical clustering has also been applied to K-means defined "supervoxels" to build individual parcellations, which were further passed to a spectral clustering algorithm for more reliable group-level parcellation (Arslan and Rueckert, 2015). However, a drawback of hierarchical clustering is its

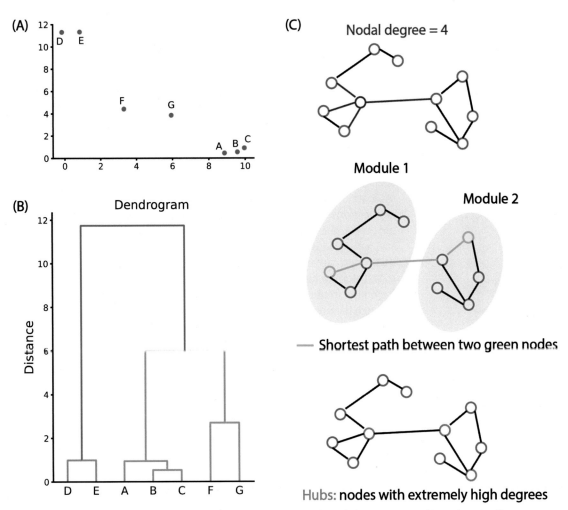

Fig. 1 (A and B) Illustration of hierarchical clustering. A dendrogram in (B) can be created based on the distance among seven data points shown in the two-dimensional space in (A). (C) Some terminologies used in graph theory.

high computational complexity, especially for large-scale datasets. For readers' interests, a detailed technical review on the computational complexity of multiple hierarchical clustering algorithms has been provided by Murtagh and colleagues (Murtagh and Contreras, 2017).

Graph theory and spectral clustering

Like other types of networks (e.g., social networks), brain networks can also be investigated using theories and methods adapted from network science, which are mainly rooted in graph theory (Bullmore and Sporns, 2009; Sporns, 2018). In the field of network science, or graph theory, data of interest are treated as a *graph* consisting of *nodes* connected by *edges* (Fig. 1C). Multiple network properties regarding how nodes are connected can then be studied, such as nodal degree, cluster coefficient, shortest path length, local/global efficiency, modularity, centrality, and small-worldness.

When graph theory is applied to fMRI data, brain regions are treated as nodes in a graph and the edges usually correspond to binarized functional connectivity indicating whether two regions are connected or not (Bassett and Bullmore, 2006). To delineate functional brain networks, multiple graph theoretical methods can be used, such as modularity maximization and Infomap algorithm (Sporns and Betzel, 2016). Intuitively speaking, modularity maximization aims to maximize a modularity quality function, which characterizes, on average, how much the observed connections between any two nodes within the same delineated module exceed the number of random connections in null models. That is to say, a module is a collection of nodes which have strong connections to each other but are loosely connected to nodes in other modules. In contrast, Infomap algorithm decomposes a network into modules depending on the frequency of visits to each node when a random walker travels through the network (Rosvall and Bergstrom, 2008). The direction of node transition in each step relies on the connection strength between each pair of nodes. Besides identifying functional networks, graph theory has also been used to study brain network alterations in disease (Hallquist and Hillary, 2018) and development (Power et al., 2010).

A substantial amount of work has used spectral clustering, which has a close relationship with graph theory, for functional mapping (Craddock et al., 2012; Shen et al., 2013; van den Heuvel et al., 2008). It partitions a graph based on the eigenvalues (i.e., spectrum) and eigenvectors of characteristic matrices of the graph such as *Laplacian matrix*. A popular spectral clustering technique is *normalized cuts*, which was first developed for natural image segmentation (Shi and Malik, 2000). This algorithm basically aims to cut a graph by breaking down the least number of edges. An advantage of spectral clustering is its ability to capture clusters with complicated shape and discontinuity (Eickhoff et al., 2015) with the potential disadvantage that the decomposed clusters tend to be equally sized (Craddock et al., 2012).

Additional methods and challenges in the field

Besides the more widely used methods introduced in previous subsections, functional atlases have also been derived using alternative approaches, such as region growing (Bellec et al., 2006; Blumensath et al., 2013) and edge detection methods (Cohen et al., 2008; Gordon et al., 2016; Nelson et al., 2010; Wig et al., 2014). Region-growing methods start from seed locations (e.g., each cortical voxel) and iteratively merge the most homogenous neighbors (i.e., those with similar fluctuations in fMRI time series) until a predefined threshold is reached. Edge detection methods draw area boundaries based on the local gradient of functional connectivity. Both types of methods generate spatially disjoint brain areas which can be further grouped into networks using other methods.

From the machine learning perspective, the models mentioned above (such as K-means) are considered *discriminative models*, which build cluster boundaries to discriminate observed data points. In contrast, *generative models* can generate new data instances by estimating the joint probabilistic distribution between the input (e.g., fMRI time series or functional connectivity) and the output (e.g., cluster label) variables. For example, a series of studies modeled functional brain organization as a mixture of multiple probabilistic distributions, i.e., using *mixture models* (Kong et al., 2019; Lashkari et al., 2010; Yeo et al., 2011). More specifically, functional connectivity profiles of each cortical region were assumed to follow a mixture of von Mises-Fisher distributions, where each distribution had a specific mean functional connectivity profile, corresponding to each cluster, which were estimated by maximum likelihood estimation. In addition, *Latent Dirichlet Allocation* (LDA) has been used to estimate cortical networks which were allowed to be spatially overlapped (Yeo et al., 2014). Hence, a single cortical region can be involved in multiple networks. LDA, first developed to solve text mining problem (Blei et al., 2003), is a three-layer model including document, topic, and word. Relationship between each pair of layers is characterized by a probabilistic distribution so that a document can contain multiple topics, which can be further composed of multiple words. In brain parcellation applications, these three layers corresponded to the entire cortex, networks, and cortical regions.

A common issue shared across almost all these methods pertains to choosing the "right" number of clusters/components. Crucial information could be neglected if only a few clusters are selected. In contrast, higher-order models might be overfitted to noise. One approach to select the optimal model order is to adopt methods from information theory, such as *Akaike's information criterion* and *Bayesian information criterion* (Calhoun et al., 2001b; Li et al., 2007; Thirion et al., 2014). These criteria calculate the trade-off between model complexity (number of clusters/components here) and goodness-of-fit of the model, taking both overfitting and underfitting into account. Alternatively, reproducibility/stability or reliability measures can also be used to select the number of clusters/components (Yeo et al., 2011). *Intraclass correlation coefficient*, a common metric for reliability, has been used to estimate the number of ICA components (Zuo et al., 2010). Reproducibility can be assessed by measures including adjusted Rand index, adjusted mutual information, and Dice overlap coefficient either on resampled data obtained by bootstrap (or cross-validation) or across independent data sets (Bellec et al., 2010; Craddock et al., 2012; Thirion et al., 2014). In the absence of "ground-truth parcellation," stability and reproducibility of delineated brain networks across different data collection and preprocessing pipelines and clustering techniques are important proxies for underlying neurobiological validity (Eickhoff et al., 2018).

For methods such as ICA and K-means, a common problem is the difficulty of generalizing individual-level components/clusters (i.e., components derived at the subject-level) to group-level components/clusters (Calhoun et al., 2009). Nonetheless, several methods have been proposed to overcome this difficulty by concatenating (or averaging) images from multiple participants before applying ICA, or averaging the ICA results across participants (Calhoun et al., 2001a,b; Guo and Pagnoni, 2008; Schmithorst and Holland, 2004). When K-means is first applied to each individual participant's data, a group analysis can be conducted by maximizing the number of voxels/regions consistently assigned to the same clusters across individuals (Golland et al., 2008). This can also be achieved by the bootstrap procedure proposed by Bellec et al. (Bellec et al., 2010), in which two regions are assigned to the same group-level cluster if the probability that they belong to the same cluster at individual level is maximized. Alternatively, K-means can also be applied directly to group-level features, e.g., a t-test map of functional connectivity across all participants (Zhang and Li, 2012).

Although references to the methods we mentioned in this section were mostly based on rs-fMRI, these methods can be used to delineate brain networks and areas from any kind of connectivity or similarity of neuroimaging features. For example, ICA and graph theoretical modeling can be also used on coactivation data provided by meta-analytic approaches (coactivation-based parcellations) (Fox et al., 2014). Furthermore, as mentioned above, ICA was first applied to task-based fMRI data for delineating task-evoked vs. task-inactive areas, then to rs-fMRI. Similarly, the multilevel bootstrap analysis based on hierarchical clustering and K-means, first developed on rs-fMRI data that we mentioned above (Bellec et al., 2010), can also be applied to derive stable clusters in task-based fMRI (Orban et al., 2015).

From evidence of individual networks to brain network atlases

Searching for canonical networks in resting-state signal

A task negative network and a task positive network

Capitalizing on the approaches described above, several networks or functional systems have been progressively disentangled. Following up on the evidence of a default-mode network, a Task Negative Network and a Task Positive Network (see Fig. 2) were sug-

Task-positive and task-negative networks

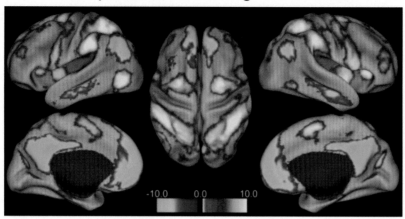

Ventral and dorsal attentional networks

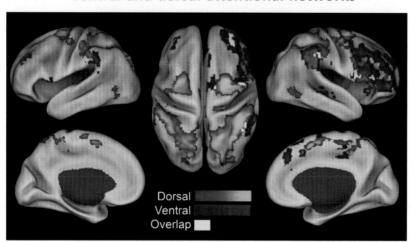

Fig. 2 The task-positive vs task negative networks and the ventral and dorsal attentional networks emerge in rs-fMRI signal. Upper part, from Fox, M. D., Snyder, A. Z., Vincent, J. L., Corbetta, M., Van Essen, D. C., Raichle, M. E., 2005. The human brain is intrinsically organized into dynamic, anticorrelated functional networks. Proc. Natl. Acad. Sci. U. S. A. 102(27), 9673–9678. https://doi.org/10.1073/pnas.0504136102. Copyright (2005) National Academy of Science and lower part, from Fox, M. D., Corbetta, M., Snyder, A. Z., Vincent, J. L., Raichle, M. E., 2006. Spontaneous neuronal activity distinguishes human dorsal and ventral attention systems. Proc. Natl. Acad. Sci. U. S. A. 103(26), 10046–10051. https://doi.org/10.1073/pnas.0604187103. Copyright (2006) National Academy of Science.

gested to be anticorrelated (Fox et al., 2005). This was found by examining correlation in resting-state signals between six predefined seed regions. Three of these regions were designated as "task-positive regions" since they usually show increases in activity during attention-demanding cognitive tasks: the intraparietal sulcus (IPS), the frontal eye field (FEF), and the middle temporal region (MT+). In contrast, the medial prefrontal cortex (MPF), the posterior cingulate cortex/precuneus (PCC) and the lateral parietal cortex (LP) were considered as "task-negative" regions as they routinely exhibit decreases in activity during attention-demanding tasks. Although the anticorrelated nature of these networks was subsequently vividly debated (Anderson et al., 2011; Fox et al., 2009; Saad et al., 2012), this seminal study paved the way toward the identification of the main functional systems that support human cognition from fMRI signals acquired at rest. It indeed demonstrated that the macroscale organization of the brain is preserved in spontaneous low-frequency BOLD signal fluctuations across the brain. Accordingly, several subsequent studies aimed to further delineate canonical networks by capitalizing on RSFC.

A ventral attentional network, a dorsal attentional network, and a frontoparietal control network

Soon after, Fox et al. (2006) further demonstrated that two different attention systems (see Fig. 2), a ventral one and a dorsal one, could be observed in rs-fMRI data of healthy participants. These two systems were previously conceptualized based on a collection of behavioral, neuroimaging, lesion, and electrophysiological studies. The derived model hence proposed that different attentional operations during sensory orienting are carried out by two separate frontoparietal systems, a bilateral dorsal attention system involved in top-down orienting of attention and a right-lateralized ventral attention system involved in reorienting attention in response to salient sensory stimuli. Fox et al. (2006) further demonstrated that these systems could be observed in spontaneous fluctuations of the signal at rest. Their study also revealed regions in the prefrontal cortex correlated with both systems, suggesting a potential mechanism for mediating the functional interaction between systems. Along the same line, Vincent et al. (2008) showed evidence for a frontoparietal control system in intrinsic functional connectivity, this network being spatially interposed between the dorsal attention system and the hippocampal-cortical memory system. Hence, slightly different, but closely related, conceptualizations of brain control and attentional networks organization emerged from the study of intrinsic connectivity across samples of healthy participants.

The default-mode subnetworks

In the same vein, the conceptualization of the default-mode network could also be further refined by examining graph-theoretic and clustering functional connection properties of the DMN regions in healthy participants. By doing so, Andrews-Hanna et al. (2010) suggested two DMN subnetworks. In line with previous studies (e.g., Buckner et al., 2009), they showed that the PCC and aMPFC (anterior medial prefrontal cortex) are crucial hubs showing high betweenness centrality, while the other DMN regions could be dissociated into two distinct subsystems. A "dorsal medial prefrontal cortex (dMPFC) subsystem" is formed by the dMPFC, temporoparietal junction (TPJ), the lateral temporal cortex (LTC), and the temporal pole (TempP), and a "medial temporal lobe (MTL) subsystem" is formed by the ventral MPFC (vMPFC), the posterior inferior parietal lobule (pIPL), the retrosplenial cortex (Rsp), the parahippocampal cortex (PHC), and the hippocampal formation (HF+). The relevance of these two subnetworks for human cognition was then further demonstrated with task-fMRI experiments.

Data-driven decomposition of RSFC into networks

Parcellation into canonical networks

By being able to confirm and further characterize core brain networks evidenced from a range of methods other than fMRI, as well as by task-fMRI, rs-fMRI established its validity for the study of brain organization. Following the first wave of relatively hypothesis-driven studies described above, several brain network atlases were developed by partitioning the cortex into a certain number of networks (Doucet et al., 2011; Gordon et al., 2016; Laumann et al., 2015). One of the most popular and widely-utilized atlases in that framework has been developed by Yeo et al. (2011). The authors identified 7 main networks (see Fig. 3) including relatively local networks, such as the visual and somatomotor networks, as well as relatively distributed, generally association, networks. Of the latter, one includes the limbic network, and others include paralimbic networks such as DMN, the dorsal attention, the ventral attention, and the frontoparietal networks, in agreement with prior reports.

In a similar view, Power et al. (2011) investigated graphs and subgraphs corresponding to functional networks in the distributed patterns of high RSFC between brain regions. They showed that the previously reported "Task Positive System" actually consists of multiple subgraphs, including the dorsal attention network, the fronto-parietal task control network, and the cingulo-opercular task control network. Hence, while using different approaches and samples, these two milestone studies show highly convergent findings, thus highlighting the

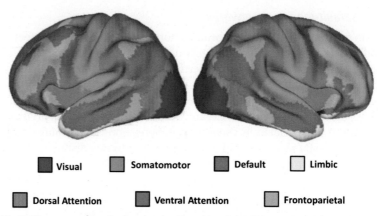

Visual   Somatomotor   Default   Limbic

Dorsal Attention   Ventral Attention   Frontoparietal

Fig. 3 The seven canonical networks. From Yeo, B. T., Krienen, F. M., Sepulcre, J., Sabuncu, M. R., Lashkari, D., Hollinshead, M., ... Buckner, R. L., 2011. The organization of the human cerebral cortex estimated by intrinsic functional connectivity. J. Neurophysiol. 106(3), 1125–1165. https://doi.org/10.1152/jn.00338.2011. Copyright © 2011 the American Physiological Society.

existence of robust functional networks in RSFC patterns with bilaterally distributed visual, sensorimotor, default-mode, and attention networks.

The neurobiological insight

These parcellations of RSFC patterns had a significant impact on the field by providing a set of canonical networks with a data-driven approach. These parcellations were not only shown to be robust through their replicability across resampling or different data sets, but they are also shown to have neurobiological validity. While this latter quality of brain parcellation remains difficult to investigate (for a review see Eickhoff et al., 2018), a common criterion of evaluation relies on comparison with features of brain organization evidenced by postmortem histological work and/or invasive in vivo techniques. Hence, the neurobiological validity of Yeo et al. and Power et al. network atlases was supported by their correspondence with well-established primary sensory-motor systems. Another criterion supporting the validity of brain parcellation based on RSFC is the convergence with patterns of brain organization shown in task fMRI, in particular, when using coactivation meta-analytic approaches on a large set of task fMRI studies (see next section for more information). Several networks delineated in RSFC by Yeo et al. and Power et al. hence demonstrated an important convergence with robust coactivation patterns revealed by decomposition approaches (Smith et al., 2009; Yeo et al., 2015a). Thus, these parcellations have proven to be not only reliable from a technical standpoint but also valid from a neurobiological standpoint.

Refining brain networks

Capitalizing on these robust atlases, follow-up investigations have provided further insight into the principles of brain organization. In addition to the 7-network partition, Yeo et al. (2011) also found a robust partition of RSFC patterns into 17 networks hence further fractionating the 7 networks into smaller networks. Many of these smaller networks were local networks, representing broad brain regions or functional modules, such as the ventral somatomotor region. Together with other studies (e.g., Craddock et al., 2012; Shen et al., 2013), Yeo et al.'s study brought evidence that decomposition approaches applied to RSFC ultimately (i.e., at a high level of partition) can be used to disentangle individual functional brain regions. Accordingly, RSFC has been used intensively in the last 15 years to parcellate the brain and/ or to parcellate some specific regions (for a review, see Eickhoff et al., 2018).

One important point of discussion that arises from those studies is that although functional organization of the brain (as for example reflected in RSFC) to a great extent corresponds to anatomical features and organization, divergence between structural and functional organizations has been frequently reported and discussed, for example, in the human hippocampus (Eickhoff et al., 2010, Genon et al., 2021). Furthermore, follow-up investigations of the network features confirmed that the main canonical networks function as relatively isolated modules. However, in contrast to the somatomotor and early visual cortices, which appeared relatively segregated, several association regions, such as the precuneus and the medial prefrontal cortex, appeared to participate in multiple networks, and hence may support large-scale integration (Yeo et al., 2014; Yeo et al., 2015a). Thus, decomposition approaches applied to RSFC importantly complement the insight provided by histological approaches in understanding brain organization.

Limitations of the unconstrained (rs-fMRI) signal to study human neurocognitive systems have been often debated (e.g., Spreng, 2012; Williamson, 2007), since, for example, brain regions that do not show signal correlation at rest may be coupled during task. Yet, RFSC has been frequently used to refine or complement the features of specific neurocognitive systems. For instance, the node and edge properties within a task-based language network of 32 brain regions have been characterized using RSFC (Labache et al., 2019). A core network (SENT_CORE) including 18 brain regions was hence identified, within which the pars triangularis of the inferior frontal gyrus and the superior temporal sulcus were identified as hubs based on their degree centrality, betweenness, and participation values. This insight provided by RFSC thus allowed the authors to hypothesize that these two hubs correspond to epicenters of sentence processing. In the same vein, RSFC was used to provide an extended spatial definition of the multidemand networks, which was further subdivided into subnetworks by using hierarchical clustering (Camilleri et al., 2018). Hence, in addition to their utility in defining canonical networks based on whole-brain decomposition, RSFC has also been used to refine and complement our understanding of specific networks representing important neurocognitive systems.

Limitations, challenges, and opportunities in brain network partitions

Due to their good reliability and important neurobiological insight, popular RSFC-based atlases have provided a referential framework for functional network analyses in a variety of studies, both in healthy and clinical populations (see next section). It should be noted, however, that these popular atlases have been derived from surface-based data, and consequently, many subcortical regions are not included in these atlases. Examples thereof are the basal ganglia and the hippocampus complex, which are particularly relevant for understanding human cognitive functioning and dysfunction (e.g., Liu et al., 2020; Strange et al., 2014). Accordingly, more recently, several studies have used high-quality data to parcellate subcortical structures partly or fully based on RSFC (e.g., Plachti et al., 2019; Tian et al., 2020) and/or to integrate subcortical structures in cortical network partitions (e.g., Ji et al., 2019).

Another important point of attention in the extensive use of canonical network atlases for neuroscientific studies in healthy and clinical populations is the degree of generalizability across the population of these "group average atlases." To address this question, the robust identification of RSFC-based networks across multiple rs-fMRI sessions in an individual has been compared to a group average partition (Laumann et al., 2015). This comparison revealed that most functional systems were grossly topologically similar in the individual and the group, hence supporting the use of group average atlases. However, some differences were also observed between the individual and the group partitions. For example, the group consensus map includes a region in the lateral occipital-temporal cortex (between the default-mode and visual systems) without clear network assignment, but in the individual, this same region showed unambiguous system affiliation (Laumann et al., 2015). These observations are consistent with reports of interindividual variability in RSFC (e.g., Mueller et al., 2013) and raised the possibility of better understanding interindividual variability in behavioral phenotype from individualized parcellation (Kong et al., 2019; Zilles and Amunts, 2013).

Evidence of interindividual variability in brain network partition, and thus in brain functional architecture, also raises the question of the need for population-specific atlases. For instance, the need for sex-specific functional networks has been suggested (Salehi et al., 2018). Along the same lines, brain network partitions in older adults may differ from those of younger adults, thus calling for age group-specific atlases (Doucet et al., 2021). Whether or not between-group variability is important enough to justify the use of population-specific atlas remains as a relatively open question. That question will be particularly relevant when considering different ethnicity and/or different geographical populations, such as European, African, and Asian populations for which population-specific structural brain templates already exist (e.g., Lee et al., 2016; Rao et al., 2017; Sivaswamy et al., 2019; Tang et al., 2010). Thus, future studies should evaluate the relevance and practical utility of population-specific functional brain atlases.

Applications of brain networks atlases

By providing a robust definition of large-scale networks and brain areas, functional atlases have been applied in a wide range of studies. Soon after their development, several major functional atlases have been used in clinical studies to identify specific dysfunction in patients (Baker et al., 2014), as well as to study brain organization in childhood (Marek et al., 2019) and functional connectivity changes in aging (Betzel et al., 2014). They have also been broadly used in cognitive neuroscience to better understand the function of brain networks and regions (for a review see Genon et al., 2018) and interindividual variability in behavior (e.g., Hearne et al., 2016). Along the same lines, they have been employed to better understand genetic loading on brain organization (Teeuw et al., 2019) and to study the influence of specific environmental conditions on brain functioning, such as the effects of sleep deprivation (Yeo et al., 2015b). In that context, it should be noted that one vital advantage brought by the use of these atlases in a wide range of studies is to offer a common framework to compare findings across studies and, more importantly, to integrate them across different fields of study.

One alternative approach to using brain network atlases for investigating RSFC involves deriving the networks from the data at hand. This is frequently done by using ICA (see second section of this chapter) (e.g., Shi et al., 2018). However, such an approach comes with the risk of double-dipping (Kriegeskorte et al., 2009). Furthermore, deriving networks from the data set at hand requires some extra work to identify relevant networks that may not straightforwardly correspond to canonical networks (see (Uddin et al., 2019) for a related discussion) and to discard potentially spurious components. Finally, the networks thereby defined often do not enjoy the stability and generalizability assessments offered by popular atlases.

Canonical networks for machine learning approaches

Deriving the networks from the same data set on which the research question is applied becomes more problematic when capitalizing on machine learning approaches with strict cross-validation. In a cross-validation setting, the data set is divided into a training and test (holdout) subsample (or set). The model (e.g., predicting intelligence score from RSFC using a linear model) is developed on the training data and tested in the holdout subsample (thereby examining the accuracy of intelligence score prediction in the unseen sample). This process of randomly splitting the data into training and test sets is usually repeated for a limited number of times. Generalizability of the model is then evaluated by summarizing (e.g., averaging) out-of-sample accuracies on the holdout sets. In such a framework, the training and test samples should absolutely not share any information. Accordingly, the networks that will serve as features for the predictive model should not be computed as premature featurization, before the cross-validation. In other words, the networks cannot be defined on the data set that will serve for testing the model. In that context, functional brain atlases represent a precious resource offering stable and well-validated networks to be used in machine learning models.

In a didactic view, two types of machine learning approaches can be distinguished: predictive approaches and multivariate associative approaches (or "doubly-multivariate approaches" (Smith and Nichols, 2018)). Brain network atlases have frequently been used for these two types of approaches. In the framework of predictive approaches, brain network atlases can be used when aiming to classify participants into groups, such as to classify patients, based on brain functional connectivity. Furthermore, several studies have employed these atlases when aiming to predict a behavioral measure from RSFC. Both applications (classification and behavioral prediction) are further developed below. Additionally, several studies have been interested in relating brain functional connectivity to a range of nonbrain variables, in particular behavioral variables, using a "doubly-multivariate approach." This latter approach is also presented in the last section below.

Connectivity-based prediction approaches in healthy and clinical populations

Predefined brain atlases and network structures, characterizing the functional organization of the human brain, offer practical tools to reduce data dimensionality for downstream analyses including the prediction of psychometric measures in healthy population and the prediction of symptoms in patients with mental disorders. A pioneering work for predicting individual behavioral scores (e.g., fluid intelligence) from RSFC in a healthy population (Finn et al., 2015) used both the Shen atlas (Shen et al., 2013) and the networks derived by Yeo et al. (2011). A behavioral prediction protocol based on connectomes computed across predefined functional areas was later proposed and became particularly popular (Shen et al., 2017). Inspired by these works, a series of studies emerged, focusing on predicting specific cognitive measures such as sustained attention and creativity (Beaty et al., 2018; Jiang et al., 2019; Rosenberg et al., 2015), investigating effects of different preprocessing strategies and algorithms on predictive ability (He et al., 2020; Li et al., 2019), exploring the feature importance for predicting different categories of psychometric measures (Chen et al., 2020; Tian and Zalesky, 2021; Wu et al., 2021), or comparing the predictive ability across rest, task and naturalistic fMRI modalities (Finn and Bandettini, 2021; Greene et al., 2020; Greene et al., 2018). Importantly, these studies capitalized on several predefined brain atlases derived from rs-fMRI (Power et al., 2011; Schaefer et al., 2018; Shen et al., 2013) or multimodal data including fMRI (Glasser et al., 2016) in order to calculate functional connectivity as brain features.

In addition, these brain atlases have played an important role not only in behavioral prediction in healthy individuals but also in predicting disease symptoms in patients. For instance, based on the Power atlas (Power et al., 2011), functional connectivity was able to provide additional prediction power of the longitudinal outcomes of autism spectrum disease (e.g., social autistic traits) beyond age and baseline behavioral scores (Plitt et al., 2015). Using functional connectivity computed from the same brain atlas, researchers have also investigated the predictability of therapeutic response in individuals with obsessive-compulsive disorder (Reggente et al., 2018). The ability to identify patients who might benefit more from a treatment has implications for personalized medicine. Furthermore, functional connectivity of dorsomedial prefrontal cortex regions, extracted from the Craddock atlas (Craddock et al., 2012), was able to predict the treatment effect of repetitive transcranial magnetic stimulation for major depressive disorder (Salomons et al., 2014). Overall, functional atlases are crucial tools for extracting functional brain features for predicting symptom severity and treatment outcomes, which could further help the development of precision medicine in the future.

Connectivity in multivariate associative approaches

Besides predictive models based on machine learning algorithms, relationships between brain and behavioral, psychometric, or demographic measures can also be explored using multivariate association models such as canonical correlation analysis (CCA) and partial least squares (PLS). Such studies can be referred to as "brain-wide association studies" (BWAS). A milestone study in this line of research was performed by applying CCA on functional connectivity among the parcels identified by a group ICA and more than 100 nonimaging subject measures. This approach revealed a positive-negative axis in the nonimaging measures that was associated with a pattern in functional connectivity that had high CCA component loadings in default-mode network (Smith et al., 2015). Across healthy participants and patients with schizophrenia, schizoaffective disorder, bipolar disorder, and attention-deficit/hyperactivity disorder, three transdiagnostic components corresponding to general psychopathology, cognitive dysfunction, and impulsivity were observed to be associated with whole-brain functional connectivity patterns calculated from the Schaefer atlas (Schaefer et al., 2018), especially and strikingly in the somatosensory-motor networks (Kebets et al., 2019) (see Fig. 4). Similarly, associations between psychometric/demographic measures and functional connectivity computed from the Glasser atlas (Glasser et al., 2016) have been investigated in healthy and clinically depressed adolescents and young adults (Mihalik et al., 2019). Furthermore, reproducibility of BWAS has been examined in studies where functional connectivity was extracted based on the Gordon atlas (Gordon et al., 2016; Marek et al., 2020). Thus, brain network atlases represent a crucial resource for representing neuroimaging features used as input in brain-wide association studies.

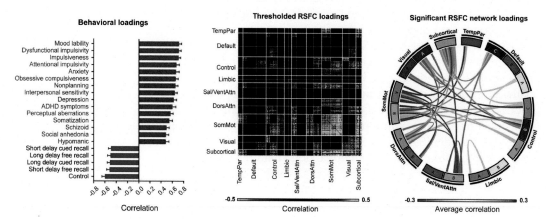

Fig. 4 Loadings of the first PLS transdiagnostic component on behavioral measures *(left)* and on RSFC *(middle)*. The component loadings on RSFC were then averaged within and between networks, shown in a circular plot *(right)*. Adapted from Kebets, V., Holmes, A. J., Orban, C., Tang, S., Li, J., Sun, N., ... Yeo, B. T., 2019. Somatosensory-motor dysconnectivity spans multiple transdiagnostic dimensions of psychopathology. Biol. Psychiatry 86(10), 779–791. https://doi.org/10.1016/j.biopsych.2019.06.013. Copyright © 2019 Society of Biological Psychiatry.

Brain network atlases for feature reduction and alternative approaches

As developed in the previous paragraphs, a range of studies employing machine learning approaches to relate functional connectivity to clinical and behavioral profiles have capitalized on functional atlases derived from rs-fMRI data. Regardless of whether these atlases represent the main canonical networks or local brain regions, they allow to represent RSFC data in a lower dimensional space than the original voxel-wise or vertex-wise data, which is crucially needed for any of the multivariate analyses developed above. In other words, they offer a feature reduction approach independent from the data in which the analysis is performed. Such feature reduction is typically required for neuroimaging data (since, for example, computing functional connectivity at the voxel level would result in a pair-wise connectivity matrix of hundreds of thousands of voxels). In addition to being computationally prohibitive, such high dimensional data would be problematic for many machine learning models. For instance, the ratio between the number of features and the number of observations (i.e., participants in a neuroimaging study) has been shown to dramatically influence the reliability of doubly multivariate approaches, such as canonical correlation analysis (Helmer et al., 2021). Thus, functional atlases are vital for machine learning studies in neuroimaging by offering a data representation whose neurobiological qualities have been extensively evaluated and that can thus serve as a reference framework across studies. Meanwhile, they also crucially address the need for independent feature reduction from a technical standpoint.

While the most popular functional atlases are based on RSFC and accordingly many studies have used canonical networks from these atlases, robust functional networks are also provided by meta-analytic approaches applied to task fMRI and PET studies. As mentioned above, meta-analytic connectivity modeling provides networks of regions that are consistently coactivated together across a range of neuroimaging studies (Langner et al., 2014). Additionally, meta-analytic approaches are also applied to neuroimaging experiments to identify sets of regions commonly activated in relation to a specific cognitive concept (Müller et al., 2018). It can hence be assumed that the resulting set of regions form the network supporting the investigated cognitive process or function (such as autobiographical memory, working memory, or motor functions). Accordingly, these networks could be, in turn, used to investigate the relationship between RSFC and behavioral phenotypes (e.g., Nostro et al., 2018; Pläschke et al., 2017). Thus, research capitalizing on meta-analytic networks can interestingly complement those capitalizing on RSFC canonical networks in better understanding the relationship between interindividual variability in RSFC and interindividual variability in behavior.

Conclusions

Several techniques have been used to delineate brain networks and parcellate the brain into functional regions based on rs-fMRI data. By well-complementing previous knowledge of local organization derived from histological work with features of macroscale organization, these techniques have significantly contributed to a better understanding of brain organization and function. Regardless of the techniques used, however, a common challenge resides in the choice of the optimal number of components into which the data should be decomposed while considering the multiscale organization of the brain. Accounting for this complexity, several atlases are now available at different levels of subdivisions. Canonical networks provided in popular functional atlases nevertheless still represent a vital resource in neuroimaging research by offering a common reference framework for investigating replicability across studies and better understanding relationship between brain and behavior in healthy and clinical populations. Furthermore, these atlases crucially provide a useful representation of brain features for the application of machine learning approaches, which in turn will contribute to the development of brain-based prediction for personalized medicine. In that framework, better understanding and addressing interindividual variability in brain functional organization, and hence in brain atlases, represent an important avenue of research for future studies.

References

Amunts, K., Lepage, C., Borgeat, L., Mohlberg, H., Dickscheid, T., Rousseau, M.-É., Oros-Peusquens, A.-M., 2013. BigBrain: an ultrahigh-resolution 3D human brain model. Science 340 (6139), 1472–1475.

Anderson, J.S., Druzgal, T.J., Lopez-Larson, M., Jeong, E.K., Desai, K., Yurgelun-Todd, D., 2011. Network anticorrelations, global regression, and phase-shifted soft tissue correction. Hum. Brain Mapp. 32 (6), 919–934.

Andrews-Hanna, J.R., Reidler, J.S., Sepulcre, J., Poulin, R., Buckner, R.L., 2010. Functional-anatomic fractionation of the brain's default network. Neuron 65 (4), 550–562. https://doi.org/10.1016/j.neuron.2010.02.005.

Arslan, S., Rueckert, D., 2015. Multi-level parcellation of the cerebral cortex using resting-state fMRI. In: Paper Presented at the Medical Image Computing and Computer-Assisted Intervention—MICCAI 2015.

Baker, J.T., Holmes, A.J., Masters, G.A., Yeo, B.T., Krienen, F., Buckner, R.L., Öngür, D., 2014. Disruption of cortical association networks in schizophrenia and psychotic bipolar disorder. JAMA Psychiat. 71 (2), 109–118.

Bassett, D.S., Bullmore, E., 2006. Small-world brain networks. Neuroscientists 12 (6), 512–523. https://doi.org/10.1177/1073858406293182.

Beall, E.B., Lowe, M.J., 2010. The non-separability of physiologic noise in functional connectivity MRI with spatial ICA at 3T. J. Neurosci. Methods 191 (2), 263–276. https://doi.org/10.1016/J.JNEUMETH.2010.06.024.

Beaty, R.E., Kenett, Y.N., Christensen, A.P., Rosenberg, M.D., Benedek, M., Chen, Q., Silvia, P.J., 2018. Robust prediction of individual creative ability from brain functional connectivity. Proc. Natl. Acad. Sci. U. S. A. 115 (5), 1087–1092. https://doi.org/10.1073/pnas.1713532115.

Beckmann, C.F., Smith, S.M., 2004. Probabilistic independent component analysis for functional magnetic resonance imaging. IEEE Trans. Med. Imaging 23 (2), 137–152. https://doi.org/10.1109/TMI.2003.822821.

Bellec, P., Perlbarg, V., Jbabdi, S., Pélégrini-Issac, M., Anton, J.L., Doyon, J., Benali, H., 2006. Identification of large-scale networks in the brain using fMRI. Neuroimage 29 (4), 1231–1243. https://doi.org/10.1016/J.NEUROIMAGE.2005.08.044.

Bellec, P., Rosa-Neto, P., Lyttelton, O.C., Benali, H., Evans, A.C., 2010. Multi-level bootstrap analysis of stable clusters in resting-state fMRI. Neuroimage 51 (3), 1126–1139. https://doi.org/10.1016/J.NEUROIMAGE.2010.02.082.

Betzel, R.F., Byrge, L., He, Y., Goñi, J., Zuo, X.-N., Sporns, O., 2014. Changes in structural and functional connectivity among resting-state networks across the human lifespan. Neuroimage 102, 345–357.

Biswal, B.B., Ulmer, J.L., 1999. Blind source separation of multiple signal sources of fMRI data sets using independent component analysis. J. Comput. Assist. Tomogr. 23 (2), 265–271. https://doi.org/10.1097/00004728-199903000-00016.

Biswal, B., Yetkin, F.Z., Haughton, V.M., Hyde, J.S., 1995. Functional connectivity in the motor cortex of resting human brain using echo-planar MRI. Magn. Reson. Med. 34 (4), 537–541. https://doi.org/10.1002/mrm.1910340409.

Blei, D.M., Ng, A.Y., Jordan, M.I., 2003. Latent Dirichlet allocation. J. Mach. Learn. Res. 3, 993–1022.

Blumensath, T., Jbabdi, S., Glasser, M.F., Van Essen, D.C., Ugurbil, K., Behrens, T.E.J., Smith, S.M., 2013. Spatially constrained hierarchical parcellation of the brain with resting-state fMRI. Neuroimage 76, 313–324. https://doi.org/10.1016/J.NEUROIMAGE.2013.03.024.

Boubela, R.N., Kalcher, K., Huf, W., Kronnerwetter, C., Filzmoser, P., Moser, E., 2013. Beyond noise: using temporal ICA to extract meaningful information from high-frequency fMRI signal fluctuations during rest. Front. Hum. Neurosci. 7 (May), 168. https://doi.org/10.3389/fnhum.2013.00168.

Buckner, R.L., Sepulcre, J., Talukdar, T., Krienen, F.M., Liu, H., Hedden, T., Johnson, K.A., 2009. Cortical hubs revealed by intrinsic functional connectivity: mapping, assessment of stability, and relation to Alzheimer's disease. J. Neurosci. 29 (6), 1860–1873. https://doi.org/10.1523/jneurosci.5062-08.2009.

Bullmore, E., Sporns, O., 2009. Complex brain networks: graph theoretical analysis of structural and functional systems. Nat. Rev. Neurosci. 10 (3), 186–198. https://doi.org/10.1038/nrn2575.

Calhoun, V.D., Adali, T., McGinty, V.B., Pekar, J.J., Watson, T.D., Pearlson, G.D., 2001a. fMRI activation in a visual-perception task: network of areas detected using the general linear model and independent components analysis. Neuroimage 14 (5), 1080–1088. https://doi.org/10.1006/NIMG.2001.0921.

Calhoun, V.D., Adali, T., Pearlson, G.D., Pekar, J.J., 2001b. A method for making group inferences from functional MRI data using independent component analysis. Hum. Brain Mapp. 14 (3), 140–151. https://doi.org/10.1002/HBM.1048.

Calhoun, V.D., Adali, T., Pearlson, G.D., Pekar, J.J., 2001c. Spatial and temporal independent component analysis of functional MRI data containing a pair of task-related waveforms. Hum. Brain Mapp. 13 (1), 43–53. https://doi.org/10.1002/HBM.1024.

Calhoun, V.D., Liu, J., Adali, T., 2009. A review of group ICA for fMRI data and ICA for joint inference of imaging, genetic, and ERP data. Neuroimage 45 (1), S163–S172. https://doi.org/10.1016/J.NEUROIMAGE.2008.10.057.

Camilleri, J.A., Müller, V.I., Fox, P., Laird, A.R., Hoffstaedter, F., Kalenscher, T., Eickhoff, S.B., 2018. Definition and characterization of an extended multiple-demand network. Neuroimage 165, 138–147. https://doi.org/10.1016/j.neuroimage.2017.10.020.

Chen, J., Tam, A., Kebets, V., Orban, C., Ooi, L.Q.R., Marek, S., Thomas Yeo, B.T., 2020. Shared and unique brain network features predict cognition, personality and mental health in childhood. BioRxiv. https://doi.org/10.1101/2020.06.24.168724. 2020.2006.2024.168724.

Cohen, A.L., Fair, D.A., Dosenbach, N.U.F., Miezin, F.M., Dierker, D., Van Essen, D.C., Petersen, S.E., 2008. Defining functional areas in individual human brains using resting functional connectivity MRI. Neuroimage 41 (1), 45–57. https://doi.org/10.1016/J.NEUROIMAGE.2008.01.066.

Cordes, D., Haughton, V., Carew, J.D., Arfanakis, K., Maravilla, K., 2002. Hierarchical clustering to measure connectivity in fMRI resting-state data. Magn. Reson. Imaging 20 (4), 305–317. https://doi.org/10.1016/S0730-725X(02)00503-9.

Craddock, R.C., James, G.A., Holtzheimer III, P.E., Hu, X.P., Mayberg, H.S., 2012. A whole brain fMRI atlas generated via spatially constrained spectral clustering. Hum. Brain Mapp. 33 (8), 1914–1928.

Doucet, G.E., Labache, L., Thompson, P.M., Joliot, M., Frangou, S., Initiative, A., s. D. N., 2021. Atlas55+: brain functional atlas of resting-state networks for late adulthood. Cereb. Cortex 31 (3), 1719–1731.

Doucet, G., Naveau, M., Petit, L., Delcroix, N., Zago, L., Crivello, F., Joliot, M., 2011. Brain activity at rest: a multiscale hierarchical functional organization. J. Neurophysiol. 105 (6), 2753–2763. https://doi.org/10.1152/jn.00895.2010.

Eickhoff, S.B., Thirion, B., Varoquaux, G., Bzdok, D., 2015. Connectivity-based parcellation: critique and implications. Hum. Brain Mapp. 36 (12), 4771–4792. https://doi.org/10.1002/HBM.22933.

Eickhoff, S.B., Yeo, B.T.T., Genon, S., 2018. Imaging-based parcellations of the human brain. Nat. Rev. Neurosci. 19 (11), 672–686. https://doi.org/10.1038/s41583-018-0071-7.

Finn, E.S., Bandettini, P.A., 2021. Movie-watching outperforms rest for functional connectivity-based prediction of behavior. Neuroimage 235. https://doi.org/10.1016/j.neuroimage.2021.117963.

Finn, E.S., Shen, X., Scheinost, D., Rosenberg, M.D., Huang, J., Chun, M.M., Constable, R.T., 2015. Functional connectome fingerprinting: identifying individuals using patterns of brain connectivity. Nat. Neurosci. 18, 1664–1671. https://doi.org/10.1038/nn.4135.

Fox, M.D., Corbetta, M., Snyder, A.Z., Vincent, J.L., Raichle, M.E., 2006. Spontaneous neuronal activity distinguishes human dorsal and ventral attention systems. Proc. Natl. Acad. Sci. U. S. A. 103 (26), 10046–10051. https://doi.org/10.1073/pnas.0604187103.

Fox, P.T., Lancaster, J.L., Laird, A.R., Eickhoff, S.B., 2014. Meta-analysis in human neuroimaging: computational modeling of large-scale databases. Annu. Rev. Neurosci. 37, 409. https://doi.org/10.1146/ANNUREV-NEURO-062012-170320.

Fox, M.D., Snyder, A.Z., Vincent, J.L., Corbetta, M., Van Essen, D.C., Raichle, M.E., 2005. The human brain is intrinsically organized into dynamic, anticorrelated functional networks. Proc. Natl. Acad. Sci. U. S. A. 102 (27), 9673–9678. https://doi.org/10.1073/pnas.0504136102.

Fox, M.D., Zhang, D., Snyder, A.Z., Raichle, M.E., 2009. The global signal and observed anticorrelated resting state brain networks. J. Neurophysiol. 101 (6), 3270–3283.

Genon, S., Bernhardt, B.C., La Joie, R., Amunts, K., Eickhoff, S.B., 2021. The many dimensions of human hippocampal organization and (dys) function. Trends Neurosci. 44, 977–989.

Genon, S., Reid, A., Langner, R., Amunts, K., Eickhoff, S.B., 2018. How to characterize the function of a brain region. Trends Cogn. Sci. 22 (4), 350–364. https://doi.org/10.1016/j.tics.2018.01.010.

Glasser, M.F., Coalson, T.S., Bijsterbosch, J.D., Harrison, S.J., Harms, M.P., Anticevic, A., Smith, S.M., 2018. Using temporal ICA to selectively remove global noise while preserving global signal in functional MRI data. Neuroimage 181, 692–717. https://doi.org/10.1016/j.neuroimage.2018.04.076.

Glasser, M.F., Coalson, T.S., Robinson, E.C., Hacker, C.D., Harwell, J., Yacoub, E., Van Essen, D.C., 2016. A multi-modal parcellation of human cerebral cortex. Nature 536, 171–178. https://doi.org/10.1038/nature18933.

Golland, Y., Golland, P., Bentin, S., Malach, R., 2008. Data-driven clustering reveals a fundamental subdivision of the human cortex into two global systems. Neuropsychologia 46 (2), 540–553. https://doi.org/10.1016/J.NEUROPSYCHOLOGIA.2007.10.003.

Gordon, E.M., Laumann, T.O., Adeyemo, B., Huckins, J.F., Kelley, W.M., Petersen, S.E., 2016. Generation and evaluation of a cortical area parcellation from resting-state correlations. Cereb. Cortex 26 (1), 288–303. https://doi.org/10.1093/cercor/bhu239.

Goutte, C., Toft, P., Rostrup, E., Nielsen, F.Å., Hansen, L.K., 1999. On clustering fMRI time series. Neuroimage 9 (3), 298–310. https://doi.org/10.1006/NIMG.1998.0391.

Greene, A.S., Gao, S., Noble, S., Scheinost, D., Constable, R.T., 2020. How tasks change whole-brain functional organization to reveal brain-phenotype relationships. Cell Rep. 32 (8). https://doi.org/10.1016/j.celrep.2020.108066.

Greene, A.S., Gao, S., Scheinost, D., Constable, R.T., 2018. Task-induced brain state manipulation improves prediction of individual traits. Nat. Commun. 9. https://doi.org/10.1038/s41467-018-04920-3.

Guo, Y., Pagnoni, G., 2008. A unified framework for group independent component analysis for multi-subject fMRI data. Neuroimage 42 (3), 1078–1093. https://doi.org/10.1016/J.NEUROIMAGE.2008.05.008.

Hallquist, M.N., Hillary, F.G., 2018. Graph theory approaches to functional network organization in brain disorders: a critique for a brave new small-world. Netw. Neurosci. 3 (1), 1–26. https://doi.org/10.1162/netn_a_00054.

He, T., Kong, R., Holmes, A.J., Nguyen, M., Sabuncu, M.R., Eickhoff, S.B., Yeo, B.T., 2020. Deep neural networks and kernel regression achieve comparable accuracies for functional connectivity prediction of behavior and demographics. Neuroimage 206. https://doi.org/10.1016/j.neuroimage.2019.116276.

Hearne, L.J., Mattingley, J.B., Cocchi, L., 2016. Functional brain networks related to individual differences in human intelligence at rest. Sci. Rep. 6 (1), 32328. https://doi.org/10.1038/srep32328.

Helmer, M., Warrington, S.D., Mohammadi-Nejad, A.-R., Ji, J.L., Howell, A., Rosand, B., Murray, J.D., 2021. On stability of canonical correlation analysis and partial least squares with application to brain-behavior associations. BioRxiv. 2020.2008.2025.265546.

van den Heuvel, M., Mandl, R., Pol, H.H., 2008. Normalized cut group clustering of resting-state fMRI data. PLoS One 3 (4), e2001. https://doi.org/10.1371/JOURNAL.PONE.0002001.

Ji, J.L., Spronk, M., Kulkarni, K., Repovš, G., Anticevic, A., Cole, M.W., 2019. Mapping the human brain's cortical-subcortical functional network organization. Neuroimage 185, 35–57.

Jiang, R., Calhoun, V.D., Fan, L., Zuo, N., Jung, R., Qi, S., Sui, J., 2019. Gender differences in connectome-based predictions of individualized intelligence quotient and subdomain scores. Cereb. Cortex 30 (3), 888–900. https://doi.org/10.1093/cercor/bhz134.

Kebets, V., Holmes, A.J., Orban, C., Tang, S., Li, J., Sun, N., Yeo, B.T., 2019. Somatosensory-motor dysconnectivity spans multiple transdiagnostic dimensions of psychopathology. Biol. Psychiatry 86 (10), 779–791. https://doi.org/10.1016/j.biopsych.2019.06.013.

Kong, R., Li, J., Orban, C., Sabuncu, M.R., Liu, H., Schaefer, A., Yeo, B.T.T., 2019. Spatial topography of individual-specific cortical networks predicts human cognition, personality, and emotion. Cereb. Cortex 29 (6), 2533–2551. https://doi.org/10.1093/cercor/bhy123.

Kriegeskorte, N., Simmons, W.K., Bellgowan, P.S.F., Baker, C.I., 2009. Circular analysis in systems neuroscience: the dangers of double dipping. Nat. Neurosci. 12 (5), 535–540. https://doi.org/10.1038/nn.2303.

Labache, L., Joliot, M., Saracco, J., Jobard, G., Hesling, I., Zago, L., Tzourio-Mazoyer, N., 2019. A SENtence Supramodal Areas AtlaS (SENSAAS) based on multiple task-induced activation mapping and graph analysis of intrinsic connectivity in 144 healthy right-handers. Brain Struct. Funct. 224, 859–882. https://doi.org/10.1007/s00429-018-1810-2.

Langner, R., Rottschy, C., Laird, A.R., Fox, P.T., Eickhoff, S.B., 2014. Meta-analytic connectivity modeling revisited: controlling for activation base rates. Neuroimage 99, 559–570. https://doi.org/10.1016/j.neuroimage.2014.06.007.

Lashkari, D., Vul, E., Kanwisher, N., Golland, P., 2010. Discovering structure in the space of fMRI selectivity profiles. Neuroimage 50, 1085–1098. https://doi.org/10.1016/j.neuroimage.2009.12.106.

Laumann, T.O., Gordon, E.M., Adeyemo, B., Snyder, A.Z., Joo, S.J., Chen, M.-Y., Dosenbach, N.U., 2015. Functional system and areal organization of a highly sampled individual human brain. Neuron 87 (3), 657–670.

Lee, H., Yoo, B.I., Han, J.W., Lee, J.J., Oh, S.Y.W., Lee, E.Y., Kim, K.W., 2016. Construction and validation of brain MRI templates from a Korean normal elderly population. Psychiatry Investig. 13 (1), 135.

Li, Y.O., Adali, T., Calhoun, V.D., 2007. Estimating the number of independent components for functional magnetic resonance imaging data. Hum. Brain Mapp. 28 (11), 1251. https://doi.org/10.1002/HBM.20359.

Li, J., Kong, R., Liégeois, R., Orban, C., Tan, Y., Sun, N., Yeo, B.T., 2019. Global signal regression strengthens association between resting-state functional connectivity and behavior. Neuroimage 196, 126–141. https://doi.org/10.1016/j.neuroimage.2019.04.016.

Liu, X., Eickhoff, S.B., Hoffstaedter, F., Genon, S., Caspers, S., Reetz, K., Patil, K.R., 2020. Joint multi-modal parcellation of the human striatum: functions and clinical relevance. Neurosci. Bull. https://doi.org/10.1007/s12264-020-00543-1.

Marek, S., Tervo-Clemmens, B., Calabro, F.J., Montez, D.F., Kay, B.P., Hatoum, A.S., Dosenbach, N.U.F., 2020. Towards reproducible brain-wide association studies. BioRxiv. https://doi.org/10.1101/2020.08.21.257758. 2020.2008.2021.257758.

Marek, S., Tervo-Clemmens, B., Nielsen, A.N., Wheelock, M.D., Miller, R.L., Laumann, T.O., Dosenbach, N.U.F., 2019. Identifying reproducible individual differences in childhood functional brain networks: an ABCD study. Dev. Cogn. Neurosci. 40, 100706. https://doi.org/10.1016/j.dcn.2019.100706.

Mckeown, M.J., Makeig, S., Brown, G.G., Jung, T.P., Kindermann, S.S., Bell, A.J., Sejnowski, T.J., 1998. Analysis of fMRI data by blind separation into independent spatial components. Hum. Brain Mapp. 6 (3), 160–188.

Mezer, A., Yovel, Y., Pasternak, O., Gorfine, T., Assaf, Y., 2009. Cluster analysis of resting-state fMRI time series. Neuroimage 45 (4), 1117–1125. https://doi.org/10.1016/J.NEUROIMAGE.2008.12.015.

Mihalik, A., Ferreira, F.S., Rosa, M.J., Moutoussis, M., Ziegler, G., Monteiro, J.M., Mourão-Miranda, J., 2019. Brain-behaviour modes of covariation in healthy and clinically depressed young people. Sci. Rep. 9. https://doi.org/10.1038/s41598-019-47277-3.

Moreno-Dominguez, D., Anwander, A., Knösche, T.R., 2014. A hierarchical method for whole-brain connectivity-based parcellation. Hum. Brain Mapp. 35, 5000–5025. https://doi.org/10.1002/HBM.22528.

Mueller, S., Wang, D., Fox, M.D., Yeo, B.T., Sepulcre, J., Sabuncu, M.R., Liu, H., 2013. Individual variability in functional connectivity architecture of the human brain. Neuron 77 (3), 586–595.

Müller, V.I., Cieslik, E.C., Laird, A.R., Fox, P.T., Radua, J., Mataix-Cols, D., Turkeltaub, P.E., 2018. Ten simple rules for neuroimaging meta-analysis. Neurosci. Biobehav. Rev. 84, 151–161.

Murtagh, F., Contreras, P., 2017. Algorithms for hierarchical clustering: an overview, II. Wiley Interdiscip. Rev. Data Min. Knowl. Discov. 7 (6), e1219. https://doi.org/10.1002/WIDM.1219.

Nelson, S.M., Cohen, A.L., Power, J.D., Wig, G.S., Miezin, F.M., Wheeler, M.E., Petersen, S.E., 2010. A parcellation scheme for human left lateral parietal cortex. Neuron 67 (1), 156–170. https://doi.org/10.1016/J.NEURON.2010.05.025.

Nostro, A.D., Müller, V.I., Varikuti, D.P., Pläschke, R.N., Hoffstaedter, F., Langner, R., Eickhoff, S.B., 2018. Predicting personality from network-based resting-state functional connectivity. Brain Struct. Funct. 223 (6), 2699–2719. https://doi.org/10.1007/s00429-018-1651-z.

Orban, P., Doyon, J., Petrides, M., Mennes, M., Hoge, R., Bellec, P., 2015. The richness of task-evoked hemodynamic responses defines a pseudohierarchy of functionally meaningful brain networks. Cereb. Cortex 25 (9), 2658–2669. https://doi.org/10.1093/CERCOR/BHU064.

Park, H.J., Friston, K., 2013. Structural and functional brain networks: from connections to cognition. Science 342 (6158). https://doi.org/10.1126/science.1238411.

Plachti, A., Eickhoff, S.B., Hoffstaedter, F., Patil, K.R., Laird, A.R., Fox, P.T., Genon, S., 2019. Multimodal Parcellations and extensive behavioral profiling tackling the Hippocampus gradient. Cereb. Cortex 29 (11), 4595–4612. https://doi.org/10.1093/cercor/bhy336.

Pläschke, R.N., Cieslik, E.C., Müller, V.I., Hoffstaedter, F., Plachti, A., Varikuti, D.P., Jockwitz, C., 2017. On the integrity of functional brain networks in schizophrenia, Parkinson's disease, and advanced age: evidence from connectivity-based single-subject classification. Hum. Brain Mapp. 38 (12), 5845–5858.

Plitt, M., Barnes, K.A., Wallace, G.L., Kenworthy, L., Martin, A., 2015. Resting-state functional connectivity predicts longitudinal change in autistic traits and adaptive functioning in autism. Proc. Natl. Acad. Sci. U. S. A. 112 (48), E6699–E6706. https://doi.org/10.1073/pnas.1510098112.

Power, J.D., 2019. Temporal ICA has not properly separated global fMRI signals: a comment on Glasser et al. (2018). Neuroimage 197, 650–651. https://doi.org/10.1016/j.neuroimage.2018.12.051.

Power, J.D., Cohen, A.L., Nelson, S.M., Wig, G.S., Barnes, K.A., Church, J.A., Petersen, S.E., 2011. Functional network organization of the human brain. Neuron 72 (4), 665–678. https://doi.org/10.1016/j.neuron.2011.09.006.

Power, J.D., Fair, D.A., Schlaggar, B.L., Petersen, S.E., 2010. The development of human functional brain networks. Neuron 67 (5), 735–748.

Rao, N.P., Jeelani, H., Achalia, R., Achalia, G., Jacob, A., Dawn Bharath, R., Yalavarthy, P.K., 2017. Population differences in brain morphology: need for population specific brain template. Psychiatry Res. Neuroimaging 265, 1–8.

Reggente, N., Moody, T.D., Morfini, F., Sheen, C., Rissman, J., O'Neill, J., Feusner, J.D., 2018. Multivariate resting-state functional connectivity predicts response to cognitive behavioral therapy in obsessive-compulsive disorder. Proc. Natl. Acad. Sci. U. S. A. 115 (9), 2222–2227. https://doi.org/10.1073/pnas.1716686115.

Rosenberg, M.D., Finn, E.S., Scheinost, D., Papademetris, X., Shen, X., Constable, R.T., Chun, M.M., 2015. A neuromarker of sustained attention from whole-brain functional connectivity. Nat. Neurosci. 19, 165–171. https://doi.org/10.1038/nn.4179.

Rosvall, M., Bergstrom, C.T., 2008. Maps of random walks on complex networks reveal community structure. Proc. Natl. Acad. Sci. U. S. A. 105 (4), 1118–1123. https://doi.org/10.1073/pnas.0706851105.

Saad, Z.S., Gotts, S.J., Murphy, K., Chen, G., Jo, H.J., Martin, A., Cox, R.W., 2012. Trouble at rest: how correlation patterns and group differences become distorted after global signal regression. Brain Connect. 2 (1), 25–32.

Salehi, M., Karbasi, A., Shen, X., Scheinost, D., Constable, R.T., 2018. An exemplar-based approach to individualized parcellation reveals the need for sex specific functional networks. Neuroimage 170, 54–67.

Salomons, T.V., Dunlop, K., Kennedy, S.H., Flint, A., Geraci, J., Giacobbe, P., Downar, J., 2014. Resting-state cortico-thalamic-striatal connectivity predicts response to dorsomedial prefrontal rTMS in major depressive disorder. Neuropsychopharmacology 39, 488–498. https://doi.org/10.1038/npp.2013.222.

Schaefer, A., Kong, R., Gordon, E.M., Laumann, T.O., Zuo, X.-N., Holmes, A.J., Yeo, B.T.T., 2018. Local-global parcellation of the human cerebral cortex from intrinsic functional connectivity MRI. Cereb. Cortex 28, 3095–3114. https://doi.org/10.1093/cercor/bhx179.

Schmithorst, V.J., Holland, S.K., 2004. Comparison of three methods for generating group statistical inferences from independent component analysis of functional magnetic resonance imaging data. J. Magn. Reson. Imaging 19 (3), 365–368. https://doi.org/10.1002/JMRI.20009.

Shen, X., Finn, E.S., Scheinost, D., Rosenberg, M.D., Chun, M.M., Papademetris, X., Constable, R.T., 2017. Using connectome-based predictive modeling to predict individual behavior from brain connectivity. Nat. Protoc. 12 (3), 506–518. https://doi.org/10.1038/nprot.2016.178.

Shen, X., Tokoglu, F., Papademetris, X., Constable, R.T., 2013. Groupwise whole-brain parcellation from resting-state fMRI data for network node identification. Neuroimage 82, 403–415.

Shi, J., Malik, J., 2000. Normalized cuts and image segmentation. IEEE Trans. Pattern Anal. Mach. Intell. 22 (8), 888–905. https://doi.org/10.1109/34.868688.

Shi, L., Sun, J., Wu, X., Wei, D., Chen, Q., Yang, W., Qiu, J., 2018. Brain networks of happiness: dynamic functional connectivity among the default, cognitive and salience networks relates to subjective well-being. Soc. Cogn. Affect. Neurosci. 13 (8), 851–862. https://doi.org/10.1093/scan/nsy059.

Sivaswamy, J., Thottupattu, A.J., Mehta, R., Sheelakumari, R., Kesavadas, C., 2019. Construction of Indian human brain atlas. Neurol. India 67 (1), 229.

Smith, S.M., Fox, P.T., Miller, K.L., Glahn, D.C., Fox, P.M., Mackay, C.E., Beckmann, C.F., 2009. Correspondence of the brain's functional architecture during activation and rest. Proc. Natl. Acad. Sci. U. S. A. 106 (31), 13040–13045. https://doi.org/10.1073/pnas.0905267106.

Smith, S.M., Nichols, T.E., 2018. Statistical challenges in "big data" human neuroimaging. Neuron 97 (2), 263–268.

Smith, S.M., Nichols, T.E., Vidaurre, D., Winkler, A.M., Behrens, T.E.J., Glasser, M.F., Miller, K.L., 2015. A positive-negative mode of population covariation links brain connectivity, demographics and behavior. Nat. Neurosci. 18, 1565–1567. https://doi.org/10.1038/nn.4125.

Sporns, O., 2018. Graph theory methods: applications in brain networks. Dialogues Clin. Neurosci. 20 (2), 121. https://doi.org/10.31887/DCNS.2018.20.2/OSPORNS.

Sporns, O., Betzel, R.F., 2016. Modular brain networks network. Annu. Rev. Psychol. 67, 613–640. https://doi.org/10.1146/annurev-psych-122414-033634.

Spreng, R.N., 2012. The fallacy of a "task-negative" network. Front. Psychol. 3, 145.

Strange, B.A., Witter, M.P., Lein, E.S., Moser, E.I., 2014. Functional organization of the hippocampal longitudinal axis. Nat. Rev. Neurosci. 15 (10), 655–669.

Tang, Y., Hojatkashani, C., Dinov, I.D., Sun, B., Fan, L., Lin, X., Toga, A.W., 2010. The construction of a Chinese MRI brain atlas: a morphometric comparison study between Chinese and Caucasian cohorts. Neuroimage 51 (1), 33–41.

Teeuw, J., Brouwer, R.M., Guimarães, J.P., Brandner, P., Koenis, M.M., Swagerman, S.C., Pol, H.E.H., 2019. Genetic and environmental influences on functional connectivity within and between canonical cortical resting-state networks throughout adolescent development in boys and girls. Neuroimage 202, 116073.

Thirion, B., Varoquaux, G., Dohmatob, E., Poline, J.-B., 2014. Which fMRI cluster-ing gives good brain parcellations? Front. Neurosci. 8. https://doi.org/10.3389/fnins.2014.00167.

Tian, Y., Margulies, D.S., Breakspear, M., Zalesky, A., 2020. Topographic organization of the human subcortex unveiled with functional connectivity gradients. Nat. Neurosci. 23 (11), 1421–1432. https://doi.org/10.1038/s41593-020-00711-6.

Tian, Y., Zalesky, A., 2021. Machine learning prediction of cognition from functional con-nectivity: are feature weights reliable? Neuroimage 245. https://doi.org/10.1016/j.neuroimage.2021.118648.

Uddin, L.Q., Yeo, B.T., Spreng, R.N., 2019. Towards a universal taxonomy of macro-scale functional human brain networks. Brain Topogr. 32 (6), 926–942.

Vincent, J.L., Kahn, I., Snyder, A.Z., Raichle, M.E., Buckner, R.L., 2008. Evidence for a frontoparietal control system revealed by intrinsic functional connectivity. J. Neurophysiol. 100 (6), 3328–3342. https://doi.org/10.1152/jn.90355.2008.

Wig, G.S., Laumann, T.O., Cohen, A.L., Power, J.D., Nelson, S.M., Glasser, M.F., Petersen, S.E., 2014. Parcellating an individual subject's cortical and subcortical brain struc-tures using snowball sampling of resting-state correlations. Cereb. Cortex 24 (8), 2036–2054. https://doi.org/10.1093/CERCOR/BHT056.

Williamson, P., 2007. Are anticorrelated networks in the brain relevant to schizophrenia? Schizophr. Bull. 33 (4), 994–1003.

Wu, J., Eickhoff, S.B., Hoffstaedter, F., Patil, K.R., Schwender, H., Yeo, B.T.T., Genon, S., 2021. A connectivity-based psychometric prediction framework for brain-behavior relationship studies. Cereb. Cortex 31 (8), 3732–3751. https://doi.org/10.1093/cercor/bhab044.

Yeo, B.T., Krienen, F.M., Chee, M.W., Buckner, R.L., 2014. Estimates of segregation and overlap of functional connectivity networks in the human cerebral cortex. Neuroimage 88, 212–227.

Yeo, B.T., Krienen, F.M., Eickhoff, S.B., Yaakub, S.N., Fox, P.T., Buckner, R.L., Chee, M.W., 2015a. Functional specialization and flexibility in human association cortex. Cereb. Cortex 25 (10), 3654–3672.

Yeo, B.T., Krienen, F.M., Sepulcre, J., Sabuncu, M.R., Lashkari, D., Hollinshead, M., Buckner, R.L., 2011. The organization of the human cerebral cortex estimated by intrinsic functional connectivity. J. Neurophysiol. 106 (3), 1125–1165. https://doi.org/10.1152/jn.00338.2011.

Yeo, B.T., Tandi, J., Chee, M.W., 2015b. Functional connectivity during rested wakeful-ness predicts vulnerability to sleep deprivation. Neuroimage 111, 147–158.

Zhang, S., Li, C.-s.R., 2012. Functional connectivity mapping of the human precuneus by resting state fMRI. Neuroimage 59 (4), 3548–3562. https://doi.org/10.1016/j.neuroimage.2011.11.023.

Zilles, K., Amunts, K., 2013. Individual variability is not noise. Trends Cogn. Sci. 17 (4), 153–155.

Zuo, X.N., Kelly, C., Adelstein, J.S., Klein, D.F., Castellanos, F.X., Milham, M.P., 2010. Reliable intrinsic connectivity networks: test-retest evaluation using ICA and dual regression approach. Neuroimage 49 (3), 2163–2177. https://doi.org/10.1016/J.NEUROIMAGE.2009.10.080.

4

Interoceptive influences on resting-state fMRI

Zhongming Liu[a,b], Xiaokai Wang[a], Ana Cecilia Saavedra Bazan[a], and Jiayue Cao[a]

[a]Department of Biomedical Engineering, University of Michigan, Ann Arbor, MI, United States, [b]Department of Electrical Engineering and Computer Science, University of Michigan, Ann Arbor, MI, United States

Introduction

As introduced in Chapter 1, resting-state functional magnetic resonance imaging (rs-fMRI) is widely used for mapping the intrinsic functional networks emerging from spontaneous brain activity (Biswal et al., 1995; Fox and Raichle, 2007). Although the resting state is defined as the absence of any overt task or external stimulus, it is underappreciated that the brain always receives internal stimuli from inside the body. Inner organs ascend visceral signals to shape brain dynamics, emotion, and cognition (Azzalini et al., 2019). The brain, in turn, descends control signals to regulate bodily states (Saper, 2002). Such bidirectional interactions allow the brain to sense, integrate, and regulate physiological conditions, which is also known as interoception (Chen et al., 2021). Interoception engages multiple levels of neural circuits to maintain the awareness of the bodily self (Craig, 2002) and the homeostasis of physiological conditions (Barrett and Simmons, 2015) across a wide range of arousal states (Wei and Van Someren, 2020). It is an intrinsic source of spontaneous brain activity, especially when the brain is disengaged from the external input in the resting state.

However, bodily signals are often considered as physiological noise in rs-fMRI (Birn, 2012). As an indirect measure of neural activity (Logothetis, 2008), fMRI is susceptible to noise or artifacts (Liu, 2016). Breathing-induced chest movement changes the magnetic field that follows the respiratory cycle (Pfeuffer et al., 2002; Raj et al., 2001). Cardiac pulsation induces an inflow artifact that follows the cardiac cycle (Dagli et al., 1999). Systemic fluctuations in end-tidal carbon dioxide (CO_2) (Golestani et al., 2015; Wise et al., 2004) and blood circulation (Tong and Frederick, 2010) alter cerebral blood volume and flow (Golestani et al., 2015; Duyn et al., 2020; Tong et al.,

Advances in Resting-State Functional MRI. https://doi.org/10.1016/B978-0-323-91688-2.00015-1

2019). Fluctuations in the respiratory depth and heart rate are also treated as noise (Birn et al., 2006; Chang et al., 2009; Shmueli et al., 2007). Their effects on the brain are widespread and often regressed out (Chang et al., 2009; Birn et al., 2008; Glover et al., 2000; Jo et al., 2010; Murphy et al., 2013)—a processing strategy that may be debatable or even undesirable (Bright and Murphy, 2015). These topics will be covered in more depth in other chapters. Removing physiological fluctuations as noise may also risk partial elimination of neuronal effects because neural control of respiratory or cardiac activity may arise from the peripheral, brainstem, subcortical, and cortical levels (Craig, 2002; Critchley and Harrison, 2013; Khalsa et al., 2018). Bodily fluctuations and their central drives have profound contributions to rs-fMRI (Chen et al., 2020; Raut et al., 2021), whereas the mechanistic understanding is currently incomplete yet important to the interpretation of fMRI.

One person's noise may be another person's signal. It is like two sides of the same coin. Neither side shows the full picture. One should exercise caution over simply taking bodily signals as the noise to the brain or entirely discarding the possibility of physiological noise (Chang et al., 2016a; Yuan et al., 2013). Given the extensive discussions about physiological noise in the literature of rs-fMRI, it is worth highlighting a relatively underdiscussed perspective that visceral influences on rs-fMRI reflect, at least in part, the network signature of interoception. To support this perspective, we discuss (1) the evidence as to how the brain engages functional networks of neural activity to sense and regulate cardiovascular, respiratory, and digestive systems; (2) how interoceptive processes contribute to resting-state networks; (3) the functional relevance and implication of interoception for arousal, cognition, emotion, and perception; and (4) the current challenges and future opportunities in this research domain.

Neural pathways of interoception

The neural pathways for interoception include an ascending (afferent) pathway, which conveys sensory signals from the viscera to the brain, and a descending (efferent) pathway, which conveys motor commands from the brain to the viscera (Chen et al., 2021). These pathways connect neurons in ganglia, brainstem, subcortical, and cortical regions to form a hierarchical network that support multiple levels of bottom-up representation and top-down regulation of bodily states. Fig. 1 shows the major brain nuclei or regions involved in interoception, which are also discussed in prior reviews (Craig, 2002; Critchley and Harrison, 2013; Khalsa et al., 2018; Berntson and Khalsa, 2021).

The sensory neurons in the nodose ganglion and the dorsal root ganglion relay visceral signals through the vagal and spinal nerves,

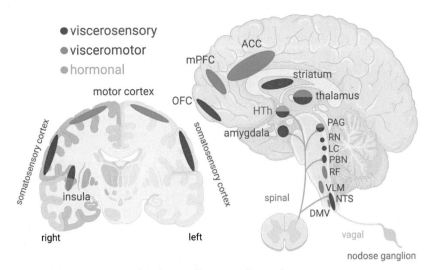

Fig. 1 Interoceptive neural pathways. The ascending pathway processes viscerosensory signals through a bottom-up cascade of brain regions shown in *red*. The descending pathway conveys visceromotor commands through a top-down cascade of brain regions shown in *blue*. Regions involved in both the ascending and descending pathways are shown in both *red* and *blue*. Note that the hypothalamus (shown in *red* and *green*) links the endocrine and nervous systems, supporting a neuroendocrine signaling pathway to guide behavior (Coll et al., 2007). The abbreviations shown refer to the orbitofrontal cortex (OFC), anterior cingulate cortex (ACC), medial prefrontal cortex (mPFC), hypothalamus (HTh), periaqueductal gray matter (PAG), raphe nucleus (RN), locus coeruleus (LC), parabrachial nucleus (PBN), reticular formation (RF), ventrolateral medulla (VLM), nucleus tractus solitarius (NTS), and dorsal motor nucleus of the vagus (DMV).

respectively (Azzalini et al., 2019; Saper, 2002). In a bottom-up hierarchical order, the vagal pathway passes through the nucleus tractus solitarius, parabrachial nucleus, cerebellum, periaqueductal gray matter, ventromedial nuclei of the thalamus, striatum, and dorsal posterior insula. The spinal pathway passes through the spinal dorsal horn, parabrachial nucleus, thalamus, hypothalamus, amygdala, and somatosensory cortex. Typically, vagal afferents carry mechanoreceptor and chemoreceptor signals related to physiological conditions, such as hunger and satiety (Mazzone and Undem, 2016; Travagli and Anselmi, 2016), and spinal afferents carry thermoreceptor and nociceptor signals related to temperature, pain, injury, and inflammation (Meacham et al., 2017; Xu and Huang, 2002). This dichotomy is, however, not absolute (Zhang et al., 2022). Different types of viscerosensory signals are integrated in the brain (Critchley and Harrison, 2013) to support interoceptive awareness (Critchley et al., 2004) and prediction (Barrett and Simmons, 2015).

In a top-down hierarchical order, the descending motor pathway includes the premotor and the motor cortex, anterior cingulate cortex, anterior insula, orbitofrontal cortex, hypothalamus, periaqueductal gray matter, reticular formation, nucleus ambiguus, ventrolateral medulla, and dorsal motor nuclei of the vagus. The ventrolateral medulla is the major source of sympathetic outflow. The dorsal motor nucleus of the vagus is the major source of parasympathetic outflow. Together, the sympathetic and parasympathetic outflows coordinate the descending regulation of various organ systems in response to bottom-up afferents (reflex mechanism) (Travagli and Anselmi, 2016; Borovikova et al., 2000; Dampney, 2016; Guyenet and Bayliss, 2015), top-down commands (central mechanism) (Saper, 2002; Barrett and Simmons, 2015; Benarroch, 1993; Cechetto, 2014), and arousal fluctuations (Wei and Van Someren, 2020; Guyenet and Bayliss, 2015). These mechanisms have distinctive, yet entangled, contributions to rs-fMRI and the networks inferred from fMRI, as will be discussed later in this chapter.

Autonomic reflexes and resting-state fMRI

The reflex mechanism is central to the homeostasis of many physiological parameters, such as blood pressure and oxygenation, both of which affect fMRI (Xu et al., 2012) and are relevant to fMRI studies using respiratory challenges (Chen and Pike, 2010). A well-understood reflex is the baroreflex (Dampney, 2016). The baroreceptors in the carotid body detect changes in blood pressure and transduce and relay the signal to the NTS through both the vagus and glossopharyngeal nerves. The NTS inhibits the cardiovascular sympathetic outflow, which increases the vascular resistance and cardiac contractility, while it also excites the cardiac parasympathetic outflow, which decreases the heart rate. As such, the baroreflex stabilizes the blood supply to the brain.

Another autonomic reflex is the chemoreflex for regulation of blood oxygenation (Dampney, 2016). Hypoxia, i.e., reduced oxygen (O_2) concentration, activates the chemoreceptors located in the carotid body and further activates a chemoreflex neural circuit in the brainstem, which increases the parasympathetic outflow to the heart and the sympathetic outflow to blood vessels, while also increasing respiratory activity. As a result, the respiratory rate and depth increase for greater oxygen uptake, while the heart rate decreases alongside peripheral vasoconstriction for greater oxygen conservation. Similarly, different chemoreflex circuits exist and respond to hypocapnia (i.e., excessive carbon dioxide) (Kara et al., 2003) and pH changes (Guyenet and Bayliss, 2015). Together, O_2- and CO_2-sensitive chemoreflexes coordinate rapid respiratory control in response to changes in the O_2 vs. CO_2 balance (Ciumas et al., 2022).

Autonomic reflexes also apply to other organs, e.g., digestive and immune systems. The vagovagal reflex regulates gastric motility and coordinates motor events across various regions of the gastrointestinal tract (Travagli and Anselmi, 2016; Powley, 2021). The vagally mediated inflammatory reflex suppresses the proinflammatory responses to injury or infection (Borovikova et al., 2000; Tracey, 2002; Ulloa, 2022). For various types of autonomic reflexes, the underlying neural circuits are mainly limited to peripheral ganglia and nerves and nuclei in the lower brainstem. Such reflex mechanisms act largely unconsciously.

Arguably, the autonomic reflexes alone can only explain a small fraction of brain-wide patterns of neural activity and interaction during a physiologically normal resting state. A greater portion of their influences on large-scale resting-state networks are more likely due to indirect vascular effects arising from the fluctuating sympathetic and parasympathetic outflows to cardiovascular and respiratory systems. The resulting physiological changes are observable in terms of blood pressure, vasomotion, heart rate, O_2 and CO_2 concentration, and respiratory rate and volume. These parameters are not independent of one another but show cofluctuations due to the shared or coordinated sympathetic and parasympathetic outflows, which drive their changes. The autonomic fluctuations underlying these parameters may collectively contribute to the changes of cerebral blood flow, volume, and oxygenation and thus affect fMRI. Relating only one of such physiological measures while ignoring other related measures may potentially mislead interpretation. It is desirable to simultaneously measure a comprehensive set of physiological parameters to characterize their collective and differential effects on rs-fMRI.

Cortical networks and interoception

Beyond the brainstem, large-scale cortical networks are also involved in interoception (Kleckner et al., 2017), linking the representations of the internal bodily states to subjective feeling (Craig, 2002; Craig, 2009), emotion (Critchley and Garfinkel, 2017), motivation (Han et al., 2018a), cognition (Azzalini et al., 2019), perception (Tallon-Baudry et al., 2018), behavior (Critchley and Harrison, 2013), and exerting central commands to the autonomic nervous system (Cechetto, 2014; Beissner et al., 2013; Levinthal and Strick, 2020). In the following, we discuss the major cortical regions involved in interoception as well as their functional anatomy and connectivity.

The insular cortex is the cortical hub for interoception (Penfield and Faulk Me, 1955), which is also known as the primary interoceptive cortex (Craig, 2002), following the convention of naming the primary cortical regions for exteroception, e.g., vision, audition, and somatosensation. The insular cortex can be differentiated into (at least) two subdivisions: the dorsal posterior insula and the ventral anterior insula (Evrard, 2019), based on their distinct laminar organizations of intra- and intercortical connectivity. The dorsal posterior insula is granular, including six differentiated layers with the ascending viscerosensory signal arriving at layer IV (also known as the granular layer), whereas the ventral anterior insula is agranular (lacking a clear granular layer). Such a laminar distinction is a hallmark circuit feature (typically shown with anterograde and retrograde tracing), which has been used to delineate different levels of cortical regions for exteroception, action, or cognition (Barbas and Rempel-Clower, 1997). Generalizing this notion to interoception, it has been suggested that the two subdivisions of the insular cortex may have different functional roles: the dorsal posterior insula is primary for viscerosensory processing, whereas the ventral anterior insula is secondary and predicts future changes of visceral states (or the so-called allostasis) and guides visceromotor control (Barrett and Simmons, 2015). Both resting-state functional connectivity in humans and anatomical tracing in monkeys suggest that the anterior and the posterior insula interact with and connect to different cortical regions, further confirming their difference in functional anatomy (Kleckner et al., 2017). The left and the right insula are also functionally distinctive as the right (but not left) insula supports conscious perception of visceral states and influences emotion (Critchley et al., 2004; Craig, 2009; Wang et al., 2019).

Paralimbic cortical regions, such as the anterior cingulate cortex, orbitofrontal cortex, and medial prefrontal cortex, are involved in interoception through their connections with the anterior insula (Critchley and Harrison, 2013; Kleckner et al., 2017; Fujimoto et al., 2021; Terasawa et al., 2013). In particular, both the anterior cingulate and anterior insula are part of the salience network (Seeley et al., 2007; Uddin, 2015). The two regions are structurally connected, as shown with diffusion tractography (Reisert et al., 2021), and are coactivated by interoceptive tasks in a number of fMRI studies (Beissner et al., 2013). Furthermore, both the anterior cingulate and dorsal amygdala are the visceromotor centers for prediction and control of interoceptive states (Barrett and Simmons, 2015; Kleckner et al., 2017).

The primary and the secondary somatosensory cortex contribute to interoception (Berntson and Khalsa, 2021; Khalsa et al., 2009). This might seem counterintuitive because the somatosensory afferents innervate the skin, instead of the viscera under the skin. In a prior study (Khalsa et al., 2009), a patient with lesions in the bilateral insular and anterior cingulate cortices could still perceive their heartbeat as accurately as healthy controls, whereas the patient could no longer do so, given pharmacological inhibition of skin afferents. It suggests that skin afferents through their ascending spinothalamocortical projection to the somatosensory cortex provide an alternative pathway for interoceptive awareness (Khalsa et al., 2009), arguably secondary to the primary pathway via the insula. To pair with this somatosensory pathway, a descending motor pathway that innervates the stomach has recently been reported with neural tracing in rodents (Levinthal and Strick, 2020). Such descending and ascending spinothalamocortical pathways, i.e., sympathetic efferents and afferents, close the loop for interoceptive sensing and regulation independent of cranial nerves (i.e., parasympathetic) that innervate visceral organs, thereby providing a shared neural pathway that subserves both interoception and exteroception.

In summary, the cortical networks of interoception have different functional roles (viscerosensory and visceromotor) and use different peripheral pathways to interact with the body, including sympathetic (spinal) and parasympathetic (vagal) pathways. In the vagal pathway, the dorsal posterior insula is the primary viscerosensory region and the anterior cingulate cortex is the primary visceromotor region. In the spinal pathway, the somatosensory cortex is the primary viscerosensory region and the motor cortex is the primary visceromotor region. Such a division is provisional and awaits confirmation by future studies.

Viscera-brain coupling

The neural or fMRI signals from brain regions involved in interoception have been demonstrated to be coupled, nearly in real-time, with cardiac and gastric rhythms. The heart and the gut have their intrinsic rhythms. The cardiac rhythm is paced by the sinoatrial node and has its normal frequency from 40 to 120 cycles per minute. The gastric rhythm is initiated by interstitial cells of Cajal (Sanders, 1996) and ranges from 3 to 7 cycles per minute. Both cardiac and gastric rhythms pace muscle contractions and generate phasic mechanoreceptor signals, which first ascend to the nucleus tractus solitarius (Powley, 2021; Moore et al., 2022) and then continue onto other subcortical and cortical regions.

The brain responds to the rhythms from the heart and the gut (Azzalini et al., 2019). Neural responses to every heartbeat elicit an event-related potential or field observable with electroencephalography (EEG) or magnetoencephalography (MEG) (Montoya et al., 1993; Park et al., 2014). When one attends to the heart or counts heartbeats, the heartbeat-evoked response is enhanced (Montoya et al., 1993; Petzschner et al., 2019), suggesting stronger interoceptive attention and awareness through the heart-brain neuroaxis (Garfinkel et al.,

2015). Cardiac interoception, indexed by the heartbeat-evoked response, can further affect the performance of exteroceptive processing and perception (Park et al., 2014; Al et al., 2020).

Although fMRI is too slow to track the responses to every heartbeat, the rhythm of the stomach (around 0.05 Hz) is right within the bandwidth of fMRI (0.01–0.1 Hz). Evidence suggests that the intrinsic gastric rhythm (or slow wave) is phase-coupled with spontaneous fMRI activity in some brain regions that process interoceptive signals as well as sensorimotor cortical regions (Rebollo et al., 2018; Rebollo and Tallon-Baudry, 2022; Choe et al., 2021; Cao et al., 2022) related to eating behaviors (Rebollo et al., 2021). This stomach-brain synchrony is primarily mediated by the ascending neural signaling from the stomach to the brain through the vagus nerve. Cutting the vagus diminishes the effect without affecting the gastric rhythm (Cao et al., 2022). Stimulating vagus afferents also increases the stomach-brain synchrony (Müller et al., 2022) and activates brain regions involved in gastric interoception (Cao et al., 2017, 2019, 2021). We hypothesize that similar ascending signaling also mediates a rhythmic coupling between the intestines and the brain, which has not been tested yet. However, this ascending mechanism is not applicable to the respiratory system since the brainstem initiates and controls the respiratory rhythm (Feldman and Del Negro, 2006). Nevertheless, the neural activity that controls respiratory activity also seems to contribute to rs-fMRI (Tu and Zhang, 2022).

The descending influences from the brain to the heart may drive the fluctuation of the heart rate or the duration of the cardiac cycle as well as the amplitude fluctuation of cardiac contraction. Heart rate variability (HRV) has been widely used to assess autonomic function (Appel et al., 1989). A common practice is to use the low-frequency component (0.04–0.15 Hz) of HRV as a proxy measure of the sympathetic tone, the high-frequency component (0.15–0.4 Hz) as a proxy measure of the parasympathetic tone, and their ratio as the balance or imbalance between sympathetic and parasympathetic activity (Ernst, 2017). The moment-by-moment correlation between the fMRI or neural signal and the HRV has been used to map and characterize the central autonomic network for sympathetic vs. parasympathetic regulation (Beissner et al., 2013; Chang et al., 2013; Kim et al., 2019). Similarly, the brain-to-stomach coupling may be assessed by evaluating the relationship between fMRI activity and the frequency or amplitude fluctuation of gastric contraction. Although such a relationship awaits affirmation, it is worth noting that gastric vagal efferents are known to innervate gastric enteric motor neurons for modulating the contraction or relaxation of gastric smooth muscles (Powley, 2021).

Given these relationships, it is plausible to assess the "causal" interaction between the brain and the viscera by evaluating the relationship between brain activity and different features of body rhythms. The phase coupling between brain activity and rhythmic events from the heart or the gut may report the ascending influence from the body to the brain. On the other hand, the coupling between brain activity and the frequency or amplitude fluctuation of cardiac or gastric rhythms may report the descending influence from the brain back to the body.

The coupling between brain and body oscillations is intriguing and seems more than just a coincidence. Instead, a binary hierarchy may govern brain and body oscillations (Klimesch, 2018). The center frequencies of canonical brain oscillations are binary multiples of one another. For delta = 2.5 Hz, theta = 5 Hz, alpha = 10 Hz, beta = 20 Hz, and gamma = 40 Hz, their ratios are 1:2, 1:4, 1:8, and 1:16, respectively. This coupling can be extended to bodily oscillations: heart rate = 1.25 Hz and breathing frequency = 0.3125 Hz, and their ratios to the delta frequency are 1:2 and 1:8, respectively. The gastric rhythm is also harmonically coupled with brain oscillations, although its ratio to the delta frequency is 1:50, which deviates from the strictly binary hierarchy. It is also intriguing that brain and body oscillations are associated to maintain rational harmonics during task execution and wakefulness, but not during sleep (Rassi et al., 2019). The functional significance of the coupling (vs. decoupling) between brain and body oscillations is not clear. It likely supports interoception and allows the bodily rhythm to entrain brain oscillations (Richter et al., 2017), providing a possible mechanism for first-person perspectives in conscious perception and behavior (Azzalini et al., 2019; Tallon-Baudry et al., 2018). In the evolutionary horizon, the enteric nervous system in the gut evolves long before the central nervous system in the brain (Furness and Stebbing, 2018; Spencer et al., 2021). Speculatively, hierarchically coupled oscillations may be the footprint of the evolutionary development of the nervous systems in both the brain and the viscera.

Implications for resting-state networks

The canonical functions of some resting-state networks should be revisited by considering the interoceptive inputs to the brain during the resting state (Kleckner et al., 2017). As shown in Fig. 2, the functional networks or systems that receive the most prominent interoceptive influences are the salience network (Seeley et al., 2007), cognitive control network (Cole and Schneider, 2007), cingulo-opercular network (Dosenbach et al., 2007), default-mode network (Raichle et al., 2001), limbic and mesolimbic systems, reticular activating system (Mesulam, 1995), sensorimotor network (Biswal et al., 1995), and the central autonomic network (Beissner et al., 2013). We discuss the functional implications of interoception to each of these networks.

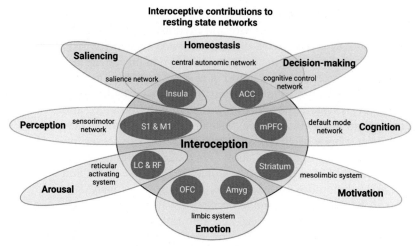

Fig. 2 Effects of interoception on resting-state networks and their canonical functions. Each *peripheral circle* describes a functional network with its name, function, and overlap with the putative network of interoception (the *center circle*). The abbreviations refer to the anterior cingulate cortex (ACC), medial prefrontal cortex (mPFC), amygdala (Amyg), orbitofrontal cortex (OFC), locus coeruleus (LC), reticular formation (RF), primary motor cortex (M1), and primary somatosensory cortex (S1).

The salience network overlaps with the interoceptive network at the anterior insular cortex—a cortical hub for switching attention across different tasks (Seeley et al., 2007), including the attention oriented toward the bodily self and the external environment (Menon and Uddin, 2010). The salience network may also play a central role in integrating internal and external inputs to guide cognition and behavior. In addition, the interoceptive network and the cingulo-opercular network share (at least) two cortical regions: the anterior insular cortex and dorsal anterior cingulate cortex (Dosenbach et al., 2007; Sadaghiani and D'Esposito, 2015), which are likely responsible for sustained and goal-driven attention and behavior to meet bodily needs. In addition, the network of interoception shares core regions with limbic and mesolimbic systems and the cognitive control network, in line with the understanding that interoception shapes emotion (Wiens, 2005), motivation (Han et al., 2018a), and decision-making (Fujimoto et al., 2021).

The default-mode network (Greicius et al., 2003) consists of regions deactivated by many externally oriented or cognitively demanding tasks (Raichle et al., 2001). Its pattern is preserved across wakefulness to sleep (Horovitz et al., 2008) or even in anesthetized or unconscious states (Vincent et al., 2007), suggesting its core function to maintain bodily function for survival. In passive but conscious states, the default-mode network, especially its midline core regions such as the medial prefrontal cortex, is involved in self-referential and affective processes functionally relevant to interoception (Andrews-Hanna et al., 2010).

The brainstem nuclei or subcortical structures involved in interoception partially overlap with those that drive arousal fluctuations, e.g., the cholinergic reticular activation system (Mesulam, 1995). Findings from recent studies suggest the profound effects of arousal on spontaneous brain activity (Chang et al., 2016b; Turchi et al., 2018). It should be noted that there are various levels of arousal. Bodily arousal refers to the systemic physiological condition. Vigilance and emotional arousal are also associated with distinct interoceptive states (Wei and Van Someren, 2020; Guyenet and Bayliss, 2015) and may be partly assessed by recording the pupil diameter, skin conductance, or heart rate variability. How interoceptive processes interplay with each type or level of arousal is not fully understood and awaits future investigations (Azzalini et al., 2019).

Last but not least, the central autonomic network observed with fMRI (Beissner et al., 2013) or diffusion MRI (Reisert et al., 2021) may be further characterized in terms of its structural and functional subdivisions that represent and regulate different visceral organs, for example, the gastric network (Rebollo et al., 2018; Choe et al., 2021; Cao et al., 2022), cardiac network (Saper, 2002; Talman, 1985), and respiratory network (Tu and Zhang, 2022).

Current challenges and future directions

Brainstem and subcortical nuclei involved in interoception are small, susceptible to artifacts and noise (Brooks et al., 2013), and are of low sensitivity in MR signal transmission and reception. This creates challenges for using fMRI to investigate interoception. Increasing availability of high-field MRI scanners, innovation in contrast mechanisms, and pulse sequences continue pushing the limit in spatial resolution and specificity. The brainstem is an underexplored part of the brain and would be an ideal test ground for technical innovation and scientific discovery.

The future of interoception research may benefit from a holistic approach (Fig. 3). The notion of interoception has transcended the boundaries of the mind and body and of central and autonomic neuroscience. It is worth taking one step further to consider the body as an interconnected system of organs as opposed to isolated organ systems. In other words, it is more desirable to investigate brain-body interactions, as opposed to brain-stomach, brain-heart, and brain-lung as parallel and independent neuro-axes. A manageable approach encompasses multiple techniques for imaging and recording the brain and the body. For example, fMRI and EEG can be recorded

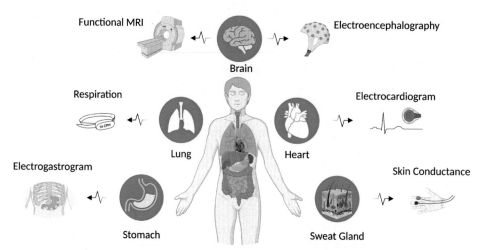

Fig. 3 Holistic approach to study interoception. Central and peripheral measures can be simultaneously collected from humans using established multimodal techniques. Assessing the relationships between the measured signals is expected to provide a more complete picture of the interoception.

concurrently to capture complementary aspects of brain activity and help identify and separate neural vs. nonneural sources (Wen and Liu, 2016). In the meantime, peripheral signals can be collected from the lung, heart, stomach, and skin to derive quantitative measures indicative of different interoceptive states. It is also desirable to perturb the interoceptive processes by using well-controlled tasks that engage different interoceptive vs. exteroceptive awareness and attention, or by using noninvasive neuromodulation, such as transcutaneous or transauricular vagus nerve stimulation, to target specific neural pathways. Cognitive or emotional assessment may further help investigate the functional significance of interoception with respect to cognitive control and emotional regulation.

Animal models provide unique opportunities to identify the cellular and molecular constituents of functional circuits for interoception (Chen et al., 2021). Viral tracing reveals structural pathways with molecular specificity and high spatial resolution, which remain unreachable with noninvasive MRI methods. Neural recordings with single-cell resolution reveal neural encoding at peripheral ganglia, brain nuclei, and regions at far greater spatial and temporal specificity than fMRI and EEG. Surgical, chemogenetic, and optogenetic tools are increasingly available and robust to perturb neural circuits more precisely than their noninvasive counterparts used for human studies.

Computational models may potentially bridge the gap between observations and theories. A compelling theory of interoception is the so-called *embodied predictive interoception coding*, (Barrett and

Simmons, 2015) extending the theory of predictive coding for perception and action over the external world (Friston, 2010). It should be noted that models of predictive coding have been explored for visual perception (Rao and Ballard, 1999) and can be scaled up to deep neural network (Lotter et al., 2017; Wen et al., 2018; Han et al., 2018b; Luczak et al., 2022). Similar models may be developed to explain how the brain predicts the states of inner organs (Petzschner et al., 2021). In addition, recent advances in representation learning of fMRI activity patterns may be used to disentangle various sources, contributing to rs-fMRI (Kim et al., 2021) or predict respiratory and cardiac signals with fMRI (Bayrak et al., 2021).

References

Al, E., Iliopoulos, F., Forschack, N., et al., 2020. Heart–brain interactions shape somatosensory perception and evoked potentials. Proc. Natl. Acad. Sci. 117 (19), 10575–10584. https://doi.org/10.1073/pnas.1915629117.

Andrews-Hanna, J.R., Reidler, J.S., Sepulcre, J., Poulin, R., Buckner, R.L., 2010. Functional-anatomic fractionation of the brain's default network. Neuron 65 (4), 550–562. https://doi.org/10.1016/j.neuron.2010.02.005.

Appel, M.L., Berger, R.D., Saul, J.P., Smith, J.M., Cohen, R.J., 1989. Beat to beat variability in cardiovascular variables: noise or music? J. Am. Coll. Cardiol. 14 (5), 1139–1148. https://doi.org/10.1016/0735-1097(89)90408-7.

Azzalini, D., Rebollo, I., Tallon-Baudry, C., 2019. Visceral signals shape brain dynamics and cognition. Trends Cogn. Sci. 23 (6), 488–509. https://doi.org/10.1016/j.tics.2019.03.007.

Barbas, H., Rempel-Clower, N., 1997. Cortical structure predicts the pattern of corticocortical connections. Cereb. Cortex 7 (7), 635–646. https://doi.org/10.1093/cercor/7.7.635.

Barrett, L.F., Simmons, W.K., 2015. Interoceptive predictions in the brain. Nat. Rev. Neurosci. 16 (7), 419–429. https://doi.org/10.1038/nrn3950.

Bayrak, R.G., Hansen, C.B., Salas, J.A., et al., 2021. From brain to body: learning low-frequency respiration and cardiac signals from fMRI dynamics. In: de Bruijne, M., Cattin, P.C., Cotin, S., et al. (Eds.), Medical Image Computing and Computer Assisted Intervention–MICCAI 2021. Lecture Notes in Computer Science, Springer International Publishing, pp. 553–563, https://doi.org/10.1007/978-3-030-87234-2_52.

Beissner, F., Meissner, K., Bär, K.J., Napadow, V., 2013. The autonomic brain: an activation likelihood estimation meta-analysis for central processing of autonomic function. J. Neurosci. 33 (25), 10503–10511. https://doi.org/10.1523/JNEUROSCI.1103-13.2013.

Benarroch, E.E., 1993. The central autonomic network: functional organization, dysfunction, and perspective. Mayo Clin. Proc. 68 (10), 988–1001. https://doi.org/10.1016/S0025-6196(12)62272-1.

Berntson, G.G., Khalsa, S.S., 2021. Neural circuits of interoception. Trends Neurosci. 44 (1), 17–28. https://doi.org/10.1016/j.tins.2020.09.011.

Birn, R.M., 2012. The role of physiological noise in resting-state functional connectivity. NeuroImage 62 (2), 864–870. https://doi.org/10.1016/j.neuroimage.2012.01.016.

Birn, R.M., Diamond, J.B., Smith, M.A., Bandettini, P.A., 2006. Separating respiratory-variation-related fluctuations from neuronal-activity-related fluctuations in fMRI. NeuroImage 31 (4), 1536–1548. https://doi.org/10.1016/j.neuroimage.2006.02.048.

Birn, R.M., Smith, M.A., Jones, T.B., Bandettini, P.A., 2008. The respiration response function: the temporal dynamics of fMRI signal fluctuations related to changes in respiration. NeuroImage 40 (2), 644–654. https://doi.org/10.1016/j.neuroimage.2007.11.059.

Biswal, B., Zerrin Yetkin, F., Haughton, V.M., Hyde, J.S., 1995. Functional connectivity in the motor cortex of resting human brain using echo-planar mri. Magn. Reson. Med. 34 (4), 537–541. https://doi.org/10.1002/mrm.1910340409.

Borovikova, L.V., Ivanova, S., Zhang, M., et al., 2000. Vagus nerve stimulation attenuates the systemic inflammatory response to endotoxin. Nature 405 (6785), 458–462. https://doi.org/10.1038/35013070.

Bright, M.G., Murphy, K., 2015. Is fMRI "noise" really noise? Resting state nuisance regressors remove variance with network structure. NeuroImage 114, 158–169. https://doi.org/10.1016/j.neuroimage.2015.03.070.

Brooks, J., Faull, O., Pattinson, K., Jenkinson, M., 2013. Physiological noise in brainstem fMRI. Front. Hum. Neurosci. 7. https://www.frontiersin.org/articles/10.3389/fnhum.2013.00623. (Accessed 2 November 2022).

Cao, J., Lu, K.H., Powley, T.L., Liu, Z., 2017. Vagal nerve stimulation triggers widespread responses and alters large-scale functional connectivity in the rat brain. PLoS One 12 (12), e0189518. https://doi.org/10.1371/journal.pone.0189518.

Cao, J., Lu, K.H., Oleson, S.T., et al., 2019. Gastric stimulation drives fast BOLD responses of neural origin. NeuroImage 197, 200–211. https://doi.org/10.1016/j.neuroimage.2019.04.064.

Cao, J., Wang, X., Powley, T.L., Liu, Z., 2021. Gastric neurons in the nucleus tractus solitarius are selective to the orientation of gastric electrical stimulation. J. Neural Eng. 18 (5), 056066. https://doi.org/10.1088/1741-2552/ac2ec6.

Cao, J., Wang, X., Chen, J., Zhang, N., Liu, Z., 2022. The vagus nerve mediates the stomach-brain coherence in rats. NeuroImage 263, 119628. https://doi.org/10.1016/j.neuroimage.2022.119628.

Cechetto, D.F., 2014. Cortical control of the autonomic nervous system. Exp. Physiol. 99 (2), 326–331. https://doi.org/10.1113/expphysiol.2013.075192.

Chang, C., Cunningham, J.P., Glover, G.H., 2009. Influence of heart rate on the BOLD signal: the cardiac response function. NeuroImage 44 (3), 857–869. https://doi.org/10.1016/j.neuroimage.2008.09.029.

Chang, C., Metzger, C.D., Glover, G.H., Duyn, J.H., Heinze, H.J., Walter, M., 2013. Association between heart rate variability and fluctuations in resting-state functional connectivity. NeuroImage 68, 93–104. https://doi.org/10.1016/j.neuroimage.2012.11.038.

Chang, C., Raven, E.P., Duyn, J.H., 2016a. Brain–heart interactions: challenges and opportunities with functional magnetic resonance imaging at ultra-high field. Philos. Trans. R Soc. Math. Phys. Eng. Sci. 374 (2067), 20150188. https://doi.org/10.1098/rsta.2015.0188.

Chang, C., Leopold, D.A., Schölvinck, M.L., et al., 2016b. Tracking brain arousal fluctuations with fMRI. Proc. Natl. Acad. Sci. U. S. A. 113 (16), 4518–4523. https://doi.org/10.1073/pnas.1520613113.

Chen, J.J., Pike, G.B., 2010. MRI measurement of the BOLD-specific flow–volume relationship during hypercapnia and hypocapnia in humans. NeuroImage 53 (2), 383–391. https://doi.org/10.1016/j.neuroimage.2010.07.003.

Chen, J.E., Lewis, L.D., Chang, C., et al., 2020. Resting-state "physiological networks". NeuroImage 213, 116707. https://doi.org/10.1016/j.neuroimage.2020.116707.

Chen, W.G., Schloesser, D., Arensdorf, A.M., et al., 2021. The emerging science of interoception: sensing, integrating, interpreting, and regulating signals within the self. Trends Neurosci. 44 (1), 3–16. https://doi.org/10.1016/j.tins.2020.10.007.

Choe, A.S., Tang, B., Smith, K.R., et al., 2021. Phase-locking of resting-state brain networks with the gastric basal electrical rhythm. PLoS One 16 (1), e0244756. https://doi.org/10.1371/journal.pone.0244756.

Ciumas, C., Rheims, S., Ryvlin, P., 2022. fMRI studies evaluating central respiratory control in humans. Front. Neural Circuits 16. https://www.frontiersin.org/articles/10.3389/fncir.2022.982963. (Accessed 31 October 2022).

Cole, M.W., Schneider, W., 2007. The cognitive control network: integrated cortical regions with dissociable functions. NeuroImage 37 (1), 343–360. https://doi.org/10.1016/j.neuroimage.2007.03.071.

Coll, A.P., Farooqi, I.S., O'Rahilly, S., 2007. The hormonal control of food intake. Cell 129 (2), 251–262. https://doi.org/10.1016/j.cell.2007.04.001.

Craig, A.D., 2002. How do you feel? Interoception: the sense of the physiological condition of the body. Nat. Rev. Neurosci. 3 (8), 655–666. https://doi.org/10.1038/nrn894.

Craig, D.A., 2009. How do you feel—now? The anterior insula and human awareness. Nat. Rev. Neurosci. 10 (1), 59–70. https://doi.org/10.1038/nrn2555.

Critchley, H.D., Garfinkel, S.N., 2017. Interoception and emotion. Curr. Opin. Psychol. 17, 7–14. https://doi.org/10.1016/j.copsyc.2017.04.020.

Critchley, H.D., Harrison, N.A., 2013. Visceral influences on brain and behavior. Neuron 77 (4), 624–638. https://doi.org/10.1016/j.neuron.2013.02.008.

Critchley, H.D., Wiens, S., Rotshtein, P., Öhman, A., Dolan, R.J., 2004. Neural systems supporting interoceptive awareness. Nat. Neurosci. 7 (2), 189–195. https://doi.org/10.1038/nn1176.

Dagli, M.S., Ingeholm, J.E., Haxby, J.V., 1999. Localization of cardiac-induced signal change in fMRI. NeuroImage 9 (4), 407–415. https://doi.org/10.1006/nimg.1998.0424.

Dampney, R.A.L., 2016. Central neural control of the cardiovascular system: current perspectives. Adv. Physiol. Educ. 40 (3), 283–296. https://doi.org/10.1152/advan.00027.2016.

Dosenbach, N.U.F., Fair, D.A., Miezin, F.M., et al., 2007. Distinct brain networks for adaptive and stable task control in humans. Proc. Natl. Acad. Sci. U. S. A. 104 (26), 11073–11078. https://doi.org/10.1073/pnas.0704320104.

Duyn, J.H., Ozbay, P.S., Chang, C., Picchioni, D., 2020. Physiological changes in sleep that affect fMRI inference. Curr. Opin. Behav. Sci. 33, 42–50. https://doi.org/10.1016/j.cobeha.2019.12.007.

Ernst, G., 2017. Heart-rate variability—more than heart beats? Front. Public Health 5. https://www.frontiersin.org/articles/10.3389/fpubh.2017.00240. (Accessed 27 October 2022).

Evrard, H.C., 2019. The organization of the primate insular cortex. Front. Neuroanat. 13. https://www.frontiersin.org/articles/10.3389/fnana.2019.00043. (Accessed 31 October 2022).

Feldman, J.L., Del Negro, C.A., 2006. Looking for inspiration: new perspectives on respiratory rhythm. Nat. Rev. Neurosci. 7 (3), 232–241. https://doi.org/10.1038/nrn1871.

Fox, M.D., Raichle, M.E., 2007. Spontaneous fluctuations in brain activity observed with functional magnetic resonance imaging. Nat. Rev. Neurosci. 8 (9), 700–711. https://doi.org/10.1038/nrn2201.

Friston, K., 2010. The free-energy principle: a unified brain theory? Nat. Rev. Neurosci. 11 (2), 127–138. https://doi.org/10.1038/nrn2787.

Fujimoto, A., Murray, E.A., Rudebeck, P.H., 2021. Interaction between decision-making and interoceptive representations of bodily arousal in frontal cortex. Proc. Natl. Acad. Sci. U. S. A. 118 (35), e2014781118. https://doi.org/10.1073/pnas.2014781118.

Furness, J.B., Stebbing, M.J., 2018. The first brain: species comparisons and evolutionary implications for the enteric and central nervous systems. Neurogastroenterol. Motil. Off. J. Eur. Gastrointest. Motil. Soc. 30 (2). https://doi.org/10.1111/nmo.13234.

Garfinkel, S.N., Seth, A.K., Barrett, A.B., Suzuki, K., Critchley, H.D., 2015. Knowing your own heart: distinguishing interoceptive accuracy from interoceptive awareness. Biol. Psychol. 104, 65–74. https://doi.org/10.1016/j.biopsycho.2014.11.004.

Glover, G.H., Li, T.Q., Ress, D., 2000. Image-based method for retrospective correction of physiological motion effects in fMRI: RETROICOR. Magn. Reson. Med. 44 (1), 162–167. https://doi.org/10.1002/1522-2594(200007)44:1<162::AID-MRM23>3.0.CO;2-E.

Golestani, A.M., Chang, C., Kwinta, J.B., Khatamian, Y.B., Jean, C.J., 2015. Mapping the end-tidal CO2 response function in the resting-state BOLD fMRI signal: spatial specificity, test–retest reliability and effect of fMRI sampling rate. NeuroImage 104, 266–277. https://doi.org/10.1016/j.neuroimage.2014.10.031.

Greicius, M.D., Krasnow, B., Reiss, A.L., Menon, V., 2003. Functional connectivity in the resting brain: a network analysis of the default mode hypothesis. Proc. Natl. Acad. Sci. U. S. A. 100 (1), 253–258. https://doi.org/10.1073/pnas.0135058100.

Guyenet, P.G., Bayliss, D.A., 2015. Neural control of breathing and CO2 homeostasis. Neuron 87 (5), 946–961. https://doi.org/10.1016/j.neuron.2015.08.001.

Han, W., Tellez, L.A., Perkins, M.H., et al., 2018a. A neural circuit for gut-induced reward. Cell 175 (3), 665–678.e23. https://doi.org/10.1016/j.cell.2018.08.049.

Han, K., Wen, H., Zhang, Y., Fu, D., Culurciello, E., Liu, Z., 2018b. Deep predictive coding network with local recurrent processing for object recognition. In: Advances in Neural Information Processing Systems. Vol. 31. Curran Associates, Inc. https://proceedings.neurips.cc/paper/2018/hash/1c63926ebcabda26b5cdb31b5cc91efb-Abstract.html. (Accessed 1 November 2022).

Horovitz, S.G., Fukunaga, M., de Zwart, J.A., et al., 2008. Low frequency BOLD fluctuations during resting wakefulness and light sleep: a simultaneous EEG-fMRI study. Hum. Brain Mapp. 29 (6), 671–682. https://doi.org/10.1002/hbm.20428.

Jo, H.J., Saad, Z.S., Simmons, W.K., Milbury, L.A., Cox, R.W., 2010. Mapping sources of correlation in resting state FMRI, with artifact detection and removal. NeuroImage 52 (2), 571–582. https://doi.org/10.1016/j.neuroimage.2010.04.246.

Kara, T., Narkiewicz, K., Somers, V.K., 2003. Chemoreflexes—physiology and clinical implications. Acta Physiol. Scand. 177 (3), 377–384. https://doi.org/10.1046/j.1365-201X.2003.01083.x.

Khalsa, S.S., Rudrauf, D., Feinstein, J.S., Tranel, D., 2009. The pathways of interoceptive awareness. Nat. Neurosci. 12 (12), 1494–1496. https://doi.org/10.1038/nn.2411.

Khalsa, S.S., Adolphs, R., Cameron, O.G., et al., 2018. Interoception and mental health: a roadmap. Biol. Psychiatry Cogn. Neurosci. Neuroimaging 3 (6), 501–513. https://doi.org/10.1016/j.bpsc.2017.12.004.

Kim, K., Ladenbauer, J., Babo-Rebelo, M., et al., 2019. Resting-state neural firing rate is linked to cardiac-cycle duration in the human cingulate and parahippocampal cortices. J. Neurosci. 39 (19), 3676–3686. https://doi.org/10.1523/JNEUROSCI.2291-18.2019.

Kim, J.H., Zhang, Y., Han, K., Wen, Z., Choi, M., Liu, Z., 2021. Representation learning of resting state fMRI with variational autoencoder. NeuroImage 241, 118423. https://doi.org/10.1016/j.neuroimage.2021.118423.

Kleckner, I.R., Zhang, J., Touroutoglou, A., et al., 2017. Evidence for a large-scale brain system supporting allostasis and interoception in humans. Nat. Hum. Behav. 1 (5), 1–14. https://doi.org/10.1038/s41562-017-0069.

Klimesch, W., 2018. The frequency architecture of brain and brain body oscillations: an analysis. Eur. J. Neurosci. 48 (7), 2431–2453. https://doi.org/10.1111/ejn.14192.

Levinthal, D.J., Strick, P.L., 2020. Multiple areas of the cerebral cortex influence the stomach. Proc. Natl. Acad. Sci. U. S. A. 117 (23), 13078–13083. https://doi.org/10.1073/pnas.2002737117.

Liu, T.T., 2016. Noise contributions to the fMRI signal: an overview. NeuroImage 143, 141–151. https://doi.org/10.1016/j.neuroimage.2016.09.008.

Logothetis, N.K., 2008. What we can do and what we cannot do with fMRI. Nature 453 (7197), 869–878. https://doi.org/10.1038/nature06976.

Lotter, W., Kreiman, G., Cox, D., 2017. Deep Predictive Coding Networks for Video Prediction and Unsupervised Learning. Published online February 28, https://doi.org/10.48550/arXiv.1605.08104.

Luczak, A., McNaughton, B.L., Kubo, Y., 2022. Neurons learn by predicting future activity. Nat. Mach. Intell. 4 (1), 62–72. https://doi.org/10.1038/s42256-021-00430-y.

Mazzone, S.B., Undem, B.J., 2016. Vagal afferent innervation of the airways in health and disease. Physiol. Rev. 96 (3), 975–1024. https://doi.org/10.1152/physrev.00039.2015.

Meacham, K., Shepherd, A., Mohapatra, D.P., Haroutounian, S., 2017. Neuropathic pain: central vs. peripheral mechanisms. Curr. Pain Headache Rep. 21 (6), 28. https://doi.org/10.1007/s11916-017-0629-5.

Menon, V., Uddin, L.Q., 2010. Saliency, switching, attention and control: a network model of insula function. Brain Struct. Funct. 214 (5), 655–667. https://doi.org/10.1007/s00429-010-0262-0.

Mesulam, M.M., 1995. Cholinergic pathways and the ascending reticular activating system of the human Braina. Ann. N. Y. Acad. Sci. 757 (1), 169–179. https://doi.org/10.1111/j.1749-6632.1995.tb17472.x.

Montoya, P., Schandry, R., Müller, A., 1993. Heartbeat evoked potentials (HEP): topography and influence of cardiac awareness and focus of attention. Electroencephalogr. Clin. Neurophysiol. 88 (3), 163–172. https://doi.org/10.1016/0168-5597(93)90001-6.

Moore, J.P., Simpson, L.L., Drinkhill, M.J., 2022. Differential contributions of cardiac, coronary and pulmonary artery vagal mechanoreceptors to reflex control of the circulation. J. Physiol. 600 (18), 4069–4087. https://doi.org/10.1113/JP282305.

Müller, S.J., Teckentrup, V., Rebollo, I., Hallschmid, M., Kroemer, N.B., 2022. Vagus nerve stimulation increases stomach-brain coupling via a vagal afferent pathway. Brain Stimul. Basic Transl. Clin. Res. Neuromodulation. https://doi.org/10.1016/j.brs.2022.08.019.

Murphy, K., Birn, R.M., Bandettini, P.A., 2013. Resting-state fMRI confounds and cleanup. NeuroImage 80, 349–359. https://doi.org/10.1016/j.neuroimage.2013.04.001.

Park, H.D., Correia, S., Ducorps, A., Tallon-Baudry, C., 2014. Spontaneous fluctuations in neural responses to heartbeats predict visual detection. Nat. Neurosci. 17 (4), 612–618. https://doi.org/10.1038/nn.3671.

Penfield, W., Faulk Me, J.R., 1955. The insula: further observations on its function1. Brain 78 (4), 445–470. https://doi.org/10.1093/brain/78.4.445.

Petzschner, F.H., Weber, L.A., Wellstein, K.V., Paolini, G., Do, C.T., Stephan, K.E., 2019. Focus of attention modulates the heartbeat evoked potential. NeuroImage 186, 595–606. https://doi.org/10.1016/j.neuroimage.2018.11.037.

Petzschner, F.H., Garfinkel, S.N., Paulus, M.P., Koch, C., Khalsa, S.S., 2021. Computational models of interoception and body regulation. Trends Neurosci. 44 (1), 63–76. https://doi.org/10.1016/j.tins.2020.09.012.

Pfeuffer, J., Van de Moortele, P.F., Ugurbil, K., Hu, X., Glover, G.H., 2002. Correction of physiologically induced global off-resonance effects in dynamic echo-planar and spiral functional imaging. Magn. Reson. Med. 47 (2), 344–353. https://doi.org/10.1002/mrm.10065.

Powley, T.L., 2021. Brain-gut communication: vagovagal reflexes interconnect the two "brains". Am. J. Physiol.-Gastrointest. Liver Physiol. 321 (5), G576–G587. https://doi.org/10.1152/ajpgi.00214.2021.

Raichle, M.E., MacLeod, A.M., Snyder, A.Z., Powers, W.J., Gusnard, D.A., Shulman, G.L., 2001. A default mode of brain function. Proc. Natl. Acad. Sci. U. S. A. 98 (2), 676–682. https://doi.org/10.1073/pnas.98.2.676.

Raj, D., Anderson, A.W., Gore, J.C., 2001. Respiratory effects in human functional magnetic resonance imaging due to bulk susceptibility changes. Phys. Med. Biol. 46 (12), 3331–3340. https://doi.org/10.1088/0031-9155/46/12/318.

Rao, R.P.N., Ballard, D.H., 1999. Predictive coding in the visual cortex: a functional interpretation of some extra-classical receptive-field effects. Nat. Neurosci. 2 (1), 79–87. https://doi.org/10.1038/4580.

Rassi, E., Dorffner, G., Gruber, W., Schabus, M., Klimesch, W., 2019. Coupling and decoupling between brain and body oscillations. Neurosci. Lett. 711, 134401. https://doi.org/10.1016/j.neulet.2019.134401.

Raut, R.V., Snyder, A.Z., Mitra, A., et al., 2021. Global waves synchronize the brain's functional systems with fluctuating arousal. Sci. Adv. 7 (30), eabf2709. https://doi.org/10.1126/sciadv.abf2709.

Rebollo, I., Tallon-Baudry, C., 2022. The sensory and motor components of the cortical hierarchy are coupled to the rhythm of the stomach during rest. J. Neurosci. 42 (11), 2205-2220. https://doi.org/10.1523/JNEUROSCI.1285-21.2021.

Rebollo, I., Devauchelle, A.D., Béranger, B., Tallon-Baudry, C., 2018. Stomach-brain synchrony reveals a novel, delayed-connectivity resting-state network in humans. Critchley H, ed. elife 7, e33321. https://doi.org/10.7554/eLife.33321.

Rebollo, I., Wolpert, N., Tallon-Baudry, C., 2021. Brain–stomach coupling: anatomy, functions, and future avenues of research. Curr. Opin. Biomed. Eng. 18, 100270. https://doi.org/10.1016/j.cobme.2021.100270.

Reisert, M., Weiller, C., Hosp, J.A., 2021. Displaying the autonomic processing network in humans—a global tractography approach. NeuroImage 231, 117852. https://doi.org/10.1016/j.neuroimage.2021.117852.

Richter, C.G., Babo-Rebelo, M., Schwartz, D., Tallon-Baudry, C., 2017. Phase-amplitude coupling at the organism level: the amplitude of spontaneous alpha rhythm fluctuations varies with the phase of the infra-slow gastric basal rhythm. NeuroImage 146, 951–958. https://doi.org/10.1016/j.neuroimage.2016.08.043.

Sadaghiani, S., D'Esposito, M., 2015. Functional characterization of the Cingulo-Opercular network in the maintenance of tonic alertness. Cereb. Cortex 25 (9), 2763–2773. https://doi.org/10.1093/cercor/bhu072.

Sanders, K., 1996. A case for interstitial cells of Cajal as pacemakers and mediators of neurotransmission in the gastrointestinal tract. Gastroenterology 111 (2), 492–515. https://doi.org/10.1053/gast.1996.v111.pm8690216.

Saper, C.B., 2002. The central autonomic nervous system: conscious visceral perception and autonomic pattern generation. Annu. Rev. Neurosci. 25 (1), 433–469. https://doi.org/10.1146/annurev.neuro.25.032502.111311.

Seeley, W.W., Menon, V., Schatzberg, A.F., et al., 2007. Dissociable intrinsic connectivity networks for salience processing and executive control. J. Neurosci. 27 (9), 2349–2356. https://doi.org/10.1523/JNEUROSCI.5587-06.2007.

Shmueli, K., van Gelderen, P., de Zwart, J.A., et al., 2007. Low-frequency fluctuations in the cardiac rate as a source of variance in the resting-state fMRI BOLD signal. NeuroImage 38 (2), 306–320. https://doi.org/10.1016/j.neuroimage.2007.07.037.

Spencer, N.J., Travis, L., Wiklendt, L., et al., 2021. Long range synchronization within the enteric nervous system underlies propulsion along the large intestine in mice. Commun. Biol. 4 (1), 1–17. https://doi.org/10.1038/s42003-021-02485-4.

Tallon-Baudry, C., Campana, F., Park, H.D., Babo-Rebelo, M., 2018. The neural monitoring of visceral inputs, rather than attention, accounts for first-person perspective in conscious vision. Cortex 102, 139–149. https://doi.org/10.1016/j.cortex.2017.05.019.

Talman, W.T., 1985. Cardiovascular regulation and lesions of the central nervous system. Ann. Neurol. 18 (1), 1–12. https://doi.org/10.1002/ana.410180102.

Terasawa, Y., Fukushima, H., Umeda, S., 2013. How does interoceptive awareness interact with the subjective experience of emotion? An fMRI study. Hum. Brain Mapp. 34 (3), 598–612. https://doi.org/10.1002/hbm.21458.

Tong, Y., Frederick, B.D., 2010. Time lag dependent multimodal processing of concurrent fMRI and near-infrared spectroscopy (NIRS) data suggests a global circulatory origin for low-frequency oscillation signals in human brain. NeuroImage 53 (2), 553–564. https://doi.org/10.1016/j.neuroimage.2010.06.049.

Tong, Y., Hocke, L.M., Frederick, B.B., 2019. Low frequency systemic hemodynamic "noise" in resting state BOLD fMRI: characteristics, causes, implications, mitigation strategies, and applications. Front. Neurosci. 13. https://www.frontiersin.org/articles/10.3389/fnins.2019.00787. (Accessed 8 September 2022).

Tracey, K.J., 2002. The inflammatory reflex. Nature 420 (6917), 853–859. https://doi.org/10.1038/nature01321.

Travagli, R.A., Anselmi, L., 2016. Vagal neurocircuitry and its influence on gastric motility. Nat. Rev. Gastroenterol. Hepatol. 13 (7), 389–401. https://doi.org/10.1038/nrgastro.2016.76.

Tu, W., Zhang, N., 2022. Neural underpinning of a respiration-associated resting-state fMRI network. Miller KL, ed. elife 11, e81555. https://doi.org/10.7554/eLife.81555.

Turchi, J., Chang, C., Ye, F.Q., et al., 2018. The basal forebrain regulates global resting-state fMRI fluctuations. Neuron 97 (4), 940–952.e4. https://doi.org/10.1016/j.neuron.2018.01.032.

Uddin, L.Q., 2015. Salience processing and insular cortical function and dysfunction. Nat. Rev. Neurosci. 16 (1), 55–61. https://doi.org/10.1038/nrn3857.

Ulloa, L., 2022. Bioelectronic neuro-immunology: neuronal networks for sympathetic-splenic and vagal-adrenal control. Neuron. https://doi.org/10.1016/j.neuron.2022.09.015.

Vincent, J.L., Patel, G.H., Fox, M.D., et al., 2007. Intrinsic functional architecture in the anaesthetized monkey brain. Nature 447 (7140), 83–86. https://doi.org/10.1038/nature05758.

Wang, X., Wu, Q., Egan, L., et al., 2019. Anterior insular cortex plays a critical role in interoceptive attention. Stephan KE, Frank MJ, Stephan KE, Faull OK, Allen M, eds. elife 8, e42265. https://doi.org/10.7554/eLife.42265.

Wei, Y., Van Someren, E.J., 2020. Interoception relates to sleep and sleep disorders. Curr. Opin. Behav. Sci. 33, 1–7. https://doi.org/10.1016/j.cobeha.2019.11.008.

Wen, H., Liu, Z., 2016. Broadband electrophysiological dynamics contribute to global resting-state fMRI signal. J. Neurosci. 36 (22), 6030–6040. https://doi.org/10.1523/JNEUROSCI.0187-16.2016.

Wen, H., Han, K., Shi, J., Zhang, Y., Culurciello, E., Liu, Z., 2018. Deep predictive coding network for object recognition. In: Proceedings of the 35th International Conference on Machine Learning. PMLR, pp. 5266–5275. https://proceedings.mlr.press/v80/wen18a.html. (Accessed 1 November 2022).

Wiens, S., 2005. Interoception in emotional experience. Curr. Opin. Neurol. 18 (4), 442–447. https://doi.org/10.1097/01.wco.0000168079.92106.99.

Wise, R.G., Ide, K., Poulin, M.J., Tracey, I., 2004. Resting fluctuations in arterial carbon dioxide induce significant low frequency variations in BOLD signal. NeuroImage 21 (4), 1652–1664. https://doi.org/10.1016/j.neuroimage.2003.11.025.

Xu, G.Y., Huang, L.Y.M., 2002. Peripheral inflammation sensitizes P2X receptor-mediated responses in rat dorsal root ganglion neurons. J. Neurosci. 22 (1), 93–102. https://doi.org/10.1523/JNEUROSCI.22-01-00093.2002.

Xu, F., Liu, P., Pascual, J.M., Xiao, G., Lu, H., 2012. Effect of hypoxia and hyperoxia on cerebral blood flow, blood oxygenation, and oxidative metabolism. J. Cereb. Blood Flow Metab. 32 (10), 1909–1918. https://doi.org/10.1038/jcbfm.2012.93.

Yuan, H., Zotev, V., Phillips, R., Bodurka, J., 2013. Correlated slow fluctuations in respiration, EEG, and BOLD fMRI. NeuroImage 79, 81–93. https://doi.org/10.1016/j.neuroimage.2013.04.068.

Zhang, T., Perkins, M.H., Chang, H., Han, W., de Araujo, I.E., 2022. An inter-organ neural circuit for appetite suppression. Cell 185 (14), 2478–2494.e28. https://doi.org/10.1016/j.cell.2022.05.007.

5

Head motion and physiological effects

Chao-Gan Yan[a] and Rasmus Birn[b]

[a]CAS Key Laboratory of Behavioral Science, Institute of Psychology, Beijing, China, [b]Department of Psychiatry, University of Wisconsin - Madison, Madison, WI, United States

The blood oxygenation level-dependent (BOLD) functional MRI (fMRI) signal is sensitive to various nonneuronal sources of fluctuations. This chapter examines the impact of two major sources of noise in resting-state fMRI: head motion, and physiological fluctuations associated with cardiac pulsations and respiration. The physics of how motion and physiological noise affect the MRI signal is first discussed. This is then followed by a discussion of the various strategies that have been developed so far to reduce the effects of motion and physiological noise. Reduction of nonneuronal sources of signal fluctuations is an important step in the processing of resting-state fMRI data in order to get more accurate measures of functional connectivity.

Introduction

The blood oxygenation level-dependent (BOLD) functional MRI (fMRI) signal is sensitive to various nonneuronal sources of fluctuations, typically referred to as "noise" in studies that are interested in mapping neuronal activity and connectivity. The most prominent noise, aside from scanner artifacts, results from the subject's head motion and physiology, in particular, cardiac and respiratory effects. Estimates of functional connectivity, such as those obtained during a resting-state scan, are particularly sensitive to this noise (Power et al., 2012a; Satterthwaite et al., 2012; Van Dijk et al., 2012). The reason has to do with the way that functional connectivity is estimated. In task fMRI, where the task timing is known, a model of the expected BOLD fMRI response is fitted to the measured signal. Sources of noise that are uncorrelated with the task are thus averaged out. In contrast, functional connectivity assesses the temporal similarity of BOLD time series between two or more regions using some metric, such as the

Advances in Resting-State Functional MRI. https://doi.org/10.1016/B978-0-323-91688-2.00013-8

Pearson's correlation coefficient (Biswal et al., 1995). Two regions with correlated nonneuronal signal variations (noise) would result in an erroneously inflated functional connectivity, while two regions with uncorrelated noise would result in reduced connectivity. These errors can be particularly problematic when the noise differs between groups, for example, when one group has higher head motion.

This chapter examines the impact of two major sources of noise in resting-state fMRI: head motion and physiology. The first sections describe the physics of how head motion affects the fMRI signal and then discusses multiple strategies to reduce the influence of motion. The following sections describe the various sources of physiological noise, discuss the prevalence of this type of noise, and review several techniques that have been developed so far to reduce the influence of physiological noise. New artifact reduction techniques are continuing to be developed, and thus, this section is not meant to be an exhaustive discussion of all techniques that have been developed but rather serve as an overview of many of the more common approaches. Finally, some important caveats about the reduction of motion and physiological noise are discussed.

The impact of head motion on RS-fMRI

Although we always ask the participants to keep their head still during fMRI scanning, head motion is inevitable during fMRI acquisition. Head motion could induce significant artifacts in resting-state fMRI signals, thus presenting a formidable challenge for RS-fMRI. This issue drew particular attention from the field in 2012. Three research groups demonstrated that the combined impact of "micro" head movements, as small as 0.1 mm from one time point to the next, can introduce systematic artifactual interindividual and group-related differences in resting-state fMRI metrics (Power et al., 2012b; Power et al., 2012a; Satterthwaite et al., 2012; Van Dijk et al., 2012). Such movements were well below quality control thresholds typically employed by most investigators, increasing the potential for artifactual findings—particularly in studies of hyperkinetic populations, such as young children or individuals with attention-deficit/hyperactivity disorder (ADHD) (Power et al., 2012b; Power et al., 2012a; Satterthwaite et al., 2012; Van Dijk et al., 2012). Since children move their heads more than adults, it is difficult to distinguish true developmental neuronal effects from higher motion-induced artifacts. Similarly, since ADHD patients move their heads much more than typically developing children, it is also difficult to distinguish the neuropathologic effect of ADHD from head motion artifacts. Since then, many resting-state fMRI publications and grant applications have been questioned by reviewers when researchers have not carefully addressed the head motion issue. Thus, assessing and minimizing the impact of head motion became increasingly urgent and important in the field.

Motion of the participant's head during an MRI scan can affect the fMRI signal in multiple ways. First, the displacement of the head can cause large signal changes, particularly at edges within the image. For example, translation of the head in one direction would cause a signal increase at one edge (where a signal that was outside the brain is now inside the brain) and a signal decrease at the opposite edge. Residual errors can occur even after image realignment due to imperfections in estimating the motion or due to interpolation effects (Grootoonk et al., 2000). Second, motion can cause spin-history effects since parts of the imaged slice may have been excited earlier or later than one repetition time (TR) ago (Friston et al., 1996). Third, the head can move into a region of B0 field inhomogeneity, and the motion itself can alter the gradient of magnetic susceptibility in the B0 direction, altering the B0 field (Jezzard and Clare, 1999). These B0 field changes can result in changes in signal dropout, spatial distortions (for echo-planar acquisitions), or blurring (for spiral acquisitions). Fourth, a large head motion can move the head into a region of altered RF-coil sensitivity (Birn et al., 2022). Finally, even motion outside of the head (e.g., movement of the shoulders or chest) can affect the signal by changing the B0 field or RF-coil loading (Yetkin et al., 1996).

Addressing the impact of head motion on resting-state fMRI

In confronting the challenge of head motion artifacts in resting-state fMRI data, several kinds of solutions exist, including (1) restraining the head motion of the participants; (2) modeling and removing head motion effects, using methods ranging from realignment parameter regression to independent component analysis and scrubbing; and (3) developing motion-optimized scanning sequences or acquisition strategies. Here we are going to discuss these solutions one by one.

Restrain the head motion

A first and obvious solution is to constrain the head motion of the participants. A common practice during MRI scanning is to use padding sponges to fill the gap between the left and right sides of the MRI head coil and the participant's head. However, this still leaves the participant with a great freedom to move their head.

Krause et al. (2019) reported a simple method, which involves applying a medical tape from one side of the head coil to the other side over the participant's forehead. The medical tape gives immediate tactile feedback to the participants when they are moving their heads. This tactile feedback significantly reduced head motion in the scanner, however with a concern that the active tactile feedback to participants may affect resting-state brain activity to some extent.

A more costly solution is to provide the participants with real-time feedback of head motion during scanning (Yang et al., 2005). The head motion parameters are analyzed in real time and displayed to the participant, which can significantly reduce their head motion. However, since participants need to extract the visual feedback information from the display, these additional cognitive resources may also change the resting-state activities.

Less common practice is to use head restraint devices, such as bite bars (Menon et al., 1997) and plaster cast head holder (Edward et al., 2000). Bite bars and plaster cast head holders could significantly reduce head motion, although sometimes it is difficult to set up and might not be feasible in some populations due to discomfort.

From the participant end, one can emphasize on the need to control head motion to the participants before scanning. For certain populations, such as children, practice sessions with MRI simulators prior to scanning (Raschle et al., 2009; de Bie et al., 2010; Lueken et al., 2012) help a lot. A mock MRI scanner could make the participants familiar with the MRI environment and reduce the stress in real scanning. The real-time feedback of head motion from the mock scanner could also train them how to better control head motion. Thus, this strategy of using a training protocol in a mock MRI scanner could lead to better-quality resting-state fMRI data for children. A more rare solution to control head motion in research studies is to sedate the participants (Lawson, 2000), but this comes with a significant concern about the occurrence of serious side effects, and ethics committees might not allow sedation in typical developing children.

During the scanning, the experimenter and scanner operating technician can oversee the head motion of the participants. Scanners can be programmed to output the head motion of the participants in real time for the experimenter to view from the control room. Once the experimenter finds that the real-time head motion of the participants exceeds some maximum head motion criteria (e.g., 3 mm and 3 degrees on any axis) during the scan, the experimenter could abort the resting-state scan and restart a new scan with additional emphasis on head motion control to the participants. This could result in discarding only the resting-state scan with big head motion without sacrificing the entire data of the participant.

Model and remove head motion related artifacts

Even with all efforts to restrain the head motion of the participants, head motion still happens during scanning. A basic correction is to realign the images to "undo" the effects of motion-induced displacement, by aligning the images across all time points so that the brain is in the same position in each image. Usually, affine registrations are used to coregister the images to a single reference volume. However, this is insufficient to remove the head motion artifacts due to partial voluming, magnetic inhomogeneity, and spin history effects (discussed below).

In a seminal study of motion effects in fMRI, Friston et al. (1996) proposed a 24-parameter autoregressive model that takes into account cumulative effects of motion on spin magnetization. When considering spin magnetization, current position is important, as the excitation of spins depends on an interaction between the local magnetic field and the Fourier transform of the slice-selective pulse (Friston et al., 1996). However, prior location can be equally important, as the spin excitation history can produce differences in local saturation. Accordingly, to account for present and past displacements, Friston et al. suggested autoregressive models that included current and past position parameters, along with the square of each parameter, i.e., the Friston 24-parameter model.

However, this model was not widely used for many years. A common practice was to regress the fMRI time series data onto the three translation and three rotation parameters for head movement estimated by motion realignment procedures (e.g., Fox et al., 2005; Vincent et al., 2007; Weissenbacher et al., 2009); the temporal derivatives of these six parameters were often included as well (e.g., Andrews-Hanna et al., 2007; Power et al., 2012a; Van Dijk et al., 2012). In 2012, three studies pointed out that these modeling-based approaches appear to be inadequate in attenuating the impact of micromovements on synchronies in the BOLD signal (Power et al., 2012a; Satterthwaite et al., 2012; Van Dijk et al., 2012).

In 2013, two studies by Satterthwaite et al. (2013) and Yan et al. (2013) systematically examined the ability of several orders of motion regression models to remove motion-related artifacts. Both studies demonstrated an advantage for higher-order models, though residual artifact remained regardless of the model used. The best performing model was based on the work by Friston et al. (1996), as the Friston 24-parameter model could account for the cumulative effects of motion on spin history.

Instead of regressing out modeled motion-related regressors derived from the realignment parameters, independent component analysis (ICA) could also be used in correcting motion artifacts.

As a data-driven approach, ICA decomposes fMRI data into components that comprise signal and structured noise. FMRIB's ICA-based X-noiseifier (ICA-FIX) can be applied to automatically separate signal from noise components but requires manual training of classifiers (Griffanti et al., 2014; Salimi-Khorshidi et al., 2014). In 2015, a data-driven method to identify and remove motion-related independent components was proposed: ICA-based Automatic Removal Of Motion Artifacts (ICA-AROMA) (Pruim et al., 2015). After ICA, ICA-AROMA detects the motion-related artifacts using four theoretical features embedded in a simple classifier: how the spatial structure of the component spatial map overlaps with the brain edge (1) and cerebrospinal fluid (2); high-frequency trend of the component time course (3); and its correlation with the realignment parameters (4). Following their identification, components determined as motion related by the classifier are removed from the fMRI data set by nonaggressive linear regression. ICA-AROMA was reported to efficiently remove head motion artifacts in benchmarking evaluations (Pruim et al., 2015; Ciric et al., 2017; Parkes et al., 2018).

An alternative approach for carrying out individual subject-level correction is to scrub contaminated volumes from time series data prior to deriving RS-fMRI metrics. This is performed variably across laboratories and typically involves the removal (Mazaika et al., 2009; Power et al., 2012a) or regressing out (Lemieux et al., 2007; Power et al., 2012b; Satterthwaite et al., 2013) of single time points characterized by a sudden, sharp movement, or segments of motion-corrupted data in an otherwise usable time series. Realizing the artifactual contributions of micromovements to resting-state fMRI findings, Power and colleagues called for the rigorous scrubbing of time frames in which micromovements occur, as well as their neighboring time points, proposing framewise displacement (FD) > 0.2mm as the threshold for frame removal (Power et al., 2012a; Power et al., 2012b). Many studies have suggested that the combination of scrubbing and modeling-based approaches brings reduction in motion-induced artifact (Power et al., 2012b; Satterthwaite et al., 2013)—this combination can be accomplished in a single, integrated regression model (i.e., by modeling motion parameters and spike regressors for each scrubbed time point). However, scrubbing approaches also have potential limitations. Scrubbing can lead to removal of large (>50%) proportions of time points from a single participant's resting-state fMRI data and can result in significant variation in the remaining numbers of time points (and therefore, degrees of freedom) from one subject to the next. Such variation can impact findings for inter-individual or group differences in resting-state fMRI metrics. From an implementation perspective, the removal of noncontiguous time points alters the underlying temporal structure of the data, precluding conventional frequency-based analyses (e.g., the fast Fourier transform [FFT]-based amplitude of low-frequency fluctuations [ALFF] measure, and fractional ALFF, [fALFF]) and requiring a more complicated and slower discrete Fourier transform (DFT) instead (Babu and Stoica, 2010).

Beyond individual-level correction approaches, correction for motion artifacts in group-level regression analyses is possible—primarily through inclusion of individual average motion estimates as a nuisance regressor (Satterthwaite et al., 2012; Van Dijk et al., 2012; Yan et al., 2013). The inclusion of individual motion estimates as a nuisance regressor at the group level permits the differentiation of interindividual differences in resting-state fMRI measures that are attributable to motion and those related to variables of interest. One limitation is that variance shared by both motion and the variable of interest will be removed—potentially leading to underestimation of relationships with the variable of interest. Additionally, sufficiently large data sets are necessary to appropriately model the impact of motion and other nuisance signals.

Head motion optimized scanning sequences or acquisition strategies

The above-mentioned modelling solutions would improve our ability to use already collected data but not completely remove motion artifacts; see reviews and benchmarking evaluations in Power et al. (2015), Ciric et al. (2017), Parkes et al. (2018). However, in prospective studies, improved fMRI scanning sequences or optimized acquisition strategies can be used.

Multi-echo fMRI (ME-fMRI) is a promising sequence to be used in removing head motion-related artifacts (Kundu et al., 2017; Power et al., 2018). Standard single-echo fMRI acquires the images at a single echo time (TE), e.g., 30 ms for a 3T scanner. ME-fMRI acquires slice images at the earliest possible TE after a normal excitation pulse. Furthermore, ME-fMRI reads out other images of the same slice at longer TEs without re-excitation. ME-fMRI can be acquired for every brain volume slice, but at little cost as the fMRI pulse sequences are idle for the early period after excitation. With ME-fMRI data, multiecho-independent component analysis (ME-ICA) could be utilized to eliminate motion artifacts. The signal decay is first modeled as a monoexponential decay: $S(TE) = S_0 \exp(-R_2^* TE)$, where S_0 is initial signal intensity and R_2^* is the decay rate. Then ME-ICA categorizes the components according to how much they reflect S_0 versus R_2^* modulation over time. Most of the S_0 related components are discarded and most of the R_2^* related components are retained following automated criteria. ME-ICA was found to effectively remove certain kinds of motion-related artifacts (Power et al., 2018).

A future promising direction could be the development of MRI acquisition strategies that are robust to motion—which already exist for structural imaging (Brown et al., 2010; White et al., 2010; Kuperman et al., 2011) and are being developed for functional imaging (Speck et al., 2006; Ooi et al., 2011; Maclaren et al., 2012). Prospective motion correction (PMC) measures the head motion in real time and prospectively updates the gradients and radio-frequency pulses accordingly (Todd et al., 2015). An MR-compatible optical camera mounted on the inside of the scanner can track a marker attached to the head of the participant. The head motion information based on the position and orientation of the marker is then used to realign the imaging field-of-view to match the participant's head movement before every radio-frequency pulse. The PMC acquisition strategy improves data quality for task fMRI (Todd et al., 2015) and resting-state fMRI (Maziero et al., 2020) affected by motion.

At the current stage in the development of head motion control solutions, one could better restrain the head motion of the participants and implement PMC data acquisition, or implement ME-fMRI sequence and utilize ME-ICA to reduce motion-related artifacts. If that is not feasible, ICA-AROMA may be a sensible approach for reducing motion artifacts, and one should at least use the Friston 24-parameter model in pre-processing to regress out head motion effects. Finally, motion covariates could be included in the group-level statistical analysis to further control motion confounding factors.

The impact of physiological noise on RS-fMRI

Fluctuations in the MRI signal over time due to physiological processes that are not related to variations in neuronal activity are typically considered to be "physiological noise" in fMRI. There are multiple sources of such physiological noise.

Cardiac pulsations

The pulsatile movement of the blood causes varying blood flow velocity during the fMRI time series acquisition, resulting in varying inflow effects. Specifically, blood flowing into the imaged slice will have been excited at a different point in time compared to the surrounding stationary tissue, leading to variations in the MRI signal (Dagli et al., 1999). As a result, large signal fluctuations at the cardiac frequency, and its harmonics, are seen in large arteries and arterioles, such as the Circle of Willis. The cardiac pulsation can also displace and deform the brain tissue, particularly in inferior regions of the brain, resulting in M0 changes (Poncelet et al., 1992). Improper spoiling of the magnetization after the acquisition of an image (e.g., by not applying strong enough crusher gradients at the end of acquisition) can also result in steady-state free precession (SSFP) signals which are modulated by cardiac fluctuations and thus lead to correlated signal changes (Zhao et al., 2000). These cardiac-related fluctuations typically occur around 1Hz, which are aliased to lower frequencies for acquisitions with a TR longer than half of the cardiac cycle. In addition, it should be kept in mind that even for short TR acquisitions, higher harmonics of the cardiac fluctuations can be aliased to lower frequencies.

Respiration

Movement of the chest wall and variations in the magnetic susceptibility in the lungs during breathing can result in B0 field changes. These B0 field changes are relatively uniform across the brain and shift the brain image in the phase-encoding direction. As a result, fluctuations at the respiratory frequency can sometimes be seen in the estimated head motion (e.g., as measured by the framewise displacement, FD) even when the head is not actually moving. Actual motion of the head during breathing can cause additional spin-history effects (Friston et al., 1996). Respiration can also affect CSF flow, resulting in in-flow effects (Feinberg and Mark, 1987). In healthy adults, these respiration-related signal changes typically occur around 0.2–0.3 Hz.

Variations in respiration depth and rate, and arterial CO_2

The above section on respiration discussed MRI signal changes associated with the primary respiration frequency (inspiration and expiration). Signal changes can also occur as a result of breath-to-breath changes in depth and rate. Variations in breathing depth and rate can cause alterations in arterial CO_2 levels. CO_2 is a potent vasodilator, and thus decreased breathing depth and/or rate leads to an increased blood flow and a resultant increase in the BOLD signal (Modarreszadeh and Bruce, 1994; Van den Aardweg and Karemaker, 2002; Wise et al., 2004; Birn et al., 2006). Similarly, hypocapnia, for example due to hyperventilation, results in reduced BOLD signal (Posse et al., 2001; Bright et al., 2009). These alterations are part of a complex feedback circuit—increased arterial CO_2 levels activate chemoreceptors that increase subsequent breathing depth and rate. This results in increased CO_2 exhalation and a reduction of arterial CO_2. Studies have shown that fluctuations of arterial CO_2 (as measured by a person's end-tidal CO_2) vary with a cycle of about 0.03 Hz (Modarreszadeh and Bruce, 1994; Van den Aardweg and Karemaker, 2002). Therefore, these variations in arterial CO_2 lead to low frequency (0.03 Hz) fluctuations in the fMRI signal.

These CO_2-related signal variations occur throughout the brain and are therefore readily identified on "carpet plots"—a two-dimensional gray-scale plot of signal intensity with voxels in one dimension and time in the other (Power, 2017). The amplitude of these fluctuations, however, is heterogeneous across the brain, with particularly strong signals in gray matter (Wise et al., 2004; Birn et al., 2006; Falahpour et al., 2013; Pinto et al., 2017). The regions with the largest CO_2-related signal variations overlap with many of the regions of the default-mode network (DMN). However, important differences are also observed, with CO_2-related fluctuations also being strong in occipital cortex, outside of the DMN (Birn et al., 2006). The latency of these CO_2-related signal changes has also been shown to vary across the brain and across subjects (Frederick et al., 2012; Falahpour et al., 2013; Pinto et al., 2017).

Changes in breathing can occur at the same time as head motion. One study that more closely examined both head motion and respiration found that some occasional head movements were associated with prolonged and relatively global signal changes after the motion (Power et al., 2018). These movements occurred at the same time as deep breaths, and thus, the prolonged global signal changes were likely the result of variations in arterial CO_2 following the respiration change. Examination of respiration and global signal changes in a large cohort of subjects (the Human Connectome Project) revealed two common types of respiration patterns—isolated occasional deep breaths, and "bursts" consisting of periodic variations in breathing depth (Lynch et al., 2020). This latter pattern was found to show significant sex differences, with males exhibiting more frequent "burst"-type breathing patterns (Lynch et al., 2020).

Heart rate

FMRI studies have also shown that the variability in heart rate is correlated with MRI signal changes throughout gray matter (Shmueli et al., 2007; Chang et al., 2009). In part, these signal fluctuations can result from changes in blood oxygenation that vary with the heart rate (Katura et al., 2006). In addition, heart rate variations are associated with activation of the sympathetic and parasympathetic nervous system. As a result, variations in heart rate are associated not only with nonneuronal MRI signal fluctuations but also with neuronal-activity associated BOLD signal changes. The heart rate can also be affected by respiration, with typically increased heart rate during inspiration and decreased heart rate during expiration. This modulation of the heart rate by the respiration results in peaks in the frequency spectrum at side-bands around the primary cardiac frequency (Raitamaa et al., 2021).

Blood pressure

Arterial blood pressure can also fluctuate during the acquisition of an RS-fMRI run. These variations result in fluctuations in cerebral blood flow (CBF) due to delays in dynamic autoregulation (Diehl et al., 1995; Lang et al., 1999). Blood pressure is under the control of the sympathetic and parasympathetic nervous systems that modulate arterial vascular tone (Failla et al., 1999) and heart rate in the low-frequency range (Akselrod et al., 1981; Katura et al., 2006). There is unfortunately very little data about the impact of blood pressure on resting-state functional connectivity in humans, likely due to the difficulty in obtaining noninvasive measures of blood pressure during an fMRI scan. Studies in rats have shown that blood pressure affects evoked fMRI responses, with transient hypertension increasing BOLD (Wang et al., 2006) and CBF (Qiao et al., 2007) signals. Increases in the amplitude of low-frequency BOLD fluctuations have been demonstrated with a drop in mean arterial pressure (Biswal and Kannurpatti, 2009).

Vasomotion

Blood vessels also exhibit spontaneous oscillations in diameter that are independent of the heart beat or respiration called vasomotion. These fluctuations typically occur at low frequencies (less than 0.1 Hz) in humans and result in low frequency oscillations in blood flow and oxygenation. While some of these fluctuations can be associated with neuronal activity and metabolism, they can also occur independently. The mechanism responsible for these oscillations is not yet completely understood, but it is believed to result from ion channels in the smooth muscle of the arterioles reacting to the oscillations in blood volume (Hudetz et al., 1998). This vasomotor activity can be modulated by internal pressure (Achakri et al., 1995), alpha- and beta-receptor agonists and antagonists (Colantuoni et al., 1984a; Filosa et al., 2004), and anesthesia (Colantuoni et al., 1984b).

Relative contributions of physiological noise

How significant are the signal variations due to physiological noise? Studies have shown that on average across gray matter, cardiac pulsations and respiratory motion (the regressors modeled by RETROICOR) account for about 5% of the variance (Bianciardi et al., 2008; Jo et al., 2010). Respiration volume and heart rate changes together account for about 16% of the variance (Chang and Glover, 2009). End-tidal CO_2 variations explain approximately 16% of the variance throughout gray matter (Wise et al., 2004), and systemic blood oxygenation changes measured peripherally by NIRS explained about 10.5% variance (compared to only 6.8% of variance explained by RETROICOR and RVT). The relative contribution of these noise sources, however, is highly heterogeneous across the brain. One study found significant cardiac fluctuations in 27.5%±8.0% of brain voxels, and removing these fluctuations resulted in a 10%–40% reduction in variance (Dagli et al., 1999). Signal changes associated with cardiac pulsations are particularly prominent in and near large vessels, such as the Circle of Willis. Furthermore, the proportion of the total signal variance due to physiological noise is greater with higher signal-to-noise ratio (SNR) and thus can vary depending on the voxel size, RF coil, and field strength (Kruger and Glover, 2001; Triantafyllou et al., 2006; Bodurka et al., 2007).

Addressing the impact of physiological noise on resting-state fMRI

Over the past three decades, a number of techniques have been developed to reduce the influence of physiological noise on the fMRI signal. Below is a description of some of the most common methods to date. However, this is still an area of active research, and thus, there will likely be new techniques that emerge as the field progresses.

Frequency filtering

Signal fluctuations associated with the cardiac and respiratory cycles (i.e., cardiac pulsation and inspiration/expiration) occur at temporal frequencies that are distinct from the fluctuations driving the neuronal-activity related functional connectivity. The neuronal-activity-related fluctuations of interest typically occur below 0.1 Hz (although evidence shows some contribution from higher frequencies as well (Chen and Glover, 2015)), while respiration occurs at around 0.3 Hz and cardiac pulsation at about 1 Hz. A relatively simple strategy to reduce the influence of cardiac and respiratory noise is to apply a low-pass filter to the data, keeping only those frequencies of interest (e.g., < 0.1 Hz). Since signal drifts and other sources of very low frequency noise are also common in fMRI, a band-pass filter can be applied (e.g., 0.01–0.1 Hz) to keep the frequencies of interest. Most current studies of RS-fMRI perform this band-pass filtering as a standard step in the analysis. However, effective reduction of cardiac and respiratory fluctuations using this technique requires that the MRI data are sampled fast enough, above the Nyquist frequency. Capturing a primary cardiac frequency at 1 Hz requires a TR shorter than 500 ms; faster heart rates would require even shorter TRs. Early RS-fMRI studies achieved this by acquiring only a limited number of slices. (This rapid acquisition was also done in early studies to demonstrate that functional connectivity was not the result of cardiac and respiratory fluctuations.) At lower sampling rates (longer TRs), the cardiac and respiratory fluctuations are aliased to lower frequencies that can overlap with the frequency range of interest. Since the cardiac- and respiratory-related signal changes are not pure sinusoids, they contain higher harmonics. These can be aliased to lower frequencies even with high sampling rates.

Modeling physiological signal fluctuations using recordings of physiology

The impact of physiological noise can be reduced by modeling the physiological fluctuations using an independent recording of the physiology acquired during the MRI scan. Cardiac pulsation, for example, can be measured using a pulse oximeter and respirations can be measured using a pneumatic belt around the participant's chest. These are commonly available on current MRI scanners, making it relatively easy to record these signals during a RS-fMRI scan. A model of the expected cardiac and respiratory signals can then be constructed using the external recordings, and this modeled signal is then fit to and subtracted from the data using multiple linear regression. One of the first implementations of this modeling was performed in k-space prior to image reconstruction (Hu et al., 1995). More common and recent implementations perform this regression in image space. RETROICOR (standing for RETROspective Image CORrection), for example, first estimates the phase of the cardiac and respiratory cycle at which each imaged slice is acquired (Glover et al., 2000). Low-order Fourier series of these phases are then fit to each voxel's time series data and removed.

Signal fluctuations due to variations in arterial CO_2 can be modeled by measuring end-tidal CO_2 during the MRI scan, e.g., using a capnograph and mask or nasal cannula. This measure of end-tidal CO_2 is typically convolved with a response function to account for delays between the CO_2 changes and associated changes in blood flow and oxygenation, as well as delays in the sampling tube (Wise et al., 2004). An estimate of the signal variations associated with breath-to-breath changes in respiratory volume and rate can also be modeled from a respiratory belt recording. Prior studies have computed either the respiration volume per time (the difference between the respiration maxima and minima, divided by the periods) (Birn et al., 2006), the windowed variance of the respiration (Chang and Glover, 2009), or the windowed envelope of the respiration waveform (Power et al., 2018). These measures are then convolved with a respiration response function (RRF), fit to the data, and then removed. Studies have shown that this respiration response function varies across the brain and across subjects and that modeled signals can be improved by estimating the RRF on a per-subject (and voxel-wise/region-wise) basis (Falahpour et al., 2013).

The RVT estimate was first developed to emulate end-tidal CO_2 removal in situations where a capnograph is unavailable. It is therefore not surprising that RVT and $etCO_2$ explain similar spatial and temporal variance. However, subsequent studies have shown that there are some spatial differences, and that while there are strong similarities, RVT and $etCO_2$ capture separate components of the noise (Chang and Glover, 2009). RVT (or any measure based purely on the respiration trace) is therefore not a perfect substitute for $etCO_2$, and both should be modeled if available.

Modeling physiological signal fluctuations from the fMRI data

The above approaches all required the physiology to be measured during the fMRI scan. However, sometimes, these recording devices are not readily available or would interfere with an fMRI task. In addition, a wealth of fMRI data exist that are openly available but do not contain associated physiological recordings. For these reasons, approaches have been developed that estimate the physiological noise, and remove/reduce it, from the acquired imaging data.

One early approach that works if the physiological fluctuations are sampled fast enough and not aliased is called IMPACT (IMage-based Physiological Artifacts estimation and Correction Technique) (Chuang and Chen, 2001). This technique first identifies voxels that show fluctuations at expected cardiac and respiratory frequencies, and then fits a Fourier series to the data re-ordered according to the estimated cardiac and respiratory phases. These estimated physiological fluctuations are then subtracted from the data.

Another physiological noise correction approach, called PESTICA, identifies physiological (cardiac and respiration)-related signal changes using temporal independent component analysis (ICA) on slice-wise data, exploiting the fact that the slices in the imaging volume are not all acquired at the same time (Beall and Lowe, 2007). This allows for a finer sampling of the physiological signals in time. This algorithm has recently been updated to account for local time-shifting of the cardiac phase (Shin et al., 2022). Recent work has also used deep-learning to estimate the physiological traces from the fMRI time course data (Aslan et al., 2019; Salas et al., 2021).

Another common approach for reducing the influence of physiological noise, particularly the more widespread signal changes associated with variations in respiration and arterial CO_2, is to derive regressors of no interest (nuisance regressors) from the fMRI data itself. The average signal over white matter, and CSF, is often used as nuisance regressor, since the neuronal-activity-related BOLD fMRI signal is generally smaller in these regions (compared to gray matter) and thus likely to contain more nonneuronal fluctuations. The temporal derivative of these average signals is also often included in the nuisance regression, since this can account for slight variations in the latency of the response across the brain. An alternative approach to averaging the signals in these regions is to combine them using principal components analysis. The first few principal components of voxel time series in these regions can then be used as nuisance regressors, a technique known as aCompCor. An alternative version of this technique, tCompCor, combines voxels from regions that have high signal variance over time (Behzadi et al., 2007). Another technique is to model the nuisance regressors on a regional basis, by including all white matter voxels within a neighborhood of each voxel as the nuisance regressors for that voxel, a technique referred to as ANATICOR (Jo et al., 2010). This requires a different set of nuisance regressors for each voxel and is thus more computationally demanding but can model variations in the noise across the brain. Nuisance regressors can also be derived from other regions where we do not expect to neuronal activation-related signal changes, such as the soft tissues in the face. PSTcor (Phase-shifted Soft-Tissue correction) uses the spatial average of each of these signals, and then shifts them in time to account for differences in the latency of these signals between the soft tissues and gray matter (Anderson et al., 2011).

Widespread signal changes, such as those associated with etCO$_2$ variations, can also be reduced by regressing out the average signal over the whole brain, also referred to as the global signal, or global signal regression (GSR). This global brain signal has been shown to be highly correlated with RVT computed from the respiration belt (Birn et al., 2006). However, this processing step is still controversial because the global signal by definition also includes the signals of interest. Simulations have shown that GSR could potentially distort estimates of functional connectivity, introducing negative correlations into areas that were not correlated and altering group differences in functional connectivity (Murphy et al., 2009; Saad et al., 2012). Others argue that GSR is a highly effective way to remove noise and that not performing this processing step would leave unwanted noise in the data (Fox et al., 2009; Power et al., 2014). The differences between these two viewpoints hinges on the relative contributions of signal vs. noise to the global signal. If the global signal contains mostly noise, then removing it can be advantageous. However, if the global signal contains enough signals of interest such that removing it would distort those signals of interest, then GSR may not be advised. The challenge is that the relative contributions of signal vs. noise to the global signal are generally unknown. Finally, studies have shown that the global signal is correlated with vigilance as measured by EEG (Wong et al., 2013). Some studies may view this as a signal of interest that should be kept in the data, while other studies may view this as a nuisance variable that should be removed. An extended discussion of this topic can be found in Murphy and Fox (2017).

A technique that extends global signal regression is APPLECOR (Affine Parameterization of Physiological Large-scale Error Correction), which considers not only an additive global noise signal, but a multiplicative one that scales with each voxel's baseline signal intensity (Marx et al., 2013). This technique has been shown to achieve more consistent maps of connectivity compared to GSR alone and offers even further improvements when combined with RVHRcor, a technique the authors call PEARCOR (the Parallel Execution of APPLECOR and RVHRcor).

Another approach to model systemic low-frequency oscillations (sLFO) of nonneuronal origin is to record blood oxygenation changes using near infrared spectroscopy (NIRS) peripherally during the fMRI scan. The recorded NIRS signal is then shifted in time for each voxel and fit to the fMRI data, a technique known as RIPTiDe (Frederick et al., 2012). More recent versions of this technique have used either the venous signal (e.g., signal from the sagittal sinus) or the global brain signal, which is advantageous in cases where a NIRS recording is not available (Erdogan et al., 2016; Tong et al., 2016; Tong et al., 2019). The average global brain signal of course includes the average of sLFOs at different delays across the brain, resulting in an average signal that is smoother than externally recorded sLFO. The average brain signal can be improved using an iterative sharpening procedure where the latencies of the sLFO across the brain are estimated and then used to provide an updated global signal. Accounting for these variations in the latency of these sLFOs across the brain can also reduce spurious correlations resulting from more conventional GSR (Tong et al., 2019).

Another class of methods to reduce the influence of physiological noise on RS-fMRI uses spatial independent component analysis (ICA) to partition the measured fMRI signal into independent spatial components. Components that are then identified as being "artifactual" (nonneuronal) are then removed. One of the first methods to use this approach to reduce physiological noise is CORSICA (CORrection of Structured noise using spatial Independent Component Analysis), which uses spatial priors for where physiological noise is typically found and a step-wise regression procedure to automatically identify physiological noise components (Perlbarg et al., 2007). A more recent approach, FIX (FMRIB's ICA-based X-noisifier), identifies artifactual components using machine learning (Griffanti et al., 2014; Salimi-Khorshidi et al., 2014). This requires first training the machine learning algorithm to classify components as either signals of interest or noise, which can be done using manual labeling in a subset of the data. These spatial ICA denoising approaches are quite good in identifying signal changes related to cardiac pulsations and apparent motion due to the B0 field changes associated with respiration, because these artifacts are spatially distinct from neuronal activity associated BOLD responses. Their success in identifying and removing signal changes related to variations in respiration volume is more mixed, likely because these artifactual signals are more global and overlap spatially with functional brain networks, such as the default mode network (Birn et al., 2008; Golestani and Chen, 2022). Instead, temporal ICA denoising approaches have been shown to more effective at reducing such noise if enough time points are available (Smith et al., 2012; Boubela et al., 2013; Glasser et al., 2018; Golestani and Chen, 2022).

Caveats

There are a number of important caveats to keep in mind regarding the various proposed methods to reduce the influence of physiological noise. First, removal of modeled or estimated physiological "noise" from the measured fMRI data assumes that this physiological noise is not correlated with, or associated with, variations in neuronal activity. This is not strictly true. A person's emotional state and arousal have been shown to be related to variations in respiration (Shea, 1996; Perl et al., 2019). In addition, there are connections between brain regions controlling respiration and arousal (Yackle et al., 2017). Studies have also shown that variations in arterial CO_2 associated with natural variations in respiration modulate neural rhythmicity and oscillatory power as measured by magneto-encephalography (Driver et al., 2016). Second, physiological noise reduction using nuisance regression can sometimes introduce correlations (Bright et al., 2017). It is also important to consider that these noise reduction approaches can result in a significant loss in the degrees of freedom.

Conclusions

Given the large number of techniques that have been developed to reduce the impact of physiological noise, how does an investigator choose which to apply? The answer to that question depends in part on who you ask—scientists developing new methods tend to prefer the methods that they have developed. In general, newer techniques have been demonstrated to perform better than older ones, and of course, new techniques are continuing to be developed. That said, the current approach that is most common is to first reduce cardiac and respiratory fluctuations using RETROICOR (or a similar approach) if physiological recordings are available. Signal fluctuations in and near pulsatile vessels can be significantly reduced using this approach, but not by regressing out more global tissue derived signals (the average white matter, CSF, global signal) or RVT (Birn et al., 2006, 2014). More widespread signal changes throughout gray matter can be significantly reduced using aCompCor. ICA-based denoising procedures can also be quite effective at reducing noise. The main challenge with these is determining which components are signal and which are noise. Multi-echo acquisitions can help with this determination for some noise sources. However, some physiological noise, such as that resulting from variations in arterial CO_2, is also echo-time dependent and therefore not easy to separate using ME-ICA. Training of classifiers using manual labeling may provide a way to distinguish signals-of-interest from noise. Reduction of nonneuronal sources of signal fluctuations is an important step in the processing of RS-fMRI data in order to get more accurate measures of functional connectivity and activation.

References

Achakri, H., Stergiopulos, N., Hoogerwerf, N., Hayoz, D., Brunner, H.R., Meister, J.J., 1995. Intraluminal pressure modulates the magnitude and the frequency of induced vasomotion in rat arteries. J Vasc Res 32 (4), 237–246.

Akselrod, S., Gordon, D., Ubel, F.A., Shannon, D.C., Berger, A.C., Cohen, R.J., 1981. Power spectrum analysis of heart rate fluctuation: a quantitative probe of beat-to-beat cardiovascular control. Science 213 (4504), 220–222.

Anderson, J.S., Druzgal, T.J., Lopez-Larson, M., Jeong, E.K., Desai, K., Yurgelun-Todd, D., 2011. Network anticorrelations, global regression, and phase-shifted soft tissue correction. Hum Brain Mapp 32 (6), 919–934.

Andrews-Hanna, J.R., Snyder, A.Z., Vincent, J.L., Lustig, C., Head, D., Raichle, M.E., Buckner, R.L., 2007. Disruption of large-scale brain systems in advanced aging. Neuron 56 (5), 924–935. 2709284.

Aslan, S., Hocke, L., Schwarz, N., Frederick, B., 2019. Extraction of the cardiac waveform from simultaneous multislice fMRI data using slice sorted averaging and a deep learning reconstruction filter. Neuroimage 198, 303–316. PMC6592732.

Babu, P., Stoica, P., 2010. Spectral analysis of nonuniformly sampled data—a review. Digital Signal Process 20 (2), 359–378.

Beall, E.B., Lowe, M.J., 2007. Isolating physiologic noise sources with independently determined spatial measures. Neuroimage 37 (4), 1286–1300.

Behzadi, Y., Restom, K., Liau, J., Liu, T.T., 2007. A component based noise correction method (CompCor) for BOLD and perfusion based fMRI. Neuroimage 37 (1), 90–101. 2214855.

Bianciardi, M., Fukunaga, M., van Gelderen, P., Horovitz, S.G., de Zwart, J., Shmueli, K., Duyn, J.H., 2008. Sources of fMRI signal variance in the human brain at rest: a 7T study. in ISMRM Workshop: High Field Systems and applications. Rome, Italy.

Birn, R.M., Cornejo, M.D., Molloy, E.K., Patriat, R., Meier, T.B., Kirk, G.R., Nair, V.A., Meyerand, M.E., Prabhakaran, V., 2014. The influence of physiological noise correction on test-retest reliability of resting-state functional connectivity. Brain Connect 4 (7), 511–522. 4146390.

Birn, R.M., Dean 3rd, D.C., Wooten, W., Planalp, E.M., Kecskemeti, S., Alexander, A.L., Goldsmith, H.H., Davidson, R.J., 2022. Reduction of motion artifacts in functional connectivity resulting from infrequent large motion. Brain Connect.

Birn, R.M., Diamond, J.B., Smith, M.A., Bandettini, P.A., 2006. Separating respiratory-variation-related fluctuations from neuronal-activity-related fluctuations in fMRI. Neuroimage 31 (4), 1536–1548.

Birn, R.M., Murphy, K., Bandettini, P.A., 2008. The effect of respiration variations on independent component analysis results of resting state functional connectivity. Hum Brain Mapp 29 (7), 740–750. 2715870.

Biswal, B.B., Kannurpatti, S.S., 2009. Resting-state functional connectivity in animal models: modulations by exsanguination. Methods Mol Biol 489, 255–274.

Biswal, B., Yetkin, F.Z., Haughton, V.M., Hyde, J.S., 1995. Functional connectivity in the motor cortex of resting human brain using echo-planar MRI. Magnetic Reson Med 34 (4), 537–541.

Bodurka, J., Ye, F., Petridou, N., Murphy, K., Bandettini, P.A., 2007. Mapping the MRI voxel volume in which thermal noise matches physiological noise—implications for fMRI. Neuroimage 34 (2), 542–549.

Boubela, R.N., Kalcher, K., Huf, W., Kronnerwetter, C., Filzmoser, P., Moser, E., 2013. Beyond noise: using temporal ICA to extract meaningful information from high-frequency fMRI signal fluctuations during rest. Front Hum Neurosci 7, 168. PMC3640215.

Bright, M.G., Bulte, D.P., Jezzard, P., Duyn, J.H., 2009. Characterization of regional heterogeneity in cerebrovascular reactivity dynamics using novel hypocapnia task and BOLD fMRI. Neuroimage 48 (1), 166–175. 2788729.

Bright, M.G., Tench, C.R., Murphy, K., 2017. Potential pitfalls when denoising resting state fMRI data using nuisance regression. Neuroimage 154, 159–168. PMC5489212.

Brown, T.T., Kuperman, J.M., Erhart, M., White, N.S., Roddey, J.C., Shankaranarayanan, A., Han, E.T., Rettmann, D., Dale, A.M., 2010. Prospective motion correction of high-resolution magnetic resonance imaging data in children. Neuroimage 53 (1), 139–145. 3146240.

Chang, C., Cunningham, J.P., Glover, G.H., 2009. Influence of heart rate on the BOLD signal: the cardiac response function. Neuroimage 44 (3), 857–869.

Chang, C., Glover, G.H., 2009. Relationship between respiration, end-tidal CO2, and BOLD signals in resting-state fMRI. Neuroimage 47 (4), 1381–1393. 2721281.

Chen, J.E., Glover, G.H., 2015. BOLD fractional contribution to resting-state functional connectivity above 0.1 Hz. Neuroimage 107, 207–218. PMC4318656.

Chuang, K.H., Chen, J.H., 2001. IMPACT: image-based physiological artifacts estimation and correction technique for functional MRI. Magn Reson Med 46 (2), 344–353.

Ciric, R., Wolf, D.H., Power, J.D., Roalf, D.R., Baum, G.L., Ruparel, K., Shinohara, R.T., Elliott, M.A., Eickhoff, S.B., Davatzikos, C., Gur, R.C., Gur, R.E., Bassett, D.S., Satterthwaite, T.D., 2017. Benchmarking of participant-level confound regression strategies for the control of motion artifact in studies of functional connectivity. Neuroimage 154, 174–187. PMC5483393.

Colantuoni, A., Bertuglia, S., Intaglietta, M., 1984a. The effects of alpha- or beta-adrenergic receptor agonists and antagonists and calcium entry blockers on the spontaneous vasomotion. Microvasc Res 28 (2), 143–158.

Colantuoni, A., Bertuglia, S., Intaglietta, M., 1984b. Quantitation of rhythmic diameter changes in arterial microcirculation. Am J Physiol 246 (4 Pt 2), H508–H517.

Dagli, M.S., Ingeholm, J.E., Haxby, J.V., 1999. Localization of cardiac-induced signal change in fMRI. Neuroimage 9 (4), 407–415.

de Bie, H.M., Boersma, M., Wattjes, M.P., Adriaanse, S., Vermeulen, R.J., Oostrom, K.J., Huisman, J., Veltman, D.J., Delemarre-Van de Waal, H.A., 2010. Preparing children with a mock scanner training protocol results in high quality structural and functional MRI scans. Eur J Pediatr 169 (9), 1079–1085. 2908445.

Diehl, R.R., Linden, D., Lucke, D., Berlit, P., 1995. Phase relationship between cerebral blood flow velocity and blood pressure. A clinical test of autoregulation. Stroke 26 (10), 1801–1804.

Driver, I.D., Whittaker, J.R., Bright, M.G., Muthukumaraswamy, S.D., murphy, k., 2016. Arterial CO2 fluctuations modulate neuronal rhythmicity: implications for MEG and FMRI studies of resting-state networks. J Neurosci 36 (33), 8541–8550. PMC4987431.

Edward, V., Windischberger, C., Cunnington, R., Erdler, M., Lanzenberger, R., Mayer, D., Endl, W., Beisteiner, R., 2000. Quantification of fMRI artifact reduction by a novel plaster cast head holder. Hum Brain Mapp 11 (3), 207–213.

Erdogan, S.B., Tong, Y., Hocke, L.M., Lindsey, K.P., de Frederick, B.F.B., 2016. Correcting for blood arrival time in global mean regression enhances functional connectivity analysis of resting state fMRI-BOLD signals. Front Hum Neurosci 10, 311. PMC4923135.

Failla, M., Grappiolo, A., Emanuelli, G., Vitale, G., Fraschini, N., Bigoni, M., Grieco, N., Denti, M., Giannattasio, C., Mancia, G., 1999. Sympathetic tone restrains arterial distensibility of healthy and atherosclerotic subjects. J Hypertens 17 (8), 1117–1123.

Falahpour, M., Refai, H., Bodurka, J., 2013. Subject specific BOLD fMRI respiratory and cardiac response functions obtained from global signal. Neuroimage 72, 252–264.

Feinberg, D.A., Mark, A.S., 1987. Human brain motion and cerebrospinal fluid circulation demonstrated with MR velocity imaging. Radiology 163 (3), 793–799.

Filosa, J.A., Bonev, A.D., Nelson, M.T., 2004. Calcium dynamics in cortical astrocytes and arterioles during neurovascular coupling. Circ Res 95 (10), e73–e81.

Fox, M.D., Snyder, A.Z., Vincent, J.L., Corbetta, M., Van Essen, D.C., Raichle, M.E., 2005. The human brain is intrinsically organized into dynamic, anticorrelated functional networks. Proc Natl Acad Sci U S A 102 (27), 9673–9678.

Fox, M.D., Zhang, D., Snyder, A.Z., Raichle, M.E., 2009. The global signal and observed anticorrelated resting state brain networks. J Neurophysiol 101 (6), 3270–3283.

Frederick, B., Nickerson, L.D., Tong, Y., 2012. Physiological denoising of BOLD fMRI data using regressor interpolation at progressive time delays (RIPTiDe) processing of concurrent fMRI and near-infrared spectroscopy (NIRS). Neuroimage 60 (3), 1913–1923. PMC3593078.

Friston, K.J., Williams, S., Howard, R., Frackowiak, R.S., Turner, R., 1996. Movement-related effects in fMRI time-series. Magnetic Resonance in Medicine 35 (3), 346–355.

Glasser, M.F., Coalson, T.S., Bijsterbosch, J.D., Harrison, S.J., Harms, M.P., Anticevic, A., Van Essen, D.C., Smith, S.M., 2018. Using temporal ICA to selectively remove global noise while preserving global signal in functional MRI data. Neuroimage 181, 692–717. PMC6237431.

Glover, G.H., Li, T.Q., Ress, D., 2000. Image-based method for retrospective correction of physiological motion effects in fMRI: RETROICOR. Magn Reson Med 44 (1), 162–167.

Golestani, A.M., Chen, J.J., 2022. Performance of temporal and spatial independent component analysis in identifying and removing low-frequency physiological and motion effects in resting-state fMRI. Front Neurosci 16, 867243. PMC9226487.

Griffanti, L., et al., 2014. ICA-based artefact removal and accelerated fMRI acquisition for improved resting state network imaging. Neuroimage 95, 232–247. 4154346.

Grootoonk, S., Hutton, C., Ashburner, J., Howseman, A.M., Josephs, O., Rees, G., Friston, K.J., Turner, R., 2000. Characterization and correction of interpolation effects in the realignment of fMRI time series. Neuroimage 11 (1), 49–57.

Hu, X., Le, T.H., Parrish, T., Erhard, P., 1995. Retrospective estimation and correction of physiological fluctuation in functional MRI. Magnetic Reson Med 34, 201–212.

Hudetz, A.G., Biswal, B.B., Shen, H., Lauer, K.K., Kampine, J.P., 1998. Spontaneous fluctuations in cerebral oxygen supply. An introduction. Adv Exp Med Biol 454, 551–559.

Jezzard, P., Clare, S., 1999. Sources of distortion in functional MRI data. Hum Brain Mapp 8 (2–3), 80–85.

Jo, H.J., Saad, Z.S., Simmons, W.K., Milbury, L.A., Cox, R.W., 2010. Mapping sources of correlation in resting state FMRI, with artifact detection and removal. Neuroimage 52 (2), 571–582. 2897154.

Katura, T., Tanaka, N., Obata, A., Sato, H., Maki, A., 2006. Quantitative evaluation of interrelations between spontaneous low-frequency oscillations in cerebral hemodynamics and systemic cardiovascular dynamics. Neuroimage 31 (4), 1592–1600.

Krause, F., Benjamins, C., Eck, J., Luhrs, M., van Hoof, R., Goebel, R., 2019. Active head motion reduction in magnetic resonance imaging using tactile feedback. Hum Brain Mapp 40 (14), 4026–4037.

Kruger, G., Glover, G.H., 2001. Physiological noise in oxygenation-sensitive magnetic resonance imaging. Magn Reson Med 46 (4), 631–637.

Kundu, P., Voon, V., Balchandani, P., Lombardo, M.V., Poser, B.A., Bandettini, P.A., 2017. Multi-echo fMRI: a review of applications in fMRI denoising and analysis of BOLD signals. Neuroimage 154, 59–80.

Kuperman, J.M., Brown, T.T., Ahmadi, M.E., Erhart, M.J., White, N.S., Roddey, J.C., Shankaranarayanan, A., Han, E.T., Rettmann, D., Dale, A.M., 2011. Prospective motion correction improves diagnostic utility of pediatric MRI scans. Pediatr Radiol 41 (12), 1578–1582.

Lang, E.W., Diehl, R.R., Timmermann, L., Baron, R., Deuschl, G., Mehdorn, H.M., Zunker, P., 1999. Spontaneous oscillations of arterial blood pressure, cerebral and peripheral blood flow in healthy and comatose subjects. Neurol Res 21 (7), 665–669.

Lawson, G.R., 2000. Controversy: sedation of children for magnetic resonance imaging. Arch Dis Child 82 (2), 150–153. PMC1718198.

Lemieux, L., Salek-Haddadi, A., Lund, T.E., Laufs, H., Carmichael, D., 2007. Modelling large motion events in fMRI studies of patients with epilepsy. Magn Reson Imag 25 (6), 894–901.

Lueken, U., Muehlhan, M., Evens, R., Wittchen, H.U., Kirschbaum, C., 2012. Within and between session changes in subjective and neuroendocrine stress parameters during magnetic resonance imaging: a controlled scanner training study. Psychoneuroendocrinology 37 (8), 1299–1308.

Lynch, C.J., Silver, B.M., Dubin, M.J., Martin, A., Voss, H.U., Jones, R.M., Power, J.D., 2020. Prevalent and sex-biased breathing patterns modify functional connectivity MRI in young adults. Nat Commun 11 (1), 5290. PMC7576607.

Maclaren, J., Herbst, M., Speck, O., Zaitsev, M., 2012. Prospective motion correction in brain imaging: a review. Magn Reson Med.

Marx, M., Pauly, K.B., Chang, C., 2013. A novel approach for global noise reduction in resting-state fMRI: APPLECOR. Neuroimage 64, 19–31. PMC3508249.

Mazaika, P., Hoeft, F., Glover, G.H., Reiss, A.L., 2009. Methods and Software for fMRI Analysis for Clinical Subjects, in Human Brain Mapping Conference.

Maziero, D., Rondinoni, C., Marins, T., Stenger, V.A., Ernst, T., 2020. Prospective motion correction of fMRI: improving the quality of resting state data affected by large head motion. Neuroimage 212, 116594. PMC7238750.

Menon, V., Lim, K.O., Anderson, J.H., Johnson, J., Pfefferbaum, A., 1997. Design and efficacy of a head-coil bite bar for reducing movement-related artifacts during functional MRI scanning. Behav Res Meth Instrum Comput 29 (4), 589–594.

Modarreszadeh, M., Bruce, E.N., 1994. Ventilatory variability induced by spontaneous variations of PaCO2 in humans. J Appl Physiol 76 (6), 2765–2775.

Murphy, K., Birn, R.M., Handwerker, D.A., Jones, T.B., Bandettini, P.A., 2009. The impact of global signal regression on resting state correlations: are anti-correlated networks introduced? Neuroimage 44 (3), 893–905. 2750906.

Murphy, K., Fox, M.D., 2017. Towards a consensus regarding global signal regression for resting state functional connectivity MRI. Neuroimage 154, 169–173. PMC5489207.

Ooi, M.B., Krueger, S., Muraskin, J., Thomas, W.J., Brown, T.R., 2011. Echo-planar imaging with prospective slice-by-slice motion correction using active markers. Magn Reson Med 66 (1), 73–81. 3122130.

Parkes, L., Fulcher, B., Yucel, M., Fornito, A., 2018. An evaluation of the efficacy, reliability, and sensitivity of motion correction strategies for resting-state functional MRI. Neuroimage 171, 415–436.

Perl, O., Ravia, A., Rubinson, M., Eisen, A., Soroka, T., Mor, N., Secundo, L., Sobel, N., 2019. Human non-olfactory cognition phase-locked with inhalation. Nat Hum Behav 3 (5), 501–512.

Perlbarg, V., Bellec, P., Anton, J.L., Pelegrini-Issac, M., Doyon, J., Benali, H., 2007. CORSICA: correction of structured noise in fMRI by automatic identification of ICA components. Magn Reson Imaging 25 (1), 35–46.

Pinto, J., Nunes, S., Bianciardi, M., Dias, A., Silveira, L.M., Wald, L.L., Figueiredo, P., 2017. Improved 7 Tesla resting-state fMRI connectivity measurements by cluster-based modeling of respiratory volume and heart rate effects. Neuroimage 153, 262–272. PMC5535271.

Poncelet, B.P., Wedeen, V.J., Weisskoff, R.M., Cohen, M.S., 1992. Brain parenchyma motion: measurement with cine echo-planar MR imaging. Radiology 185 (3), 645–651.

Posse, S., Kemna, L.J., Elghahwagi, B., Wiese, S., Kiselev, V.G., 2001. Effect of graded hypo- and hypercapnia on fMRI contrast in visual cortex: quantification of $T(*)(2)$ changes by multiecho EPI. Magn Reson Med 46 (2), 264–271.

Power, J.D., 2017. A simple but useful way to assess fMRI scan qualities. Neuroimage 154, 150–158. PMC5296400.

Power, J.D., Barnes, K.A., Snyder, A.Z., Schlaggar, B.L., Petersen, S.E., 2012a. Spurious but systematic correlations in functional connectivity MRI networks arise from subject motion. Neuroimage 59 (3), 2142–2154. 3254728.

Power, J.D., Barnes, K.A., Snyder, A.Z., Schlaggar, B.L., Petersen, S.E., 2012b. Steps toward optimizing motion artifact removal in functional connectivity MRI; a reply to Carp. Neuroimage.

Power, J.D., Mitra, A., Laumann, T.O., Snyder, A.Z., Schlaggar, B.L., Petersen, S.E., 2014. Methods to detect, characterize, and remove motion artifact in resting state fMRI. Neuroimage 84, 320–341. 3849338.

Power, J.D., Plitt, M., Gotts, S.J., Kundu, P., Voon, V., Bandettini, P.A., Martin, A., 2018. Ridding fMRI data of motion-related influences: removal of signals with distinct spatial and physical bases in multiecho data. Proc Natl Acad Sci U S A 115 (9), E2105–E2114. PMC5834724.

Power, J.D., Schlaggar, B.L., Petersen, S.E., 2015. Recent progress and outstanding issues in motion correction in resting state fMRI. Neuroimage 105, 536–551. PMC4262543.

Pruim, R.H., Mennes, M., Buitelaar, J.K., Beckmann, C.F., 2015. Evaluation of ICA-AROMA and alternative strategies for motion artifact removal in resting state fMRI. Neuroimage.

Pruim, R.H., Mennes, M., van Rooij, D., Llera, A., Buitelaar, J.K., Beckmann, C.F., 2015. ICA-AROMA: a robust ICA-based strategy for removing motion artifacts from fMRI data. Neuroimage.

Qiao, M., Rushforth, D., Wang, R., Shaw, R.A., Tomanek, B., Dunn, J.F., Tuor, U.I., 2007. Blood-oxygen-level-dependent magnetic resonance signal and cerebral oxygenation responses to brain activation are enhanced by concurrent transient hypertension in rats. J Cereb Blood Flow Metab 27 (6), 1280–1289.

Raitamaa, L., Huotari, N., Korhonen, V., Helakari, H., Koivula, A., Kananen, J., Kiviniemi, V., 2021. Spectral analysis of physiological brain pulsations affecting the BOLD signal. Hum Brain Mapp 42 (13), 4298–4313. PMC8356994.

Raschle, N.M., Lee, M., Buechler, R., Christodoulou, J.A., Chang, M., Vakil, M., Stering, P.L., Gaab, N., 2009. Making MR imaging child's play—pediatric neuroimaging protocol, guidelines and procedure. J Vis Exp 29, 3148936.

Saad, Z.S., Gotts, S.J., Murphy, K., Chen, G., Jo, H.J., Martin, A., Cox, R.W., 2012. Trouble at rest: how correlation patterns and group differences become distorted after global signal regression. Brain Connect 2 (1), 25–32.

Salas, J.A., Bayrak, R.G., Huo, Y., Chang, C., 2021. Reconstruction of respiratory variation signals from fMRI data. Neuroimage 225, 117459. PMC7868104.

Salimi-Khorshidi, G., Douaud, G., Beckmann, C.F., Glasser, M.F., Griffanti, L., Smith, S.M., 2014. Automatic denoising of functional MRI data: combining independent component analysis and hierarchical fusion of classifiers. Neuroimage 90, 449–468. 4019210.

Satterthwaite, T.D., Elliott, M.A., Gerraty, R.T., Ruparel, K., Loughead, J., Calkins, M.E., Eickhoff, S.B., Hakonarson, H., Gur, R.C., Gur, R.E., Wolf, D.H., 2013. An improved framework for confound regression and filtering for control of motion artifact in the preprocessing of resting-state functional connectivity data. Neuroimage 64, 240–256. PMC3811142.

Satterthwaite, T.D., Wolf, D.H., Loughead, J., Ruparel, K., Elliott, M.A., Hakonarson, H., Gur, R.C., Gur, R.E., 2012. Impact of in-scanner head motion on multiple measures of functional connectivity: relevance for studies of neurodevelopment in youth. Neuroimage 60 (1), 623–632.

Shea, S.A., 1996. Behavioural and arousal-related influences on breathing in humans. Exp Physiol 81 (1), 1–26.

Shin, W., Koenig, K.A., Lowe, M.J., 2022. A comprehensive investigation of physiologic noise modeling in resting state fMRI; time shifted cardiac noise in EPI and its removal without external physiologic signal measures. Neuroimage 254, 119136.

Shmueli, K., van Gelderen, P., de Zwart, J.A., Horovitz, S.G., Fukunaga, M., Jansma, J.M., Duyn, J.H., 2007. Low-frequency fluctuations in the cardiac rate as a source of variance in the resting-state fMRI BOLD signal. Neuroimage 38 (2), 306–320.

Smith, S.M., Miller, K.L., Moeller, S., Xu, J., Auerbach, E.J., Woolrich, M.W., Beckmann, C.F., Jenkinson, M., Andersson, J., Glasser, M.F., Van Essen, D.C., Feinberg, D.A., Yacoub, E.S., Ugurbil, K., 2012. Temporally-independent functional modes of spontaneous brain activity. Proc Natl Acad Sci U S A 109 (8), 3131–3136. PMC3286957.

Speck, O., Hennig, J., Zaitsev, M., 2006. Prospective real-time slice-by-slice motion correction for fMRI in freely moving subjects. Magma 19 (2), 55–61.

Todd, N., Josephs, O., Callaghan, M.F., Lutti, A., Weiskopf, N., 2015. Prospective motion correction of 3D echo-planar imaging data for functional MRI using optical tracking. Neuroimage 113, 1–12. 4441089.

Tong, Y., Hocke, L.M., Frederick, B.B., 2019. Low frequency systemic hemodynamic "noise" in resting state BOLD fMRI: characteristics, causes, implications, mitigation strategies, and applications. Front Neurosci 13, 787. PMC6702789.

Tong, Y., Hocke, L.M., Lindsey, K.P., Erdogan, S.B., Vitaliano, G., Caine, C.E., Frederick, B., 2016. Systemic low-frequency oscillations in BOLD signal vary with tissue type. Front Neurosci 10, 313. PMC4928460.

Triantafyllou, C., Hoge, R.D., Wald, L.L., 2006. Effect of spatial smoothing on physiological noise in high-resolution fMRI. Neuroimage 32 (2), 551–557.

Van den Aardweg, J.G., Karemaker, J.M., 2002. Influence of chemoreflexes on respiratory variability in healthy subjects. Am J Respir Crit Care Med 165 (8), 1041–1047.

Van Dijk, K.R., Sabuncu, M.R., Buckner, R.L., 2012. The influence of head motion on intrinsic functional connectivity MRI. Neuroimage 59 (1), 431–438.

Vincent, J.L., Patel, G.H., Fox, M.D., Snyder, A.Z., Baker, J.T., Van Essen, D.C., Zempel, J.M., Snyder, L.H., Corbetta, M., Raichle, M.E., 2007. Intrinsic functional architecture in the anaesthetized monkey brain. Nature 447 (7140), 83–86.

Wang, R., Foniok, T., Wamsteeker, J.I., Qiao, M., Tomanek, B., Vivanco, R.A., Tuor, U.I., 2006. Transient blood pressure changes affect the functional magnetic resonance imaging detection of cerebral activation. Neuroimage 31 (1), 1–11.

Weissenbacher, A., Kasess, C., Gerstl, F., Lanzenberger, R., Moser, E., Windischberger, C., 2009. Correlations and anticorrelations in resting-state functional connectivity MRI: a quantitative comparison of preprocessing strategies. Neuroimage 47 (4), 1408–1416.

White, N., Roddey, C., Shankaranarayanan, A., Han, E., Rettmann, D., Santos, J., Kuperman, J., Dale, A., 2010. PROMO: real-time prospective motion correction in MRI using image-based tracking. Magn Reson Med 63 (1), 91–105. 2892665.

Wise, R.G., Ide, K., Poulin, M.J., Tracey, I., 2004. Resting fluctuations in arterial carbon dioxide induce significant low frequency variations in BOLD signal. Neuroimage 21 (4), 1652–1664.

Wong, C.W., Olafsson, V., Tal, O., Liu, T.T., 2013. The amplitude of the resting-state fMRI global signal is related to EEG vigilance measures. Neuroimage 83, 983–990. 3815994.

Yackle, K., Schwarz, L.A., Kam, K., Sorokin, J.M., Huguenard, J.R., Feldman, J.L., Luo, L., Krasnow, M.A., 2017. Breathing control center neurons that promote arousal in mice. Science 355 (6332), 1411–1415. PMC5505554.

Yan, C.G., Cheung, B., Kelly, C., Colcombe, S., Craddock, R.C., Di Martino, A., Li, Q., Zuo, X.N., Castellanos, F.X., Milham, M.P., 2013. A comprehensive assessment of regional variation in the impact of head micromovements on functional connectomics. Neuroimage 76, 183–201.

Yang, S., Ross, T.J., Zhang, Y., Stein, E.A., Yang, Y., 2005. Head motion suppression using real-time feedback of motion information and its effects on task performance in fMRI. Neuroimage 27 (1), 153–162.

Yetkin, F.Z., Haughton, V.M., Cox, R.W., Hyde, J., Birn, R.M., Wong, E.C., Prost, R., 1996. Effect of motion outside the field of view on functional MR. AJNR Am J Neuroradiol 17 (6), 1005–1009.

Zhao, X., Bodurka, J., Jesmanowicz, A., Li, S.J., 2000. B(0)-fluctuation-induced temporal variation in EPI image series due to the disturbance of steady-state free precession. Magn Reson Med 44 (5), 758–765.

6

Physiological brain pulsations

Vesa Kiviniemi[a,b]

[a]Oulu Functional NeuroImaging, Health Science and Technology (HST), Oulu University, Oulu, Finland, [b]Diagnostic Imaging, Medical Research Center (MRC), Oulu University Hospital, Oulu, Finland

Historical perspective

The earliest report of discernible physical reactions in human brain was based on observations by Angelo Mosso in patients with open skull lesions, who showed increased brain pulsatility during performance of cued visual and cognitive tasks (Mosso, 1881). A decade later, hyperemic reactions and marked pulsatile responses during direct electric brain stimulus were reported in an open skull animal model experiment by Roy and Sherrington. They also noted marked overall pulsatility of the open brain that *"...can be seen the changes in volume of the brain which occur when the vaso-motor centre causes the rhythmic variations of the blood-pressure which are generally called 'Traube-Hering' waves. When these undulations appear during an experiment on the cerebral circulation (and in our experience they are especially frequently met with in such experiments), the brain expands with each rise of the blood-pressure and contracts with each successive fall, as is well shewn in"* (Roy and Sherrington, 1890).

Just over a decade later in 1901, Prof. Hans Berger, based on his direct interoperative observations of brain surgery patients, described three forms of intracranial pressure pulsations: "eine pulsatorische, eine respiratorische und vasomotorische Bewenung" (Berger, 1901), cf. Fig. 1 for a novel illustration of these waves. Berger found the pulsations while he was pursuing his studies of new forms of brain activity that finally led to the invention of electroencephalography (EEG).

The EEG remained a leading clinical tool for investigating the brain function till 1960s when positron emission tomography (PET) enabled the detection of brain blood flow and metabolism changes. In 1985, PET studies showed how during brain activation regional blood flow elevates local blood oxygenation more than what glucose metabolism would demand (Fox and Raichle, 1986).

Some years later in 1990, Ogawa discovered, based on Faraday's notes and Pauling's research (Pauling and Coryell, 1936) on magnetic

Advances in Resting-State Functional MRI. https://doi.org/10.1016/B978-0-323-91688-2.00012-6

properties of blood, that the use of repetitive functional magnetic resonance imaging (fMRI) susceptible to oxygen level of the brain enabled the detection of this peculiar nonstoichiometric brain activation change (Ogawa et al., 1990). The discrepant increase in regional blood oxygenation in activated regions could be detected with T2*-weighted scanned every few seconds, and Ogawa named the contrast blood oxygenation level-dependent contrast (BOLD) (Ogawa et al., 1990). Two years later, several groups simultaneously found out that regional neuronal activation leads to increase in BOLD contrast with a 3–5 s latency depending on activated area (Bandettini et al., 1992; Kwong et al., 1992; Ogawa et al., 1992). This led into an explosion of neuroscientific discoveries on human and animal brain function with unprecedented spatiotemporal accuracy.

The current literature holds that the activated neurons stimulate the regional neurovascular unit, which triggers a dilation of regional precapillary arteriole sphincters. The vasomotor-tone relaxation leads to vasodilation and increase in pulsatile blood flow after a response delay of 1–2 s (Drew et al., 2011; Grubb et al., 2020; Huotari et al., 2022; Ma et al., 2016). The inflowing oxygenated blood balloons the cortical veins draining the activated area 2–3 s later depending on the activated area (Biswal et al., 2003; Buxton, 2012; Huotari et al., 2022) (cf. Fig. 2A and B). In terms of the MR spin changes, the increased blood flow reduces local paramagnetic deoxyhemoglobin concentration, in conjunction with increasing local blood volume (Buxton, 2012).

The increasing oxygenated venous blood reduces dephasing of regional (peri)vascular water proton spins in fMRI data, which is detectable as an increase in the susceptibility weighted T2* BOLD signal intensity level downstream from activated areas (Bandettini et al., 1992; Buxton, 2012; Kwong et al., 1992; Ogawa et al., 1992), cf. Fig. 2A. Ultrafast functional $MREG_{BOLD}$ (Huotari et al., 2022) and a pioneering ultrasound study by Kucewizc (Kucewicz et al., 2007) both provide noninvasive evidence pointing to the original theme of Musso (Mosso, 1881), which is the localization of brain activation was based on the detection of increased arterial pulsatility in the activated brain area, Fig. 2A vs. B.

An important factor connected to the mixing of BOLD fMRI signal sources has previously been the low temporal resolution. BOLD data sampling that fulfills the Nyquist theorem requirement of >2 Hz sampling is needed in order to avoid the mixing of the cardiovascular with other brain pulsations due to aliasing (Huotari et al., 2019). Recent advances in 3D fMRI, however, offer 20–40 times faster simultaneous sampling of BOLD changes with other hydrodynamic brain pulsations, and importantly enables the separation of all the three physiological brain pulsation mechanisms, i.e., vasomotor, respiratory, and cardiovascular pulsations and their interactions without signal

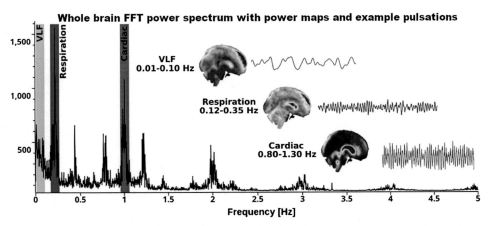

Fig. 1 Whole brain MREG signal FFT spectrum illustrating frequency peaks all three physiological brain pulsations originally described by Berger in 1903. The <0.1 Hz very-low-frequency (VLF) vasomotor waves are shown in green, 0.3 Hz respiratory pulsation in blue and cardiovascular pulsations in red with frequency peaks, and corresponding time signals on right side. VLF (a.k.a. ALFF), respiratory, and 1 Hz cardiac power maps show midline spatial distribution of the brain pulsations. Please note that the 5 Hz MREG$_{BOLD}$ spectral window shows 4 harmonic cardiac peaks and heterodyne power peaks on symmetrically on both sides of the 1 Hz cardiac power peak reflecting respiratory amplitude modulation of the cardiovascular pulsations (Raitamaa et al., 2021).

aliasing (Hennig et al., 2007, 2021; Lin et al., 2012; Posse et al., 2013) (cf. Fig. 1). These three pulsations function as main hydrodynamic drivers of the brain. While the pulsations have independent myogenic pressure sources, they all are interactive. The pulsations are all modulated by autonomous nervous system, and, the slower pulsations tend to modulate the faster ones by pressure effects. For instance, respiration modulates cardiovascular amplitude envelope (CREM) especially in CSF (Raitamaa et al., 2021), while neurovascularly coupled vasomotor dilations lead to increased cardiovascular pulse amplitude, i.e., CHE signal in active brain areas, cf. Fig. 2 (Huotari et al., 2022).

Intracranial hydrodynamic physiology

Before the mechanisms by which the physiological pulsations function, some novel elements of the intracranial physiology need to be reviewed from a hydrodynamic perspective. The central nervous system is guarded with a strong cranial bone and further insulated by meninges from its surrounding environment in order to give the neuronal tissue the stable conditions they require to perform its multifaceted tasks even while the brain is moving even in accelerating conditions (Wagshul et al., 2011).

The brain itself consists of 70%–80% of water, which involves three basic fluid compartments intertwined with each other (Natali et al., 2013; Ringstad and Eide, 2020). The brain (and spinal cord below) is embedded within cerebrospinal fluid (CSF) that has been for long thought to function mostly as a cushion dampening external impulses on the sensitive neuronal structures. Recently, however, the CSF has increasingly been regarded as an active part of the CNS that actually uses the physical impulses to filter water and solutes both along CSF conduits from aquaducts to perivascular Virchow-Robins spaces and finally through the CNS tissue structures (Gouveia-Freitas and Bastos-Leite, 2021; Jessen et al., 2015). CSF movement is therefore a vital ingredient in maintaining homeostatic balance that supports the underlying neuronal activity (Wagshul et al., 2011).

In addition to relatively uniform CSF, there are other two fluid compartments in the central nervous system that can be further divided into subcompartments (Vladić et al., 2009). The circulating blood can be divided into arterial, capillary, and venous blood based on both anatomy and its gaseous/metabolite content. The third compartment is the interstitial fluid (ISF) that divides into extracellular and intracellular compartments. Water molecules traverse between these relatively isolated compartments mediated via pores in barrier tissue (meninges, glia limitans) and specific aquaporin water channels (AQP) found in cellular membranes depending on the location and scale of hydrodynamic activity.

The CSF is impeded within arachnoid space between the fibrous pia mater brain surface and tight dura mater bordering the bone structures around the brain. In the centrally located CSF ventricles and other conduits, the pia mater attaches to ependymal cells and basically everywhere else the pia attaches to astrocytic end feet (Gouveia-Freitas and Bastos-Leite, 2021; Zhang et al., 1990). In the arachnoid space, arteries enter and veins exit the brain tissue, and there, the pia forms a special CSF-filled perivascular Virchow-Robin space reaching all the way into capillary level, which recently was discovered to function as a conduit of CSF solutes both in and out of the brain as driven by physiological forces replacing the lymphatic vessels seen elsewhere in the body (Jessen et al., 2015; Wagshul et al., 2011; Zhang et al., 1990).

The blood compartment is also tightly separated from the brain interstitial tissue by the three-layer blood-brain barrier (BBB): the endothelial mono-cellular layer, the basement membrane, and a sheet of astrocyte end-feet layer (a.k.a. the glia limitans) of the BBB (Hladky and Barrand, 2014). Furthermore, occasionally, pericytes are also impeded in the BBB wall. The endothelial cells are firmly sewn together by tight junctions that make the BBB impenetrable to any polar molecules. Thus, active transport from the blood into the interstitium is required for substances like glucose and proteins.

Astrocytic end feet in the glia limitans is tightly coupled to each other but may have 20–40 nm pores or interwoven clefts between adjacent end feet, which can mediate some solute transport in and out of the interstitium in conjunction with the water flowing through AQP4 water channels covering the luminal edge of astrocytes, cf. Fig. 3. Of note, no other organ has this kind of tight border except the brain and spinal cord, and as the original discovery of the BBB states, water-soluble dyes were injected in the blood color in every other tissue except the CNS.

The brain itself is thus guarded by the bone, meninges, and cushioning CSF, and the interstitial brain tissue, i.e., neuronal and glial cells, are further insulated with BBB and pia mater. One reason for the heavy packaging is that the neurons require rather precise concentrations of metabolites, neurotransmitters, and electrolytes in order to maintain stable baseline activity of functional networks with the ability to spread action potentials across their axons over to other neurons (Hladky and Barrand, 2014; Jessen et al., 2015; Wagshul et al., 2011). As much as 20%–25% of the whole-body glucose metabolism is *continuously* used to maintain the required substrate concentrations over all these borders, especially the BBB and cellular membranes. Maintenance requires both influx and efflux, which sounds like a double effort.

For example, even small regional decreases in the concentration and/or water dynamics of the electrolyte solute concentrations can locally slow down neuronal activity to 2–4 Hz delta activity, i.e., sleep. Conversely, an increase in solute concentrations can awaken sleeping neurons (Ding et al., 2016). Further increase is known to lead to pathological hyperactive states including epileptic seizures as K+ concentrations may be elevated due to several reasons like regional water and/or K+ levels (Patel et al., 2019).

As the brain homeostasis requires a lot of energy, that has not been as abundant over the (Darwinian) aeons as it is nowadays; nature has evolved fascinating ways to utilize all available energy resources to facilitate the transport of solutes over the insulating borders to neuronal tissue. For instance, even *voluntary* treadmill exercise facilitates better CSF kinetics alone ((Rasmussen et al., 2019; von Holstein-Rathlou et al., 2018). Moreover, the energy used to drive blood flow and gas exchange is mediated into the brain cavity by the hydraulic water pressure effects via blood vessels and CSF canals (Dreha-Kulaczewski et al., 2017; Kiviniemi et al., 2016; Raitamaa et al., 2021; Ringstad and Eide, 2020). Every CSF impulse is further transformed into microscale water and solute movement in the poro-elastic perivascular spaces, which at the same time dampen also the physical exercise impacts of the body. As many forms of physical impulses are harvested as intracranial CSF solute movement, the stinging fact seems to be that every move you make, every breath you take, it will be flushing your brain.

Physiological forces driving solute transport

The unique incompressible surroundings of the CNS offer several characteristic physical features. In addition to the safeguarding nature that prevents direct harm to the elastic CNS structures, any kind of physical impulses asserted in and on the body affects the brain hydrodynamics. The CSF water is actually very susceptible to any physical power exerted on the body, especially within the incompressible cranial bones that surround the brain with only a few exceptions of foramen magnum and some canals for blood vessels and cranial nerves.

An increase in the volume (or pressure) in any of the intracranial fluid compartments inside the cranial vault leads to immediate loss of fluid volume that is required to be filled. In other words, for any intracranial pressure impulse, something needs to give in to dampen the impulse and recover the volume loss. This is called the Monro-Kellie doctrine (Dreha-Kulaczewski et al., 2017; Wagshul et al., 2011; Wilson, 2016). Previously (and unfortunately even now still due to the difficulty of imaging the spinal canal), the intracranial space alone was considered in the doctrine. However, more recent view on the Monro-Kellie doctrine emphasizes the important nature of the spinal canal as an compensatory fluid volume reservoir (Wilson, 2016) that may also facilitate solute removal (Cai et al., 2019; Jacob et al., 2019).

In healthy brains, the reciprocal fluid flow driven by the Monro-Kellie doctrine is used to drive not only brain metabolites to desired regions but also remove metabolic end-products as solutes of the CSF along perivascular spaces. The physiological brain pulsations detected some 100 years ago introduce driving pressure impulses reaching all brain fluid compartments. Cardiovascular impulses are mediated into the brain via high-pressure arteries, while respiratory inhalation empties low-pressure cranial veins and CSF (Dreha-Kulaczewski et al., 2015, 2017; Ringstad and Eide, 2020; Santisakultarm et al., 2012; Wagshul et al., 2011). Importantly, both cardiac and respiratory pulsations induce simultaneously occurring CSF counter-pulsations from the spinal cord that reach the brain tissue via the incompressible CSF paravascular spaces as dictated by the Monro-Kellie doctrine.

Both systemic blood pressure and more regional arterial vasomotor tone waves further regulate brain tissue cerebral blood flow (CBF) and volume (CBV) depending on the arousal state (Chang et al., 2016; Liu et al., 2018). The vasomotor waves introduce pressure and fluid volume gradients inside the brain and simultaneously also modulate the amplitudes of the cardiorespiratory pulsatility locally (Huotari et al., 2022). In addition, all these physiological pulsations are interactive, respond to autonomous nervous system control, and importantly, also directly modulate each other (Raitamaa et al., 2021). Finally, also the overall body position during sleep and movement status during the day, as the impulses from physical exercise also move CSF and have an effect on the total CNS solute transportation (Jessen et al., 2015; Lee et al., 2015; von Holstein-Rathlou et al., 2018). All these physiological pulsation drivers form a multilevel, backed-up system for hydrodynamics solute transportation within the CNS.

Cardiovascular brain pulsation

In addition to the enormous task of providing continuously metabolites for the energy consuming brain tissue by perfusing the capillaries present for each neuron, the physical pulsatility of the arteries itself is also utilized as a driving force of the brain water between fluid compartments at macroscopic and microscopic levels. Each heart beat brings 6–10 ml of blood into the cranium and this macroscopic volume increase needs to be compensated as dictated by the Monro-Kellie doctrine. Some 0.1 sec after the cardiac impulse arrival to the brain, the sagittal sinus venous outflow increases, partially dictated by the blood-volume change (Rivera-Rivera et al., 2017). Also, CSF flow in ventricles and aqueducts reacts similarly to compensate the fast intracranial volume change of the cardiovascular impulse (Dreha-Kulaczewski et al., 2015, 2017).

From another perspective, the perivascular structures also have a microscopical cushioning effect that turns the original impulse of the heart into a widespread water movement along the cerebral arterial tree and into the brain interstitium via AQP4 water channels. In other words, brain tissue as a whole has evolved into a poroelastic form that absorbs motion energy to facilitate CSF solute transport.

Microscopic studies in rodent brain have shown that parenchymal cardiovascular pulsatility is the main contributor to periarterial CSF water convection in the pial vessels (Iliff et al., 2013). As discovered by Nedergaard in 2013, these cardiovascular pulsations play a key role in maintaining brain tissue homeostasis by driving the CSF-mediated convection of solutes through the brain parenchyma as mediated by AQP4 channels and small pores in the astrocytic endfeet sheaths of glia limitans (Iliff et al., 2013; Nedergaard, 2013)). The cardiovascular brain pulsations are directly proportional to the convective efficiency for transporting CSF solutes along perivascular CSF conduits (Mestre et al., 2018).

Macroscopically, cardiac pulsatility has been shown to be a main driver of the CSF in aqueducts and over foramen magnum between the spinal canal reservoir and intracranial space (Dreha-Kulaczewski et al., 2015, 2017; Kiviniemi et al., 2016; Raitamaa et al., 2021; Ringstad and Eide, 2020; Wagshul et al., 2011). The whole brain level power analysis shows that similar to microscopic level, the periarterial areas are dominated by cardiac pulsation power and also

CSF spaces in both ventricular and frontal cortex (Mestre et al., 2018; Santisakultarm et al., 2012). However, both the respiratory and vaso-motor pulsations are more dominant than cardiovascular pulsation *deeper within* the gray matter; white matter, on the other hand, is especially dominated by respiratory pulsations in the human brain (Huotari et al., 2019; Kiviniemi et al., 2016; Raitamaa et al., 2021; Windischberger et al., 2002).

Neuronal activation leading to vasomotor tone relaxation also in-creases the amplitude of the low-frequency envelope of cardiovascu-lar pulsatility in activated areas some 1.3 ($\pm$2.2) seconds before BOLD response, cf. in Fig. 2B. The pulsatile nature of the blood flow has itself been shown to increase nitric-oxide (NO)-mediated vasodilation,

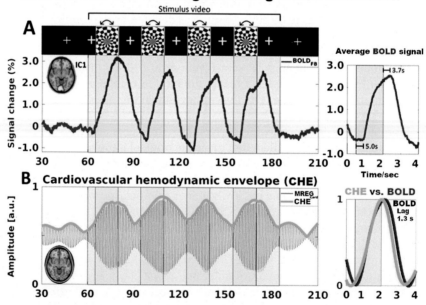

Fig. 2 Ultrafast MREG scanning captures simultaneously two different functional contrasts. (A) The classic susceptibility-based BOLD changes and, (B) arterial pulsation increase in activated visual areas (Huotari et al., 2022). The green cardio hemodynamic envelope (CHe) set over each cardiovascular impulse reflects time domain changes in arterial pulse amplitude (blue signal capturing spin phase changes from accelerating flow). The green CHE signal precedes the classical red BOLD signal on an average 1.3 s as the arteries dilate first to increase their pulsatile flow. After the arterial dilation, the classical susceptibility-based BOLD response in the hyperemic veins shows increase in signal intensity. Modified from Huotari, N., Tuunanen, J., Raitamaa, L., Raatikainen, V., Kananen, J., Helakari, H., Tuovinen, T., Järvelä, M., Kiviniemi, V., Korhonen, V., 2022. Cardiovascular pulsatility increases in visual cortex before blood oxygen level dependent response during stimulus. Front. Neurosci. 16, 836378. https://doi.org/10.3389/fnins.2022.836378.

which further facilitates the downstream vasodilatory brain responses in the vascular trees (Nakano et al., 2000). Increased arterial pulsation further augments the metabolite flow in perivascular areas during activity (Mestre et al., 2018; Xie et al., 2013).

Cardiovascular fMRI signal

The accelerative cardiovascular impulses are known to introduce spin phase changes and steady-state spin-coherence perturbations in gradient recalled sequences since 1980s and they have different properties depending on the flow with respect to gradient readout directions (Duyn, 1997; von Schulthess and Higgins, 1985). Novel ultrafast 3D VEPI & MREG sequences have also detected the same drop in arterial signal intensity upon the arrival of cardiovascular impulses that traverse the arterial tree (Kiviniemi et al., 2016; Posse et al., 2013). The arterial signal drop is strong near arteries but overall in the brain it is less powerful than the other two brain pulsations in the human brain (Ra itamaa et al., 2021). Fig. 2 illustrates an example of the arterial impulses that precede BOLD signal by some 1.3 sec during brain activation.

Respiratory brain pulsation

Respiratory brain pulsations are a most dominant driving force of movement in the intracranial fluid compartments (Raitamaa et al., 2021). During inspiration, the reduced intrathoracic pressure draws venous blood from the brain and this volume loss is counterbalanced by an inward movement of CSF from the spinal canal as dictated by the Monro-Kellie doctrine (Dreha-Kulaczewski et al., 2017), cf. Fig. 3.

Venous outflow from the brain also has unique characteristics, namely, it travels through the only existing incompressible veins in the body—the dural venous sinuses. All other veins in the body collapse from the marked negative pressures but the dura holds the venous sinuses open and mediates the negative pressures during inhalation high into the cortical level (Dreha-Kulaczewski et al., 2017; Vinje et al., 2019). Furthermore, the negative pressure within the dural sinuses seems to be connected to the accumulation of solutes close to the brain vertex next to the dural lymphatic vessels found only next to venous sinuses (Ringstad and Eide, 2020). This mechanism is also bound to affect the perivenous CSF space in the cortex, which together with counterphase venous blood volume changes creates a perivenous CSF pump. In theory this kind of pumping activity has the capability to remove perivenous CSF solutes into the subarachnoid CSF space for further removal from the CNS.

The original research findings of the glymphatic system by Iliff & Nedergaard in the pial surface indicate that the CSF and its solutes are driven by cardiovascular brain pulsations that push CSF from the

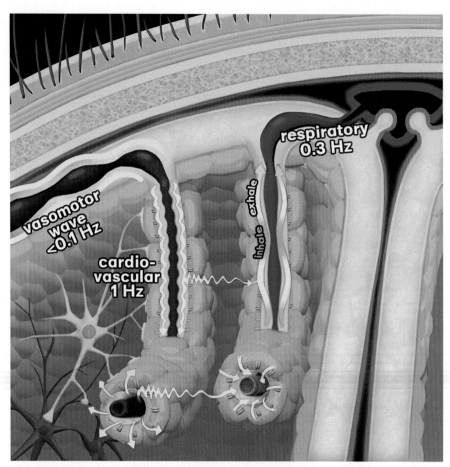

Fig. 3 Theoretical illustration of the physiological brain pulsations in the brain parenchyma. Vasomotor waves control the regional cerebral blood flow by flow resistance and also modulate the cardiorespiratory pulsation amplitude. Arterial pulsation pushes blood inside the vessel and solutes along the periarterial CSF space and into the brain interstitium via AQP4 channels and glia limitans pores. Inhalation draws venous blood from the cortical veins and compensatory CSF flows inward into the perivenous space, countering the venous blood volume loss. During exhalation, the intrathoracic pressure slows down the flow in cerebral veins. The continuous blood flow from the capillaries then balloons the cortical veins during exhalation with increased deoxygenated blood volume. The ballooning venous walls then push out the CSF from the perivenous space along with the CSF solutes. This results in a pumping effect within the perivenous space like the overall CSF/venous pulsation (Dreha-Kulaczewski et al., 2015; Wagshul et al., 2011).

glia limitans of the BBB through brain tissue into the perivascular space and then out of the brain (Iliff et al., 2013; Nedergaard, 2013) (see Fig. 3 for the mention of fast arterial pulsations). The glymphatic water convection has been demonstrated to carry both interstitial and perivascular CSF in a highly similar ways in both mice and human brain, despite the 3000-fold difference in size between the two (Eide and Ringstad, 2019; Ringstad and Eide, 2020). Importantly, the injected CSF tracers tend to fill only periarterial rather than perivenous spaces, which together with the know differences in pulsators of these vessels (Mestre et al., 2018; Santisakultarm et al., 2012), suggests that the perivenous routes may only serve as outflow conduits in the cortex, while in the white matter veins tend to flow toward the internal cavernous sinuses depending on hydrostatic effects of body posture.

The cardiorespiratory pulsations are the two main drivers of the perivascular CSF convection (Mestre et al., 2018; Santisakultarm et al., 2012), and these pressure pulsations extend from CSF into perivascular CSF spaces throughout the brain inside the incompressible cranial vault in accordance with the Monro-Kellie doctrine to replace the volume changes. On a macroscopic scale, respiration waves may function as a driver of the glymphatic convection in addition to the effects of slow vasomotor (neurogenic, systemic and intrinsic arterial) waves and faster cardiovascular pulsations all around the brain, cf. Fig. 3 (Dreha-Kulaczewski et al., 2017; Helakari et al., 2022; Wagshul et al., 2011).

Respiratory fMRI signal

In T2* weighted fMRI data, the respiration induces propagating waves of BOLD signal changes along venous tracks: inhalation increases and exhalation decreases the BOLD signal dominantly the brain cortex but also in the white matter (Birn et al., 2009; Kiviniemi et al., 2016; Raitamaa et al., 2021; Wise et al., 2004; Yuan et al., 2013). The signal is formed by the classical susceptibility effect, as inhalation reduces deoxyhemoglobin volume in veins; and the volume loss is replaced by CSF in perivascular spaces abundant in each voxels and increases the T2* weighted signal. Exhalation does the opposite as the veins re-filling from capillary flow balloon of deoxygenated blood, which then increases water proton dephasing resulting in reduced T2* signal intensity. Detection of these changes requires ultrafast T2* signal scanning in order to avoid aliasing between physiological signal sources.

Vasomotor waves in awake brain

Cerebral blood flow is regionally controlled by flow resistance, where the caliber of the vessel is regulated by the vasomotor tone of the myocytes encircling the arterial walls. Vasomotor tone is under several systemic and regional control factors that then integratively form the baseline tone that directs the regional blood flow.

The most famous vasomotor action is the activation-induced regional hyperemia of the BOLD response, as described earlier in this chapter. In awake conditions, the neuronal activity is tightly coupled to regional blood flow during both cued task activations (Bandettini et al., 1992) as well as spontaneous brain activity fluctuations (Biswal et al., 1995) within functional connected, independent brain networks (Beckmann and Smith, 2004; Biswal et al., 1995; Fox and Raichle, 2007;

Kiviniemi et al., 2003; Smith et al., 2009). Recently, the functional activity fluctuations are also reported to propagate in quasiperiodic waves over the independent functional networks (Abbas et al., 2019; Kiviniemi et al., 2016), and even functionally connected white matter tracks have been shown to exhibit low-frequency signals along the connected tracks (Gore et al., 2019).

The early works of spontaneous functional connectivity relate the functionally connected VLF fluctuations to slow vasomotor waves (Biswal et al., 1995; Kiviniemi et al., 2000). The spontaneous vasomotor waves were first noted by Hales in 1733 after exsanguinating a horse with a continuous blood pressure measurement. These waves are often divided into two characteristic waves, the slower 0.03 Hz Mayer waves and somewhat faster 0.1 Hz Traube-Hering waves, although some researchers think there may be several different low frequency oscillators (Miyakawa et al., 1984; Kiviniemi, 2004).

In principle, the two types of waves reflect the origin of the oscillatory activity; the slower Mayerian waves are autorhythmic cellular fluctuations in myocytes modulated by direct neuronal control from autonomous system and the Traube-Hering waves stem from mechanistic interactions of cardiorespiratory pulsations from the thoracic area. Recent studies on VLF brain activity have actually detected three coexisting phenomena that can form both functionally connected standing and propagating waves in the cortex (Bolt et al., 2022).

In addition, systemically there are slower hours and long whole body rhythms affecting also the brain vasculature and hydrodynamics including the circadian rhythm, hormonal control mechanisms (e.g., renin-angiotensin), and gastro-intestinal blood volume changes for instance. The multiple key factors of blood flow control act simultaneously, which results into a 1/f-type fractal behavior in the measured signal, where fast changes in one factor may be relatively small in amplitude but larger changes involving multiple control mechanisms require more time. The identification of unique waves may be difficult also due to their interaction with other control mechanisms. Thus, research results tend to give different frequency ranges depending on underlying physiological state or pathology.

The autonomous nervous system, systemic blood volume, neurohumoral, hormonal, circadian, regional feedback loops of chemosensory, and direct neuronal control mechanisms engage continuous waves of changes in tissue vasculature all over the body (Miyakawa et al., 1984; Preiss and Polosa, 1974; Kiviniemi, 2004). In turn, cerebral blood flow control is mediated by vasomotor tone, which is under integrated action of multiple physiological controllers that maintain a stable inflow of required metabolites and outflow of metabolic products (Clarke and Sokoloff, 1999).

Sleep changes all physiological brain pulsations

As previously mentioned, in the awake state, cerebral blood flow is tightly related to baseline neuronal activity (Bandettini et al., 1992; Logothetis et al., 2001; Smith et al., 2009). In anesthetized and sleeping mice, the neurovascular coupling is present but becomes overshadowed by a dominant 0.04 Hz vasomotor waves (Ma et al., 2016). Human fMRI data corroborates that cerebral blood flow becomes overpowered by physiological pulsations like vasomotor waves around 0.03–0.05 Hz in sleep or under anesthesia (Chang et al., 2016; Helakari et al., 2022; Kiviniemi et al., 2000, 2005; Liu et al., 2018).

In the human brain, increased very-low-frequency BOLD fluctuations in the posterior brain regions have been detected during light N1-N2 sleep and episodes of low-awake vigilance (Chow et al., 2013; Fukunaga et al., 2006; Horovitz et al., 2008). Deactivation cycles of the cholinergic nucleus basalis precede widely distributed hemodynamic BOLD signal increase in the posterior brain regions (Liu et al., 2018). K-complexes formed in the basal ganglia lead to increased vasomotor tone over the peripheral vasomotor waves and are then enter the sleeping brain and mask the neurovascular activity (Fultz et al., 2019; Özbay et al., 2019; Picchioni et al., 2022). The CSF pulsations are strongly linked to the vasomotor waves and k-complexes, and currently, it seems that it is unclear which impulse follows the other.

During sleep, the glymphatic brain solute transport and water convection increase markedly (Jessen et al., 2015; Nedergaard, 2013; Xie et al., 2013). Areas undergoing the increased hydrodynamics in sleep have reduced intraparenchymal electrolyte concentrations that slow down the neuronal activity from awake fast rhythms to classical sleep delta power (Ding et al., 2016; Xie et al., 2013). The slow wave sleep areas of human sleep delta activity were shown to present marked increase in all three physiological brain pulsations in reversed order of frequency; strong increases were seen in both vasomotor and respiratory pulsations while the cardiovascular pulsations were less obvious (Helakari et al., 2022).

Recent study also indicates that the vasomotor waves move injected CSF tracers along the perivascular space, thus serving directly as a glymphatic driver (van Veluw et al., 2020). The vasomotor tone waves also regulate amplitude of the cardiorespiratory pulsations as the tonus of the arterial wall (card) and venous blood volume (respectively) determine their downstream pulse amplitude (Hadaczek et al., 2006; Helakari et al., 2022; Kananen et al., 2020; Mestre et al., 2018; Santisakultarm et al., 2012). The altering frequency distributions over a range of frequencies introduce a reduction in the spectral entropy of both EEG (Mahon et al., 2008) and BOLD signal alike (Helakari et al., 2022).

During sleep, upper-airway resistance increases (Malik et al., 2012; Sowho et al., 2014) and this may be related to the increased venous/CSF pulsations as increased respiratory frequency power of the brain fMRI signal (Helakari et al., 2022). The supine position in sleep redirects the venous flow to the jugular vein from the cavernous sinus and paraspinal venous plexus that dominates in the prone position, and this may influence further the respiratory brain pulsatility (Gisolf et al., 2004). The breathing is also more stable during sleep (Malik et al., 2012), which partially explains the reduced respiratory entropy changes in sleep (Helakari et al., 2022). The increased and stabilized vasomotor and respiratory pulsations in sleep would likely increase the CSF flow toward the neuropil, which might inflate the extracellular space, as has been shown to occur during sleep/wake transitions (Xie et al., 2013).

Pathological alterations in brain pulsations

There are several micro and macroscopical studies indicating that altered interstitial and CSF convection driven by physiological pulsations is strongly connected to brain pathology (Dreha-Kulaczewski et al., 2015; Eide and Ringstad, 2019; Gouveia-Freitas and Bastos-Leite, 2021; Jahanian et al., 2014; Jessen et al., 2015; Kananen et al., 2020; Mestre et al., 2018, 2020; Nedergaard, 2013; Nedergaard and Goldman, 2020; Ringstad and Eide, 2020; Tuovinen et al., 2020), for further reading (Arbel-Ornath et al., 2013). The amplitude of (very) low-frequency BOLD signal fluctuations (ALFF) developed by Yu-Feng Zang (Yu-Feng et al., 2007) has become an important and robust (Biswal et al., 2010) metric in detecting brain pathology. $BOLD_{ALFF}$ has been highly successful in detecting disease-related changes in several major neuropsychiatric disease based on the high number of meta-analyses of over 1370 articles so far on BOLD ALFF on PubMed. Diseases such as attention-deficit hyperactivity disorder (ADHD) (Yu-Feng et al., 2007), major depression (Gong et al., 2020b), Alzheimer's disease (AD), mild-cognitive impairment (MCI) (Mascali et al., 2015; Yang et al., 2018; Zhang et al., 2021), Parkinson's disease (Wang et al., 2018), schizophrenia (Gong et al., 2020a), and stroke (Tsai et al., 2014) have shown significant changes in the brain BOLD signal amplitude to name a few. Brain $BOLD_{ALFF}$ can be clinically useful in spinal pathology like cervical myelopathy (Takenaka et al., 2020), discogenic pain (Ma et al., 2020), or in lower back pain as well (Zhou et al., 2018). The $BOLD_{ALFF}$ competes with respiratory pulsation amplitude in cortex, and the high variance in ALFF may be due to aliased respiration over ALFF if BOLD data are sampled noncritically with respect to faster cardiorespiratory pulsations (Mao et al., 2015; Raitamaa et al., 2021).

In close resemblance to ALFF, the coefficient of variation ($BOLD_{CV}$ = BOLD signal STD/mean) has been shown to be altered in ageing, AD, and small vessel vasculopathies (Jahanian et al., 2014; Makedonov et al., 2013a,b; Makedonov et al., 2016). The ultra-fast $BOLD_{CV}$ showed that the variance is mostly altered in cardiovascular frequency in AD (Tuovinen et al., 2020). Further optical flow analysis of the propagation cardiovascular brain impulses detected mostly increased speed but even reversed cardiovascular impulse propagation in AD in areas recently shown to have leaking BBB (Nation et al., 2019; Rajna et al., 2021).

Pathological mechanism can alter the physical brain pulsations driving both blood and CSF in different time scales. Changes may be sudden blood/CSF flow cessation like stroke, which leads to brain edema as the interstitium flows with paravascular CSF water within seconds because metabolism fails (Mestre et al., 2020). Changes can also occur slowly over decades; blood pressure increases as elastin and collagen loss stiffens arteries with advancing age, which reduces cardiovascular pulsatility (Hussein et al., 2020) and increases BBB permeability (Nation et al., 2019). Increased blood pressure is a risk factor for both AD and stroke, as it reduces vessel-wall pulsatility along with perivascular CSF convection (Hussein et al., 2020; Mestre et al., 2018). Also, AQP4 water channels decline with age and allows amyloid to accumulate into perivascular structures, decreasing cognition and leading in part to AD ((Nedergaard and Goldman, 2020; Zeppenfeld et al., 2017).

Intractable epilepsy has also been shown to be related to AQP4 water channel absence from the astrocytic end feet lining the perivascular space (Eid 2005 PNAS). Intracranial EcoG measurements in intractable epilepsy patients have also demonstrated strong drive of respiratory brain pulsations (Herrero et al., 2018; Zelano et al., 2016). Indeed, respiratory brain signal variance, amplitude, and coherence were shown to be altered in intractable epilepsy patients in $MREG_{BOLD}$ signal (Kananen et al., 2018, 2020). Ultrafast $MREG_{BOLD}$ enables localization of individual abnormalities of brain pulsation variance in a two-center study > 6 standard deviations above age matched control population mean values (Kananen et al., 2020).

Within the spinal and brain tissue aqueducts the cardiac pulsatility is the dominating force moving the CSF, only at deep breaths the CSF increasingly enters these central CSF spaces (Dreha-Kulazewski et al., 2017). On the brain and spinal cord surface, however, the respiration effects dominate as CSF signal source (Kiviniemi et al., 2016; Raitamaa et al., 2021). The inspiration induces a strong cortical propagating wave starting from the infratentorial regions extending forward and traversing into the perivenous spaces (Santisakultarm et al., 2012; Kiviniemi et al., 2016; Raitamaa et al., 2021) in the brain cortex. The cyclic CSF in/out flow with countering venous outflow volume changes driven also by inspiration can introduce a net flow wave that pushes CSF out from the perivenous spaces (Kiviniemi et al., 2016, Elabasy et al., 2023). The respiration is also the most powerful signal in the white matter (Raitamaa et al., 2021), but there the flow trajectories are towards the brain center identical to the local venous paths (Kiviniemi et al., 2016; Elabasy et al., 2023).

Summary

To conclude, physiological brain pulsations have been long regarded as noise in resting-state fMRI, but now, their importance has finally started to dawn due to increased understanding of the key elements of the brain electrohydrodynamic physiology. Another important milestone is the advances in fMRI scanning technology such as the increase in sampling rates stemming from combined parallel imaging and advanced k-space undersampling. Due to these ingredients, it seems finally plausible that individual-level diagnostics can be made accurately based on detecting directly the effects of microscopic disease mechanisms on the human brain noninvasively.

References

Abbas, A., Belloy, M., Kashyap, A., Billings, J., Nezafati, M., Schumacher, E.H., Keilholz, S., 2019. Quasi-periodic patterns contribute to functional connectivity in the brain. Neuroimage 191, 193–204. https://doi.org/10.1016/j.neuroimage.2019.01.076.

Arbel-Ornath, M., Hudry, E., Eikermann-Haerter, K., Hou, S., Gregory, J.L., Zhao, L., Betensky, R.A., Frosch, M.P., Greenberg, S.M., Bacskai, B.J., 2013. Interstitial fluid drainage is impaired in ischemic stroke and Alzheimer's disease mouse models. Acta Neuropathol. (Berl.) 126, 353–364.

Bandettini, P.A., Wong, E.C., Hinks, R.S., Tikofsky, R.S., Hyde, J.S., 1992. Time course EPI of human brain function during task activation. Magn. Reson. Med. 25, 390–397. https://doi.org/10.1002/mrm.1910250220.

Beckmann, C.F., Smith, S.M., 2004. Probabilistic Independent Component Analysis for Functional Magnetic Resonance Imaging. IEEE Trans. Med. Imaging 23, 137–152.

Berger, H., 1901. Lehre Von Der Blutzirkulation in Der Schadelhohle Des Menschen Namentlich Unter Dem Einfluss Von Medikamenten.

Birn, R.M., Murphy, K., Handwerker, D.A., Bandettini, P.A., 2009. fMRI in the presence of task-correlated breathing variations. Neuroimage 47, 1092–1104. https://doi.org/10.1016/j.neuroimage.2009.05.030.

Biswal, B., Yetkin, F.Z., Haughton, V.M., Hyde, J.S., 1995. Functional connectivity in the motor cortex of resting human brain using echo-planar MRI. Magn. Reson. Med. Off. J. Soc. Magn. Reson. Med. Soc. Magn. Reson. Med. 34, 537–541.

Biswal, B.B., Pathak, A.P., Ulmer, J.L., Hudetz, A.G., 2003. Decoupling of the hemodynamic and activation-induced delays in functional magnetic resonance imaging. J. Comput. Assist. Tomogr. 27, 219–225.

Biswal, B.B., Mennes, M., Zuo, X.N., Gohel, S., Kelly, C., Smith, S.M., Beckmann, C.F., Adelstein, J.S., Buckner, R.L., Colcombe, S., Dogonowski, A.M., Ernst, M., Fair, D.,

Hampson, M., Hoptman, M.J., Hyde, J.S., Kiviniemi, V.J., Kotter, R., Li, S.J., Lin, C.P., Lowe, M.J., Mackay, C., Madden, D.J., Madsen, K.H., Margulies, D.S., Mayberg, H.S., McMahon, K., Monk, C.S., Mostofsky, S.H., Nagel, B.J., Pekar, J.J., Peltier, S.J., Petersen, S.E., Riedl, V., Rombouts, S.A., Rypma, B., Schlaggar, B.L., Schmidt, S., Seidler, R.D., Siegle, G.J., Sorg, C., Teng, G.J., Veijola, J., Villringer, A., Walter, M., Wang, L., Weng, X.C., Whitfield-Gabrieli, S., Williamson, P., Windischberger, C., Zang, Y.F., Zhang, H.Y., Castellanos, F.X., Milham, M.P., 2010. Toward discovery science of human brain function. Proc. Natl. Acad. Sci. U. S. A. 107, 4734–4739. https://doi.org/10.1073/pnas.0911855107.

Bolt, T., Nomi, J.S., Bzdok, D., Salas, J.A., Chang, C., Thomas Yeo, B.T., Uddin, L.Q., Keilholz, S.D., 2022. A parsimonious description of global functional brain organization in three spatiotemporal patterns. Nat. Neurosci. 25 (8), 1093–1103. https://doi.org/10.1038/s41593-022-01118-1.

Buxton, R.B., 2012. Dynamic models of BOLD contrast. Neuroimage 62, 953–961. https://doi.org/10.1016/j.neuroimage.2012.01.012.

Cai, R., Pan, C., Ghasemigharagoz, A., Todorov, M.I., Förstera, B., Zhao, S., Bhatia, H.S., Parra-Damas, A., Mrowka, L., Theodorou, D., 2019. Panoptic imaging of transparent mice reveals whole-body neuronal projections and skull-meninges connections. Nat. Neurosci. 22, 317–327.

Chang, C., Leopold, D.A., Scholvinck, M.L., Mandelkow, H., Picchioni, D., Liu, X., Ye, F.Q., Turchi, J.N., Duyn, J.H., 2016. Tracking brain arousal fluctuations with fMRI. Proc. Natl. Acad. Sci. U. S. A. 113, 4518–4523. https://doi.org/10.1073/pnas.1520613113.

Chow, H.M., Horovitz, S.G., Carr, W.S., Picchioni, D., Coddington, N., Fukunaga, M., Xu, Y., Balkin, T.J., Duyn, J.H., Braun, A.R., 2013. Rhythmic alternating patterns of brain activity distinguish rapid eye movement sleep from other states of consciousness. Proc. Natl. Acad. Sci. U. S. A. 110, 10300–10305. https://doi.org/10.1073/pnas.1217691110.

Clarke, D.D., Sokoloff, L., 1999. Substrates of cerebral metabolism. In: Siegel, G.J., Agranoff, B.W., Albers, R.W., et al. (Eds.), Basic Neurochemistry: Molecular, Cellular and Medical Aspects, sixth ed. Lippincott-Raven, Philadelphia.

Ding, F., O'Donnell, J., Xu, Q., Kang, N., Goldman, N., Nedergaard, M., 2016. Changes in the composition of brain interstitial ions control the sleep-wake cycle. Science 352, 550–555. https://doi.org/10.1126/science.aad4821.

Dreha-Kulaczewski, S., Joseph, A.A., Merboldt, K.-D., Ludwig, H.-C., Gärtner, J., Frahm, J., 2015. Inspiration is the major regulator of human CSF flow. J. Neurosci. Off. J. Soc. Neurosci. 35, 2485–2491. https://doi.org/10.1523/JNEUROSCI.3246-14.2015.

Dreha-Kulaczewski, S., Joseph, A.A., Merboldt, K.-D., Ludwig, H.-C., Gärtner, J., Frahm, J., 2017. Identification of the upward movement of human CSF in vivo and its relation to the brain venous system. J. Neurosci. Off. J. Soc. Neurosci. 37, 2395–2402. https://doi.org/10.1523/JNEUROSCI.2754-16.2017.

Drew, P.J., Shih, A.Y., Kleinfeld, D., 2011. Fluctuating and sensory-induced vasodynamics in rodent cortex extend arteriole capacity. Proc. Natl. Acad. Sci. U. S. A. 108, 8473–8478. https://doi.org/10.1073/pnas.1100428108.

Duyn, J.H., 1997. Steady state effects in fast gradient echo magnetic resonance imaging. Magn. Reson. Med. 37, 559–568. https://doi.org/10.1002/mrm.1910370414.

Eide, P.K., Ringstad, G., 2019. Delayed clearance of cerebrospinal fluid tracer from entorhinal cortex in idiopathic normal pressure hydrocephalus: a glymphatic magnetic resonance imaging study. J. Cereb. Blood Flow Metab. 39, 1355–1368.

Elabasy, A., Suhonen, M., Rajna, Z., Hosni, Y., Kananen, J., Annunen, J., Ansakorpi, H., Korhonen, V., Seppänen, T., Kiviniemi, V., 2023. Respiratory brain impulse propagation in focal epilepsy. Sci. Rep. 13 (1), 5222. https://doi.org/10.1038/s41598-023-32271-7. PMID: 36997658; PMCID: PMC10063583.

Fox, P.T., Raichle, M.E., 1986. Focal physiological uncoupling of cerebral blood flow and oxidative metabolism during somatosensory stimulation in human subjects. Proc. Natl. Acad. Sci. U. S. A. 83, 1140–1144.

Fox, M.D., Raichle, M.E., 2007. Spontaneous fluctuations in brain activity observed with functional magnetic resonance imaging. Nat. Rev. Neurosci. 8, 700–711. https://doi.org/10.1038/nrn2201.

Fukunaga, M., Horovitz, S.G., van Gelderen, P., de Zwart, J.A., Jansma, J.M., Ikonomidou, V.N., Chu, R., Deckers, R.H., Leopold, D.A., Duyn, J.H., 2006. Large-amplitude, spatially correlated fluctuations in BOLD fMRI signals during extended rest and early sleep stages. Magn. Reson. Imaging 24, 979–992.

Fultz, N.E., Bonmassar, G., Setsompop, K., Stickgold, R.A., Rosen, B.R., Polimeni, J.R., Lewis, L.D., 2019. Coupled electrophysiological, hemodynamic, and cerebrospinal fluid oscillations in human sleep. Science 366, 628–631.

Gisolf, J., van Lieshout, J.J., van Heusden, K., Pott, F., Stok, W.J., Karemaker, J.M., 2004. Human cerebral venous outflow pathway depends on posture and central venous pressure. J. Physiol. 560, 317–327. https://doi.org/10.1113/jphysiol.2004.070409.

Gong, J., Wang, J., Luo, X., Chen, G., Huang, H., Huang, R., Huang, L., Wang, Y., 2020a. Abnormalities of intrinsic regional brain activity in first-episode and chronic schizophrenia: a meta-analysis of resting-state functional MRI. J. Psychiatry Neurosci. JPN 45, 55–68. https://doi.org/10.1503/jpn.180245.

Gong, J., Wang, J., Qiu, S., Chen, P., Luo, Z., Wang, J., Huang, L., Wang, Y., 2020b. Common and distinct patterns of intrinsic brain activity alterations in major depression and bipolar disorder: voxel-based meta-analysis. Transl. Psychiatry 10, 353. https://doi.org/10.1038/s41398-020-01036-5.

Gore, J.C., Li, M., Gao, Y., Wu, T.-L., Schilling, K.G., Huang, Y., Mishra, A., Newton, A.T., Rogers, B.P., Chen, L.M., Anderson, A.W., Ding, Z., 2019. Functional MRI and resting state connectivity in white matter—a mini-review. Magn. Reson. Imaging 63, 1–11. https://doi.org/10.1016/j.mri.2019.07.017.

Gouveia-Freitas, K., Bastos-Leite, A.J., 2021. Perivascular spaces and brain waste clearance systems: relevance for neurodegenerative and cerebrovascular pathology. Neuroradiology 63, 1581–1597. https://doi.org/10.1007/s00234-021-02718-7.

Grubb, S., Cai, C., Hald, B.O., Khennouf, L., Murmu, R.P., Jensen, A.G.K., Fordsmann, J., Zambach, S., Lauritzen, M., 2020. Precapillary sphincters maintain perfusion in the cerebral cortex. Nat. Commun. 11, 395. https://doi.org/10.1038/s41467-020-14330-z.

Hadaczek, P., Yamashita, Y., Mirek, H., Tamas, L., Bohn, M.C., Noble, C., Park, J.W., Bankiewicz, K., 2006. The "perivascular pump" driven by arterial pulsation is a powerful mechanism for the distribution of therapeutic molecules within the brain. Mol. Ther. J. Am. Soc. Gene Ther. 14, 69–78. https://doi.org/10.1016/j.ymthe.2006.02.018.

Hennig, J., Zhong, K., Speck, O., 2007. MR-Encephalography: fast multi-channel monitoring of brain physiology with magnetic resonance. Neuroimage 34, 212–219.

Helakari, H., Korhonen, V., Holst, S.C., Piispala, J., Kallio, M., Vayrynen, T., Huotari, N., Raitamaa, L., Kananen, J., Jarvela, M., et al., 2022. Sleep-specific changes in physiological brain pulsations. J. Neurosci. 42 (12), 2503–2515. https://doi.org/10.1523/JNEUROSCI.0934-21.2022.

Hennig, J., Kiviniemi, V., Riemenschneider, B., Barghoorn, A., Akin, B., Wang, F., LeVan, P., 2021. 15 Years MR-encephalography. Magn. Reson. Mater. Phys. Biol. Med. 34, 85–108. https://doi.org/10.1007/s10334-020-00891-z.

Herrero, J.L., Khuvis, S., Yeagle, E., Cerf, M., Mehta, A.D., 2018. Breathing above the brain stem: volitional control and attentional modulation in humans. J. Neurophysiol. 119, 145–159.

Hladky, S.B., Barrand, M.A., 2014. Mechanisms of fluid movement into, through and out of the brain: evaluation of the evidence. Fluids Barriers CNS 11, 26.

Horovitz, S.G., Fukunaga, M., de Zwart, J.A., van Gelderen, P., Fulton, S.C., Balkin, T.J., Duyn, J.H., 2008. Low frequency BOLD fluctuations during resting wakefulness and light sleep: a simultaneous EEG-fMRI study. Hum. Brain Mapp. 29, 671–682. https://doi.org/10.1002/hbm.20428.

Huotari, N., Raitamaa, L., Helakari, H., Kananen, J., Raatikainen, V., Rasila, A., Tuovinen, T., Kantola, J., Borchardt, V., Kiviniemi, V.J., Korhonen, V.O., 2019. Sampling Rate Effects on Resting State fMRI Metrics. Front. Neurosci. 13, 279. https://doi.org/10.3389/fnins.2019.00279.

Huotari, N., Tuunanen, J., Raitamaa, L., Raatikainen, V., Kananen, J., Helakari, H., Tuovinen, T., Järvelä, M., Kiviniemi, V., Korhonen, V., 2022. Cardiovascular pulsatility increases in visual cortex before blood oxygen level dependent response during stimulus. Front. Neurosci. 16, 836378. https://doi.org/10.3389/fnins.2022.836378.

Hussein, A., Matthews, J.L., Syme, C., Macgowan, C., MacIntosh, B.J., Shirzadi, Z., Pausova, Z., Paus, T., Chen, J.J., 2020. The association between resting-state functional magnetic resonance imaging and aortic pulse-wave velocity in healthy adults. Hum. Brain Mapp. 41, 2121–2135. https://doi.org/10.1002/hbm.24934.

Iliff, J.J., Wang, M., Zeppenfeld, D.M., Venkataraman, A., Plog, B.A., Liao, Y., Deane, R., Nedergaard, M., 2013. Cerebral arterial pulsation drives paravascular CSF-interstitial fluid exchange in the murine brain. J. Neurosci. Off. J. Soc. Neurosci. 33, 18190–18199. https://doi.org/10.1523/JNEUROSCI.1592-13.2013.

Jacob, L., Boisserand, L.S.B., Geraldo, L.H.M., de Brito Neto, J., Mathivet, T., Antila, S., Barka, B., Xu, Y., Thomas, J.-M., Pestel, J., 2019. Anatomy and function of the vertebral column lymphatic network in mice. Nat. Commun. 10, 1–16.

Jahanian, H., Ni, W.W., Christen, T., Moseley, M.E., Tamura, M.K., Zaharchuk, G., 2014. Spontaneous BOLD Signal Fluctuations in Young Healthy Subjects and Elderly Patients with Chronic Kidney Disease. PLoS One 9. https://doi.org/10.1371/journal.pone.0092539.

Jessen, N.A., Munk, A.S.F., Lundgaard, I., Nedergaard, M., 2015. The glymphatic system: a beginner's guide. Neurochem. Res. 40, 2583–2599. https://doi.org/10.1007/s11064-015-1581-6.

Kananen, J., Tuovinen, T., Ansakorpi, H., Rytky, S., Helakari, H., Huotari, N., Raitamaa, L., Raatikainen, V., Rasila, A., Borchardt, V., Korhonen, V., LeVan, P., Nedergaard, M., Kiviniemi, V., 2018. Altered physiological brain variation in drug-resistant epilepsy. Brain Behav. 8, e01090. https://doi.org/10.1002/brb3.1090.

Kananen, J., Helakari, H., Korhonen, V., Huotari, N., Järvelä, M., Raitamaa, L., Raatikainen, V., Rajna, Z., Tuovinen, T., Nedergaard, M., 2020. Respiratory-related brain pulsations are increased in epilepsy—a two-centre functional MRI study. Brain Commun. 2, fcaa076.

Kiviniemi, V., Jauhiainen, J., Tervonen, O., Pääkkö, E., Oikarinen, J., Vainionpää, V., Rantala, H., Biswal, B., 2000. Slow vasomotor fluctuation in fMRI of anesthetized child brain. Magn. Reson. Med. Off. J. Soc. Magn. Reson. Med. Soc. Magn. Reson. Med. 44, 373–378.

Kiviniemi, V., Kantola, J.-H., Jauhiainen, J., Hyvärinen, A., Tervonen, O., 2003. Independent component analysis of nondeterministic fMRI signal sources. Neuroimage 19, 253–260.

Kiviniemi, V., 2004. Spontaneous Blood Oxygen Fluctuation in Awake and Sedated Brain Cortex – A BOLD fMRI Study. http://urn.fi/urn:isbn:9514273885. (Accessed 18 June 2004).

Kiviniemi, V., Haanpää, H., Kantola, J.-H., Jauhiainen, J., Vainionpää, V., Alahuhta, S., Tervonen, O., 2005. Midazolam sedation increases fluctuation and synchrony of the resting brain BOLD signal. Magn. Reson. Imaging 23, 531–537. https://doi.org/10.1016/j.mri.2005.02.009.

Kiviniemi, V., Wang, X., Korhonen, V., Keinänen, T., Tuovinen, T., Autio, J., LeVan, P., Keilholz, S., Zang, Y.-F., Hennig, J., Nedergaard, M., 2016. Ultra-fast magnetic resonance encephalography of physiological brain activity—Glymphatic pulsation mechanisms? J. Cereb. Blood Flow Metab. 36, 1033–1045. https://doi.org/10.1177/0271678X15622047.

Kucewicz, J.C., Dunmire, B., Leotta, D.F., Panagiotides, H., Paun, M., Beach, K.W., 2007. Functional tissue pulsatility imaging of the brain during visual stimulation. Ultrasound Med. Biol. 33, 681–690. https://doi.org/10.1016/j.ultrasmedbio.2006.11.008.

Kwong, K.K., Belliveau, J.W., Chesler, D.A., Goldberg, I.E., Weisskoff, R.M., Poncelet, B.P., Kennedy, D.N., Hoppel, B.E., Cohen, M.S., Turner, R., 1992. Dynamic magnetic resonance imaging of human brain activity during primary sensory stimulation. Proc. Natl. Acad. Sci. U. S. A. 89, 5675–5679.

Lee, H., Xie, L., Yu, M., Kang, H., Feng, T., Deane, R., Logan, J., Nedergaard, M., Benveniste, H., 2015. The effect of body posture on brain glymphatic transport. J. Neurosci. Off. J. Soc. Neurosci. 35, 11034–11044. https://doi.org/10.1523/JNEUROSCI.1625-15.2015.

Lin, F.-H., Tsai, K.W.K., Chu, Y.-H., Witzel, T., Nummenmaa, A., Raij, T., Ahveninen, J., Kuo, W.-J., Belliveau, J.W., 2012. Ultrafast inverse imaging techniques for fMRI. Neuroimage 62, 699–705.

Liu, X., de Zwart, J.A., Schölvinck, M.L., Chang, C., Frank, Q.Y., Leopold, D.A., Duyn, J.H., 2018. Subcortical evidence for a contribution of arousal to fMRI studies of brain activity. Nat. Commun. 9, 395.

Logothetis, N.K., Pauls, J., Augath, M., Trinath, T., Oeltermann, A., 2001. Neurophysiological investigation of the basis of the fMRI signal. Nature 412, 150–157.

Ma, Y., Shaik, M.A., Kozberg, M.G., Kim, S.H., Portes, J.P., Timerman, D., Hillman, E.M., 2016. Resting-state hemodynamics are spatiotemporally coupled to synchronized and symmetric neural activity in excitatory neurons. Proc. Natl. Acad. Sci. U. S. A. 113, E8463–E8471. https://doi.org/10.1073/pnas.1525369113.

Ma, M., Zhang, H., Liu, R., Liu, H., Yang, X., Yin, X., Chen, S., Wu, X., 2020. Static and dynamic changes of amplitude of low-frequency fluctuations in cervical discogenic pain. Front. Neurosci. 14, 733. https://doi.org/10.3389/fnins.2020.00733.

Mahon, P., Greene, B.R., Lynch, E.M., McNamara, B., Shorten, G.D., 2008. Can state or response entropy be used as a measure of sleep depth? Anaesthesia 63, 1309–1313.

Makedonov, I., Black, S.E., Macintosh, B.J., 2013a. BOLD fMRI in the white matter as a marker of aging and small vessel disease. PLoS One 8, e67652. https://doi.org/10.1371/journal.pone.0067652.

Makedonov, I., Black, S.E., MacIntosh, B.J., 2013b. Cerebral small vessel disease in aging and Alzheimer's disease: a comparative study using MRI and SPECT. Eur. J. Neurol. 20, 243–250.

Makedonov, I., Chen, J.J., Masellis, M., MacIntosh, B.J., Alzheimer's Disease Neuroimaging Initiative, 2016. Physiological fluctuations in white matter are increased in Alzheimer's disease and correlate with neuroimaging and cognitive biomarkers. Neurobiol. Aging 37, 12–18. https://doi.org/10.1016/j.neurobiolaging.2015.09.010.

Malik, V., Smith, D., Lee-Chiong, T., 2012. Respiratory physiology during sleep. Sleep Med. Clin. 7, 497–505. https://doi.org/10.1016/j.jsmc.2012.06.011.

Mao, D., Ding, Z., Jia, W., Liao, W., Li, X., Huang, H., Yuan, J., Zang, Y.-F., Zhang, H., 2015. Low-frequency fluctuations of the resting brain: high magnitude does not equal high reliability. PLoS One 10, e0128117. https://doi.org/10.1371/journal.pone.0128117.

Mascali, D., DiNuzzo, M., Gili, T., Moraschi, M., Fratini, M., Maraviglia, B., Serra, L., Bozzali, M., Giove, F., 2015. Intrinsic patterns of coupling between correlation and amplitude of low-frequency fMRI fluctuations are disrupted in degenerative dementia mainly due to functional disconnection. PLoS One 10, e0120988. https://doi.org/10.1371/journal.pone.0120988.

Mestre, H., Du, T., Sweeney, A.M., Liu, G., Samson, A.J., Peng, W., Mortensen, K.N., Stæger, F.F., Bork, P.A.R., Bashford, L., Toro, E.R., Tithof, J., Kelley, D.H., Thomas, J.H., Hjorth, P.G., Martens, E.A., Mehta, R.I., Solis, O., Blinder, P., Kleinfeld, D., Hirase, H., Mori, Y., Nedergaard, M., 2020. Cerebrospinal fluid influx drives acute ischemic tissue swelling. Science 367 (6483), eaax7171. https://doi.org/10.1126/science.aax7171. Epub 2020 Jan 30. PMID: 32001524; PMCID: PMC7375109.

Mestre, H., Tithof, J., Du, T., Song, W., Peng, W., Sweeney, A.M., Olveda, G., Thomas, J.H., Nedergaard, M., Kelley, D.H., 2018. Flow of cerebrospinal fluid is driven by arterial pulsations and is reduced in hypertension. Nat. Commun. 9, 4878. https://doi.org/10.1038/s41467-018-07318-3.

Miyakawa, K., Koepchen, H., Polosa, C., 1984. Mechanisms of Blood Pressure Waves. Japan Scientific Societies Press, Japan, pp. 1–360.

Mosso, A., 1881. Concerning the Circulation of the Blood in the Human Brain. Verlag von Viet & Company, Leipzig.

Nakano, T., Tominaga, R., Nagano, I., Okabe, H., Yasui, H., 2000. Pulsatile flow enhances endothelium-derived nitric oxide release in the peripheral vasculature. Am. J. Physiol.-Heart Circ. Physiol. 278, H1098–H1104. https://doi.org/10.1152/ajpheart.2000.278.4.H1098.

Natali, F., Dolce, C., Peters, J., Gerelli, Y., Stelletta, C., Leduc, G., 2013. Water dynamics in neural tissue. J. Physical Soc. Japan 82, SA017. https://doi.org/10.7566/JPSJS.82SA.SA017.

Nation, D.A., Sweeney, M.D., Montagne, A., Sagare, A.P., D'Orazio, L.M., Pachicano, M., Sepehrband, F., Nelson, A.R., Buennagel, D.P., Harrington, M.G., 2019. Blood–brain barrier breakdown is an early biomarker of human cognitive dysfunction. Nat. Med. 25, 270–276.

Nedergaard, M., 2013. Neuroscience. Garbage truck of the brain. Science 340, 1529–1530. https://doi.org/10.1126/science.1240514.

Nedergaard, M., Goldman, S.A., 2020. Glymphatic failure as a final common pathway to dementia. Science 370, 50–56.

Ogawa, S., Lee, T.M., Kay, A.R., Tank, D.W., 1990. Brain magnetic resonance imaging with contrast dependent on blood oxygenation. Proc. Natl. Acad. Sci. U. S. A. 87, 9868–9872.

Ogawa, S., Tank, D.W., Menon, R., Ellermann, J.M., Kim, S.-G., Merkle, H., Ugurbil, K., 1992. Intrinsic signal changes accompanying sensory stimulation: functional brain mapping with magnetic resonance imaging. Proc. Natl. Acad. Sci. U. S. A. 89, 5951–5955.

Özbay, P.S., Chang, C., Picchioni, D., Mandelkow, H., Chappel-Farley, M.G., van Gelderen, P., de Zwart, J.A., Duyn, J., 2019. Sympathetic activity contributes to the fMRI signal. Commun. Biol. 2, 421. https://doi.org/10.1038/s42003-019-0659-0.

Patel, D.C., Tewari, B.P., Chaunsali, L., Sontheimer, H., 2019. Neuron-glia interactions in the pathophysiology of epilepsy. Nat. Rev. Neurosci. 20, 282–297.

Pauling, L., Coryell, C.D., 1936. The magnetic properties and structure of hemoglobin, oxyhemoglobin and carbonmonoxyhemoglobin. Proc. Natl. Acad. Sci. U. S. A. 22 (4), 210–216. https://doi.org/10.1073/pnas.22.4.210.

Picchioni, D., Özbay, P.S., Mandelkow, H., de Zwart, J.A., Wang, Y., van Gelderen, P., Duyn, J.H., 2022. Autonomic arousals contribute to brain fluid pulsations during sleep. Neuroimage 249, 118888. https://doi.org/10.1016/j.neuroimage.2022.118888.

Posse, S., Ackley, E., Mutihac, R., Zhang, T., Hummatov, R., Akhtari, M., Chohan, M., Fisch, B., Yonas, H., 2013. High-speed real-time resting-state FMRI using multi-slab echo-volumar imaging. Front. Hum. Neurosci. 7, 479. https://doi.org/10.3389/fnhum.2013.00479.

Preiss, G., Polosa, C., 1974. Patterns of sympathetic neuron activity associated with Mayer waves. Am. J. Physiol.-Leg. Content 226, 724–730.

Raitamaa, L., Huotari, N., Korhonen, V., Helakari, H., Koivula, A., Kananen, J., Kiviniemi, V., 2021. Spectral analysis of physiological brain pulsations affecting the BOLD signal. Hum. Brain Mapp. 42 (13), 4298–4313. https://doi.org/10.1002/hbm.25547.

Rajna, Z., Mattila, H., Huotari, N., Tuovinen, T., Krüger, J., Holst, S.C., Korhonen, V., Remes, A.M., Seppänen, T., Hennig, J., 2021. Cardiovascular brain impulses in Alzheimer's disease. Brain 144 (7), 2214–2226. https://doi.org/10.1093/brain/awab144.

Rasmussen, R., Nicholas, E., Petersen, N.C., Dietz, A.G., Xu, Q., Sun, Q., Nedergaard, M., 2019. Cortex-wide changes in extracellular potassium ions parallel brain state transitions in awake behaving mice. Cell Rep. 28, 1182–1194.e4. https://doi.org/10.1016/j.celrep.2019.06.082.

Ringstad, G., Eide, P.K., 2020. Cerebrospinal fluid tracer efflux to parasagittal dura in humans. Nat. Commun. 11, 1–9.

Rivera-Rivera, L.A., Schubert, T., Turski, P., Johnson, K.M., Berman, S.E., Rowley, H.A., Carlsson, C.M., Johnson, S.C., Wieben, O., 2017. Changes in intracranial venous blood flow and pulsatility in Alzheimer's disease: a 4D flow MRI study. J. Cereb. Blood Flow Metab. 37, 2149–2158. https://doi.org/10.1177/0271678X16661340.

Roy, C.S., Sherrington, C.S., 1890. On the regulation of the blood-supply of the brain. J. Physiol. 11. 85–158.17.

Santisakultarm, T.P., Cornelius, N.R., Nishimura, N., Schafer, A.I., Silver, R.T., Doerschuk, P.C., Olbricht, W.L., Schaffer, C.B., 2012. In vivo two-photon excited fluorescence microscopy reveals cardiac- and respiration-dependent pulsatile blood flow in cortical blood vessels in mice. Am. J. Physiol. Heart Circ. Physiol. 302, H1367–H1377. https://doi.org/10.1152/ajpheart.00417.2011.

Smith, S.M., Fox, P.T., Miller, K.L., Glahn, D.C., Fox, P.M., Mackay, C.E., Filippini, N., Watkins, K.E., Toro, R., Laird, A.R., Beckmann, C.F., 2009. Correspondence of the brain's functional architecture during activation and rest. Proc. Natl. Acad. Sci. U. S. A. 106, 13040–13045.

Sowho, M., Amatoury, J., Kirkness, J.P., Patil, S.P., 2014. Sleep and respiratory physiology in adults. Clin. Chest Med. 35, 469–481. https://doi.org/10.1016/j.ccm.2014.06.002.

Takenaka, S., Kan, S., Seymour, B., Makino, T., Sakai, Y., Kushioka, J., Tanaka, H., Watanabe, Y., Shibata, M., Yoshikawa, H., Kaito, T., 2020. Resting-state amplitude of low-frequency fluctuation is a potentially useful prognostic functional biomarker in cervical myelopathy. Clin. Orthop. 478, 1667–1680. https://doi.org/10.1097/CORR.0000000000001157.

Tsai, Y.-H., Yuan, R., Huang, Y.-C., Weng, H.-H., Yeh, M.-Y., Lin, C.-P., Biswal, B.B., 2014. Altered resting-state fMRI signals in acute stroke patients with ischemic penumbra. PLoS One 9, e105117. https://doi.org/10.1371/journal.pone.0105117.

Tuovinen, T., Kananen, J., Rajna, Z., Lieslehto, J., Korhonen, V., Rytty, R., Mattila, H., Huotari, N., Raitamaa, L., Helakari, H., 2020. The variability of functional MRI brain signal increases in Alzheimer's disease at cardiorespiratory frequencies. Sci. Rep. 10, 1–11.

van Veluw, S.J., Hou, S.S., Calvo-Rodriguez, M., Arbel-Ornath, M., Snyder, A.C., Frosch, M.P., Greenberg, S.M., Bacskai, B.J., 2020. Vasomotion as a driving force for paravascular clearance in the awake mouse brain. Neuron 105. 549–561.e5.

Vinje, V., Ringstad, G., Lindstrøm, E.K., Valnes, L.M., Rognes, M.E., Eide, P.K., Mardal, K.-A., 2019. Respiratory influence on cerebrospinal fluid flow–a computational study based on long-term intracranial pressure measurements. Sci. Rep. 9, 1–13.

Vladić, A., Klarica, M., Bulat, M., 2009. Dynamics of distribution of 3H-inulin between the cerebrospinal fluid compartments. Brain Res. 1248, 127–135. https://doi.org/10.1016/j.brainres.2008.10.044.

von Holstein-Rathlou, S., Petersen, N.C., Nedergaard, M., 2018. Voluntary running enhances glymphatic influx in awake behaving, young mice. Neurosci. Lett. 662, 253–258. https://doi.org/10.1016/j.neulet.2017.10.035.

von Schulthess, G.K., Higgins, C.B., 1985. Blood flow imaging with MR: spin-phase phenomena. Radiology 157, 687–695. https://doi.org/10.1148/radiology.157.3.2997836.

Wagshul, M.E., Eide, P.K., Madsen, J.R., 2011. The pulsating brain: a review of experimental and clinical studies of intracranial pulsatility. Fluids Barriers CNS 8, 5. https://doi.org/10.1186/2045-8118-8-5.

Wang, J., Zhang, J.-R., Zang, Y.-F., Wu, T., 2018. Consistent decreased activity in the putamen in Parkinson's disease: a meta-analysis and an independent validation of resting-state fMRI. Giga. Science 7, giy071. https://doi.org/10.1093/gigascience/giy071.

Wilson, M.H., 2016. Monro-Kellie 2.0: the dynamic vascular and venous pathophysiological components of intracranial pressure. J. Cereb. Blood Flow Metab. 36, 1338–1350. https://doi.org/10.1177/0271678X16648711.

Windischberger, C., Langenberger, H., Sycha, T., Tschernko, E.M., Fuchsjäger-Mayerl, G., Schmetterer, L., Moser, E., 2002. On the origin of respiratory artifacts in BOLD-EPI of the human brain. Magn. Reson. Imaging 20, 575–582. https://doi.org/10.1016/S0730-725X(02)00563-5.

Wise, R.G., Ide, K., Poulin, M.J., Tracey, I., 2004. Resting fluctuations in arterial carbon dioxide induce significant low frequency variations in BOLD signal. Neuroimage 21, 1652–1664. https://doi.org/10.1016/j.neuroimage.2003.11.025.

Xie, L., Kang, H., Xu, Q., Chen, M.J., Liao, Y., Thiyagarajan, M., O'Donnell, J., Christensen, D.J., Nicholson, C., Iliff, J.J., Takano, T., Deane, R., Nedergaard, M., 2013. Sleep drives metabolite clearance from the adult brain. Science 342, 373–377. https://doi.org/10.1126/science.1241224.

Yang, L., Yan, Y., Wang, Y., Hu, X., Lu, J., Chan, P., Yan, T., Han, Y., 2018. Gradual disturbances of the amplitude of low-frequency fluctuations (ALFF) and fractional ALFF in Alzheimer spectrum. Front. Neurosci. 12, 975. https://doi.org/10.3389/fnins.2018.00975.

Yuan, H., Zotev, V., Phillips, R., Bodurka, J., 2013. Correlated slow fluctuations in respiration, EEG, and BOLD fMRI. Neuroimage 79, 81–93. https://doi.org/10.1016/j.neuroimage.2013.04.068.

Yu-Feng, Z., Yong, H., Chao-Zhe, Z., Qing-Jiu, C., Man-Qiu, S., Meng, L., Li-Xia, T., Tian-Zi, J., Yu-Feng, W., 2007. Altered baseline brain activity in children with ADHD revealed by resting-state functional MRI. Brain Dev. 29, 83–91. https://doi.org/10.1016/j.braindev.2006.07.002.

Zelano, C., Jiang, H., Zhou, G., Arora, N., Schuele, S., Rosenow, J., Gottfried, J.A., 2016. Nasal respiration entrains human limbic oscillations and modulates cognitive function. J. Neurosci. Off. J. Soc. Neurosci. 36, 12448–12467. https://doi.org/10.1523/JNEUROSCI.2586-16.2016.

Zeppenfeld, D.M., Simon, M., Haswell, J.D., D'Abreo, D., Murchison, C., Quinn, J.F., Grafe, M.R., Woltjer, R.L., Kaye, J., Iliff, J.J., 2017. Association of perivascular localization of aquaporin-4 with cognition and Alzheimer disease in aging brains. JAMA Neurol. 74, 91–99. https://doi.org/10.1001/jamaneurol.2016.4370.

Zhang, E.T., Inman, C.B., Weller, R.O., 1990. Interrelationships of the pia mater and the perivascular (Virchow-Robin) spaces in the human cerebrum. J. Anat. 170, 111–123.

Zhang, X., Xue, C., Cao, X., Yuan, Q., Qi, W., Xu, W., Zhang, S., Huang, Q., 2021. Altered patterns of amplitude of low-frequency fluctuations and fractional amplitude of low-frequency fluctuations between amnestic and vascular mild cognitive impairment: an ALE-based comparative meta-analysis. Front. Aging Neurosci. 13, 711023. https://doi.org/10.3389/fnagi.2021.711023.

Zhou, F., Gu, L., Hong, S., Liu, J., Jiang, J., Huang, M., Zhang, Y., Gong, H., 2018. Altered low-frequency oscillation amplitude of resting state-fMRI in patients with discogenic low-back and leg pain. J. Pain Res. 11, 165–176. https://doi.org/10.2147/JPR.S151562.

Systemic low-frequency oscillations in resting-state fMRI

Yunjie Tong[a] and Lia M. Hocke[b]
[a]Weldon School of Biomedical Engineering, Purdue University, West Lafayette, IN, United States, [b]McLean Hospital, Harvard Medical School, Belmont, MA, United States

Characteristics of low-frequency oscillations (LFOs)

Low-frequency oscillations (LFOs) in blood-oxygenation-level-dependent (BOLD) (Dreha-Kulaczewski et al., 2015) fMRI have been studied extensively, but their origin and even definition are still debated. In this chapter, we will therefore apply two criteria to our definition of LFOs: the frequency band of the signal and whether the signal is moving instead of being stationary (immobile). **Frequency content**: The first criterion is simply a definition. LFOs are signals that occur in the brain (and in some cases throughout the body) that have frequencies between ~0.009 and 0.2 Hz. The exact endpoints of this band are extremely variable in the literature. The original paper on resting-state connectivity used 0.01–0.1 Hz (Biswal et al., 1995), but later papers expanded this range. For the purpose of this discussion, we will use the range of 0.01–0.15 Hz. Using 0.15 Hz as the top of the range holds a particular significance, as this is traditionally the highest expected frequency in neuronally generated hemodynamic signals (based on the shape of the canonical hemodynamic response function (Josephs and Henson, 1999)). Even though this assumption has been recently challenged by Chen et al. (2021), this band defines the frequency range where spectral filtering cannot be used to remove nonneuronal signals from the expected neuronal ones. **Moving vs stationary**: Compared to stationary LFOs that are confined to particular brain region/regions (e.g., visual stimulus causes activation in the visual cortex), some LFO signals are nonstationary. These signals can stem from neuronal activation. For example, dynamic functional connectivity analysis (Chang and Glover, 2010) has been developed to catch the spatiotemporal dynamics of the resting-state neuronal activation. In this chapter,

Advances in Resting-State Functional MRI. https://doi.org/10.1016/B978-0-323-91688-2.00004-7

we will focus on one unique physiological moving LFO signal, which has the following characteristics: (1) global, as presented in the brain and the periphery, with a likely origin outside of the brain, and (2) nonstationary, as moving in the whole brain. Because of these features, LFOs are referred to as systemic low-frequency oscillations (sLFOs), to be distinguished from other physiological and neuronal oscillations. In this chapter, we will discuss these features in detail, and the possible physiological origins of sLFOs and their applications in resting-state fMRI studies.

Features

Global oscillation

sLFOs are not confined to the brain. The same sLFO signal can be found in the peripheries (i.e., fingertips and toes) by concurrent near infrared spectroscopy (NIRS). For example, in early studies (see Fig. 1) (Tong et al., 2012, 2013), healthy subjects' resting-state fMRI (rs-fMRI) data were acquired, while oxy- and deoxy-hemoglobin concentration ([HbO] and [Hb]) oscillations were obtained at fingertips and toes simultaneously using NIRS. High cross-correlation coefficients were found between brain BOLD and NIRS signals and among NIRS signals at various peripheral sites. More interestingly, the delays (i.e., corresponding to the maximum cross-correlation coefficient) relative to one peripheral site reflect anatomy. First, no clear delays were found in

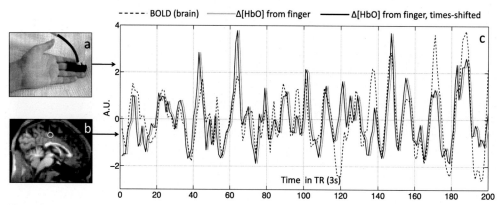

Fig. 1 Concurrent resting-state fMRI and NIRS study. NIRS probe was placed on the finger, as shown in (A). The resting-state fMRI is shown in (B) with an example voxel marked by a *circle*. The simultaneously acquired NIRS signal (i.e., changes in oxyhemoglobin concentration Δ[HbO]) and the BOLD signal from the example voxel are shown in (C). The *dark line* represents the shifted Δ[HbO] signal, which matches the BOLD signal temporally. Modified from Tong, Y., Hocke, L.M., Licata, S.C., Frederick, B., 2012. Low-frequency oscillations measured in the periphery with near-infrared spectroscopy are strongly correlated with blood oxygen level-dependent functional magnetic resonance imaging signals. J. Biomed. Opt. 17(10), 106004.

the sLFO signals from symmetric peripheral sites (left fingertip vs right fingertip), while 2–3 s delays were found between sLFO signals from the fingertip and toe (fingertip sLFO is leading). More importantly, arrival times of the same sLFO wave (taken from the fingertips and toes) vary in the brain by voxel location in a reproducible spatial pattern in the brain (Tong et al., 2013). Since BOLD contrast in fMRI and [HbO]/[Hb] in NIRS are both blood-based (Cui et al., 2011; Strangman et al., 2002; Sassaroli et al., 2006), the sLFO signal must be closely related to blood (i.e., blood flow, volume, etc.), suggesting that sLFOs travel systemically with the blood through the body.

Based on the fMRI/fNIRS literature, the delay differences of sLFO at various anatomic positions support the theory that the sLFO stem from the cardiopulmonary system and travel to different sites along the vascular pathway. The vascular system is complex, with numerous pathways leading to various arrival times of sLFOs throughout the body. For instance, the vascular pathways are symmetric across the midline of the body, which explains why similar LFO arrival times are found between left and right fingertips or toes. On the same note, the pathway from the cardiopulmonary system to fingertips is shorter than that from the cardiopulmonary system to toes, which explains the later arrival time at the toes (Li et al., 2018, 2020). fMRI studies of the large vessels, including arteries and veins in the neck, further support the theory by showing that the sLFO appears in the internal carotid arteries (ICAs) seconds before it does in the brain tissue (represented by the fMRI global mean).

Additionally, regional hemodynamic oscillations along different pathways can modulate the sLFO's shape. For example, the waveform of the sLFO signal detected at the toe is slightly different from that detected at the fingertip (in addition to the delay), which indicates the different "noise" added along these two pathways. Similarly, as the sLFOs move through the cerebrovasculature, the most complex vascular system in the body, regional neuronal/physiological oscillations in the brain will modulate the waveform of sLFO, leading to different BOLD signals (i.e., dynamic signals) at different voxels. Here, we use "dynamic sLFO signal" to represent the traveling wave with modulated waveform.

Dynamic sLFO and blood flow-like patterns

- **Spatiotemporal moving patterns of sLFO**
  In rs-fMRI studies, extensive effort has been made to understand the moving patterns of sLFOs in various populations (Tong et al., 2015; Tong and Frederick, 2010, 2014; Fitzgerald et al., 2021; Wu et al., 2017; Siegel et al., 2016; Ni et al., 2017; Lv et al., 2013, 2019; Christen et al., 2015; Chen et al., 2019a; Khalil et al., 2017, 2020). Research into low-frequency physiological noise in fMRI has established

that a significant (20%–70%) fraction of the low-frequency variance in rs-fMRI data (Liu et al., 2017) can be modeled quite effectively as a single sLFO with varying arrival delays (i.e., the "arrival time" of the sLFO signal at a given voxel) across the brain. With this unique single low-frequency signal, cross-correlation analysis can be carried out with each voxel's low-frequency BOLD signal. The spatiotemporal pattern of the delay is similar to that of blood flow propagating throughout the brain, leading us to hypothesize that the sLFO, not only is related to the blood but also moves with it. To test this hypothesis, a dynamic susceptibility-contrast MRI (DSC) scan was acquired in addition to the rs-fMRI in the same cohort (Chen et al., 2019a; Khalil et al., 2017, 2020; Tong et al., 2017; Tanritanir et al., 2020). The blood arrival map (i.e., delay map) from DSC data was compared to that derived from rs-fMRI data. Largely similar moving patterns are found between these sets of maps. It seems that the sLFO signal acts like an intrinsic bolus that moves through the cerebrovasculature with the blood, much like gadolinium in the DSC case.

These findings support the hypothesis that the sLFO signal at least partly enters the brain with the blood inflow and moves with the blood. However, there is a discrepancy between the DSC and rs-fMRI-based delay maps in some studies, especially in the voxels with earlier arrival times. In other words, the delay map from rs-fMRI is more accurate in assessing blood arrival time for the capillaries, venules, and veins. One possible explanation is that BOLD contrast relies on the concentration of deoxyhemoglobin, which is widely present in the capillaries, venules, and veins. For example, He et al. (2018) found higher correlations among venous voxels in the resting-state BOLD correlation maps, and among arterial voxels in the resting-state CBV-weighted (not BOLD contrast) correlation maps. In the arterioles and arteries, deoxyhemoglobin concentration is low, leading to low signal-to-noise ratio (SNR) for sLFO contrast and rendering less reliable delay values.

The other crucial factor in calculating accurate delay maps is to identify the low-frequency reference signal, to which all BOLD signals are correlated. This LFO signal should be systemic and uncontaminated by neuronal oscillations. Based on these requirements, LFOs extracted from the (1) global mean fMRI signal (GS, which does contain neuronal information, but the reason for its use is the lack of spatial differentiation) (Strangman et al., 2002); (2) averaged fMRI signal in the superior sagittal sinus (SSS); and the (3) concurrent NIRS peripheral signal have been used. However, a more refined and robust procedure has been developed for this purpose (see the Rapidtide Package, 2022).

Here the moving patterns of sLFO are associated with perfusion and blood flow. However, a large body of evidence also associates the propagation patterns with arousal regulation and modulated across different states of vigilance (Gu et al., 2021), while some are associated with dynamic functional connectivity and infraslow electrical activity (Thompson et al., 2014; Abbas et al., 2019).

- **Transit time of sLFO**
  One way to quantify the cerebral blood transit time is to assess the time it takes for the blood to travel from the ICAs to a major draining vein such as the SSS. For healthy subjects, this time is about 6 s, as can be verified by many other imaging modalities (Schreiber et al., 2002, 2005; Gilroy et al., 1963), including contrast-enhanced computed tomography (CT). Under the hypothesis that the sLFOs travel with the blood, it is natural to assume that the delay between sLFO signals observed at ICA and at the SSS would reflect the same value. To assess the feasibility of this approach, studies have been carried out on Myconnectome (Poldrack et al., 2015) data (Tong et al., 2019a) (brain changes over a year from one, 45 year old subject), in which the ICA, SSS, and internal jugular veins (IJV) were identified in rs-fMRI with the help of the T1- and T2-weighted anatomical images. The results revealed the clear pathway of the sLFOs moving through the cerebrovasculature, i.e., ICA → brain (GS) → SSS → IJV, similar to an extrinsic contrast bolus. Later, this finding was confirmed on randomly selected subjects from the Adolescent Brain Cognitive Development (ABCD) (Casey et al., 2018) and Human Connectome Project (HCP) (Van Essen et al., 2013) data (Yao et al., 2019; Wu et al., 2016).

  Interestingly, the sLFO signals found in the ICAs are negatively correlated with those from the brain tissue and big veins. Extra caution was taken to ensure that the negative correlation was not due to the pseudo-periodic nature of the sLFO (see "Using sLFOs to denoising the rs-fMRI data" section). Since there is little BOLD contrast in arterial blood, there must be other contrast in fMRI signals obtained from the arteries (Weisskoff and Kiihne, 1992; Gao and Liu, 2012). Multiple explanations are given for the fMRI contrast in arteries, which are beyond the scope of this chapter (e.g., partial-volume effects, hematocrit fluctuations, and in-flow effects (von Schulthess and Higgins, 1985)).

  Notwithstanding, the physiological mechanisms, which underlie the sLFO generation from the cardiopulmonary system (see "Physiological origins of sLFOs" section), remain unclear. However, the intrinsic, detectable, and traveling oscillations are of great interest to many, since the sLFO can serve as a natural biomarker/bolus for assessing global circulation.

Physiological origins of sLFOs

Physiological contributors to systemic LFOs in fMRI are manifold, including variations in sympathetic nervous system tone, partial pressure of carbon dioxide (CO_2) fluctuations modulated by respiratory and cardiac processes, blood-pressure regulation, low-frequency neuronal "waves," and neuronal waves following motion and gastric motility. However, the source of sLFOs and its mechanism of production and propagation in the body have not been conclusively determined and may in fact be a combination of multiple, independent signals with distinct sources. We will review the current theories and the evidence supporting them below. Previous publications have already reviewed various possible sources (Tong et al., 2019b; Hocke et al., 2016; Sassaroli et al., 2012), so here we will concentrate on signals fitting the characteristics of the sLFOs discussed above including (1) that the signal is nonneuronal; (2) nonstationary (moving); (3) global, as in, present in the brain and the periphery; and (4) that it originates outside of the brain.

Respiration and cardiac pulsation

For a long time, fMRI was sampled at frequencies too slow to resolve the high-frequency signals of cardiac (~1 Hz) and respiratory waveforms (0.3–0.4 Hz) and their harmonics. Part of the signal, acquired with long TRs, was therefore aliased into the low-frequency range (Dagli et al., 1999; Raj et al., 2001; Glover et al., 2000). Advances in fMRI acquisition and hardware (see Chapter 11) have improved to the point that the aliasing of these signals is less an issue regarding the fundamental frequency. However, effects of aliasing cannot be completely neglected due to the higher order harmonics of the waveforms (Chen et al., 2019b). The spatial extent of these aliased signals is mostly confined to large vessels and edge effects (Chen et al., 2019b) and, therefore, overlaps only partially with the spatial distribution of the sLFO described above. Moreover, Hocke et al. have shown in various studies that the sLFO persists with short TR fMRI (0.4 s) acquisition which would capture all but the second harmonics of cardiac waves (>0.5 Hz) as well as NIRS studies which sample above 5 Hz (e.g., Hocke et al., 2016; Sutoko et al., 2019).

Nevertheless, cardiac and respiratory effects can still contribute to the global signal, especially since respiratory challenges create circulatory turbulences, which match the gross vascular anatomy and delay structure of the sLFO. Reproducibility of the delay structure is indeed diminished by removal of either the neuronal or the nonneuronal (e.g., cardiac, respiratory) components by independent component analysis-based denoising (Aso et al., 2017). Numerous variation models have been proposed. The most popular ones, RETROICOR

(Glover et al., 2000), variations of the respiration variation and respiration volume per time (RVT) model by Birn and Chang (Birn et al., 2006, 2008) and variations in heart-rate regressors (Chang et al., 2013), have previously been shown to explain additive variation in data with longer TRs. These models typically incorporate a delay of several seconds or voxel-wise convolutions to match the modeled noise waveforms with the fMRI data, suggesting a moving component to these signals. However, Hocke et al. have found that with short TR (400 ms), the latter variation models only accounted for a very small portion of the low-frequency band in fMRI and are spatially and temporally distinct from the sLFOs, which can be captured in the periphery (Hocke et al., 2016). Moreover, it has been shown that all of the models and variations only account for a very small portion (~27%) of the variance in global rs-fMRI signal fluctuation attributed to respiration and/or motion, with the most explanatory variables being motion estimates and the standard deviation and mean of heart rate (Power et al., 2017). These studies do not exclude the influence of cardiac and respiration on the sLFO; however, there may be other models, through which, cardiac and respiration can contribute to the sLFOs (Liu, 2016).

Carbon dioxide

Carbon dioxide (CO_2) is an end product of aerobic cellular respiration, often measured with a capnograph as end-tidal partial pressure of CO_2 (ETCO$_2$), which is the level of CO_2 that is released at the end of an exhaled breath (typically reported in units of millimeter mercury (mm Hg)). It is highly related to cardiac output and pulmonary blood flow, partially sharing the same mechanism by inducing changes in cerebral blood flow and volume due to CO_2's role as a potent vasodilator (Crystal, 2015; Ogoh, 2019; Gisolf et al., 2004; Shibutani et al., 1994). It has been shown that RVT (especially when convolved with the "respiration response function") is highly correlated to ETCO$_2$, both temporally (negative) and spatially (Chang and Glover, 2009). However, here we focus on the direct effects of spontaneous fluctuations in ETCO$_2$, especially in the LF band of the fMRI BOLD signal rather than the indirect effect from respiration and/or cardiac. Wise et al. (2004) found that ETCO$_2$ in the 0–0.05 Hz band was correlated significantly with both an increased BOLD fMRI signal in gray and white matter and an increased middle cerebral artery (MCA) blood velocity (measured with transcranial Doppler ultrasound). This correlation is attributed to the vasodilatory effect of CO_2, resulting in increases in arterial diameter and blood volume with higher CO_2. Consequently, fluctuations in ETCO$_2$ can clearly contribute to sLFOs and move with the blood. Wise found that at least 28% of the variance in the low-frequency BOLD signal could be explained by ETCO$_2$ variations. This percentage is likely to be an underestimation of actual contribution, since only a single delay (6.3 s) was used for all voxels and subsequent studies have shown optimal delay differences, including for eyes open and closed (Peng et al., 2013).

Yao et al. conducted a gas-challenge fMRI study (Yao et al., 2021), in which elevated CO_2 was used in inspired gas to induce a global increase in BOLD signal (due to vessel dilation). During this global increase, the sLFO signal does not disappear but simply superimposes the CO_2-related BOLD signal variations, suggesting that CO_2 and sLFO are modulating the BOLD fMRI simultaneously, with CO_2 effects being the dominant signal. In addition, the spatiotemporal pattern of the sLFO signal moving through the brain, after the CO_2 effect is minimized in the BOLD fMRI, is similar to that found in the resting state, though with a shorter transit time on account of the increased flow velocity due to the CO_2. This pattern shows that sLFOs and other neuronal/physiological processes can modulate the BOLD signal simultaneously and independently. There may still be variations of the CO_2 signal that might contribute to sLFO such as breath-by-breath O_2-CO_2 exchange ratio (bER), which has recently been suggested to be a more robust regressor than $ETCO_2$ in the analysis of cerebral hemodynamic fluctuations at rest and cerebrovascular responses to breath hold challenge (Chan et al., 2020a,b); however, more research is needed.

Vasomotion

Vasomotion is a spontaneous oscillation between 0.01 and 0.3 Hz in **vascular tone**, i.e., the degree of constriction in a blood vessel relative to its maximally dilated state. Vasomotor oscillations (Nilsson and Aalkjaer, 2003) can be observed in isolated vessels in various species—including humans—and in blood vessels from different sites in the body, including the periphery (e.g., Fagrell et al., 1980) and the brain (e.g., Mayhew et al., 1996). Believed to be independent of respiration, cardiac pulsation, and neuronal activity, the oscillations are modulated by vascular smooth muscles and the endothelium, and linked to various signaling mechanisms including oscillatory intracellular calcium (overview Aalkjaer et al. (2011) and Loh et al. (2018)). Laser Doppler studies on endothelially dependent vasomotion in the human peripheries have reported peak frequencies in the low-frequency band around 0.01 Hz (Kvernmo et al., 1999; Stefanovska et al., 1999). The oscillations can influence blood pressure and flow, resulting in periodic oscillations in the local blood flow. This would imply a very localized (also referred to as stationary in the introduction) nature of origin of these signals; however, the vascular variations induced would then propagate with the blood, resulting in a mix of stationary and moving signals.

Recent research has revised the understanding regarding vaso-motion in rs-fMRI from a purely localized signal, to the one that has more spatially distributed effects. Ultra-large two-photon microscopy (e.g., precise measurements for a field of view of up to 7 mm) on rodents showed co-fluctuations in low-frequency arteriole vasomotion below <0.2 Hz with a peak frequency of 0.1 Hz in distant mirrored transhemispheric sites as a link between the envelope of the γ band (gamma-band oscillations refer to rhythmic brain activity, typically around 30–80 Hz) and changes in CBV and later in tissue oxygenation (BOLD effect). This effect could be reproduced even in the absence of neuronal activity (Winder et al., 2017; Mateo et al., 2017; Drew et al., 2020). Similar results to that of the ultra-large photon microscopy results could be found in high-resolution rodent fMRI studies (He et al., 2018). Complementary human visual cortex data (He et al., 2018) showed high and widespread correlations among sulcus veins and specified intracortical veins of the visual cortex, similar to the observed frequencies in sLFO (<0.1 Hz in venous voxels, instead of 0.01–0.04 Hz as seen in rats). The authors could spatially link these semilocalized sLFO waves propagating along the cortical gradient of hierarchy, with activity sensitive to the brain arousal levels (Gu et al., 2021), matching previous research on arousal (Chang et al., 2016).

Gastric waves

Gastric waves or the gastric rhythm refers to the synchronized gut motions which oscillate with a slow pseudoperiodic signal at around 0.05 Hz (1 cycle/20 s) (Azzalini et al., 2019). These oscillations can be measured with electrogastrograms (EGG) via electrodes placed on the abdomen. The EGG signal is a mixture of the more transient smooth muscle activity generating gastric peristaltic contractions in tandem with the slow electrical gastric rhythm, constantly generated in the stomach wall by Intersial Cells of Cajal (Suzuki et al., 1986; Wolpert et al., 2020). The latter gastric rhythm is even present when muscular contractions are absent (Bozler, 1945), or when the stomach is completely disconnected from the central nervous system (Suzuki et al., 1986). Interestingly, a link also exists between arousal, sleep, and gut homeostasis (Richter et al., 2017; Pigarev, 2014; Vaccaro et al., 2020). As this field is relatively new, many questions remain, including that of LFO directionality. Rebollo et al. (2018) found the rs-fMRI signal in all motor and sensory (vision, audition, touch and interoception and olfaction) regions (Rebollo and Tallon-Baudry, 2022) to be highly synchronized with delays to the gastric rhythm, with delays of 3.3 s between the earliest (somatosensory cortices) and the latest (dorsal precuneus and extrastriate body area) nodes of the proposed "gastric network." These regions of greater delays lie in close proximity to vessels at the end of the vascular path through the brain (superior sagittal and transverse sinuses). However, the observations above are consistent with a hemodynamic perturbation generated in the stomach and propagating through the cerebral vasculature. Previous observations (Mohamed Yacin et al., 2011) in the peripheries support the global nature of the gastric signal. Yacin et al. used fingertip photoplethysmogram to reconstruct the gastric slow wave. The strong relationship ($R \geq 0.9$) between gastric activity and systemic LFOs in the periphery suggests that the gastric signal contributes to a significant portion of the sLFO variance observable in the periphery.

Application and impact
sLFOs delay-based analysis in clinical populations

Irrespective of the origins of sLFOs, various clinical applications have emerged from the resulting delay structure (as a measure of flow) and correlation coefficients (as a measure of perfusion) (Khalil et al., 2017, 2020; Nishida et al., 2019; Aso et al., 2020). Delay analysis has been extensively evaluated in healthy subjects in order to quantify typical spatiotemporal blood flow patterns as described herein. During the last decade, studies have further demonstrated the application of delay analysis for compromised circulation due to arterial occlusion from atherosclerosis, moyamoya disease, or stroke, especially as an alternative to DSC (see "An alternative to DSC" section). These pathologies can lead to extremely long delay times (≥ 10 s), where standard noninvasive flow measures, such as arterial spin labeling (ASL), are not able to quantify.

- **Stroke**

  A stroke occurs when blood clots or other particles, such as fatty deposits called plaques, build up in the blood vessels and block the blood flow to the brain (ischemic stroke) or when an artery in the brain ruptures (hemorrhagic stroke). Clinically, DSC MRI is the most common method to assess brain perfusion in acute ischemic stroke to guide clinical decision-making. However, as mentioned earlier, due to the requirement of the gadolinium-based contrast in DSC-MRI and its possible side effects, as well as cumulative effects in the brain (McDonald et al., 2015; Gulani et al., 2017), alternatives such as BOLD fMRI delay analysis are highly desirable. Numerous studies have used hemodynamic delays in MRI analysis to investigate hypoperfusion of acute (Lv et al., 2013; Chen et al., 2019a; Khalil et al., 2017, 2020; Qian et al., 2015; Amemiya et al., 2014), subacute (Siegel et al., 2016; Ni et al., 2017), and chronic stages of stroke (Siegel et al., 2016; Amemiya et al., 2014; Zhao et al., 2018). These studies have shown that with a rs-fMRI scans of only 204 s (Tanritanir et al., 2020), significant delays in BOLD signal correspond to areas of hypoperfusion can be quantified that are in agreement with DSC-MRI (e.g., Lv et al., 2013; Amemiya et al., 2014), even with cases of total perfusion-diffusion mismatch (Khalil et al., 2021). High correlations were found between delay maps and time-to-peak (TTP) maps derived from DSC-MRI in ischemic stroke (Khalil et al., 2017) with superior agreement when the SSS was chosen as a reference signal instead of the global signal (Christen et al., 2015). Similar results were found when comparing BOLD delay to ASL, with increased delay strongly correlated with a decrease in blood flow assessed by ASL (Siegel et al., 2016).

- **Moyamoya disease**

  Another pathology benefiting from delay analysis is moyamoya disease. Moyamoya disease is a rare progressive cerebrovascular disorder, which is caused by the blockage of the arteries in the basal ganglia. The tiny vessels, which form at the base of brain in order to compensate for the blockage, create a tangle, which has lent the disease its Japanese name "moyamoya," meaning "a puff of smoke." Medications as well as revascularization are typically used to treat the disease. Donahue et al. have revealed and accurately quantitated extremely long delays in vivo (>21 s) in Moyamoya patients, distinguishable from those in healthy controls (~8 s) (Donahue et al., 2016). Moreover, they demonstrated longitudinally that delays in regions that co-localized with surgical revascularization were subsequently reduced (see Fig. 2). They did so by using an exogenous CO_2 manipulation; however, the exogenous CO_2 manipulation is not necessary—the hemodynamic delays arising from moyamoya disease can also be quantified using the endogenous sLFO signal from rsfMRI alone as a probe (Wu et al., 2017; Christen et al., 2015; Jahanian et al., 2018).

- **Other pathologies**

  LFOs in various other pathologies have been studied. For instance, Nishida et al. (2019) studied arterial occlusive disease, in which arteries throughout the body become narrowed, demonstrating that delay maps were highly correlated with CVR maps of SPECT. Coloigner et al. (2016) studied delays in sickle cell disease, an inherited blood disorder marked by deformed hemoglobin (sickle cell instead of donut shaped), which rupture easily and tend to stick together, leading to vaso-occlusion, ischemia, and infarct. Patients are also chronically anemic, leading to increased blood flow (Borzage et al., 2016). Coloigner et al. (2016) found that pattern confirmed in delay analysis when changes in the areas vascularized by anterior, middle, and posterior cerebral arteries were almost instantaneous and shorter in sickle cell disease patients in comparison to controls. In addition, the area vascularized by an artery with stenosis lagged behind the other areas. Yan et al. (2019) successfully detected perfusion deficits (from BOLD delay) in

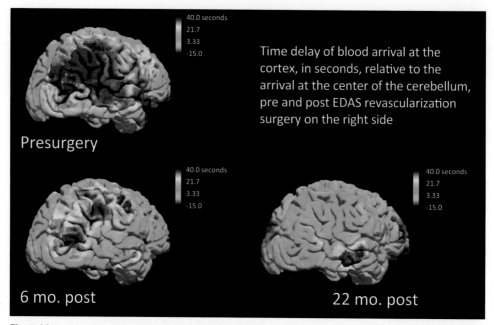

Fig. 2 Moyamoya hemodynamic changes following revascularization. The cortical time-to-peak (TTP) maps in a 57-year-old female subject with severe bilateral moyamoya disease are shown before and after (6 and 22 month) right indirect revascularization. The markedly improved cortical time to peak in the right hemisphere following surgery is clearly visible. Modified from Donahue, M.J., Strother, M.K., Lindsey, K.P., Hocke, L.M., Tong, Y., Frederick, B.D., 2016. Time delay processing of hypercapnic fMRI allows quantitative parameterization of cerebrovascular reactivity and blood flow delays. J. Cereb. Blood Flow Metab. 36(10), 1767–1779.

patients with Alzheimer's disease and mild cognitive impairment. Delay maps have even been considered a biomarker for aging (Satow et al., 2017) with Aso (Aso et al., 2020) detecting age-related delay maps with aging of the venous drainage. These various studies demonstrate the breath of pathologies on which delay analysis can shine a new light on.

Assessing LFO in cerebrospinal fluid (CSF) movement

The glymphatic system plays an important role in the clearance of metabolic waste products in the brain, especially in sleep (Jessen et al., 2015). Glymphatic clearance is achieved by the circulation of CSF within and out of the brain. Without its own engine, CSF flow is assumed to be mostly facilitated by the hemodynamic oscillations in the brain. In several studies, the derivative of global LFO signal (in rs-fMRI) is found to be highly correlated with (Raitamaa et al., 2021; Attarpour et al., 2021) and leading (by a few seconds) CSF movement measured at the third and fourth ventricles (Fultz et al., 2019; Yang et al., 2022). Since the temporal derivative of global LFO signal represents the cumulative change in CBV, changes in CBV might drive CSF movement according to the Monro-Kellie doctrine (Mokri, 2001). Thus, sLFO might also hold the key in understanding the dynamics of and vascular coupling with CSF movement.

An alternative to DSC

DSC has been the golden standard in imaging the blood flow in the brain. However, it is invasive with severe side effects induced in some patients. Aso et al. have proposed that sLFOs may behave like a contrast agent (Aso et al., 2019), allowing the measurement of a broad range of perfusion information, such as the arterial-venous delay (Tong et al., 2017; Satow et al., 2017; Nishida et al., 2019). However, spontaneous sLFOs in deoxyhemoglobin concentration might be hard to track due to spurious correlations. Recently, Poublanc et al. has further enhanced endogenous deoxyhemoglobin with the help of RespirAct (Poublanc et al., 2021). They found great consistency between their result and that of DSC.

Using sLFOs to denoising the rs-fMRI data

Calculating functional networks that are based on regional neuronal activations is the ultimate goal of most rs-fMRI studies. Given the need to distinguish the neuronal signal from many noise sources, including physiological processes and motion artifacts, many "denoising" methods have been developed over the past few decades (see Chapters 6 and 10). Among them, several focus on the low-frequency noise explicitly. Global signal regression (GSR) is one of the most widespread methods in this category. The major drawback of this method is the creation of spurious, negative correlations between brain regions (Murphy et al., 2009; Carbonell et al., 2011). The global mean signal is a summation of many copies of the sLFO signal with a range of delays. However, the global signal is not aligned temporally with the BOLD signal from individual voxels of the brain, leading to spurious regression outcomes. One obvious solution is dynamic global-signal regression (Aso et al., 2017), in which different copies of the sLFOs, each shifted by the appropriate time delays, are used (Erdogan et al., 2016). The key prerequisite for this approach is to have the accurate sLFO, which could be obtained from concurrently measured peripheral NIRS data, BOLD signals extracted from big veins or the GS using a recursive analytical procedure (Aso et al., 2017; Tong et al., 2011). As a word of caution, demonstrated by Hocke et al. (2016) it is important to understand some statistical pitfalls when evaluating the optimal delay and correlation. The combination of low-pass filtering and a delay estimation based on cross-correlations can distort the correlation distribution so severely that the significance of the correlation has to be evaluated using specialized approaches.

References

Aalkjaer, C., Boedtkjer, D., Matchkov, V., 2011. Vasomotion—what is currently thought? Acta Physiol (Oxford) 202 (3), 253–269.

Abbas, A., Belloy, M., Kashyap, A., et al., 2019. Quasi-periodic patterns contribute to functional connectivity in the brain. NeuroImage 191, 193–204.

Amemiya, S., Kunimatsu, A., Saito, N., Ohtomo, K., 2014. Cerebral hemodynamic impairment: assessment with resting-state functional MR imaging. Radiology 270 (2), 548–555.

Aso, T., Jiang, G., Urayama, S.I., Fukuyama, H., 2017. A resilient, non-neuronal source of the spatiotemporal lag structure detected by BOLD signal-based blood flow tracking. Front. Neurosci. 11, 256.

Aso, T., Urayama, S., Fukuyama, H., Murai, T., 2019. Axial variation of deoxyhemoglobin density as a source of the low-frequency time lag structure in blood oxygenation level-dependent signals. PLoS One 14 (9), e0222787.

Aso, T., Sugihara, G., Murai, T., et al., 2020. A venous mechanism of ventriculomegaly shared between traumatic brain injury and normal ageing. Brain 143 (6), 1843–1856.

Attarpour, A., Ward, J., Chen, J.J., 2021. Vascular origins of low-frequency oscillations in the cerebrospinal fluid signal in resting-state fMRI: interpretation using photoplethysmography. Hum. Brain Mapp. 42 (8), 2606–2622.

Azzalini, D., Rebollo, I., Tallon-Baudry, C., 2019. Visceral signals shape brain dynamics and cognition. Trends Cogn. Sci. 23 (6), 488–509.

Birn, R.M., Diamond, J.B., Smith, M.A., Bandettini, P.A., 2006. Separating respiratory-variation-related fluctuations from neuronal-activity-related fluctuations in fMRI. NeuroImage 31 (4), 1536–1548.

Birn, R.M., Smith, M.A., Jones, T.B., Bandettini, P.A., 2008. The respiration response function: the temporal dynamics of fMRI signal fluctuations related to changes in respiration. NeuroImage 40 (2), 644–654.

Biswal, B., Yetkin, F.Z., Haughton, V.M., Hyde, J.S., 1995. Functional connectivity in the motor cortex of resting human brain using echo-planar MRI. Magn. Reson. Med. 34 (4), 537–541.

Borzage, M.T., Bush, A.M., Choi, S., et al., 2016. Predictors of cerebral blood flow in patients with and without anemia. J. Appl. Physiol. (1985) 120 (8), 976–981.

Bozler, E., 1945. The action potentials of the stomach. Am. J. Physiol.-Legacy Content 144 (5), 693–700.

Carbonell, F., Bellec, P., Shmuel, A., 2011. Global and system-specific resting-state fMRI fluctuations are uncorrelated: principal component analysis reveals anti-correlated networks. Brain Connect. 1 (6), 496–510.

Casey, B.J., Cannonier, T., Conley, M.I., et al., 2018. The adolescent brain cognitive development (ABCD) study: Imaging acquisition across 21 sites. Dev. Cogn. Neurosci. 32, 43–54.

Chan, S.T., Evans, K.C., Song, T.Y., et al., 2020a. Cerebrovascular reactivity assessment with O_2-CO_2 exchange ratio under brief breath hold challenge. PLoS One 15 (3), e0225915.

Chan, S.T., Evans, K.C., Song, T.Y., et al., 2020b. Dynamic brain-body coupling of breath-by-breath O_2-CO_2 exchange ratio with resting state cerebral hemodynamic fluctuations. PLoS One 15 (9), e0238946.

Chang, C., Glover, G.H., 2009. Relationship between respiration, end-tidal CO_2, and BOLD signals in resting-state fMRI. NeuroImage 47 (4), 1381–1393.

Chang, C., Glover, G.H., 2010. Time-frequency dynamics of resting-state brain connectivity measured with fMRI. NeuroImage 50 (1), 81–98.

Chang, C., Metzger, C.D., Glover, G.H., Duyn, J.H., Heinze, H.J., Walter, M., 2013. Association between heart rate variability and fluctuations in resting-state functional connectivity. NeuroImage 68, 93–104.

Chang, C., Leopold, D.A., Scholvinck, M.L., et al., 2016. Tracking brain arousal fluctuations with fMRI. Proc. Natl. Acad. Sci. U. S. A. 113 (16), 4518–4523.

Chen, Q., Zhou, J., Zhang, H., et al., 2019a. One-step analysis of brain perfusion and function for acute stroke patients after reperfusion: a resting-state fMRI study. J. Magn. Reson. Imaging 50 (1), 221–229.

Chen, J.E., Polimeni, J.R., Bollmann, S., Glover, G.H., 2019b. On the analysis of rapidly sampled fMRI data. NeuroImage 188, 807–820.

Chen, J.E., Glover, G.H., Fultz, N.E., Rosen, B.R., Polimeni, J.R., Lewis, L.D., 2021. Investigating mechanisms of fast BOLD responses: the effects of stimulus intensity and of spatial heterogeneity of hemodynamics. NeuroImage 245, 118658.

Christen, T., Jahanian, H., Ni, W.W., Qiu, D., Moseley, M.E., Zaharchuk, G., 2015. Noncontrast mapping of arterial delay and functional connectivity using resting-state functional MRI: a study in Moyamoya patients. J. Magn. Reson. Imaging 41 (2), 424–430.

Coloigner, J., Vu, C., Bush, A., et al., 2016. BOLD delay times using group delay in sickle cell disease. Proc. SPIE Int. Soc. Opt. Eng., 9784.

Crystal, G.J., 2015. Carbon dioxide and the heart: physiology and clinical implications. Anesth. Analg. 121 (3), 610–623.

Cui, X., Bray, S., Bryant, D.M., Glover, G.H., Reiss, A.L., 2011. A quantitative comparison of NIRS and fMRI across multiple cognitive tasks. NeuroImage 54 (4), 2808–2821.

Dagli, M.S., Ingeholm, J.E., Haxby, J.V., 1999. Localization of cardiac-induced signal change in fMRI. NeuroImage 9 (4), 407–415.

Donahue, M.J., Strother, M.K., Lindsey, K.P., Hocke, L.M., Tong, Y., Frederick, B.D., 2016. Time delay processing of hypercapnic fMRI allows quantitative parameterization of cerebrovascular reactivity and blood flow delays. J. Cereb. Blood Flow Metab. 36 (10), 1767–1779.

Dreha-Kulaczewski, S., Joseph, A.A., Merboldt, K.D., Ludwig, H.C., Gartner, J., Frahm, J., 2015. Inspiration is the major regulator of human CSF flow. J. Neurosci. 35 (6), 2485–2491.

Drew, P.J., Mateo, C., Turner, K.L., Yu, X., Kleinfeld, D., 2020. Ultra-slow oscillations in fMRI and resting-state connectivity: neuronal and vascular contributions and technical confounds. Neuron 107 (5), 782–804.

Erdogan, S.B., Tong, Y., Hocke, L.M., Lindsey, K.P., deB Frederick, B., 2016. Correcting for blood arrival time in global mean regression enhances functional connectivity analysis of resting state fMRI-BOLD signals. Front. Hum. Neurosci. 10, 311.

Fagrell, B., Intaglietta, M., Ostergren, J., 1980. Relative hematocrit in human skin capillaries and its relation to capillary blood flow velocity. Microvasc. Res. 20 (3), 327–335.

Fitzgerald, B., Yao, J.F., Talavage, T.M., Hocke, L.M., Frederick, B.D., Tong, Y., 2021. Using carpet plots to analyze transit times of low frequency oscillations in resting state fMRI. Sci. Rep. 11 (1), 7011.

Fultz, N.E., Bonmassar, G., Setsompop, K., et al., 2019. Coupled electrophysiological, hemodynamic, and cerebrospinal fluid oscillations in human sleep. Science 366 (6465), 628–631.

Gao, J.H., Liu, H.L., 2012. Inflow effects on functional MRI. NeuroImage 62 (2), 1035–1039.

Gilroy, J., Bauer, R.B., Krabbenhoft, K.L., Meyer, J.S., 1963. Cerebral circulation time in cerebral vascular disease measured by serial angiography. Am. J. Roentgenol. Radium Therapy, Nucl. Med. 90, 490–505.

Gisolf, J., Wilders, R., Immink, R.V., van Lieshout, J.J., Karemaker, J.M., 2004. Tidal volume, cardiac output and functional residual capacity determine end-tidal CO_2 transient during standing up in humans. J. Physiol. 554 (Pt 2), 579–590.

Glover, G.H., Li, T.Q., Ress, D., 2000. Image-based method for retrospective correction of physiological motion effects in fMRI: RETROICOR. Magn. Reson. Med. 44 (1), 162–167.

Gu, Y., Sainburg, L.E., Kuang, S., et al., 2021. Brain activity fluctuations propagate as waves traversing the cortical hierarchy. Cereb. Cortex 31 (9), 3986–4005.

Gulani, V., Calamante, F., Shellock, F.G., Kanal, E., Reeder, S.B., International Society for Magnetic Resonance in M, 2017. Gadolinium deposition in the brain: summary of evidence and recommendations. Lancet Neurol. 16 (7), 564–570.

He, Y., Wang, M., Chen, X., et al., 2018. Ultra-slow single-vessel BOLD and CBV-based fMRI spatiotemporal dynamics and their correlation with neuronal intracellular calcium signals. Neuron 97 (4), 925–939 e925.

Hocke, L.M., Tong, Y., Lindsey, K.P., deB Frederick, B., 2016. Comparison of peripheral near-infrared spectroscopy low-frequency oscillations to other denoising methods in resting state functional MRI with ultrahigh temporal resolution. Magn. Reson. Med. 76 (6), 1697–1707.

Jahanian, H., Christen, T., Moseley, M.E., Zaharchuk, G., 2018. Erroneous resting-state fMRI connectivity maps due to prolonged arterial arrival time and how to fix them. Brain Connect. 8 (6), 362–370.

Jessen, N.A., Munk, A.S., Lundgaard, I., Nedergaard, M., 2015. The glymphatic system: a beginner's guide. Neurochem. Res. 40 (12), 2583–2599.

Josephs, O., Henson, R.N., 1999. Event-related functional magnetic resonance imaging: modelling, inference and optimization. Philos. Trans. R. Soc. Lond. Ser. B Biol. Sci. 354 (1387), 1215–1228.

Khalil, A.A., Ostwaldt, A.C., Nierhaus, T., et al., 2017. Relationship between changes in the temporal dynamics of the blood-oxygen-level-dependent signal and hypoperfusion in acute ischemic stroke. Stroke 48 (4), 925–931.

Khalil, A.A., Villringer, K., Fillebock, V., et al., 2020. Non-invasive monitoring of longitudinal changes in cerebral hemodynamics in acute ischemic stroke using BOLD signal delay. J. Cereb. Blood Flow Metab. 40 (1), 23–34.

Khalil, A., Röhrs, K., Nolte, C.H., Galinovic, I., 2021. Total perfusion-diffusion mismatch detected using resting-state functional MRI. BJR Case Rep. 7 (5), 20210056.

Kvernmo, H.D., Stefanovska, A., Kirkeboen, K.A., Kvernebo, K., 1999. Oscillations in the human cutaneous blood perfusion signal modified by endothelium-dependent and endothelium-independent vasodilators. Microvasc. Res. 57 (3), 298–309.

Li, Y., Zhang, H., Yu, M., Yu, W., Frederick, B.D., Tong, Y., 2018. Systemic low-frequency oscillations observed in the periphery of healthy human subjects. J. Biomed. Opt. 23 (5), 1–11.

Li, Y., Ma, Y., Ma, S., et al., 2020. Asymmetry of peripheral vascular biomarkers in ischemic stroke patients, assessed using NIRS. J. Biomed. Opt. 25 (6), 1–16.

Liu, T.T., 2016. Noise contributions to the fMRI signal: an overview. NeuroImage 143, 141–151.

Liu, T.T., Nalci, A., Falahpour, M., 2017. The global signal in fMRI: nuisance or information? NeuroImage 150, 213–229.

Loh, Y.C., Tan, C.S., Ch'ng, Y.S., Yeap, Z.Q., Ng, C.H., Yam, M.F., 2018. Overview of the microenvironment of vasculature in vascular tone regulation. Int. J. Mol. Sci. 19 (1).

Lv, Y., Margulies, D.S., Cameron Craddock, R., et al., 2013. Identifying the perfusion deficit in acute stroke with resting-state functional magnetic resonance imaging. Ann. Neurol. 73 (1), 136–140.

Lv, Y., Wei, W., Song, Y., et al., 2019. Non-invasive evaluation of cerebral perfusion in patients with transient ischemic attack: an fMRI study. J. Neurol. 266 (1), 157–164.

Mateo, C., Knutsen, P.M., Tsai, P.S., Shih, A.Y., Kleinfeld, D., 2017. Entrainment of arteriole vasomotor fluctuations by neural activity is a basis of blood-oxygenation-level-dependent "resting-state" connectivity. Neuron 96 (4), 936–948 e933.

Mayhew, J.E., Askew, S., Zheng, Y., et al., 1996. Cerebral vasomotion: a 0.1-Hz oscillation in reflected light imaging of neural activity. NeuroImage 4 (3 Pt 1), 183–193.

McDonald, R.J., McDonald, J.S., Kallmes, D.F., et al., 2015. Intracranial gadolinium deposition after contrast-enhanced MR imaging. Radiology 275 (3), 772–782.

Mohamed Yacin, S., Srinivasa Chakravarthy, V., Manivannan, M., 2011. Reconstruction of gastric slow wave from finger photoplethysmographic signal using radial basis function neural network. Med. Biol. Eng. Comput. 49 (11), 1241–1247.

Mokri, B., 2001. The Monro-Kellie hypothesis: applications in CSF volume depletion. Neurology 56 (12), 1746–1748.

Murphy, K., Birn, R.M., Handwerker, D.A., Jones, T.B., Bandettini, P.A., 2009. The impact of global signal regression on resting state correlations: are anti-correlated networks introduced? NeuroImage 44 (3), 893–905.

Ni, L., Li, J., Li, W., et al., 2017. The value of resting-state functional MRI in subacute ischemic stroke: comparison with dynamic susceptibility contrast-enhanced perfusion MRI. Sci. Rep. 7, 41586.

Nilsson, H., Aalkjaer, C., 2003. Vasomotion: mechanisms and physiological importance. Mol. Interv. 3 (2), 79–89. 51.

Nishida, S., Aso, T., Takaya, S., et al., 2019. Resting-state functional magnetic resonance imaging identifies cerebrovascular reactivity impairment in patients with arterial occlusive diseases: a pilot study. Neurosurgery 85 (5), 680–688.

Ogoh, S., 2019. Interaction between the respiratory system and cerebral blood flow regulation. J. Appl. Physiol. 127 (5), 1197–1205.

Peng, T., Niazy, R., Payne, S.J., Wise, R.G., 2013. The effects of respiratory CO_2 fluctuations in the resting-state BOLD signal differ between eyes open and eyes closed. Magn. Reson. Imaging 31 (3), 336–345.

Pigarev, I., 2014. The visceral theory of sleep. Neurosci. Behav. Physiol. 44 (4), 421–434.

Poldrack, R.A., Laumann, T.O., Koyejo, O., et al., 2015. Long-term neural and physiological phenotyping of a single human. Nat. Commun. 6, 8885.

Poublanc, J., Sobczyk, O., Shafi, R., et al., 2021. Perfusion MRI using endogenous deoxyhemoglobin as a contrast agent: preliminary data. Magn. Reson. Med. 86 (6), 3012–3021.

Power, J.D., Plitt, M., Laumann, T.O., Martin, A., 2017. Sources and implications of whole-brain fMRI signals in humans. NeuroImage 146, 609–625.

Qian, T., Wang, Z., Gao, P., 2015. Measuring the timing information of blood flow in acute stroke with the "background noise"of BOLD signal. ISMRM, Toronto.

Raitamaa, L., Huotari, N., Korhonen, V., et al., 2021. Spectral analysis of physiological brain pulsations affecting the BOLD signal. Hum. Brain Mapp. 42 (13), 4298–4313.

Raj, D., Anderson, A.W., Gore, J.C., 2001. Respiratory effects in human functional magnetic resonance imaging due to bulk susceptibility changes. Phys. Med. Biol. 46 (12), 3331–3340.

Rapidtide [Computer Program].

Rebollo, I., Tallon-Baudry, C., 2022. The sensory and motor components of the cortical hierarchy are coupled to the rhythm of the stomach during rest. J. Neurosci. 42 (11), 2205–2220.

Rebollo, I., Devauchelle, A.-D., Béranger, B., Tallon-Baudry, C., 2018. Stomach-brain synchrony reveals a novel, delayed-connectivity resting-state network in humans. elife 7, e33321.

Richter, C.G., Babo-Rebelo, M., Schwartz, D., Tallon-Baudry, C., 2017. Phase-amplitude coupling at the organism level: the amplitude of spontaneous alpha rhythm fluctuations varies with the phase of the infra-slow gastric basal rhythm. NeuroImage 146, 951–958.

Sassaroli, A., deB Frederick, B., Tong, Y., Renshaw, P.F., Fantini, S., 2006. Spatially weighted BOLD signal for comparison of functional magnetic resonance imaging and near-infrared imaging of the brain. NeuroImage 33 (2), 505–514.

Sassaroli, A., Pierro, M., Bergethon, P.R., Fantini, S., 2012. Low-frequency spontaneous oscillations of cerebral hemodynamics investigated with near-infrared spectroscopy: a review. IEEE J. Sel. Top. Quantum Electron. 18 (4), 1478–1492.

Satow, T., Aso, T., Nishida, S., et al., 2017. Alteration of venous drainage route in idiopathic normal pressure hydrocephalus and normal aging. Front. Aging Neurosci. 9, 387.

Schreiber, S.J., Franke, U., Doepp, F., Staccioli, E., Uludag, K., Valdueza, J.M., 2002. Dopplersonographic measurement of global cerebral circulation time using echo contrast-enhanced ultrasound in normal individuals and patients with arteriovenous malformations. Ultrasound Med. Biol. 28 (4), 453–458.

Schreiber, S.J., Doepp, F., Spruth, E., Kopp, U.A., Valdueza, J.M., 2005. Ultrasonographic measurement of cerebral blood flow, cerebral circulation time and cerebral blood volume in vascular and Alzheimer's dementia. J. Neurol. 252 (10), 1171–1177.

Shibutani, K., Muraoka, M., Shirasaki, S., Kubal, K., Sanchala, V.T., Gupte, P., 1994. Do changes in end-tidal PCO_2 quantitatively reflect changes in cardiac output? Anesth. Analg. 79 (5), 829–833.

Siegel, J.S., Snyder, A.Z., Ramsey, L., Shulman, G.L., Corbetta, M., 2016. The effects of hemodynamic lag on functional connectivity and behavior after stroke. J. Cereb. Blood Flow Metab. 36 (12), 2162–2176.

Stefanovska, A., Bracic, M., Kvernmo, H.D., 1999. Wavelet analysis of oscillations in the peripheral blood circulation measured by laser doppler technique. IEEE Trans. Biomed. Eng. 46 (10), 1230–1239.

Strangman, G., Culver, J.P., Thompson, J.H., Boas, D.A., 2002. A quantitative comparison of simultaneous BOLD fMRI and NIRS recordings during functional brain activation. NeuroImage 17 (2), 719–731.

Sutoko, S., Chan, Y.L., Obata, A., et al., 2019. Denoising of neuronal signal from mixed systemic low-frequency oscillation using peripheral measurement as noise regressor in near-infrared imaging. Neurophotonics 6 (1), 015001.

Suzuki, N., Prosser, C.L., Dahms, V., 1986. Boundary cells between longitudinal and circular layers: essential for electrical slow waves in cat intestine. Am. J. Physiol. 250 (3 Pt 1), G287–G294.

Tanritanir, A.C., Villringer, K., Galinovic, I., et al., 2020. The effect of scan length on the assessment of BOLD delay in ischemic stroke. Front. Neurol. 11, 381.

Thompson, G.J., Pan, W.J., Magnuson, M.E., Jaeger, D., Keilholz, S.D., 2014. Quasi-periodic patterns (QPP): large-scale dynamics in resting state fMRI that correlate with local infraslow electrical activity. NeuroImage 84, 1018–1031.

Tong, Y., Frederick, B.D., 2010. Time lag dependent multimodal processing of concurrent fMRI and near-infrared spectroscopy (NIRS) data suggests a global circulatory origin for low-frequency oscillation signals in human brain. NeuroImage 53 (2), 553–564.

Tong, Y., Frederick, B., 2014. Tracking cerebral blood flow in BOLD fMRI using recursively generated regressors. Hum. Brain Mapp. 35 (11), 5471–5485.

Tong, Y., Bergethon, P.R., Frederick, B.D., 2011. An improved method for mapping cerebrovascular reserve using concurrent fMRI and near-infrared spectroscopy with regressor interpolation at progressive time delays (RIPTiDe). NeuroImage 56 (4), 2047–2057.

Tong, Y., Hocke, L.M., Licata, S.C., Frederick, B., 2012. Low-frequency oscillations measured in the periphery with near-infrared spectroscopy are strongly correlated with blood oxygen level-dependent functional magnetic resonance imaging signals. J. Biomed. Opt. 17 (10), 106004.

Tong, Y., Hocke, L.M., Nickerson, L.D., Licata, S.C., Lindsey, K.P., Frederick, B., 2013. Evaluating the effects of systemic low frequency oscillations measured in the periphery on the independent component analysis results of resting state networks. NeuroImage 76, 202–215.

Tong, Y., Hocke, L.M., Fan, X., Janes, A.C., Frederick, B., 2015. Can apparent resting state connectivity arise from systemic fluctuations? Front. Hum. Neurosci. 9, 285.

Tong, Y., Lindsey, K.P., Hocke, L.M., Vitaliano, G., Mintzopoulos, D., Frederick, B.D., 2017. Perfusion information extracted from resting state functional magnetic resonance imaging. J. Cereb. Blood Flow Metab. 37 (2), 564–576.

Tong, Y., Yao, J.F., Chen, J.J., Frederick, B.D., 2019a. The resting-state fMRI arterial signal predicts differential blood transit time through the brain. J. Cereb. Blood Flow Metab. 39 (6), 1148–1160.

Tong, Y., Hocke, L.M., Frederick, B.B., 2019b. Low frequency systemic hemodynamic "noise" in resting state BOLD fMRI: characteristics, causes, implications, mitigation strategies, and applications. Front. Neurosci. 13, 787.

Vaccaro, A., Kaplan Dor, Y., Nambara, K., et al., 2020. Sleep loss can cause death through accumulation of reactive oxygen species in the gut. Cell 181 (6), 1307–1328 e1315.

Van Essen, D.C., Smith, S.M., Barch, D.M., et al., 2013. The WU-Minn human connectome project: an overview. NeuroImage 80, 62–79.

von Schulthess, G.K., Higgins, C.B., 1985. Blood flow imaging with MR: spin-phase phenomena. Radiology 157 (3), 687–695.

Weisskoff, R.M., Kiihne, S., 1992. MRI susceptometry: image-based measurement of absolute susceptibility of MR contrast agents and human blood. Magn. Reson. Med. 24 (2), 375–383.

Winder, A.T., Echagarruga, C., Zhang, Q., Drew, P.J., 2017. Weak correlations between hemodynamic signals and ongoing neural activity during the resting state. Nat. Neurosci. 20 (12), 1761–1769.

Wise, R.G., Ide, K., Poulin, M.J., Tracey, I., 2004. Resting fluctuations in arterial carbon dioxide induce significant low frequency variations in BOLD signal. NeuroImage 21 (4), 1652–1664.

Wolpert, N., Rebollo, I., Tallon-Baudry, C., 2020. Electrogastrography for psychophysiological research: practical considerations, analysis pipeline, and normative data in a large sample. Psychophysiology 57 (9), e13599.

Wu, C., Honarmand, A.R., Schnell, S., et al., 2016. Age-related changes of normal cerebral and cardiac blood flow in children and adults aged 7 months to 61 years. J. Am. Heart Assoc. 5 (1).

Wu, J., Dehkharghani, S., Nahab, F., Allen, J., Qiu, D., 2017. The effects of acetazolamide on the evaluation of cerebral hemodynamics and functional connectivity using blood oxygen level-dependent MR imaging in patients with chronic steno-occlusive disease of the anterior circulation. AJNR Am. J. Neuroradiol. 38 (1), 139–145.

Yan, S., Qi, Z., An, Y., Zhang, M., Qian, T., Lu, J., 2019. Detecting perfusion deficit in Alzheimer's disease and mild cognitive impairment patients by resting-state fMRI. J. Magn. Reson. Imaging 49 (4), 1099–1104.

Yang, H.S., Inglis, B., Talavage, T.M., et al., 2022. Coupling between cerebrovascular oscillations and CSF flow fluctuations during wakefulness: an fMRI study. J. Cereb. Blood Flow Metab. 42 (6), 1091–1103.

Yao, J., Wang, J.H., Yang, H.C., et al., 2019. Cerebral circulation time derived from fMRI signals in large blood vessels. J. Magn. Reson. Imaging 50 (5), 1504–1513.

Yao, J.F., Yang, H.S., Wang, J.H., et al., 2021. A novel method of quantifying hemodynamic delays to improve hemodynamic response, and CVR estimates in CO_2 challenge fMRI. J. Cereb. Blood Flow Metab. 41 (8), 1886–1898.

Zhao, Y., Lambon Ralph, M.A., Halai, A.D., 2018. Relating resting-state hemodynamic changes to the variable language profiles in post-stroke aphasia. Neuroimage Clin. 20, 611–619.

8

The role of vigilance in resting-state functional MRI

Thomas T. Liu[a,b,c,d]

[a]Center for Functional MRI, University of California San Diego, La Jolla, CA,
United States, [b]Department of Radiology, University of California San Diego,
La Jolla, CA, United States, [c]Department of Psychiatry, University of California
San Diego, La Jolla, CA, United States, [d]Department of Bioengineering,
University of California San Diego, La Jolla, CA, United States

Introduction

One of the key features of resting-state fMRI (rsfMRI) experiments is that the subjects are scanned while they are at "rest" for a period typically ranging from 5 to 15 min or more. However, the definition of rest can vary considerably across studies. In some rsfMRI studies, subjects are asked to rest quietly with their eyes closed. In other studies, subjects are asked to keep their eyes open and may be asked to fixate on a target, such as a small cross located at the center of a blank screen. In the absence of an engaging task, subjects often report difficulty in maintaining a constant level of vigilance or wakefulness during rsfMRI scans. Indeed, when Tagliazucchi and Laufs (2014) examined data from 1100 subjects scanned by research groups from across the world, they found that about a third of participants lost wakefulness within the first 3 min of a resting-state scan and that half of the participants had fallen asleep after 10 min. The investigators did not find a significant difference in the ability to stay awake between subjects who kept their eyes closed versus those who kept their eyes open (without fixation), but did find that the probability of staying awake was greater when there was a fixation challenge.

The unconstrained nature of the rsFMRI experiment has led to considerable debate and discussion about what the rsfMRI signal actually represents (Gonzalez-Castillo et al., 2021). Over the past decade, there has been a growing appreciation that fluctuations and differences in vigilance and arousal are likely to be a significant contributor to the rsfMRI signal and should be considered when analyzing and interpreting rsfMRI studies (Liu and Falahpour, 2020). There can be pronounced fluctuations in vigilance within a scan, as well as major

Advances in Resting-State Functional MRI. https://doi.org/10.1016/B978-0-323-91688-2.00005-9

differences both in mean vigilance levels and the magnitude of vigilance fluctuations between subjects, scans, and experimental groups and conditions. Indeed, there is a wide range of factors that can affect vigilance, including time of day, diet, sleep history, medication use, disease state, and anxiety levels. Many rsfMRI studies implicitly assume that all participants are in similar states of wakefulness or vigilance, but the validity of this assumption is rarely assessed in either the experimental design phase or the analysis and interpretation phase.

The goal of this chapter is to provide the reader with an appreciation of the various ways in which changes and differences in vigilance can affect the rsfMRI signal and an understanding of potential approaches for addressing these effects. We begin by considering various approaches for assessing vigilance in rsfMRI studies. We then review the evidence from a wide range of studies that have contributed to our understanding of the effect of vigilance on the amplitude of the rsfMRI signal and derived metrics such as measures of both static and dynamic functional connectivity. Next we consider data-driven approaches for estimating vigilance effects, examine the relation between vigilance and disease, and consider future directions and challenges.

Assessing vigilance

While we all have an intuitive understanding of what it means to be in different states of vigilance, such as awake, drowsy, or asleep, it is useful to review some of the terms and definitions that have been used in prior studies. Terms such as vigilance, wakefulness, or arousal are frequently used to refer to the level to which a subject is awake and alert (Oken et al., 2006; Jobert et al., 1994; Olbrich et al., 2009; Matejcek, 1982; Tagliazucchi and Laufs, 2014; Satpute et al., 2019). Related terms include cortical arousal, sustained attention, and tonic alertness (Oken et al., 2006; Sadaghiani et al., 2010; Olbrich et al., 2011).

EEG-based measures

Quantitative measures of vigilance based on electroencephalographic (EEG) measures have been developed and applied by many investigators over the course of the last century (Oken et al., 2006; Olbrich et al., 2009; Knaut et al., 2019). For differentiating wakefulness from sleep and characterizing different sleep stages, EEG-based metrics have been standardized by the American Academy of Sleep Medicine (AASM, 2009), with a sleep stage score assigned to each 30 s epoch. There are five stages: a wake (W) stage, three stages (N1, N2, N3) of nonrapid eye movement (NREM) sleep, and a rapid eye movement (R) stage. The three NREM stages are as follows: N1 for the state that occurs between sleep and wakefulness, N2 for light sleep, and N3 for deep sleep or slow-wave sleep.

For rsFMRI studies, we are interested in fluctuations that primarily occur in the bandwidth of 0.01–0.1 Hz, corresponding to periods of 10–100 s. Thus, there is interest in measures that have a temporal resolution finer than the 30 s epochs used to define AASM sleep stages. In particular, since most rsfMRI studies assume that subjects are awake during the scan, we would especially like to characterize temporal fluctuations in arousal and vigilance that occur between the wake and N1 stages. A number of relevant EEG-based metrics have been proposed and are summarized in Table 1. Although the specific details vary, these metrics typically compare the EEG power in middle-frequency bands (e.g., α and β bands) that are associated with increased wakefulness to the power in low-frequency bands (e.g., δ and θ) that are associated with decreased wakefulness (Klimesch,

Table 1 Examples of vigilance metrics.

Name	Description	Electrodes	Period	References
Inverse index of wakefulness	$\sqrt{\dfrac{P_{\delta,\theta[2-7\,\text{Hz}]}}{P_{\alpha[8-12\,\text{Hz}]}}}$	C3,4; P3,4	120 s	Horovitz et al. (2008)
Alpha slow wave index (1)	$\dfrac{P_{\alpha[8-11.5\,\text{Hz}]}}{P_{\delta,\theta[2-8\,\text{Hz}]}}$	Cz	30 s	Jobert et al. (1994)
Alpha slow wave index (2)	$\dfrac{P_{\alpha[8-12\,\text{Hz}]}}{P_{\delta,\theta[1-8\,\text{Hz}]}}$	C3	30 s	Larson-Prior et al. (2009)
EEG vigilance (1)	$\dfrac{P_{\alpha[8-12\,\text{Hz}]}}{P_{\delta,\theta,\alpha[2-12\,\text{Hz}]}}$	F3,4; O1,2	3 s	Olbrich et al. (2009)
EEG vigilance (2)	$\sqrt{\dfrac{P_{\alpha[7-13\,\text{Hz}]}}{P_{\delta,\theta[1-7\,\text{Hz}]}}}$	All	1.8 s	Wong et al. (2013)
EEG wakefulness index	$\dfrac{P_{\theta_f,\alpha_o,\sigma_o,\beta_f[4-30\,\text{Hz}]}}{P_{\delta_f,\theta_o,\alpha_f,\sigma_c,\beta_f[0.5-30\,\text{Hz}]}}$	F3,4; O1,2; C3,4	2 s	Knaut et al. (2019)

Continued

Table 1 Examples of vigilance metrics—cont'd

Name	Description	Electrodes	Period	References
EEG alertness index	$\sqrt{\dfrac{P_{\alpha[8-12\,Hz]}}{P_{\theta[3-7\,Hz]}}}$	P3,4,z; O1,2,z	2.1 s	Goodale et al. (2021)
LFP arousal index	$\sqrt{\dfrac{P_{\beta[15-25\,Hz]}}{P_{\theta[3-7\,Hz]}}}$	Intracranial: V1,V2, F, P	2.6 s	Chang et al. (2016)
Frontal orbital beta power	$P_{\beta:[13-30\,Hz]}$	F3,4,z; O1,2,z	4 s	Mayeli et al. (2020)
Pupillometry	Pupil diameter	NA	>20 ms	Schwalm and Rosales Jubal (2017)
Behavioral arousal index	% Eyelid opening	NA	2.6 s	Chang et al. (2016)

The notation $P_{\delta,\theta[f_1-f_2\,Hz]}$ indicates power in the indicated EEG frequency bands (e.g., δ and θ bands), as well as the minimum f_1 and maximum f_2 frequencies covered by the collection of indicated bands. EEG electrode locations are specified with the standard notation of F, O, C, and P for frontal, occipital, central, and parietal regions, respectively. For metrics where the band powers in the definition are limited to certain regions, these constraints are indicated as subscripts, with the subscripts f, o, and c referring to frontal, occipital, and central regions, respectively. For example θ_f indicates θ band power from the frontal region. With the exception of the Inverse Index of Wakefulness, all of the metrics are designed to increase with vigilance. For the vigilance metric proposed by Olbrich et al. (2009), the expression provided in the table is used to define two major stages of vigilance, each of which has three substages as defined in the cited paper. In the work of Wong et al. (2013), variants of the EEG vigilance metric defined using regional subsets of electrodes were also used. For the metrics in Chang et al. (2016), the electrode locations refer to the placement of intracranial electrodes. This is an updated version of a table from Liu and Falahpour (2020).

1999; Oken et al., 2006). For example, in Jobert et al. (1994), Larson-Prior et al. (2009), and Wong et al. (2013), the proposed metrics have the form of either the ratio of the power in the alpha band to the power in the delta and theta bands or the square root of this ratio. This ratio is sometimes referred to as the alpha slow-wave index. Other variations include using the inverse of the square root of the alpha slow-wave ratio as an inverse index of wakefulness (Horovitz et al., 2008) or computing the ratio of the power in the alpha band to the power in the delta, theta, and alpha bands (Olbrich et al., 2009). While most metrics are computed using a predefined subset of the EEG electrodes, Knaut et al. (2019) recently proposed an EEG wakefulness index that is a ratio of powers that depends on both the EEG frequency band and topography. Examples of EEG spectra obtained in different vigilance states are shown in the lower row of Fig. 1.

While measuring EEG activity in the MRI environment is currently the gold standard for quantitative assessment of vigilance in rsfMRI studies, it is very rarely done due to the significant cost and technical and logistical challenges associated with the process. Simultaneous

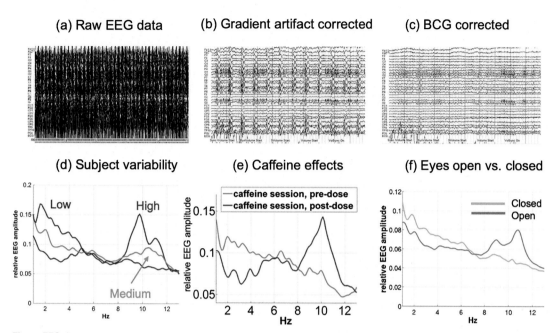

Fig. 1 EEG data and spectra. Upper row: EEG data acquired in the MRI environment. (A) Raw EEG data prior to correction for artifacts; (B) after correction for gradient artifacts; (C) after additional correction for ballistocardiogram (BCG) artifacts. Lower row: Examples of EEG spectra for different vigilance states. (D) Spectra from scans with low, medium, and high vigilance levels; (E) spectra before and after a 200 mg caffeine dose; and (F) spectra in the eyes-open and eyes-closed states. Data in panels (D, E) from Wong et al. (2013); data in panel (F) from Wong et al. (2015).

EEG and fMRI measurements require specialized MRI-compatible EEG equipment, and the acquired EEG data are highly contaminated by the time-varying electric fields generated by the switching of the MRI gradient fields, cardiac-induced motion of the EEG wires within the strong main magnetic field (e.g., 3 Tesla) used for most rsfMRI studies (known as ballistocardiogram artifact), and subject motion within the magnetic field. While tremendous advances have been made in both software and hardware solutions to minimize the contamination of the EEG data, careful and time-intensive processing of the data is still required to achieve high-quality data for use in further analysis. The upper row in Fig. 1 shows examples of EEG data acquired before and after correction for gradient and ballistocardiogram (BCG) effects. In addition to the processing challenges, the time and effort required to place the EEG cap on the subject's head prior to the fMRI scan are often not compatible with the tight time and budget constraints of many rsfMRI studies. Due to these various factors, relatively few rsfMRI studies currently acquire simultaneous EEG and fMRI measurements.

Measures of pupil size and eye closure

For rsfMRI studies where the subjects are instructed to keep their eyes open, vigilance can be assessed with measures of pupil diameter or eyelid closure. While these measures are generally easier to obtain than EEG measures, they still have their challenges. Pupillometry has the advantage of high temporal resolution but the need for specialized equipment and the time and expertise required to set up the equipment and obtain reliable measures in the MRI environment have limited its wider adoption (Yellin et al., 2015; Schneider et al., 2016; Breeden et al., 2017). In a rsfMRI study that used both pupillometry and EEG, Mayeli et al. (2020) found a significant temporal correlation between pupil size measures and EEG-based measures of frontal and occipital beta power.

Measures of eye closure can be obtained with an MRI-compatible video camera and have been used to assess drowsiness and the presence of microsleeps (Poudel et al., 2014; Chang et al., 2016; Wang et al., 2016). The main challenge associated with this approach lies in the processing of the video data to obtain reliable measures of eyelid closure, which typically requires some level of manual assessment and correction. However, the development and validation of machine learning-based approaches may minimize this concern. As the cost of an MRI-compatible video camera is considerably lower than that of either MRI-compatible EEG or pupillometry equipment, it has the potential to be more widely adopted in rsfMRI studies. One of the limitations of both the pupillometry and eyelid closure approaches is that they do not readily provide a measure of vigilance when a subject closes their eyes. This challenge may be partially addressed in the future with advanced image processing approaches that can estimate eyeball motion even when the eyes are closed.

Physiology-based measures

Although not always obtained, measures of respiratory and cardiac activity can be readily acquired with the standard peripheral equipment and software that is provided with modern MRI systems (as further discussed in Chapter 10). Additional measures, such as galvanic skin resistance or end-tidal carbon dioxide levels, can be obtained via additional MRI-compatible equipment. As discussed in more detail below, these physiological measures have been shown to be related to measures of vigilance (Yuan et al., 2013; Özbay et al., 2019) and are sometimes used as a proxy for vigilance. For example, Raut et al. (2021) used a measure of respiratory variation as an indicator of arousal. However, physiological activity typically reflects a host of additional factors that may not be directly related to vigilance, and the use of physiological measures as independent indicators of vigilance should therefore be viewed with some caution.

Because of the challenges associated with acquiring independent measures of vigilance during rsfMRI studies, there has been growing interest in data-driven approaches that use fMRI data to estimate vigilance levels. This is especially relevant in studies where there is limited time available for additional set-up procedures. We will describe these data-driven approaches below in the section on "fMRI-based vigilance estimates".

Links between vigilance and the rsfMRI signal

Our understanding of the role of vigilance in rsfMRI has grown steadily over the past two decades and benefited from findings from a broad range of experimental studies. The approaches used in these studies include long duration scans to examine the transition between wake and sleep, manipulation of the vigilance state with pharmacological agents, scanning subjects with their eyes opened or closed, studying subjects after sleep deprivation, and making use of the naturally occurring variability in vigilance across subjects and time of day.

Long-duration scans

Key insights into vigilance effects in rsfMRI have emerged from studies that have used relatively long-duration (e.g., 30 min) resting-state scans that facilitate observation of changes in the rsfMRI signal that occur as subjects fluctuate between sleep stages. These studies have reported changes in the amplitude of the rsfMRI BOLD signal and both static and dynamic measures of functional connectivity (FC).

Defining amplitude as the standard deviation of the BOLD time course, Fukunaga et al. (2006) found that the mean BOLD signal amplitude in the visual cortex increased during early sleep stages, with amplitudes that were comparable to those observed with visual stimulation. A follow-up study by the same group confirmed the earlier findings and also reported that increases in BOLD signal amplitude were related to increases in the Inverse Index of Wakefulness (see Table 1) in multiple brain regions, including the visual cortex, auditory cortex, and precuneus (Horovitz et al., 2008). An example of this increase is shown in Fig. 2C. Using a different approach, Olbrich et al. (2009) regressed the BOLD signal onto measures of vigilance state (see Table 1) and examined the amplitudes of the regression fit coefficients as a function of state. They found that decreases in vigilance were associated with an increase in the BOLD signal amplitude in the occipital cortex, the anterior cingulate, the frontal cortex, the parietal cortices,

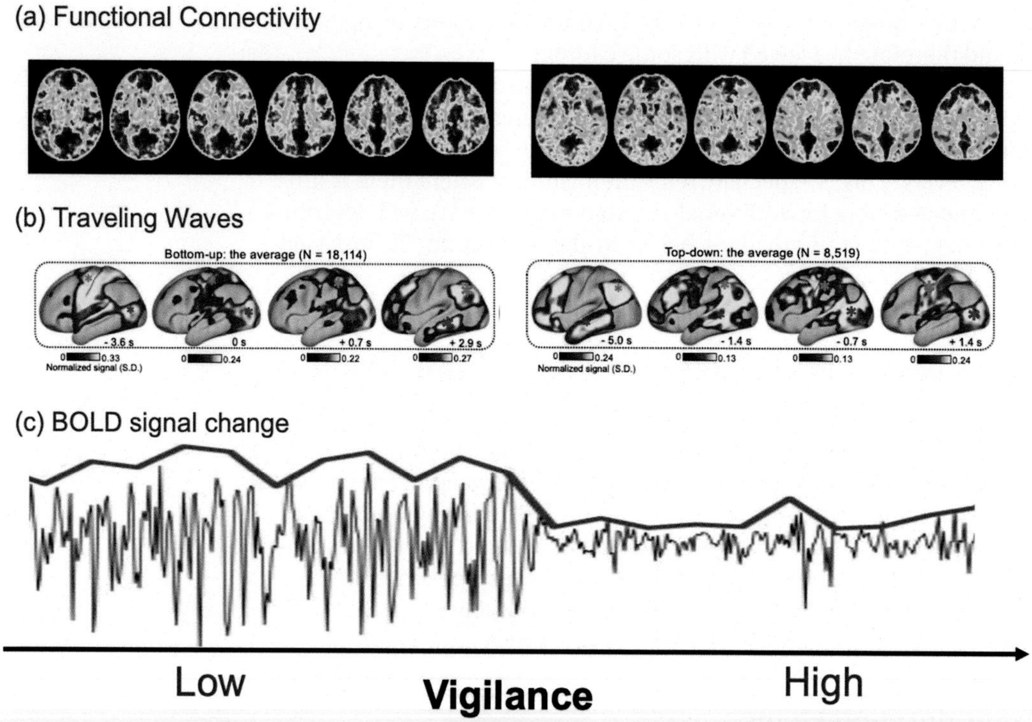

Fig. 2 Relationships between vigilance and the rsfMRI signal, with vigilance increasing from left to right. (A) Functional connectivity maps obtained in a low vigilance state prior to caffeine (lefthand side) and high vigilance state after caffeine (right-hand side) from Wong et al. (2012). (B) Snapshots of bottom-up traveling wave (left-hand side) and top-down traveling wave (right-hand side) from Gu et al. (2021; used with permission). There is a greater prevalence of bottom-up waves when subjects are in a low vigilance state. (C) An example of rsfMRI signal fluctuations in the visual cortex of a subject transitioning from a state of low vigilance to a state of high vigilance (Horovitz et al., 2008; used with permission). The line above the signal indicates the standard deviation of the signal computed over 2-min intervals.

and the temporal cortices and a decrease in BOLD signal amplitude in the thalamus and frontal regions. Similarly, Bijsterbosch et al. (2017) reported that the BOLD signal amplitude in primary sensory and motor areas increased over the course of a 15-min rsfMRI scan, presumably as subjects became less alert.

In addition to regional effects, investigators have considered global effects, often with a focus on the rsfMRI global signal, defined as the average of the BOLD signals in either the entire brain or gray matter regions. Larson-Prior et al. (2009) found that the global signal spectral power significantly increased during light sleep as compared with awake states, accompanied by a general trend toward significant power increases in individual regions of interest. Similarly, McAvoy et al. (2019) reported that the amplitude of the global mean signal increased with sleep depth and concluded that the increase in global signal amplitude reflected a proportionally greater decrease in oxygen consumption with sleep as compared to the sleep-related decrease in blood flow. Using a series of 20-min long rsfMRI scans, Soon et al. (2021) described the presence of large widespread modulations of the BOLD signal in response to microsleep incidents. The key features of the response were consistent across microsleep durations and were tightly linked with changes in respiration and heart rate. Taken as a whole, the long duration studies suggest a strong overall link between decreases in vigilance and increases in BOLD signal amplitude.

Measures of FC are of primary interest for many rsfMRI studies, and a number of studies have examined the relation between FC and vigilance with long-duration studies. Both Larson-Prior et al. (2011) and Sämann et al. (2011) found decreases in the extent of anticorrelations between the default mode network (DMN) and the task-positive network (TPN) during the transition to light sleep. Laumann et al. (2017) found that a multivariate measure of kurtosis was significantly correlated with an index of sleep, suggesting the temporal variability of FC measures increased with sleep. As a demonstration of the link between vigilance and FC, machine learning approaches have been used to classify sleep stages using FC matrices computed over relatively short-time (e.g., 60 s) windows, with reliable classifications obtained with both nonlinear and linear support vector machines (Tagliazucchi et al., 2012a; Altmann et al., 2016). Clustering of windowed FC estimates represents one approach for identifying dynamic FC states, and it has been shown that the estimated states roughly correspond to the average FC states found in different sleep stages, suggesting that variations in dynamic FC states are associated with fluctuations in wakefulness (Haimovici et al., 2017; Zou et al., 2019). Stevner et al. (2019) used a hidden Markov model (HMM) approach to associate multiple FC states with each sleep stage and characterized the transitions between the states, describing a level of complexity of brain activity beyond what can be characterized with standard sleep-stage scoring.

Pharmacological manipulation

Pharmacological agents can be used effectively to probe the relationship between vigilance and the rsfMRI signal. Of particular interest are the effects of caffeine, which is a widely used stimulant that can lead to acute increases in vigilance in subjects enrolled in rsfMRI studies. In a pair of related papers, Wong and colleagues found that a 200 mg dose of caffeine significantly reduced the amplitude of the rsfMRI global signal and later demonstrated that increases in vigilance due to caffeine were significantly correlated with decreases in the amplitude of the global signal (Wong et al., 2012, 2013). Fig. 1E shows example EEG spectra obtained before and after the caffeine dose. With regards to FC, the same investigative team showed that caffeine induces spatially widespread decreases in rsfMRI FC measures (Wong et al., 2012) and demonstrated that caffeine-induced increases in EEG vigilance were significantly correlated with increases in the anticorrelation between nodes of the DMN and TPN (Wong et al., 2013). Examples of the caffeine-induced FC reductions are shown in Fig. 2A. The rsfMRI observations were shown to be consistent with magnetoencephalography-based observations of decreased resting-state connectivity due to caffeine administration in the same sample of subjects (Tal et al., 2013). The increased presence of anticorrelations was largely attributed to the caffeine-related reduction in resting-state global activity (Wong et al., 2012, 2013). Furthermore, the caffeine-induced increases in anticorrelation were consistent with the decreases in anticorrelations observed in the transition to light sleep, as discussed above (Larson-Prior et al., 2011; Sämann et al., 2011).

Studies using pharmacological agents that reduce vigilance have largely found effects consistent with those observed in the transition to sleep in the long duration studies previously discussed. For example, Kiviniemi et al. (2005) found that the sedative midazolam led to both an increase in the spectral power of low frequency BOLD fluctuations and an increase in FC within the sensory-motor network. In a later study, Greicius et al. (2008) also reported a midazolam-related increase in FC in the sensory-motor network but reported a decrease in FC in the DMN. Similarly, Licata et al. (2013) reported that the sedative zolpidem resulted in a drug-related increase in FC in a number of sensory, motor, and limbic networks. In a study investigating the effects of the depressant alcohol, Esposito et al. (2010) found that alcohol led to an increase in spontaneous BOLD fluctuations in the visual cortex.

Eyes open versus eyes closed

A major decision in the design of rsfMRI studies is whether to scan subjects in the eyes-closed (EC) or eyes-open (EO) condition. While there has historically been quite a bit of variability in the choice of condition, there is a growing trend to scan subjects in the EO state, usually with fixation, as this experimental condition can provide greater consistency in resting-state metrics (Patriat et al., 2013). Indeed, subjects scanned in the EC state tend to have lower vigilance and are more likely to fall asleep during a scan (Wong et al., 2015; Tagliazucchi and Laufs, 2014).

In general, prior studies have found that the amplitude of the resting-state BOLD signal is decreased in the EO condition as compared to the EC condition (Bianciardi et al., 2009; Jao et al., 2013; McAvoy et al., 2008, 2012; Patriat et al., 2013; Xu et al., 2014; Yan et al., 2009; Yang et al., 2007; Yuan et al., 2014; Zou et al., 2009). For example,

Jao et al. (2013) found that the average variance of the BOLD signal was significantly lower in the EO condition. However, there is variability in the findings, with regional resting-state activity sometimes found to be higher in the EO condition. These differences likely reflect variations in processing approaches, such as the use of global signal regression and physiological noise reduction in some studies and not others.

In a study utilizing simultaneous EEG fMRI, Wong et al. (2015) reported increases in EEG vigilance in the EO state versus the EC state and found a significant negative correlation (across subjects) between the observed increases in vigilance and changes in global signal amplitude. The observed negative relation had nearly the same slope as that of a previously reported relationship between caffeine-induced changes in vigilance and global signal amplitude (Wong et al., 2013), suggesting that the basic mechanisms underlying the vigilance and global signal amplitude relationship may be somewhat independent of the experimental manipulation. Fig. 1F shows example EEG spectra in the EO and EC states.

Studies examining the effects of eye state on functional connectivity have found generally lower FC values in the EO state as compared to the EC state (Bianciardi et al., 2009; Xu et al., 2014; Zou et al., 2009; McAvoy et al., 2008) that are consistent with decreased global activity and increased vigilance in the EO state. It is also likely that decreases in global signal amplitude and increases in vigilance may account for the increased reliability of connectivity measures that have been observed in the EO state as compared to the EC state (Patriat et al., 2013). However, is important to note that effects may be network-dependent. For example, Costumero et al. (2020) found higher FC between the visual cortex and DMN and sensorimotor networks for subjects scanned in the EC state and higher FC between the visual network and the salience network for another group of subjects scanned in the EO state.

Allen et al. (2018) examined dynamic FC (DFC) states in both the EO and EC conditions. They identified a DFC state related to increased drowsiness (lower alpha and higher delta and theta power) in which there were high levels of FC in the sensorimotor and visual regions and the increased presence of anticorrelations between the thalamus and these regions. This state was found in both EC and EO conditions but occurred more frequently in the EC state. Weng et al. (2020) reported evidence for a hyperconnected DFC state that was more prevalent in the EC state and a hypoconnected DFC state that was more prevalent in the EO state.

Sleep deprivation

Scanning subjects who have been deprived of sleep constitutes a powerful paradigm for examining the effects of reduced vigilance on the rs-fMRI signal (Chee and Zhou, 2019). In a study of subjects who were scanned both after a night of regular sleep and after one night of sleep deprivation (SD), De Havas et al. (2012) found sleep deprivation-related reductions in both DMN functional connectivity and the degree of anticorrelation between the DMN and other regions. In a follow-up study, Yeo et al. (2015) reported sleep deprivation related increases in the amplitude of the global signal and found that subjects who exhibited less vigilance declines after sleep deprivation showed stronger anticorrelations among several networks when global signal regression was used. However, when global signal regression was not applied, the authors observed a spatially widespread increase in functional connectivity with sleep deprivation (Yeo et al., 2015). Other studies have also reported spatially widespread increases in FC with sleep deprivation (Wirsich et al., 2018; Kaufmann et al., 2016), increases between the cerebellum and bilateral caudate (Zhang et al., 2019), and increases in FC density in sensory integration and arousal regulating areas, such as the thalamus (Yang et al., 2018). However, some networks show the opposite effect with FC decreases observed between the cerebellum and a number of brain regions (Zhang et al., 2019) and FC density decreases in the posterior cingulate cortex and precuneus (Yang et al., 2018). The variability in the findings most likely reflects both biological differences in the response of various functional networks to sleep deprivation and methodological differences in experimental paradigms (e.g., hours of sleep deprivation), the functional networks examined, and specific details of the processing and analysis approaches adopted.

Further insights into the effects of sleep deprivation have been gained by examining the behavior of the rs-fMRI signal around certain events, such as spontaneous eye closures and microsleeps. When examining spontaneous eye closures in sleep deprived subjects, Ong et al. (2015) found reductions in the FC in the DMN and DAN that exceeded those that had been previously associated with sleep deprivation. In related work, Wang et al. (2016) identified a low-arousal DFC state associated with spontaneous eye closures and another high-arousal state associated with periods of the eyes remaining wide open. A follow-up study found that subjects with a greater fraction of high arousal states showed higher levels of vigilance, working memory, and processing speed after sleep restriction (Patanaik et al., 2018). In a study of microsleeps, Poudel et al. (2018) observed spatially widespread increases in the BOLD signal coincident with microsleep events after both normal rest and sleep-deprivation conditions.

Naturally occurring variations across scans and time of day

Our understanding of the role of vigilance in rs-fMRI studies has also been enhanced by considering variations in vigilance that occur across subjects, scans, and time of day. Wong et al. (2013) found a strong and significant negative correlation between the amplitude of the rs-fMRI global signal and EEG vigilance measures across scans and subjects in the EC condition, with a weaker and nearly significant correlation observed in the EO condition. Scans for which the subjects exhibited relatively higher vigilance levels had lower global signal amplitudes, while scans with relatively lower vigilance levels were associated with higher global signal amplitudes. Fig. 1D shows example EEG spectra from scans with low, medium, and high vigilance levels.

In comparing rs-fMRI activity across two 15-min rs-fMRI runs acquired within the same session, Bijsterbosch et al. (2017) found increased sensorimotor rs-fMRI signal amplitude in the second scan, possibly reflecting a reduction in vigilance. In addition, a large number of FC changes were observed between scans and these changes could be accounted for by changes in the signal amplitudes of the rs-fMRI networks.

In considering the time of day effects, Cordani et al. (2018) found that the rs-fMRI BOLD signal amplitude in the sensory cortices decreased at times corresponding to dawn and dusk, possibly reflecting an anticipatory mechanism in which vigilance is increased and spontaneous activity is reduced in order to improve visual perception during times associated with low light levels. In other studies of the time of day effects, regionally dependent increases and decreases were found in both measures of low-frequency amplitude and FC (Jiang et al., 2016; Steel et al., 2019). Orban et al. (2020) hypothesized that rs-fMRI global signal amplitude would follow the known circadian rhythm of arousal and be lowest in the morning, increase in the midafternoon, and dip in the early evening. Paradoxically, they observed a steady decrease in global signal amplitude and FC with time of day, possibly reflecting the effects of additional mechanisms, such as anxiety or arousal related to the anticipation of a scan session. Consistent with prior findings (e.g., Bijsterbosch et al. (2017)), the authors did find an increase in global signal amplitude with increased time in the scanner, suggesting that the link between arousal and the global signal may depend on timescale (Orban et al., 2020).

Dynamics of vigilance effects

So far, we have primarily considered vigilance effects at time scales greater than or equal to the standard sleep staging epoch of 30 s. However, from firsthand experience, we know that there are moment-to-moment variations in vigilance occurring at much shorter time scales. In considering the relation between vigilance and the rs-fMRI signal at finer time scales, we are typically limited by the temporal resolution of fMRI data acquisition, which is currently on the order of 1 s for whole brain acquisitions, and not by the temporal resolution of vigilance estimates, such as EEG, which are on the order of a millisecond. Thus, vigilance estimates are usually smoothed and downsampled to the temporal resolution of the fMRI data prior to computing metrics, such as temporal correlation, relating the two signals.

Temporal fluctuations in vigilance

A number of simultaneous EEG fMRI studies have reported that EEG alpha power is negatively correlated with fMRI signals in widespread regions of the brain, including the visual and frontoparietal cortices (Goldman et al., 2002; Laufs et al., 2003a, b; Moosmann et al., 2003), but positively correlated in other regions such as the thalamus, insula, and anterior cingulate (Goldman et al., 2002; Feige et al., 2005; Difrancesco et al., 2008; Sadaghiani et al., 2010; Moosmann et al., 2003). Given the relationship between EEG alpha power and vigilance, these studies were instrumental in establishing a link between moment-to-moment variations in vigilance and the fMRI signal. In later work, Falahpour et al. (2018a) found similar spatial patterns of correlation obtained with the metric of EEG vigilance used in Wong et al. (2013) and further demonstrated a significant negative correlation between the rs-fMRI global signal and EEG vigilance time series. An example of this negative correlation is shown in Fig. 3A, where the plot in the middle row shows the vigilance time series in blue and the global signal (inverted for display) in red. The top row shows the rs-fMRI images obtained by averaging over time points corresponding to the top 10% of vigilance values. The appearance of positive signal values in the thalamus and negative values in sensory areas is consistent with the prior literature. In the bottom row, images obtained by averaging over time points with the lowest 10% of vigilance values shows the opposite pattern, with negative signal values in the thalamus and positive values in sensory areas.

The findings obtained with simultaneous EEG-fMRI are largely consistent with those reported in studies that use pupillometry and eyelid closure. Schneider et al. (2016) reported that spontaneous pupil dilations (an indicator of increased vigilance) were associated with increased BOLD activity in the salience network, thalamus, and frontoparietal regions. In contrast, spontaneous pupil constrictions were associated with increased BOLD activity in the visual and sensorimotor areas. Several other studies have reported similar findings of a positive correlation between pupil size and the rs-fMRI BOLD signal in regions comprising cingulo-opercular, default mode, and frontoparietal networks and a negative correlation in the visual and sensorimotor regions (Yellin et al., 2015; Breeden et al., 2017; DiNuzzo et al., 2019). Related work has demonstrated a positive correlation between pupil size and BOLD activity in the locus coeruleus (Murphy et al., 2014; DiNuzzo et al., 2019), a brainstem structure containing

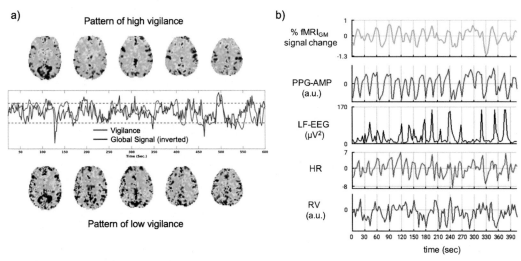

Fig. 3 (A) Example of the negative correlation ($r = -0.33$) between vigilance time course (blue) and the rs-fMRI global signal (*red*; inverted for display) from Liu and Falahpour (2020). Average rs-fMRI images from time points corresponding to the top and lower 10% of vigilance values are shown above and below, respectively, the time courses. (B) Example from Özbay et al. (2019) of the relationship between rs-fMRI global signal (computed within gray matter) and low-frequency EEG power (LF-EEG), amplitude of the photoplethysmography signal (PPG-AMP) as a measure of vascular tone, respiratory variation (RV), and heart rate (HR) during N2 sleep. Increases in LF-EEG (an indicator of subcortical arousal related to the occurrence of K-complexes) are associated with decreases in both the global signal and PPG-AMP.

norepinephrine neurons that are thought to modulate pupil size (Joshi et al., 2016). Using infrared video showing the status of the eyes in nonhuman primates, Chang et al. (2016) found widespread negative correlations between eyelid closure and the BOLD signal, but reported positive correlations for the thalamus and cerebellum. The thalamic findings were extended in human studies performed by Iidaka (2021), who reported that among the major subthalamic nuclei, the greatest association with arousal was found in the mediodorsal nucleus.

In light of the observed correlation between vigilance and the rs-fMRI BOLD signal, we should also expect to find that vigilance fluctuations will influence rs-fMRI-derived dynamic FC estimates. Indeed, Tagliazucchi et al. (2012b) found that time-varying increases in alpha power were correlated with decreases in functional connectivity in awake subjects, consistent with a reduction in BOLD signal amplitude with increased vigilance. In similar work, Scheeringa et al. (2012) reported that FC in the visual system decreased with increases in alpha power and Chang et al. (2013) observed that the time-varying strength of connectivity between the DMN and default attention network (DAN) was inversely proportional to the alpha power measured within the same time window (40-s window length).

Transient events

Prior work has shown that key features of the rs-fMRI signal and derived metrics are largely driven by transient events with large signal excursions (Liu and Duyn, 2013; Tagliazucchi et al., 2016; Nalci et al., 2017). In an early study examining the role of transient activity in EEG-BOLD correlations, Poudel et al. (2014) observed transient changes in BOLD activity associated with microsleeps during the performance of a continuous tracking task. Furthermore, in subjects who exhibited a higher occurrence of microsleeps, the authors found that postcentral EEG theta was positively correlated with the BOLD signal in the thalamus, basal forebrain, visual, posterior parietal, and prefrontal cortices. Similarly, Soon et al. (2021) observed large BOLD signal transients at both microsleep onset and awakening and found that these were associated with EEG vigilance measures. Han et al. (2019) hypothesized that the observed correlations between the EEG and rs-fMRI signals reflected sequential spectral transition (SST) events originally reported in large-scale electrocorticography (ECoG) recordings performed in monkeys (Liu et al., 2015) and later shown to be coupled with peaks in the rs-fMRI global signal (Liu et al., 2018). SST events are thought to be linked to transient modulations in arousal and last for 10–20s with an initial decrease in mid-band (alpha and beta; 8–30 Hz) activity followed by an increase in low-frequency (delta and theta; <8 Hz) activity and a burst of high-frequency broadband gamma activity (>30 Hz) (Liu et al., 2015).

Traveling waves

There is a growing appreciation that traveling waves may play a key role in the dynamics of the brain at multiple spatial and temporal scales (Muller et al., 2018; see also Chapter 12). In the area of rs-fMRI, quasi-periodic patterns (QPP) of activity that move across the brain were observed by Majeed et al. (2009, 2011) and later shown to be related to electrophysiological measures of resting-state activity in rats (Thompson et al., 2014). These quasiperiodic patterns appear to be related to the global signal, with the type of QPP depending on the amplitude of the global signal (Liu et al., 2017; Yousefi et al., 2017). Gu et al. (2021) examined time segments associated with large global signal amplitudes and found evidence for waves propagating along the principal gradient (PG) of rs-fMRI connectivity, with both bottom-up (sensorimotor to DMN) and top-down (DMN to sensorimotor) propagation observed. The authors further found a dependence of the relative proportion of the bottom-up and top-down waves on vigilance state and demonstrated an association between subcortical activity and the occurrence of the propagating waves. Examples of these waves are shown in Fig. 2B. Liu et al. (2021) adapted the analytical approach from Gu et al. (2021) to identify global traveling waves in the mouse brain that were closely linked to physiological measures of arousal, such as pupil diameter. Using a measure of low frequency respiratory variability as an indicator of arousal, Raut et al. (2021) presented evidence from human rs-fMRI data for arousal-related traveling waves propagating along the PG and also demonstrated traveling waves associated with SST events in nonhuman primates. Recently, Gu et al. (2022) reported that SST-like events in human EEG data were associated with an orderly spatiotemporal pattern of fMRI deactivations and activations lasting for about 20 s.

The global signal

As touched upon in previous sections, the role of vigilance in rs-fMRI is tightly tied to the global signal, an entity which has received considerable attention in the rs-fMRI literature (Liu et al., 2017) and is discussed further in Chapter 5. Much of this attention has focused on a commonly used preprocessing step known as global signal regression (GSR) in which the global signal is regressed out of rs-fMRI data prior to the computation of various metrics of interest, such as measures of FC. While GSR has gained widespread adoption, its use has sparked a great deal of controversy, with some investigators touting its advantages while others argue strongly against its use (Liu et al., 2017). Nalci et al. (2017) demonstrated that GSR can be approximated as a temporal downweighting process that effectively censors time points with large global signal amplitudes. Given the relation between the global signal and vigilance fluctuations, GSR will also tend to censor out time points associated with large vigilance-related fluctuations and can thereby reduce the effects of these fluctuations on the rs-fMRI signal. As a consequence, GSR can have a significant effect on studies examining the relation between vigilance and the rs-fMRI signal. For example, Falahpour et al. (2018b) noted that prior studies that did not use GSR generally found a negative correlation between EEG alpha power (or vigilance) and the BOLD signal in widespread regions of the brain, including the lingual gyrus, posterior cingulate, cuneus, and precuneus (Goldman et al., 2002; Laufs et al., 2003b; Falahpour et al., 2018a). In contrast, in a study that used GSR, Sadaghiani et al. (2010) found positive correlations in additional areas not reported in prior studies, including the dorsal anterior cingulate cortex, the anterior insula, and the anterior prefrontal cortex, possibly reflecting a GSR-related positive shift in the correlation between EEG vigilance and the rs-fMRI signal (Falahpour et al., 2018b).

fMRI-based vigilance estimates

As discussed in the section "Assessing vigilance", the MRI environment can pose significant challenges to assessing vigilance with standard approaches such as EEG. These challenges have led to increased interest in methods for estimating vigilance from the fMRI data alone. The seminal work of Tagliazucchi et al. (2012a) demonstrated that sleep-stage estimates could be obtained with nonlinear support vector machines applied to windowed rs-fMRI connectivity estimates. This approach was later extended by Altmann et al. (2016) with the use of linear support vector machines.

Chang et al. (2016) studied arousal fluctuations in awake monkeys sitting in complete darkness and introduced a template-based approach to predict the fluctuations. To generate a spatial template, the authors computed on a per-voxel basis the correlation between the fMRI data and an eye-based metric of arousal. They then correlated the resulting spatial template with each volume of an independent fMRI data set to form an estimate of arousal for each timepoint in the test data set. This template-based approach was later extended to predict EEG-based estimates of vigilance acquired in humans (Falahpour et al., 2018a). The authors found that predictive performance was related to the standard deviation of the vigilance fluctuations, with higher performance obtained when subjects showed more variability in their vigilance fluctuations across the course of a scan. Furthermore, the template-based approach could be used to estimate the standard deviation of the vigilance fluctuations. Gu et al. (2020) used a global co-activation map as a template and demonstrated that the resulting estimates were comparable to those obtained using the template in (Falahpour et al., 2018a). A graphical summary of the template-based approach is provided in Fig. 4.

Applying the template-based approach to an auditory stimulus task, Goodale et al. (2021) showed that predictions of prestimulus alertness could account for a significant amount of the trial-to-trial variability in the response to stimuli. In addition, the authors showed that high performance could be achieved with as few as 1% of the voxels in the spatial map, as long as high magnitude voxels with both positive and negative signs were included in the reduced template. Similarly, Iidaka (2021) demonstrated that a template with voxels confined to the thalamus could be used to predict arousal. In a modified template-based approach, Poudel et al. (2021) first generated a template based on correlations between fMRI data and EEG theta power. For each timepoint, this template was then correlated with a map of beta coefficients obtained through temporal deconvolution of the fMRI data with a hemodynamic response function. Large values in the resulting correlation time series were identified as sleep-like events in a sample of subjects performing a cognitive task after partial sleep deprivation.

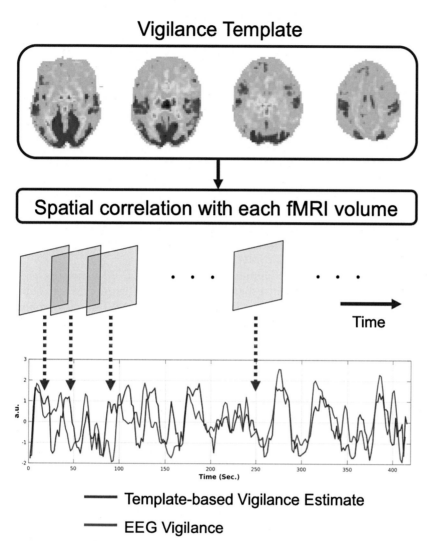

Fig. 4 Graphical summary from Liu and Falahpour (2020) of the template-based approach for the prediction of vigilance fluctuations. The vigilance template is obtained by correlating the fMRI data with an estimate of EEG vigilance and then applied to an independent simultaneous EEG-fMRI data set. For each timepoint in the fMRI data, the spatial correlation between the template and the fMRI volume is computed to form an estimate of vigilance *(red line)*. For this example, the estimate is highly correlated ($r = 0.51$) with the EEG-based measure of vigilance *(blue line)*.

Vigilance and disease

Because of its ease of use in clinical populations, there has been widespread interest in the use of rs-fMRI for the study of disease-related alterations in brain activity and connectivity. For example, rs-fMRI studies of schizophrenia have reported disease-related differences in functional connectivity and signal variance (Wang et al., 2015; Calhoun et al., 2011; Yang et al., 2014). Given the importance of vigilance for normal cognitive function, it is not surprising to find that alterations in vigilance are associated with a broad range of mental disorders such as depression, anxiety, autism, and schizophrenia (Razavi et al., 2013; Boutros et al., 2008; Jawinski et al., 2019; Sander et al., 2016; Ulke et al., 2018). For example, Jawinski et al. (2019) found an association between resting-state vigilance levels (as assessed with EEG) and genetic markers for major depressive disorder, autism spectrum disorder, and Alzheimer's disease.

In light of the overlapping links between rs-fMRI measures, vigilance, and disease, investigators should exert caution when both designing and interpreting the findings of rs-fMRI studies where disease-related vigilance effects may be at play. As an example, Yang et al. (2014) reported that the variance of the global signal was significantly higher in patients with schizophrenia as compared to normal controls and concluded that the differences reflected an increase in neural coupling. However, given the link between the global signal amplitude and vigilance (Wong et al., 2013) and evidence for decreases in EEG vigilance with schizophrenia (Razavi et al., 2013; Boutros et al., 2008), it is also possible that disease-related vigilance effects may have contributed to the observed differences, a potential confound acknowledged by the authors (Yang et al., 2014). Ideally, rs-fMRI studies would include measures of vigilance acquired during the acquisition of the rs-fMRI data or in a session that is temporally adjacent to the scan, such as a psychomotor vigilance task performed before or after the scans (Yeo et al., 2015). The fMRI-based vigilance estimation approaches described in the section "fMRI-based vigilance estimates" represent promising options that are likely to become more widely adopted, especially with further development and validation of the approaches. Recently, Tu et al. (2020) used the vigilance template from Falahpour et al. (2018a) to show that vigilance levels did not differ between a cohort of subjects with chronic lower back pain and a cohort of healthy controls, and were thus unlikely to account for group differences in brain dynamics. rs-fMRI studies would also benefit from self-reports of vigilance states, such as the estimated time spent in sleep and other mental states (Soehner et al., 2019; Gonzalez-Castillo et al., 2021).

The impaired clearance of metabolic waste products from the brain through the glymphatic system is receiving increasing interest as a potential mechanism for a range of neurodegenerative diseases (Rasmussen et al., 2018). A key component of the glymphatic system is the flow of cerebrospinal fluid (CSF). Fultz et al. (2019) demonstrated that rs-fMRI global signal measures were correlated with MRI-based measures of CSF flow in healthy subjects studied during non-REM sleep. While the underlying mechanisms are still under investigation, recent work suggests a key role for transient autonomic arousal events that do not require the presence of sleep (Picchioni et al., 2022). Han et al. (2021b) hypothesized that an impairment of the glymphatic system would reduce the coupling between CSF flow and arousal-related variations in the global signal and found that decreases in coupling were associated with increased risk for Alzheimer's disease and markers of pathology. Using a similar approach, Han et al. (2021a) reported that reduced coupling between CSF flow and the global signal was related to Parkinson's disease cognitive impairment.

Challenges and future directions

One of the key challenges faced in trying to advance our understanding of the role of vigilance in rs-fMRI lies in the interconnected nature of the sources that contribute to the rs-fMRI signal. In addition to intrinsic fluctuations in brain activity that are related to cortical functions (including vigilance), the rs-fMRI signal reflects system-related instabilities, subject motion, and physiological fluctuations (Liu, 2016). Interpretation of the rs-fMRI signal is complicated by the fact that factors such as subject motion, respiration, and cardiac activity not only have their origins in the brain networks that control these functions but also are related to the brain's state of arousal (Iacovella and Hasson, 2011). For example, Yuan et al. (2013) reported significant correlations between measures of EEG alpha power, respiratory activity, and the rs-fMRI BOLD signal. Gu et al. (2020) examined motion-related changes in FC measures and found evidence that these changes may primarily reflect arousal-related contributions as opposed to pure motion-related effects on the MRI signal.

The standard general linear model (GLM) approach for the processing of rs-fMRI data typically involves the inclusion of nuisance regressors to represent confounds that are not of interest (Liu, 2016). A common strategy is to investigate the relationship that is of interest (e.g., between vigilance and rs-fMRI signal) both before and after regressing out signal components that can be explained by the nuisance regressors. As examples, Yuan et al. (2013) demonstrated that the correlation between EEG alpha power and rs-fMRI signal was reduced after respiratory and cardiac nuisance regressors were projected out of the rs-fMRI data, and Patanaik et al. (2018) found that the relation between vigilance and the global signal was reduced when motion was used as a covariate. However, as GLM-based approaches do not fully take into account the complexity of the relationship between the rs-fMRI signal, vigilance, and physiological processes, it is likely that the resulting estimates are biased and potentially misleading. A more accurate understanding will depend on the development of models with biological validity and effective methods for estimating the parameters of such models. Advances in computational models that can model various vigilance states will also be beneficial (Hahn et al., 2021). The development and validation of improved models and methods will require high-quality data and well-designed experimental approaches, and the support and use of open multimodal neuroimaging databases and standardized approaches (Poldrack et al., 2017; Babayan et al., 2019) will be critical.

Noninvasive studies in human subjects have greatly contributed to our understanding of vigilance effects in rs-fMRI and will continue to play a key role. As the rs-fMRI signal is a complex reflection of neural, metabolic, and vascular factors (Liu, 2013), the use of complementary approaches such as noninvasive MRI measures of cerebral blood flow and metabolism and additional measures of physiology, such as skin vascular tone (Özbay et al., 2019), can help to advance our understanding (see also Fig. 3B). Studies of arousal in task-based fMRI studies can also provide insights (Li et al., 2021), and innovations in the design and analysis of both task-based and rs-fMRI experiments are likely to be important for advancing the field. In parallel, invasive studies in animal models are uniquely poised to offer additional insights and will continue to play a critical role in elucidating the mechanisms underlying the relationship between vigilance fluctuations and the rs-fMRI signal. For example, in a study with awake macaques, Turchi et al. (2018) used injections of the $GABA_A$ agonist muscimol to inactive two regions of the nucleus basalis of Meynert, a group of neurons in the basal forebrain with widespread arousal-related modulatory projections to the cortex. The injections led to reduced global signal fluctuations, with greater reductions observed in periods of drowsiness as compared to periods of alertness. A potential drawback of animal studies is the widespread use of anesthesia (Matsui et al., 2016; Zerbi et al., 2019), which can limit the relevance of these studies for understanding the dynamics of the nonanesthetized awake brain. This concern can be addressed by the continued development and application of methods for conducting rs-fMRI studies with awake animals (Ma et al., 2020).

Conclusion

A quote commonly attributed to the German philosopher Arthur Schopenhauer states that "All truth passes through three stages. First, it is ridiculed. Second, it is violently opposed. Third, it is accepted as being self-evident." Since the initial report of resting-state functional connectivity nearly three decades ago by Biswal et al. (1995), the field of rs-fMRI has gone through similar stages and is now firmly in the third stage where rs-fMRI is considered a standard approach that has been adopted by a wide range of studies, including a number of large-scale imaging projects. Yet despite the broad adoption of rs-fMRI and numerous studies investigating the characteristics of the rs-fMRI signal, the origins and mechanisms underlying the rs-fMRI signal are still not very well understood. As we have described in this chapter, there is now substantial evidence that vigilance variations can have a profound effect on the rs-fMRI signal. Furthermore, given the pronounced variations in vigilance that can occur across subjects and experimental conditions, these effects can have a significant impact on the interpretation of rs-fMRI studies, especially those involving populations where there are pronounced alterations of vigilance states. The continued development of approaches to better estimate and account for vigilance effects is thus likely to play a key role in improving the interpretation of rs-fMRI data in both clinical and research settings.

References

AASM, 2009. The AASM Manual for the Scoring of Sleep and Associated Events: Rules, Terminology and Technical Specification.

Allen, E.A., Damaraju, E., Eichele, T., Wu, L., Calhoun, V.D., 2018. EEG signatures of dynamic functional network connectivity states. Brain Topogr. 31 (1), 101–116.

Altmann, A., Schröter, M.S., Spoormaker, V.I., Kiem, S.A., Jordan, D., Ilg, R., Bullmore, E.T., Greicius, M.D., Czisch, M., Sämann, P.G., 2016. Validation of non-rem sleep stage decoding from resting state fMRI using linear support vector machines. NeuroImage 125, 544–555.

Babayan, A., Erbey, M., Kumral, D., Reinelt, J.D., Reiter, A.M.F., Röbbig, J., Schaare, H.L., Uhlig, M., Anwander, A., Bazin, P.-L., Horstmann, A., Lampe, L., Nikulin, V.V., Okon-Singer, H., Preusser, S., Pampel, A., Rohr, C.S., Sacher, J., Thöne-Otto, A., Trapp, S., Nierhaus, T., Altmann, D., Arelin, K., Blöchl, M., Bongartz, E., Breig, P., Cesnaite, E., Chen, S., Cozatl, R., Czerwonatis, S., Dambrauskaite, G., Dreyer, M., Enders, J., Engelhardt, M., Fischer, M.M., Forschack, N., Golchert, J., Golz, L., Guran, C.A., Hedrich, S., Hentschel, N., Hoffmann, D.I., Huntenburg, J.M., Jost, R., Kosatschek, A., Kunzendorf, S., Lammers, H., Lauckner, M.E., Mahjoory, K., Kanaan, A.S., Mendes, N., Menger, R., Morino, E., Näthe, K., Neubauer, J., Noyan, H., Oligschläger, S., Panczyszyn-Trzewik, P., Poehlchen, D., Putzke, N., Roski, S., Schaller, M.-C., Schieferbein, A., Schlaak, B., Schmidt, R., Gorgolewski, K.J., Schmidt, H.M., Schrimpf, A., Stasch, S., Voss, M., Wiedemann, A., Margulies, D.S., Gaebler, M., Villringer, A., 2019. A mind-brain-body dataset of MRI, EEG, cognition, emotion, and peripheral physiology in young and old adults. Scientific Data 6, 180308.

Bianciardi, M., Fukunaga, M., van Gelderen, P., Horovitz, S., de Zwart, J., Duyn, J., 2009. Modulation of spontaneous fMRI activity in human visual cortex by behavioral state. NeuroImage 45, 160–168.

Bijsterbosch, J., Harrison, S., Duff, E., Alfaro-Almagro, F., Woolrich, M., Smith, S., 2017. Investigations into within- and between-subject resting-state amplitude variations. NeuroImage 159, 57–69.

Biswal, B., Yetkin, F.Z., Haughton, V.M., Hyde, J.S., 1995. Functional connectivity in the motor cortex of resting human brain using echo-planar MRI. Magn. Reson. Med. 34 (4), 537–541.

Boutros, N.N., Arfken, C., Galderisi, S., Warrick, J., Pratt, G., Iacono, W., 2008. The status of spectral EEG abnormality as a diagnostic test for schizophrenia. Schizophr. Res. 99, 225–237.

Breeden, A.L., Siegle, G.J., Norr, M.E., Gordon, E.M., Vaidya, C.J., 2017. Coupling between spontaneous pupillary fluctuations and brain activity relates to inattentiveness. Eur. J. Neurosci. 45 (2), 260–266.

Calhoun, V.D., Sui, J., Kiehl, K., Turner, J., Allen, E., Pearlson, G., 2011. Exploring the psychosis functional connectome: aberrant intrinsic networks in schizophrenia and bipolar disorder. Front. Psychiatry 2, 75.

Chang, C., Liu, Z., Chen, M.C., Liu, X., Duyn, J.H., 2013. EEG correlates of time-varying BOLD functional connectivity. NeuroImage 72, 227–236.

Chang, C., Leopold, D.A., Schölvinck, M.L., Mandelkow, H., Picchioni, D., Liu, X., Ye, F.Q., Turchi, J.N., Duyn, J.H., 2016. Tracking brain arousal fluctuations with fMRI. Proc. Natl. Acad. Sci. 113 (16), 4518–4523.

Chee, M.W.L., Zhou, J., 2019. Functional connectivity and the sleep-deprived brain. Prog. Brain Res. 246, 159–176.

Cordani, L., Tagliazucchi, E., Vetter, C., Hassemer, C., Roenneberg, T., Stehle, J.H., Kell, C.A., 2018. Endogenous modulation of human visual cortex activity improves perception at twilight. Nat. Commun. 9 (1), 1274.

Costumero, V., Bueichekú, E., Adrián-Ventura, J., Ávila, C., 2020. Opening or closing eyes at rest modulates the functional connectivity of V1 with default and salience networks. Sci. Rep. 10 (1), 9137.

De Havas, J.A., Parimal, S., Soon, C.S., Chee, M.W.L., 2012. Sleep deprivation reduces default mode network connectivity and anti-correlation during rest and task performance. NeuroImage 59 (2), 1745–1751.

Difrancesco, M.W., Holland, S.K., Szaflarski, J.P., 2008. Simultaneous EEG/functional magnetic resonance imaging at 4 Tesla: correlates of brain activity to spontaneous alpha rhythm during relaxation. J. Clin. Neurophysiol.: Off. Publ. Am. Electroencephalogr. Soc. 25 (5), 255–264.

DiNuzzo, M., Mascali, D., Moraschi, M., Bussu, G., Maugeri, L., Mangini, F., Fratini, M., Giove, F., 2019. Brain networks underlying eye's pupil dynamics. Front. Neurosci. 13, 965.

Esposito, F., Pignataro, G., Renzo, G.D., Spinali, A., Paccone, A., Tedeschi, G., Annunziato, L., 2010. Alcohol increases spontaneous BOLD signal fluctuations in the visual network. NeuroImage 53, 534–543.

Falahpour, M., Chang, C., Wong, C.W., Liu, T.T., 2018a. Template-based prediction of vigilance fluctuations in resting-state fMRI. NeuroImage 174, 317–327.

Falahpour, M., Nalci, A., Liu, T.T., 2018b. The effects of global signal regression on estimates of resting-state blood oxygen-level-dependent functional magnetic resonance imaging and electroencephalogram vigilance correlations. Brain Connectivity 8 (10), 618–627.

Feige, B., Scheffler, K., Esposito, F., Di Salle, F., Hennig, J., Seifritz, E., 2005. Cortical and subcortical correlates of electroencephalographic alpha rhythm modulation. J. Neurophysiol. 93 (5), 2864–2872.

Fukunaga, M., Horovitz, S.G., Van Gelderen, P., De Zwart, J.A., Jansma, J.M., Ikonomidou, V.N., Chu, R., Deckers, R.H.R., Leopold, D.A., Duyn, J.H., 2006. Large-amplitude, spatially correlated fluctuations in BOLD fMRI signals during extended rest and early sleep stages. Magn. Reson. Imaging 24 (8), 979–992.

Fultz, N.E., Bonmassar, G., Setsompop, K., Stickgold, R.A., Rosen, B.R., Polimeni, J.R., Lewis, L.D., 2019. Coupled electrophysiological, hemodynamic, and cerebrospinal fluid oscillations in human sleep. Science 366 (6465), 628–631.

Goldman, R.I., Stern, J.M., Engel, J., Cohen, M.S., 2002. Simultaneous EEG and fMRI of the alpha rhythm. NeuroReport 13 (18), 2487–2492.

Gonzalez-Castillo, J., Kam, J.W.Y., Hoy, C.W., Bandettini, P.A., 2021. How to interpret resting-state fMRI: ask your participants. J. Neurosci. 41 (6), 1130–1141.

Goodale, S.E., Ahmed, N., Zhao, C., Zwart, J.A.D., Özbay, P.S., Picchioni, D., Duyn, J., Englot, D.J., Morgan, V.L., Chang, C., 2021. fMRI-based detection of alertness predicts behavioral response variability. eLife 10, e62376.

Greicius, M.D., Kiviniemi, V., Tervonen, O., Vainionpää, V., Alahuhta, S., Reiss, A.L., Menon, V., 2008. Persistent default-mode network connectivity during light sedation. Hum. Brain Mapp. 29 (7), 839–847.

Gu, Y., Han, F., Sainburg, L.E., Liu, X., 2020. Transient arousal modulations contribute to resting-state functional connectivity changes associated with head motion parameters. Cereb. Cortex 30 (10), 5242–5256.

Gu, Y., Sainburg, L.E., Kuang, S., Han, F., Williams, J.W., Liu, Y., Zhang, N., Zhang, X., Leopold, D.A., Liu, X., 2021. Brain activity fluctuations propagate as waves traversing the cortical hierarchy. Cereb. Cortex 31 (9), 3986–4005.

Gu, Y., Han, F., Sainburg, L.E., Schade, M.M., Buxton, O.M., Duyn, J.H., Liu, X., 2022. An orderly sequence of autonomic and neural events at transient arousal changes. bioRxiv. https://doi.org/10.1101/2022.02.05.479238.

Hahn, G., Zamora-López, G., Uhrig, L., Tagliazucchi, E., Laufs, H., Mantini, D., Kringelbach, M.L., Jarraya, B., Deco, G., 2021. Signature of consciousness in brain-wide synchronization patterns of monkey and human fMRI signals. NeuroImage 226, 117470.

Haimovici, A., Tagliazucchi, E., Balenzuela, P., Laufs, H., 2017. On wakefulness fluctuations as a source of BOLD functional connectivity dynamics. Sci. Rep. 7, 5908.

Han, F., Gu, Y., Liu, X., 2019. A neurophysiological event of arousal modulation may underlie fMRI-EEG correlations. Front. Neurosci. 13, 823.

Han, F., Brown, G.L., Zhu, Y., Belkin-Rosen, A.E., Lewis, M.M., Du, G., Gu, Y., Eslinger, P.J., Mailman, R.B., Huang, X., Liu, X., 2021a. Decoupling of global brain activity and cerebrospinal fluid flow in Parkinson's disease cognitive decline. Mov. Disord. 36 (9), 2066–2076.

Han, F., Chen, J., Belkin-Rosen, A., Gu, Y., Luo, L., Buxton, O.M., Liu, X., Initiative, A.D.N., 2021b. Reduced coupling between cerebrospinal fluid flow and global brain activity is linked to Alzheimer disease-related pathology. PLoS Biol. 19 (6), e3001233.

Horovitz, S.G., Fukunaga, M., De Zwart, J.A., Van Gelderen, P., Fulton, S.C., Balkin, T.J., Duyn, J.H., 2008. Low frequency BOLD fluctuations during resting wakefulness and light sleep: a simultaneous EEG-fMRI study. Hum. Brain Mapp. 29 (6), 671–682.

Iacovella, V., Hasson, U., 2011. The relationship between BOLD signal and autonomic nervous system functions: implications for processing of "physiological noise". Magn. Reson. Imaging 29 (10), 1338–1345.

Iidaka, T., 2021. Fluctuations in arousal correlate with neural activity in the human thalamus. Cerebral Cortex Commun. 2 (3), tgab055.

Jao, T., Vértes, P.E., Alexander-Bloch, A.F., Tang, I.-N., Yu, Y.-C., Chen, J.-H., Bullmore, E.T., 2013. Volitional eyes opening perturbs brain dynamics and functional connectivity regardless of light input. NeuroImage 69, 21–34.

Jawinski, P., Kirsten, H., Sander, C., Spada, J., Ulke, C., Huang, J., Burkhardt, R., Scholz, M., Hensch, T., Hegerl, U., 2019. Human brain arousal in the resting state: a genome-wide association study. Mol. Psychiatry 24, 1599–1609.

Jiang, C., Yi, L., Su, S., Shi, C., Long, X., Xie, G., Zhang, L., 2016. Diurnal variations in neural activity of healthy human brain decoded with resting-state blood oxygen level dependent fMRI. Front. Hum. Neurosci. 10, 634.

Jobert, M., Schulz, H., Jahnig, P., Tismer, C., Bes, F., Escola, H., 1994. A computerized method for detecting episodes of wakefulness during sleep based on the alpha slow-wave index. Sleep 17, 37–46.

Joshi, S., Li, Y., Kalwani, R.M., Gold, J.I., 2016. Relationships between pupil diameter and neuronal activity in the locus coeruleus, colliculi, and cingulate cortex. Neuron 89 (1), 221–234.

Kaufmann, T., Elvsashagen, T., Alnæs, D., Zak, N., Pedersen, P.Ø., Norbom, L.B., Quraishi, S.H., Tagliazucchi, E., Laufs, H., Bjørnerud, A., Malt, U.F., Andreassen, O.A., Roussos, E., Duff, E.P., Smith, S.M., Groote, I.R., Westlye, L.T., 2016. The brain functional connectome is robustly altered by lack of sleep. NeuroImage 127, 324–332.

Kiviniemi, V.J., Haanpää, H., Kantola, J.-H., Jauhiainen, J., Vainionpää, V., Alahuhta, S., Tervonen, O., 2005. Midazolam sedation increases fluctuation and synchrony of the resting brain BOLD signal. Magn. Reson. Imaging 23 (4), 531–537.

Klimesch, W., 1999. EEG alpha and theta oscillations reflect cognitive and memory performance: a review and analysis. Brain Res. Brain Res. Rev. 29 (2–3), 169–195.

Knaut, P., von Wegner, F., Morzelewski, A., Laufs, H., 2019. EEG-correlated fMRI of human alpha (de-)synchronization. Clin. Neurophysiol.: Off. J. Int. Federat. Clin. Neurophysiol. 130 (8), 1375–1386.

Larson-Prior, L.J., Zempel, J.M., Nolan, T.S., Prior, F.W., Snyder, A.Z., Raichle, M.E., 2009. Cortical network functional connectivity in the descent to sleep. Proc. Natl. Acad. Sci. U. S. A. 106 (11), 4489–4494.

Larson-Prior, L.J., Power, J.D., Vincent, J.L., Nolan, T.S., Coalson, R.S., Zempel, J., Snyder, A.Z., Schlaggar, B.L., Raichle, M.E., Petersen, S.E., 2011. Modulation of the brain's functional network architecture in the transition from wake to sleep. Prog. Brain Res. 193, 277–294.

Laufs, H., Kleinschmidt, A., Beyerle, A., Eger, E., Salek-Haddadi, A., Preibisch, C., Krakow, K., 2003a. EEG-correlated fMRI of human alpha activity. NeuroImage 19 (4), 1463–1476.

Laufs, H., Krakow, K., Sterzer, P., Eger, E., Beyerle, A., Salek-Haddadi, A., Kleinschmidt, A., 2003b. Electroencephalographic signatures of attentional and cognitive default modes in spontaneous brain activity fluctuations at rest. Proc. Natl. Acad. Sci. 100, 1053–1058.

Laumann, T.O., Snyder, A.Z., Mitra, A., Gordon, E.M., Gratton, C., Adeyemo, B., Gilmore, A.W., Nelson, S.M., Berg, J.J., Greene, D.J., McCarthy, J.E., Tagliazucchi, E., Laufs, H., Schlaggar, B.L., Dosenbach, N.U.F., Petersen, S.E., 2017. On the stability of BOLD fMRI correlations. Cereb. Cortex 27, 4719–4732.

Li, R., Ryu, J.H., Vincent, P., Springer, M., Kluger, D., Levinsohn, E.A., Chen, Y., Chen, H., Blu-menfeld, H., 2021. The pulse: transient fMRI signal increases in subcortical arousal systems during transitions in attention. NeuroImage 232, 117873.

Licata, S.C., Nickerson, L.D., Lowen, S.B., Trksak, G.H., Maclean, R.R., Lukas, S.E., 2013. The hypnotic zolpidem increases the synchrony of BOLD signal fluctuations in widespread brain networks during a resting paradigm. NeuroImage 70, 211–222.

Liu, T.T., 2013. Neurovascular factors in resting-state functional MRI. NeuroImage 80, 339–348.

Liu, T.T., 2016. Noise contributions to the fMRI signal: an overview. NeuroImage 143, 141–151.

Liu, X., Duyn, J.H., 2013. Time-varying functional network information extracted from brief instances of spontaneous brain activity. Proc. Natl. Acad. Sci. U. S. A. 110 (11), 4392–4397.

Liu, T.T., Falahpour, M., 2020. Vigilance effects in resting-state fMRI. Front. Neurosci. 14, 321.

Liu, X., Yanagawa, T., Leopold, D., Chang, C., Ishida, H., Fuji, I.N., Duyn, J., 2015. Arousal transitions in sleep, wakefulness, and anesthesia are characterized by an orderly sequence of cortical events. NeuroImage 116, 222–231.

Liu, T.T., Nalci, A., Falahpour, M., 2017. The global signal in fMRI: nuisance or information? NeuroImage 150, 213–229.

Liu, X., De Zwart, J.A., Schölvinck, M.L., Chang, C., Ye, F.Q., Leopold, D.A., Duyn, J.H., 2018. Subcortical evidence for a contribution of arousal to fMRI studies of brain activity. Nat. Commun. 9 (1), 395.

Liu, X., Leopold, D.A., Yang, Y., 2021. Single-neuron firing cascades underlie global spontaneous brain events. Proc. Natl. Acad. Sci. 118 (47), e2105395118.

Ma, Y., Ma, Z., Liang, Z., Neuberger, T., Zhang, N., 2020. Global brain signal in awake rats. Brain Struct. Funct. 225 (1), 227–240.

Majeed, W., Magnuson, M., Keilholz, S.D., 2009. Spatiotemporal dynamics of low frequency fluctuations in BOLD fMRI of the rat. J. Magn. Reson. Imaging: JMRI 30 (2), 384–393.

Majeed, W., Magnuson, M., Hasenkamp, W., Schwarb, H., Schumacher, E.H., Barsalou, L., Keilholz, S.D., 2011. Spatiotemporal dynamics of low frequency BOLD fluctuations in rats and humans. NeuroImage 54 (2), 1140–1150.

Matejcek, M., 1982. Vigilance and the EEG. In: Herrmann, W. (Ed.), Electroencephalography in Drug Research. Gustav Fischer, Stuttgart, pp. 405–441.

Matsui, T., Murakami, T., Ohki, K., 2016. Transient neuronal coactivations embedded in globally propagating waves underlie resting-state functional connectivity. Proc. Natl. Acad. Sci. U. S. A. 113 (23), 6556–6561.

Mayeli, A., Zoubi, O.A., Misaki, M., Stewart, J.L., Zotev, V., Luo, Q., Phillips, R., Fischer, S., Götz, M., Paulus, M.P., Refai, H., Bodurka, J., 2020. Integration of simultaneous resting-state electroencephalography, functional magnetic resonance imaging, and eye-tracker methods to determine and verify electroencephalography vigilance measure. Brain Connectivity 10 (10), 535–546.

McAvoy, M., Larson-Prior, L., Nolan, T., Vaishnavi, S., Raichle, M., d'Avossa, G., 2008. Resting states affect spontaneous bold oscillations in sensory and paralimbic cortex. J. Neurophysiol. 100, 922–931.

McAvoy, M., Larson-Prior, L., Ludwikow, M., Zhang, D., Snyder, A.Z., Gusnard, D.L., Raichle, M.E., d'Avossa, G., 2012. Dissociated mean and functional connectivity BOLD signals in visual cortex during eyes closed and fixation. J. Neurophysiol. 108, 2363–2372.

McAvoy, M.P., Tagliazucchi, E., Laufs, H., Raichle, M.E., 2019. Human non-REM sleep and the mean global BOLD signal. J. Cereb. Blood Flow Metab.: Off. J. Int. Soc. Cereb. Blood Flow Metab. 39 (11), 2210–2222.

Moosmann, M., Ritter, P., Krastel, I., Brink, A., Thees, S., Blankenburg, F., Taskin, B., Obrig, H., Villringer, A., 2003. Correlates of alpha rhythm in functional magnetic resonance imaging and near infrared spectroscopy. NeuroImage 20 (1), 145–158.

Muller, L., Chavane, F., Reynolds, J., Sejnowski, T.J., 2018. Cortical travelling waves: mechanisms and computational principles. Nat. Rev. Neurosci. 19 (5), 255–268.

Murphy, P.R., O'Connell, R.G., O'Sullivan, M., Robertson, I.H., Balsters, J.H., 2014. Pupil diameter covaries with BOLD activity in human locus coeruleus. Hum. Brain Mapp. 35 (8), 4140–4154.

Nalci, A., Rao, B.D., Liu, T.T., 2017. Global signal regression acts as a temporal downweighting process in resting-state fMRI. NeuroImage 152, 602–618.

Oken, B., Salinsky, M., Elsas, S., 2006. Vigilance, alertness, or sustained attention: physiological basis and measurement. Clin. Neurophysiol. 117, 1885–1901.

Olbrich, S., Mulert, C., Karch, S., Trenner, M., Leicht, G., Pogarell, O., Hegerl, U., 2009. EEG-vigilance and BOLD effect during simultaneous EEG/fMRI measurement. NeuroImage 45 (2), 319–332.

Olbrich, S., Sander, C., Matschinger, H., Mergl, R., Trenner, M., Schönknecht, P., Hegerl, U., 2011. Brain and body. J. Psychophysiol. 25 (4), 190–200.

Ong, J.L., Kong, D., Chia, T.T.Y., Tandi, J., Thomas Yeo, B.T., Chee, M.W.L., 2015. Coactivated yet disconnected-neural correlates of eye closures when trying to stay awake. NeuroImage 118, 553–562.

Orban, C., Kong, R., Li, J., Chee, M.W.L., Yeo, B.T.T., 2020. Time of day is associated with paradoxical reductions in global signal fluctuation and functional connectivity. PLoS Biol. 18 (2), e3000602.

Özbay, P.S., Chang, C., Picchioni, D., Mandelkow, H., Chappel-Farley, M.G., Gelderen, P., v., Zwart, J.A. d., Duyn, J., 2019. Sympathetic activity contributes to the fMRI signal. Commun. Biol. 2 (1), 421.

Patanaik, A., Tandi, J., Ong, J.L., Wang, C., Zhou, J., Chee, M.W.L., 2018. Dynamic functional connectivity and its behavioral correlates beyond vigilance. NeuroImage 177, 1–10.

Patriat, R., Molloy, E., Meier, T., Kirk, G., Nair, V., Meyerand, M., Prabhakaran, V., Birn, R.M., 2013. The effect of resting condition on resting-state fMRI reliability and consistency: A comparison between resting with eyes open, closed, and fixated. NeuroImage 78, 463–473.

Picchioni, D., Özbay, P.S., Mandelkow, H., Zwart, J.A.D., Wang, Y., Gelderen, P.V., Duyn, J.H., 2022. Autonomic arousals contribute to brain fluid pulsations during sleep. NeuroImage 249, 118888.

Poldrack, R.A., Baker, C.I., Durnez, J., Gorgolewski, K.J., Matthews, P.M., Munafò, M.R., Nichols, T.E., Poline, J.-B., Vul, E., Yarkoni, T., 2017. Scanning the horizon: towards transparent and reproducible neuroimaging research. Nat. Rev. Neurosci. 18 (2), 115–126.

Poudel, G.R., Innes, C.R.H., Bones, P.J., Watts, R., Jones, R.D., 2014. Losing the struggle to stay awake: divergent thalamic and cortical activity during microsleeps. Hum. Brain Mapp. 35 (1), 257–269.

Poudel, G.R., Innes, C.R.H., Jones, R.D., 2018. Temporal evolution of neural activity and connectivity during microsleeps when rested and following sleep restriction. NeuroImage 174, 263–273.

Poudel, G.R., Hawes, S., Innes, C.R.H., Parsons, N., Drummond, S.P.A., Caeyensberghs, K., Jones, R.D., 2021. RoWDI: rolling window detection of sleep intrusions in the awake brain using fMRI. J. Neural Eng. 18 (5), 056063.

Rasmussen, M.K., Mestre, H., Nedergaard, M., 2018. The glymphatic pathway in neurological disorders. Lancet Neurol. 17 (11), 1016–1024.

Raut, R.V., Snyder, A.Z., Mitra, A., Yellin, D., Fujii, N., Malach, R., Raichle, M.E., 2021. Global waves synchronize the brain's functional systems with fluctuating arousal. Sci. Adv. 7 (30), eabf2709.

Razavi, N., Jann, K., Koenig, T., Kottlow, M., Hauf, M., Strik, W., Dierks, T., 2013. Shifted coupling of EEG driving frequencies and fMRI resting state networks in schizophrenia spectrum disorders. PLoS ONE 8, e76604.

Sadaghiani, S., Scheeringa, R., Lehongre, K., Morillon, B., Giraud, A.-L., Kleinschmidt, A., 2010. Intrinsic connectivity networks, alpha oscillations, and tonic alertness: a simultaneous electroencephalography/functional magnetic resonance imaging study. J. Neurosci. 30 (30), 10243–10250.

Sämann, P.G., Wehrle, R., Hoehn, D., Spoormaker, V.I., Peters, H., Tully, C., Holsboer, F., Czisch, M., 2011. Development of the brain's default mode network from wakefulness to slow wave sleep. Cereb. Cortex 21 (9), 2082–2093.

Sander, C., Hensch, T., Wittekind, D.A., Böttger, D., Hegerl, U., 2016. Assessment of wakefulness and brain arousal regulation in psychiatric research. Neuropsychobiology 72, 195–205.

Satpute, A.B., Kragel, P.A., Barrett, L.F., Wager, T.D., Bianciardi, M., 2019. Deconstructing arousal into wakeful, autonomic and affective varieties. Neurosci. Lett. 693, 19–28.

Scheeringa, R., Petersson, K.M., Kleinschmidt, A., Jensen, O., Bastiaansen, M.C.M., 2012. EEG alpha power modulation of fMRI resting-state connectivity. Brain Connect. 2 (5), 254–264.

Schneider, M., Hathway, P., Leuchs, L., Sämann, P.G., Czisch, M., Spoormaker, V.I., 2016. Spontaneous pupil dilations during the resting state are associated with activation of the salience network. NeuroImage 139, 189–201.

Schwalm, M., Rosales Jubal, E., 2017. Back to pupillometry: how cortical network state fluctuations tracked by pupil dynamics could explain neural signal variability in human cognitive neuroscience. eNeuro 4 (6). https://doi.org/10.1523/ENEURO.0293-16.2017.

Soehner, A.M., Chase, H.W., Bertocci, M., Greenberg, T., Stiffler, R., Lockovich, J.C., Aslam, H.A., Graur, S., Bebko, G., Phillips, M.L., 2019. Unstable wakefulness during resting-state fMRI and its associations with network connectivity and affective psychopathology in young adults. J. Affect. Disord. 258, 125–132.

Soon, C.S., Vinogradova, K., Ong, J.L., Calhoun, V.D., Liu, T., Helen, Z.J., Ng, K.K., Chee, M.W., 2021. Respiratory, cardiac, EEG, BOLD signals and functional connectivity over multiple microsleep episodes. NeuroImage 118129.

Steel, A., Thomas, C., Trefler, A., Chen, G., Baker, C.I., 2019. Finding the baby in the bath water—evidence for task-specific changes in resting state functional connectivity evoked by training. NeuroImage 188, 524–538.

Stevner, A.B.A., Vidaurre, D., Cabral, J., Rapuano, K., Nielsen, S.F.V., Tagliazucchi, E., Laufs, H., Vuust, P., Deco, G., Woolrich, M.W., Van Someren, E., Kringelbach, M.L., 2019. Discovery of key whole-brain transitions and dynamics during human wakefulness and non-REM sleep. Nat. Commun. 10 (1), 1035.

Tagliazucchi, E., Laufs, H., 2014. Decoding wakefulness levels from typical fMRI resting-state data reveals reliable drifts between wakefulness and sleep. Neuron 82, 695–708.

Tagliazucchi, E., von Wegner, F., Morzelewski, A., Borisov, S., Jahnke, K., Laufs, H., 2012a. Automatic sleep staging using fMRI functional connectivity data. NeuroImage 63 (1), 63–72.

Tagliazucchi, E., von Wegner, F., Morzelewski, A., Brodbeck, V., Laufs, H., 2012b. Dynamic BOLD functional connectivity in humans and its electrophysiological correlates. Front. Hum. Neurosci. 6, 339.

Tagliazucchi, E., Siniatchkin, M., Laufs, H., Chialvo, D.R., 2016. The voxel-wise functional connectome can be efficiently derived from co-activations in a sparse spatio-temporal point-process. Front. Neurosci. 10, 381.

Tal, O., Diwakar, M., Wong, C.W., Olafsson, V., Lee, R., Huang, M.-X., Liu, T.T., 2013. Caffeine-induced global reductions in resting-state BOLD connectivity reflect widespread decreases in MEG connectivity. Front. Hum. Neurosci. 7, 63.

Thompson, G.J., Pan, W.-J., Magnuson, M.E., Jaeger, D., Keilholz, S.D., 2014. Quasi-periodic patterns (QPP): large-scale dynamics in resting state fMRI that correlate with local infraslow electrical activity. NeuroImage 84 (C), 1018–1031.

Tu, Y., Fu, Z., Mao, C., Falahpour, M., Gollub, R.L., Park, J., Wilson, G., Napadow, V., Gerber, J., Chan, S.-T., Edwards, R.R., Kaptchuk, T.J., Liu, T., Calhoun, V., Rosen, B., Kong, J., 2020. Distinct thalamocortical network dynamics are associated with the pathophysiology of chronic low back pain. Nat. Commun. 11 (1), 3948.

Turchi, J., Chang, C., Ye, F.Q., Russ, B.E., Yu, D.K., Cortes, C.R., Monosov, I.E., Duyn, J.H., Leopold, D.A., 2018. The basal forebrain regulates global resting-state fMRI fluctuations. Neuron 97 (4), 940–952.e4.

Ulke, C., Tenke, C.E., Kayser, J., Sander, C., Böttger, D., Wong, L.Y.X., Alvarenga, J.E., Fava, M., McGrath, P.J., Deldin, P.J., Mcinnis, M.G., Trivedi, M.H., Weissman, M.M., Pizzagalli, D.A., Hegerl, U., Bruder, G.E., 2018. Resting EEG measures of brain arousal in a multisite study of major depression. Clin. EEG Neurosci. 50 (1), 3–12.

Wang, H.-L.S., Rau, C.-L., Li, Y.-M., Chen, Y.-P., Yu, R., 2015. Disrupted thalamic resting-state functional networks in schizophrenia. Front. Behav. Neurosci. 9, 45.

Wang, C., Ong, J.L., Patanaik, A., Zhou, J., Chee, M.W.L., 2016. Spontaneous eyelid closures link vigilance fluctuation with fMRI dynamic connectivity states. Proc. Natl. Acad. Sci. U. S. A. 113 (34), 9653–9658.

Weng, Y., Liu, X., Hu, H., Huang, H., Zheng, S., Chen, Q., Song, J., Cao, B., Wang, J., Wang, S., Huang, R., 2020. Open eyes and closed eyes elicit different temporal properties of brain functional networks. NeuroImage 222, 117230.

Wirsich, J., Rey, M., Guye, M., Bénar, C., Lanteaume, L., Ridley, B., Confort-Gouny, S., Cassé-Perrot, C., Soulier, E., Viout, P., Rouby, F., Lefebvre, M.-N., Audebert, C., Truillet, R., Jouve, E., Payoux, P., Bartrés-Faz, D., Bordet, R., Richardson, J.C., Babiloni, C., Rossini, P.M., Micallef, J., Blin, O., Ranjeva, J.-P., Pharmacog Consortium, 2018. Brain networks are independently modulated by donepezil, sleep, and sleep deprivation. Brain Topogr. 31 (3), 380–391.

Wong, C.W., Olafsson, V., Tal, O., Liu, T.T., 2012. Anti-correlated networks, global signal regression, and the effects of caffeine in resting-state functional MRI. NeuroImage 63 (1), 356–364.

Wong, C.W., Olafsson, V., Tal, O., Liu, T.T., 2013. The amplitude of the resting-state fMRI global signal is related to EEG vigilance measures. NeuroImage 83, 983–990.

Wong, C.W., DeYoung, P.N., Liu, T.T., 2015. Differences in the resting-state fMRI global signal amplitude between the eyes open and eyes closed states are related to changes in EEG vigilance. NeuroImage 124 (Pt A), 24–31.

Xu, P., Huang, R., Wang, J., Van Dam, N.T., Xie, T., Dong, Z., Chen, C., Gu, R., Zang, Y.-F., He, Y., Fan, J., Luo, Y.-J., 2014. Different topological organization of human brain functional networks with eyes open versus eyes closed. NeuroImage 90, 246–255.

Yan, C., Liu, D., He, Y., Zou, Q., Zhu, C., Zuo, X., Long, X., Zang, Y., 2009. Spontaneous brain activity in the default mode network is sensitive to different resting-state conditions with limited cognitive load. PLoS One 4, e5743.

Yang, H., Long, X.-Y., Yang, Y., Yan, H., Zhu, C.-Z., Zhou, X.-P., Zang, Y.-F., Gong, Q.-Y., 2007. Amplitude of low frequency fluctuation within visual areas revealed by resting-state functional MRI. NeuroImage 36 (1), 144–152.

Yang, G.J., Murray, J.D., Repovs, G., Cole, M.W., Savic, A., Glasser, M.F., Pittenger, C., Krystal, J.H., Wang, X.-J., Pearlson, G.D., Glahn, D.C., Anticevic, A., 2014. Altered global brain signal in schizophrenia. Proc. Natl. Acad. Sci. U. S. A. 111 (20), 7438–7443.

Yang, L., Lei, Y., Wang, L., Chen, P., Cheng, S., Chen, S., Sun, J., Li, Y., Wang, Y., Hu, W., Yang, Z., 2018. Abnormal functional connectivity density in sleep-deprived subjects. Brain Imaging Behav. 12 (6), 1650–1657.

Yellin, D., Berkovich-Ohana, A., Malach, R., 2015. Coupling between pupil fluctuations and resting-state fMRI uncovers a slow build-up of antagonistic responses in the human cortex. NeuroImage 106, 414–427.

Yeo, B.T.T., Tandi, J., Chee, M.W.L., 2015. Functional connectivity during rested wakefulness predicts vulnerability to sleep deprivation. NeuroImage 111, 147–158.

Yousefi, B., Shin, J., Schumacher, E.H., Keilholz, S.D., 2017. Quasi-periodic patterns of intrinsic brain activity in individuals and their relationship to global signal. NeuroImage 167, 297–308.

Yuan, H., Zotev, V., Phillips, R., Bodurka, J., 2013. Correlated slow fluctuations in respiration, EEG, and BOLD fMRI. NeuroImage 79, 81–93.

Yuan, B.-K., Wang, J., Zang, Y.-F., Liu, D.-Q., 2014. Amplitude differences in high-frequency fMRI signals between eyes open and eyes closed resting states. Front. Hum. Neurosci. 8, 503.

Zerbi, V., Floriou-Servou, A., Markicevic, M., Vermeiren, Y., Sturman, O., Privitera, M., von Ziegler, L., Ferrari, K.D., Weber, B., De Deyn, P.P., Wenderoth, N., Bohacek, J., 2019. Rapid reconfiguration of the functional connectome after chemogenetic locus coeruleus activation. Neuron 103 (4), 702–718.

Zhang, Y., Yang, Y., Yang, Y., Li, J., Xin, W., Huang, Y., Shao, Y., Zhang, X., 2019. Alterations in cerebellar functional connectivity are correlated with decreased psychomotor vigilance following total sleep deprivation. Front. Neurosci. 13, 134.

Zou, Q., Long, X., Zuo, X., Yan, C., Zhu, C., Yang, Y., Liu, D., He, Y., Zang, Y., 2009. Functional connectivity between the thalamus and visual cortex under eyes closed and eyes open conditions: a resting-state fMRI study. Hum. Brain Mapp. 30 (9), 3066–3078.

Zou, G., Xu, J., Zhou, S., Liu, J., Su, Z.H., Zou, Q., Gao, J.-H., 2019. Functional MRI of arousals in non-rapid eye movement sleep. Sleep.

9

Multimodal methods to help interpret resting-state fMRI

Xiaoqing Alice Zhou, Yuanyuan Jiang, Weitao Man, and Xin Yu

Athinoula A. Martinos Center for Biomedical Imaging, Massachusetts General Hospital and Harvard Medical School, Charlestown, MA, United States

Introduction

Resting-state fMRI (rs-fMRI) presents a unique tool to map the functional connectivity and categorize specific resting-state networks. The rs-fMRI network designates the spatial patterns of the voxel-wise correlation of blood-oxygenation-level-dependent (BOLD) signals, presenting a $< 0.1\,\mathrm{Hz}$ low-frequency fluctuation (LFF) (Biswal et al., 1997, 1995; Cordes et al., 2001; Fukunaga et al., 2006; Obrig et al., 2000). There are two basic assumptions: (i) the specific rs-fMRI network presents underlying neuronal connections; (ii) synchronized (associated) neuronal oscillations lead to the low-frequency rs-fMRI signal fluctuation. To support these assumptions, concurrent neuronal and fMRI signals need to be acquired to validate causality in combination with the circuit-specific modulatory schemes. To fulfill this need, multimodal brain recording/mapping platforms have been developed.

Before introducing the multimodal neurotechniques with fMRI when interpreting rs-fMRI-based functional connectivity, it is important to compare rs-fMRI as a unique brain-function mapping tool with other neurotechniques and discuss how rs-fMRI provides a different angle to understand brain function. Typical methods to study brain function include electrophysiological recordings and optical imaging, both of which enable brain dynamic signal acquisition and analysis with respective temporal and spatial characteristics.

Electrophysiological techniques (Ephys) have shown a typical cross-scale neuronal activity recording capability, i.e., from single neurons (action potential), multiunit recording, and to populational neuronal encoding (e.g., local field potential, or global EEG mapping), presenting a gold standard approach to link brain function to behavior. For example, in vitro Ephys applies a patch-clamp scheme to target select neurons and measures synchronized synaptic activity

Advances in Resting-State Functional MRI. https://doi.org/10.1016/B978-0-323-91688-2.00007-2

changes, e.g., long-term potentiation (LTP) or long-term depression (LTD), which can be used to define the connectivity strength between neurons in brain slices (Malinow and Tsien, 1990; Lev-Ram et al., 2003; Zhang et al., 1998). Besides cell–cell interactions, multiarray or laminar electrodes have also been applied to perform single-unit or multiunit recordings from a number of cells in the local area or across cortical layers, as well as local field potential recordings for neuronal populational recordings in hundreds of neurons across sub-millimeter spatial scale (Logothetis et al., 2001; Cardoso et al., 2012).

The latest technological advancement of optical sensors and imaging schemes has extended micro-to-mesoscale brain functional mapping optically. In the past decade, genetic tools have enabled the development of genetically encoded fluorescent indicators with high kinetics and sensitivity to sensing neurotransmitters, e.g., calcium (Ca^{2+}) and glutamate (Glu), and neuromodulators, e.g., dopamine (DA), norepinephrine (NE), and acetylcholine (Ach), as well as membrane voltage (Chen et al., 2013; Marvin et al., 2013; Sun et al., 2018; Patriarchi et al., 2018; Feng et al., 2019; Borden et al., 2020; Jing et al., 2018; Gong et al., 2015; Tian et al., 2009). The implementation of these sensors has significantly extended the capability of optical imaging methods with cellular resolution. Due to spectral penetrating limitation of photons with visual light wavelength, conventional single-photon optical imaging only allows for the cortical surface functional imaging. Multiphoton imaging (including the three-photon imaging using near-infrared light) has extended the penetration depth to 1.2–1.3 mm, allowing for mapping hippocampal activity in the dorsal CA1 region of mice without invasive surgery (Ouzounov et al., 2017; Wang et al., 2018b). Besides the effort to extend penetration depth, wide-field two-photon imaging has extended the field of view (FOV) from millimeters to sub-centimeters across the cortex (Mateo et al., 2017). It should be noted that both imaging schemes are still limited in sampling rate at around 20–50 Hz. Until now, kilohertz sample rate optical imaging schemes have been reported to record the millisecond- to tens-of-millisecond-scale neuronal responses through Ca^{2+}/Glu or voltage sensors (Zhang et al., 2019; Wu et al., 2020). To achieve ultrafast sampling rate, fiber photometry and micro-lens-based imaging schemes have also been developed to target the deep brain regions with a few hundred-micron diameters (Jennings et al., 2015; Hamel et al., 2015; Gulati et al., 2017; Ghosh et al., 2011). Interestingly, similar to the in vivo electrophysiology, fiber photometry provides a neuronal-populational recoding scheme attributing to the fast kinetics of genetically encoded biosensor with the added benefit of sensing a variety of neurotransmitters and neuromodulators with cellular specificity.

As mentioned earlier, rs-fMRI mainly measures LFF (< 0.1 Hz) (Biswal et al., 1997; Cordes et al., 2001; Fukunaga et al., 2006; Obrig et al., 2000) of the BOLD signal across cerebrovasculature, estimating coordinated neuronal activity of brain networks based on the assumed neurovascular coupling (Raichle et al., 2001; Greicius et al., 2003; Hampson et al., 2006; Cabral et al., 2014). Thus, rs-fMRI does not provide direct neuronal activity measurements but rather the neurovascular coupled signal from large voxels (in the millimeter scale for human brains and submillimeter scale for animal brains). However, rs-fMRI provides a noninvasive, spatially extensive detection of correlated hemodynamic signal changes (e.g., BOLD signal). These spatiotemporal patterns are considered to encompass neuronal oscillatory patterns during normal cognitive processing as well as in pathological brain states to specify impaired functional circuits from brain-wide vascular functional abnormalities (Sun et al., 2011; Wang et al., 2020; Smith et al., 2013; Deco et al., 2011; Allen et al., 2014; Sheline and Raichle, 2013; Palacios et al., 2013; Andrews-Hanna et al., 2007; Di Martino et al., 2014). To better understand the underlying mechanisms of LFF-based rs-fMRI connectivity, numerous efforts have been made combining fMRI with other techniques. In particular, besides synchronized activity based on well-defined neuronal circuits, e.g., callosal interhemispheric or thalamocortical projections, emerging evidence suggests that LFF correlation relies on brain-state fluctuation through global neuromodulatory projections, as well as the autonomic regulation of cerebrovascular tones (Chang et al., 2016; Duyn, 2011; Turchi et al., 2018; Fox and Raichle, 2007; Pais-Roldan et al., 2019, 2020; Wang et al., 2018a; Duyn et al., 2020). These neuronal regulatory sources converge through neurovascular coupling (including astrocyte function), i.e., the neuro-glio-vascular (NGV) signaling, contributing to the low-frequency rs-fMRI signal fluctuation. The goal of this chapter is to introduce the existing multimodal neuroimaging methodologies targeting NGV signaling in animals, which have provided crucial insights for translational functional connectivity mapping with rs-fMRI.

The significance of animal fMRI to interpret rs-fMRI

Animal fMRI has been somewhat overshadowed by the translational power of fMRI directly applied to human brains, given fMRI's noninvasive and large-scale imaging capability. However, given safety and technological limitations in human studies, it remains challenging to better interpret the fMRI signal in both task and resting conditions. In contrast, with animal fMRI, advanced multimodal imaging is more feasible, presenting a great potential to decipher the underlying NGV signaling of rs-fMRI across different spatial and temporal scales. Animal fMRI applies high-field MRI to achieve higher spatial resolution than human MR when mapping the function dynamic circuits, e.g., the NGV circuit. To evaluate the spatial advantage of animal fMRI, one critical issue is to compare the typical size of different parameters related to rodents and human brains. Although the human brain is at least 10 times larger in one dimension ($\sim$10–15 cm long) than the rodent brain (1–1.5 cm) (DeSilva et al., 2021; Badea et al., 2007), their neuronal soma sizes are highly comparable to those of rodents (human neuronal soma: 10–25 μm vs. mouse neuronal soma: 8–20 μm) (Hodge et al., 2019). Also, the human red blood cell has a diameter of $\sim$7–8 μm in comparison with mouse red blood cells with a diameter of 5–6 μm, presenting similar microscopic scales for microvessels supplying brain parenchyma in both species (O'Connell et al., 2015; Diez-Silva et al., 2010). Although neuronal metabolism, cerebrovascular structure (e.g., the number distribution of arteriole and venules), and oxygen extraction efficiency (mediated by many factors, e.g., the hemoglobin oxygen affinity) vary between species (Li et al., 2021; Hirsch et al., 2012; Wearing et al., 2021; Webb et al., 2022), the basic unit of NGV signaling is at the similar spatial scale (Kugler et al., 2021; Zisis et al., 2021).

For high-field human fMRI (7 T and above), the highest spatial resolution is near 0.5–0.7 mm isotropic resolution (He et al., 2018; Berman et al., 2021; Feinberg et al., 2018). For a sample application of human laminar fMRI studies (using 750 × 750 um resolution), given the typical cortical thickness of 1–4 mm (Fischl and Dale, 2000), only 5–6 voxels can be sampled across the thickest human cortex, e.g., motor cortex (Huber et al., 2017, 2021). In contrast, high-field animal fMRI ($>$11.7 T) can push the spatial resolution to a few hundreds of microns (even down to 50–100 μm in-plane resolution) (Pais-Roldan et al., 2020; Chen et al., 2020; Yu et al., 2016). Since the rat cortex has a typical 2 mm cortical thickness, detailed layer-specific signaling can be achieved with more than 20 voxels (Yu et al., 2014; Nunes et al., 2019).

In addition, animal fMRI studies can be readily combined with electrophysiological recording, optical imaging, including fiber photometry, and brain stimulation with electrodes or fiber-mediated optogenetics and Designer Receptors Exclusively Activated by Designer Drugs (DREADD)-based chemogenetic tools (Roth, 2016; Zerbi et al., 2019). This unique multimodal neuroimaging platform highlights the significant potential for animal fMRI to interpret rs-fMRI when aiming to specify its NGV signaling.

Multimodal neuroimaging to interpret rs-fMRI

Investigate neurovascular coupling using simultaneous electrophysiological recordings with fMRI

In 2001, Logothetis et al. established their protocol for simultaneous fMRI with in vivo electrophysiological recordings of monkey brains, providing simultaneous measurements of concurrent neuronal and fMRI signals from the visual cortex (Logothetis et al., 2001). This work revealed that the BOLD fMRI signal is marginally more correlated with local field potential (LFP) than spiking activity. Although more detailed cerebral metabolic rate of oxygen ($CMRO_2$) measurements based on calibrated BOLD in anesthetized rats have indicated a stronger cerebral energetic correlation with spiking activity (Hyder and Rothman Douglas, 2010; Smith et al., 2002), the strong coupling between LFP and BOLD signal remains to be a major focus of neuronal origin studies of fMRI signal. One critical observation based on simultaneous recording of the high-frequency LFP with optical-imaging-based hemodynamic responses is the synchronized gamma power and hemodynamic signal fluctuation (Niessing et al., 2005). This work led to extensive efforts to verify the coupling between BOLD signal variance and LFP gamma power fluctuation in both animal and human brains (Lachaux et al., 2007; Nir et al., 2007; Shmuel and Leopold, 2008; Schölvinck et al., 2010; Keilholz, 2014; Goense and Logothetis, 2008; He and Raichle, 2009). These studies directly analyze the concurrent neuronal and fMRI signals to investigate neurovascular coupling underlying the fMRI signal acquired in both task-related and resting conditions.

One typical analytical approach to explore the neurovascular coupling mechanism underlying rs-fMRI signal fluctuation is cross-correlation analysis of the concurrent LFP and fMRI signals. Schölvinck et al. performed rs-fMRI with simultaneous LFP recordings across different cortices, and correlated the LFP power fluctuation at different frequencies with the concurrent rs-fMRI signals (Schölvinck et al., 2010). The correlation coefficients were calculated by introducing different lag times between the two time series. As shown in Fig. 1, the positive peak correlation coefficient was detected at the lag time 7–8 s, indicating that the neuronal oscillation potentially drives the rs-fMRI signal low-frequency fluctuation. Interestingly, the most robust positive correlation is detected at 40–80 Hz LFP signal across animals, suggesting the strong coupling between gamma power and rs-fMRI LFF.

It should be noted that the coupling of the fMRI signal and the beta/alpha-band oscillatory features has shown much larger variability across subjects in both animal and human studies (Scheeringa et al., 2011; Schölvinck et al., 2010; Keilholz, 2014). One plausible explanation is that the altered coupling of global neuronal oscillation with rs-fMRI signals relies on vigilant states of the brain (Özbay et al., 2019; Chang et al., 2016), which involve different NGV coupling mechanisms. The complexity of neurovascular coupling can be exemplified by the hippocampal ripple activity-based event-related fMRI study (Logothetis et al., 2012). Correlating the fMRI response to sharp-wave ripples with higher frequency (100–250 Hz) than fast gamma oscillation in the hippocampus as a triggering signal, the results show both positive and

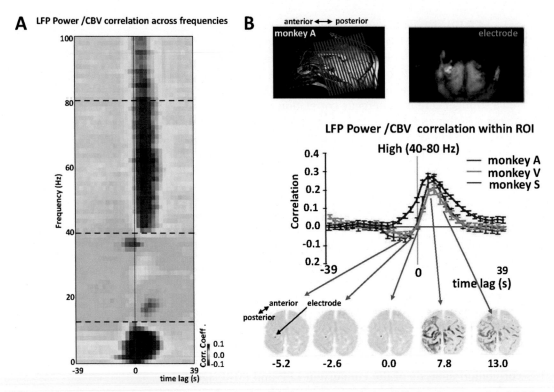

Fig. 1 Simultaneous LFP and rs-fMRI in monkey brains. (A) The cross-correlation (CC) plots between global fMRI signal and LFP frequency-specific power profiles. Positive peak CC is detected at the positive lag time (5–8 s), indicating the neuronal oscillation drives global BOLD signal fluctuation. (B) Upper panel shows the multislice echo-planar imaging (EPI) fMRI geometry to cover the posterior monkey brain and electrode location. The lower panel shows the cross-correlation plots of the high gamma power fluctuation (40–80 Hz) with the rs-fMRI signal from three monkeys, showing the peak positive CC at lag time of 7.8 s (the time-lapsed correlation maps are shown in the bottom panel from −5.2-s to 13-s lag time). Adapted from Schölvinck, M.L., Maier, A., Ye, F.Q., Duyn, J.H., Leopold, D.A., 2010. Neural basis of global resting-state fMRI activity. Proc. Natl. Acad. Sci. U. S. A. 107, 10238–10243.

negative correlation patterns with varied temporal dynamic features of BOLD signals across different brain regions. This study suggests that conventional hemodynamic response function (HRF) with fixed temporal dynamics would introduce biased weighting for neuronal activation toward a presumed HRF. Also, BOLD fMRI signal correlation patterns may differentiate between function-specific processing and neuromodulation, between bottom-up and top-down functional signals, and present variable signs toward excitation and inhibition, depending on the circuit-dependent direct or indirect nature of the local neural activity (Logothetis, 2008; Lauritzen, 2008, 2001;

Devor et al., 2007; Chaigneau et al., 2007; Harel et al., 2002; Anenberg et al., 2015). The major challenge when interpreting the rs-fMRI signal is how to specify the linkage of hemodynamic responses through cerebrovasculature and the underlying neural activity patterns, as the true HRF may be unclear.

Vasomotion, to bridge gamma power and rs-fMRI signal fluctuation

One emerging theory has proposed that vasomotion near 0.1 Hz serves as a critical mediator to bridge neuronal oscillation and low-frequency rs-fMRI signal fluctuation. Vasomotion is a spontaneous rhythmic oscillation in vessel diameter, mainly occurring in arterioles and arteries controlled by smooth muscle cell contraction due to synchronized Ca^{2+} oscillation (Arciero and Secomb, 2012; Koenigsberger et al., 2005). Using the intraoperative intrinsic optical imaging method and postoperative fMRI, Rashubskiy et al. reported high-amplitude ~0.1 Hz sinusoidal oscillations of the cortical hemodynamics in human brains, of which the wave-like propagation pattern was caused by the oscillation of diameters of pial arteries (Rayshubskiy et al., 2014). This human brain-mapping result suggests vasomotion contribution to rs-fMRI signal, but not simply caused by cardiac pulsation and systemic blood pressure oscillation. Using a wide-field two-photon imaging with simultaneous LFP recordings of awake mice, Mateo et al. reported that the arteriole diameter changes covary with the gamma power fluctuation across two hemispheres in awake mice (Mateo et al., 2017). Also, the detected arteriole diameter oscillation is centered at 0.1 Hz with broader bandwidth from 0.05 to 0.15 Hz, presenting a strengthened bilateral spectral coherence pattern in mirrored cortices of two hemispheres (Choi et al., 2022). This phenomenon mimics the bilateral rs-fMRI correlation patterns as reported by Biswal et al. in their original rs-fMRI study in human motor cortices (Biswal et al., 1995). Drew et al. have written a thorough review to discuss vasomotion as a contributing physiological source of the rs-fMRI signal (Drew et al., 2020). However, it should be noted that how vasomotion-specific diameter changes are transferred to the BOLD signal fluctuation remains to be elucidated when assigning to specific neuronal circuit-based functional connectivity. Tong et al. also reported consistent delay times of the systemic low-frequency oscillation from the internal carotid artery and global/venous resting-state BOLD signal fluctuation in human brains, presenting a potential strategy to bridge vasomotion and BOLD signal oscillation by specifying the blood transit time through the cerebrovascular tree (Tong et al., 2019). Rather than only measuring the BOLD signal fluctuation with rs-fMRI, it is also important to retrieve vasomotion oscillatory signals directly with fMRI by measuring arteriole diameter changes.

Arteriole diameter changes are involved in active hyperemia responses underlying the hemodynamic changes coupled to neuronal activity. Neuronal activation-driven arteriole dilation can be in part measured with cerebral blood volume (CBV) fMRI. CBV fMRI method was originally developed when using gadolinium as an exogeneous agent to estimate the arteriole-based CBV changes in the human visual cortex by Belliveau et al. in 1991 (Belliveau et al., 1991). Later, iron-based nanoparticles became routinely used as T2*-weighted contrast agents to measure the arteriole diameter changes with CBV-fMRI (Mandeville, 2012). The abovementioned study on global LFP-correlated whole-brain rs-fMRI signal fluctuation (Chang et al., 2016) is based on CBV signal changes after an administration of iron nanoparticles in monkeys. For human CBV fMRI, vascular space occupancy (VASO) was developed to measure the null blood signal from T1-weighted images as endogenous CBV estimates (Lu et al., 2003). The BOLD and CBV rs-fMRI-based functional connectivity has been compared in both rodent and human brains (He et al., 2018; Zhang et al., 2018; Magnuson et al., 2010). These studies are mainly focusing on improving the spatial specificity of brain networks given the different contributions of vascular components. It has been considered that CBV fMRI from arterioles and capillaries would be spatially more specific to neuronal sources than BOLD signals that could be mislocalized by large draining veins using gradient-echo echo-planar-imaging (EPI) sequence (Kim et al., 2013; Oliveira et al., 2022; Kim and Kim, 2011).

To further investigate the dynamic contribution of vasomotion to the BOLD rs-fMRI signal fluctuation, high-field fMRI methods have been developed to directly map vasodynamics from microvessels with high spatiotemporal resolution (Yu et al., 2016, 2012; Moon et al., 2013; Fukuda et al., 2021). In particular, Yu et al. established a single-vessel mapping method to identify the venule-dominated BOLD versus arteriole-dominated CBV fMRI signals from penetrating microvessels through cortical layer V (Yu et al., 2016) and the hippocampus (Chen et al., 2019) in rat brains. Ultra-slow rs-fMRI signal fluctuations (<0.04 Hz) have also been detected with the single-vessel fMRI from penetrating venules and arterioles with different spatial scales in anesthetized rat brains (He et al., 2018), of which different temporal dynamic lag times are detected based on cross-correlation analysis with anesthetics-induced global neuronal oscillation (Fig. 2). This single-vessel CBV fMRI serves as a critical step to bridge 2-PM-based microscopic vasodynamic measurements with the mesoscale penetrating arteriole dilation across the whole hemisphere (Mateo et al., 2017). This single-vessel fMRI method has provided a promising

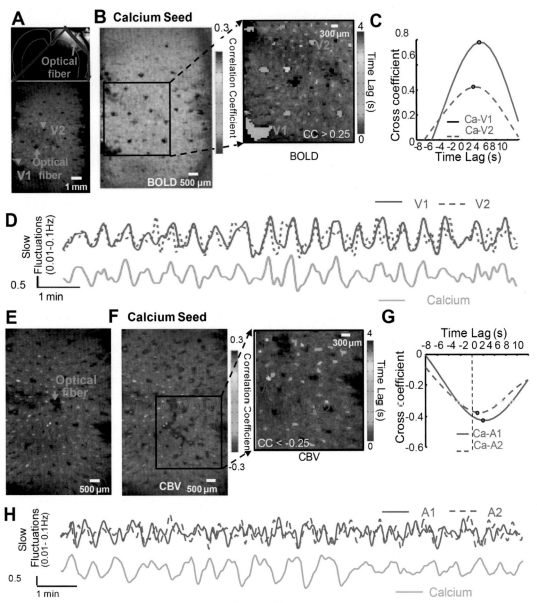

Fig. 2 Simultaneous single-vessel BOLD/CBV fMRI and Ca²⁺-based fiber photometry. (A) The localization of the optical fiber tip in the MR images and arteriole-venule (A-V) map of the 2D slice to detect penetrating cortical vessels. (B) Ca²⁺-based single-vessel BOLD-based rs-fMRI correlation shows peak correlation coefficients located primarily at penetrating venules (dark dots, e.g., V1, V2). The right panel shows the lag-time map for individual venules with CC > 0.25. (C) The cross-correlation plots of two presentative venule voxels with Ca²⁺ signals show that Ca²⁺ leads venule BOLD fluctuation with lag time from 2 to 4 s. (D) The representative time course of the Ca²⁺ and venule BOLD fMRI signal fluctuation (bandpass filter 0.01–0.1 Hz). (E) The A-V map to show the representative arterioles as bright dots. (F) Ca²⁺-based single-vessel CBV-based rs-fMRI correlation shows peak correlation coefficients located primarily at penetrating arterioles (bright dots, e.g., A1, A2). The right panel shows the lag-time map for individual arterioles with CC > 0.25. (G) The cross-correlation plots of two presentative arteriole voxels with Ca²⁺ signals show that Ca²⁺ leads arteriole fluctuation with lag time from 1.5 to 2.5 s. (H) The representative time course of the Ca²⁺ and arteriole CBV fMRI signal fluctuation (bandpass filter 0.01–0.1 Hz). Adapted from He, Y., Wang, M., Chen, X., Pohmann, R., Polimeni, J. R., Scheffler, K., Rosen, B.R., Kleinfeld, D., Yu, X., 2018. Ultra-slow single-vessel BOLD and CBV-based fMRI spatiotemporal dynamics and their correlation with neuronal intracellular calcium signals. Neuron 97, 925–939e5.

platform to directly map ~0.1 Hz vasomotion underlying the BOLD rs-fMRI correlation patterns in awake animal brains.

Resting-state cerebral blood flow/velocity mapping with arteriole spin labeling (ASL) and phase-contrast imaging

Besides the CBV-based rs-fMRI studies to elucidate potential vasomotion contributions (Fig. 2) to functional connectivity mapping, there are a series of studies to map the resting-state cerebral blood flow changes in combination with electrophysiological recordings. The CBF fMRI method was originally proposed by Kwong et al. in 1992 (Kwong et al., 1992). Local blood flow changes upon neuronal activation were measured using arterial spins, which could be polarized by an external field (Detre et al., 1992; Williams et al., 1992). The ASL-based CBF fMRI method detects local changes in the flow of blood containing labeled protons through brain tissue (Detre and Wang, 2002). Resting-state ASL-based fMRI has been shown sensitive to and able to reveal resting-state networks similar to the BOLD-based rs-fMRI networks (De Luca et al., 2006). For example, by reducing the BOLD contamination in the resting state, CBF signal fluctuation has also reported ASL-based functional connectivity of the sensorimotor network (Chuang et al., 2008). Moreover, a recent effort by Chen et al. has demonstrated that the resting-state blood velocity dynamic changes can be measured from individual arteriole and venules, using a high-resolution phase-contrast (PC) MR imaging (Chen et al., 2021). In contrast to the single-vessel BOLD or CBV, which highlights either individual arteriole or venules, this PC-based single-vessel mapping method detects large-scale blood velocity changes across the cerebrovascular network. An emerging trend of high-resolution rs-fMRI in animal models is to identify resting-state vasodynamic changes at the single-vessel level in deep brain regions in the context of brain-state fluctuation. This is a key step to decipher the NGV signaling contribution to resting-state fMRI.

Simultaneous fiber photometry-based Ca^{2+} recording and fMRI

In 2012, Schulz et al. applied the simultaneous fMRI and cell-type-specific calcium recordings in the rodent cortex using implanted optical fibers (Schulz et al., 2012). The calcium indicators, e.g., OGB-1 or Rhod-2, were used to detect the intracellular Ca^{2+} transients in both neurons and astrocytes through the implanted optical fiber. This study presented the feasibility to combine fiber photometry with animal fMRI to decipher distinct neuronal and astrocytic activity contributions to evoked fMRI signals. In contrast to electrophysiological recordings, fiber photometry has no electromagnetic interference with RF pulsing and fast gradient switches during MR scanning. The diamagnetic property of the optic fiber also leads to little field distortion in MR images. Meanwhile, fiber photometry enables fast sampling of the fluorescent signal from deep brain regions with minimally invasive insertion. In contrast to multiphoton microscopic imaging, e.g., three-photon with 1–1.3 mm penetration depth (Ouzounov et al., 2017; Wang et al., 2018b), the fiber implantation can be used to target much deep brain structures, where optical fiber bundle and micro-lens with sub-millimeter diameter have also been implanted to measure fluorescent neuronal signals based on a variety of indicators (Jennings et al., 2015; Hamel et al., 2015; Gulati et al., 2017; Ghosh et al., 2011). This unique multi-modal imaging scheme has led to the interpretation of rs-fMRI signals in the context of NGV coupling across different spatial scales.

To investigate the linkage between neuronal Ca^{2+} and rs-fMRI signal, Schwalm et al. have applied OGB-based fiber photometry to reveal the brain-wide global correlation of slow oscillation-associated neuronal calcium transients with the resting-state fMRI signals in anesthetized rat brains (Schwalm et al., 2017). One advanced experimental application is to replace exogenous Ca^{2+} indicator with genetically encoded calcium indicators (GECIs), e.g., the GCaMP. Using different promoters, GCaMP can be specifically expressed in either neurons or astrocytes, enabling the detection of evoked cell-type-specific Ca^{2+} transients and the concurrent BOLD signal from the fiber-targeted brain regions (Wang et al., 2018a; He et al., 2018; Albers et al., 2018b; Liang et al., 2017). Wang et al. has reported distinct coupling feature of concurrent astrocytic Ca^{2+} transients and fMRI signals. In contrast to the typical positive correlation in the evoked condition, i.e., evoked astrocytic Ca^{2+} transients coincide with the positive BOLD signal in the activated brain region, spontaneous intrinsic Ca^{2+} transients with higher amplitude are negatively correlated with global BOLD signal changes during rest (Fig. 3). Interestingly, this negative coupling event coincides with specific activation in the central thalamic regions, as well as the ascending arousal midbrain regions. Thalamocortical oscillation is a major source to regulate the brain state-dependent rs-fMRI low-frequency fluctuation, which has been reported across different species (Chang et al., 2016; He et al., 2018; Liu et al., 2018). Interestingly, when an unanesthetized monkey is experiencing vigilant-state changes during scanning, the eye-open/close-based arousal index is positively correlated with a few subcortical nuclei along the arousal pathway, including thalamic regions, but showing a broad negative correlation throughout the cortex, which is similar to the spontaneous astrocyte-coupled rs-fMRI signal pattern (Fig. 4) (Chang et al., 2016). Meanwhile, astrocytes have also been reported to be directly involved in regulating brain states (Poskanzer and Yuste, 2011, 2016) as well as brain state-dependent vasodynamic changes

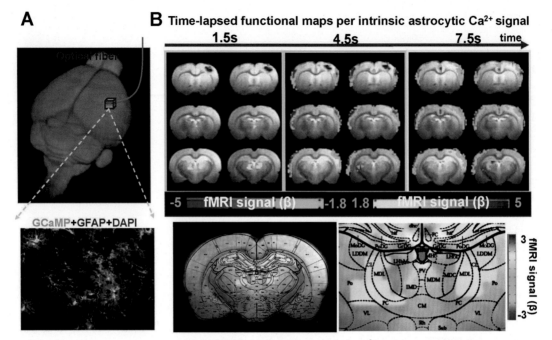

Fig. 3 The spatio-temporal relation between BOLD and intrinsic astrocytic Ca²⁺ signals during forepaw stimulation. (A) The schematic plan for simultaneous fMRI and fiber optic astrocytic Ca²⁺ recording. (B) The time-lapsed whole-brain fMRI maps. Positive BOLD in the central lateral thalamus (CL) and midbrain reticular formation (MRF) upon stimulation, followed by the negative BOLD in the whole cortex (lower panel shows the atlas location of the activated CL and MRF). Adapted from Wang, M., He, Y., Sejnowski, T.J., Yu, X., 2018. Brain-state dependent astrocytic Ca(2+) signals are coupled to both positive and negative BOLD-fMRI signals. Proc. Natl. Acad. Sci. U. S. A. 115, E1647–E1656.

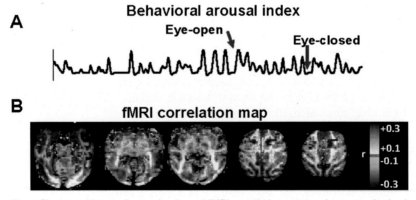

Fig. 4 The arousal-state fluctuation-based fMRI correlation patterns in unanesthetized monkeys. (A) The arousal index time course of the eye open and close based on the eye size measurement. (B) The arousal index-based fMRI correlation maps show a negative correlation across the whole cortex but a positive correlation in the thalamus. Adapted from Chang, C., Leopold, D.A., Scholvinck, M.L., Mandelkow, H., Picchioni, D., Liu, X., Ye, F.Q., Turchi, J.N., Duyn, J.H., 2016. Tracking brain arousal fluctuations with FMRI. Proc. Natl. Acad. Sci. U. S. A. 113, 4518–4523.

(He et al., 2018; Wang et al., 2018a). In the latest study by Jiang et al., the fiber photometry-based Glu and Ca^{2+} signal recordings were used to decipher the BOLD fMRI signal (Jiang et al., 2020). These studies have revealed the pivotal role of astrocytes underlying the global rs-fMRI signal fluctuation.

One caveat of existing multimodal animal neuroimaging studies is the usage of anesthetics, which confounds the interpretation of the rs-fMRI signals. In particular, astrocyte and vasodynamic function can be altered by anesthetic effects. Although the astrocyte-mediated global rs-fMRI signal fluctuation has been reported in animals treated with different anesthetics (Pradier et al., 2021; Wang et al., 2018a), the more profound brain-state changes underlying the rs-fMRI signal fluctuation remain to be elucidated in awake or nonanesthetized (e.g., during sleep) animals with the multimodal fMRI platform. Ma et al. have applied the optical imaging method to acquire concurrent neuronal Ca^{2+} and hemoglobin-based intrinsic optical imaging signals in awake mice to acquire spatiotemporal convolution patterns at the cortical surface. This mapping scheme has recently been applied inside the MR scanner for cortical rs-fMRI correlation analysis (Lake et al., 2020). The fiber photometry-based Ca^{2+} recording has also been applied for awake rodent fMRI studies (Liang et al., 2017). One critical direction of awake animal rs-fMRI with the multimodal neuroimaging platform is to address the stress-related confounds through neuromodulation. Also, it remains challenging to achieve high-resolution fMRI images of awake mice given the motion artifacts and the typical air-tissue interface-induced image distortion in high-field scanners.

Besides GECIs, genetic engineering advancement has developed a series of fluorescent biosensors with fast kinetics in a sub-millisecond time scale. These genetically encoded biosensors can be readily implemented into the multimodal neuroimaging platform. The membrane-voltage sensitive indicator (e.g., ASAP3) has been used to record action potentials of neurons with kHz sampling rates (Zhang et al., 2019; Wu et al., 2020). Also, Jiang et al. have expressed Glu sensor (iGluSnFR) at the membrane of neurons to detect both synaptic Glu spike per stimulation and baseline extracellular Glu signal clearance through astrocytes with simultaneous fMRI (Jiang et al., 2020). The genetically encoded dopamine (Patriarchi et al., 2018), acetylcholine (Jing et al., 2018; Borden et al., 2020), norepinephrine (Feng et al., 2019), and oxytocin (Muller et al., 2014; Xiong et al., 2021) can also be recorded with concurrent rs-fMRI signal to further interpret the neuromodulatory contribution to rs-fMRI signal fluctuations in animal brains.

Circuit-specific modulation to interpret rs-fMRI

One remaining challenge to interpret the rs-fMRI signal is to distinguish the circuit-specific functional connectivity from global correlation patterns caused by either direct brain-wide neuromodulatory effects or autonomic regulation (Özbay et al., 2019; Pais-Roldán et al., 2021; Drew et al., 2020). Since Biswal et al. reported the original interhemispheric correlation of rs-fMRI signal detected in bilateral motor cortices (Biswal et al., 1995), transcallosal projection is considered to underlie interhemispheric connectivity. rs-fMRI studies on human subjects with callosotomy or acallosal patients, as well as in animal models (Magnuson et al., 2014; O'Reilly et al., 2013), have reported altered bilateral correlation patterns (Uddin et al., 2008; Tyszka et al., 2011; Owen et al., 2013; Khanna et al., 2012; Tovar-Moll et al., 2014), but the remaining interhemispheric connectivity independent of callosal projections also suggests alterative commissural circuits or global subcortical projections as additional contributing sources (Tyszka et al., 2011; Uddin et al., 2008; Khanna et al., 2012; Tovar-Moll et al., 2014; Corsi-Cabrera et al., 2012). Multimodal fMRI studies have implemented circuit-specific modulatory schemes, e.g., optogenetics or DREADDS, to investigate the cellular and circuit contributions to rs-fMRI signals in animal models.

Optogenetic studies rely on genetically encoded light-sensitive ion channels, channelrhodopsin-2 (ChR2) (Nagel et al., 2003), which can be expressed specifically into neuronal subtypes or through anterograde or retrograde neuronal projections to target specific neural circuits in animal models (Zhang et al., 2007; Boyden et al., 2005). Instead of conventional task-related experimental design for either bottom-up or top-down neuronal network investigation, optogenetics delivers the modulation on individual relay nuclei or certain cell types along neuronal networks through light exposure to targeted brain regions. Optical fiber can be used to deliver light pulses when targeting deep brain nuclei, similar to the abovementioned fiber photometry studies. Lee et al. have reported the optogenetically driven fMRI signal in the mouse brain (Lee et al., 2010). Instead of the correlation analysis between concurrent oscillatory neuronal and rs-fMRI signals, optogenetic tools enable causality analysis to specify cellular or circuit-specific contributions to modulate the rs-fMRI signal fluctuation. Jung et al. have specifically labeled the GABAergic neurons with ChR2 to inhibit the local circuit and suppress the downstream subcortical network (Jung Won et al., 2022). By comparing correlation maps with and without optogenetic silencing, ipsilateral thalamocortical and corticocortical circuit-specific contributions to the rs-fMRI signal fluctuation can be well identified. Besides the intra-hemispheric connectivity studies, Chen et al. has labeled the callosal fiber with ChR2 and combined both fiber photometry-based neuronal Ca^{2+} recording and fMRI to examine interhemispheric fMRI responses directly mediated by optogenetically driven callosal activation (Chen et al., 2019). An intriguing application of the callosal circuit-specific optogenetic fMRI study is to implement a line-scanning fMRI method applied by Yu et al. in 2014 (Yu et al., 2014) (Fig. 5). The laminar-specific fMRI signal can be mapped with ultra-high spatiotemporal resolution across different cortical layers across 50 μm thickness every 50 ms. This method has been used to specify layer-specific BOLD signal changes upon direct optogenetic stimulation of the somatosensory cortex (Albers et al., 2018a). Lately, Chen et al. implanted an inductive coupling coil to boost the signal-to-noise ratio (SNR) to 2–3 times, allowing the laminar-specific correlation analysis of resting-state line-scanning fMRI signal (Chen et al., 2022). Choil et al. further developed a multiscale line-scanning fMRI method to detect the laminar-specific rs-fMRI

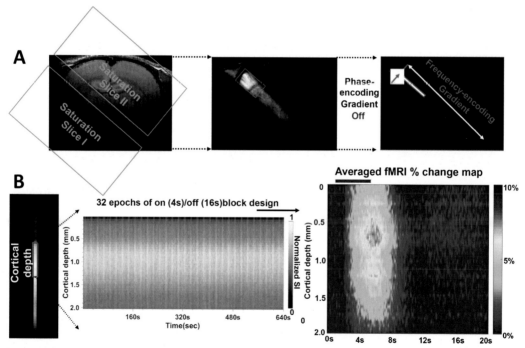

Fig. 5 (A) The schematic of line-scanning fMRI. Two saturation slices were localized to dampen the MR signal from brains regions outside of the region of interest. The FOV was aligned with frequency-encoding direction parallel to the cortical radial axis. After turning off the phase-encoding gradient, the single-line profile was acquired. (B) The single-line profile was collected with TR at 50 ms with a block design (4 s on and 16 s off for 32 epochs). The bright stripes across cortical layers (50 um per voxel) presented the repeated patterns of the evoked BOLD signal. The averaged fMRI map can be created by averaging all epochs to show the high spatiotemporal resolution line-scanning fMRI percentage map (x-axis is cortical depth, and y-axis is time). Adapted from Yu, X., Qian, C., Chen, D.Y., Dodd, S.J., Koretsky, A.P., 2014. Deciphering laminar-specific neural inputs with line-scanning fMRI. Nat. Methods 11, 55–58.

signal correlation across adjacent cortical regions (Choi et al., 2022). The latest study by Choi et al. has applied the bilateral line-scanning fMRI method to characterize the laminar-specific correlation in the somatosensory cortex between two hemispheres, showing a stronger coherence pattern at the 0.08–0.1 Hz oscillatory bandwidth at Layer II-III, where the callosal projection neurons are primarily located (Choi et al., 2022). Interestingly, the bilateral line-scanning-based laminar rs-fMRI also distinguishes 0.01–0.04 Hz coherence patterns across all layers, indicating the global slow neuronal oscillation in anesthetized rat brains. These studies provide a critical platform to decipher the cellular and circuit-specific contribution to the different spectral dynamics of the rs-fMRI signal across different brain regions.

DREADDs are modified muscarinic G-protein-coupled receptors (GPCRs). The DREADD-based receptors can be specifically expressed in desired neuronal subtypes using viral vectors, similar to optogenetic ChR2, aiming to target specific neural circuits. To enhance neuronal firing and activate Gq signaling in neurons or across specific neural circuits, a hM3Dq DREADD (human muscarinic receptor) was originally developed to be expressed in rodent brains (Alexander et al., 2009; Armbruster et al., 2007). The hM3Dq receptor agonist, clozapine-N-oxide (CNO), is a pharmacologically inert metabolite of clozapine, of which the recommended doses (generally 0.1–3 mg/kg) do not show pharmacological and behavioral impact in rodents (Alexander et al., 2009; Krashes et al., 2011; Farrell et al., 2013; Guettier et al., 2009; Urban and Roth, 2015; Zhu et al., 2014; Ferguson et al., 2011, 2013). CNO can also be used to target Gi-DREADDs (Armbruster et al., 2007). hM4Di is the typically used inhibitory Gi-DREADD (for review, see (Urban and Roth, 2015)) by different groups to silence neurons and mediate correlated animal behaviors (Armbruster et al., 2007; Ferguson et al., 2011; Kozorovitskiy et al., 2012; Ray et al., 2011; Carter et al., 2013; Teissier et al., 2015; Bourane et al., 2015). In addition, the κ-opioid-derived DREADD (KORD) is another chemogenetic GPCR to inhibit neuronal activity with a pharmacological inert compound, salvinorin B (Marchant et al., 2016; Vardy et al., 2015; Denis et al., 2015).

Chemogenetics provides a noninvasive neural circuit modulation method, allowing the dissection of circuit-specific contributions to global functional connectivity. rs-fMRI has been performed to characterize altered low-frequency rs-fMRI signal fluctuation in animal brains with chemogenetic activation of D1 neurons in the dorsal striatum (Nakamura et al., 2020) or the mesolimbic projection from the ventral tegmental area (VTA) to nucleus accumbens (Roelofs et al., 2017). The Gi-DREADD has also been used to modulate inhibitory neuronal activity in the somatosensory cortex (Markicevic et al., 2020) or prefrontal cortex (mPFC) (Rocchi et al., 2022), to identify the altered connectivity strength due to imbalanced excitatory/inhibitory circuitry. Interestingly, Rocchi et al. reported an increased functional connectivity following chemogenetic-mediated inhibitory effect. Despite the decreased neuronal spiking rates mediated by inhibitory hM4Di DREADD, the increased low-frequency (0.1–4 Hz) EEG oscillatory power caused by chemogenetic inhibition could directly contribute to this increase. This novel observation nicely illustrates the power of chemogenetics when combining rs-fMRI to elucidate the neuronal basis of the functional connectivity.

It should be noted that chemogenetic inhibition relies on diffusion of the administered drug in the local brain area. In contrast, optogenetic inhibition needs to deliver sufficient light exposure to cover the targeted functional nuclei. Due to the limited photon penetration, high-power and long-term light exposure can cause heating-induced MR signal offsets, confounding the rs-fMRI signal fluctuation. On the other hand, Gi-DREADD-based inhibition shows a more robust and consistent regulatory effect to targeted functional nuclei or circuits and is better suited to global functional connectivity mapping with rs-fMRI.

Other multimodal methodology in animal rs-fMRI

One intriguing method to further specify the cellular/circuit-specific contribution to rs-fMRI signal fluctuation is "hemogenetic" fMRI (Ghosh et al., 2022). Jasanoff's lab has developed a genetically encoded molecular probe to transfer the cellular calcium activity to hemodynamic responses, enabling the global mapping of the functional connectivity patterns from genetically targeted cells and circuits. The hemogenetic probe, NOSTICS, is a chimeric enzyme that combines the catalytic domain of inducible nitric oxide synthase (iNOS) with the CaM-binding and reductase domains of neuronal nitric oxide synthase (nNOS). NOSTICS retains calcium sensitivity and can be selectively manipulated by drugs, e.g., the iNOS-selective inhibitor 1400 W, which do not disturb nNOS function for normal hemodynamic responses. Thus, by measuring differences in the rs-fMRI correlation maps of animals administered with and without the drug treatment, the NOSTICS-boosted hemodynamic coupling circuit will be specifically identified.

Also, both functional ultrasound (fUS) and optoacoustic (fOA) methods have been developed to map animal brain function. The fUS detects power Doppler signal changes related to the cerebral blood volume (CBV) and flow (CBF), which has shown a unique advantage to map neuronal activity-coupled vascular hemodynamic function in freely behaving rodents (Boido et al., 2019; Deffieux et al., 2018; Mace et al., 2011), birds (Rau et al., 2018), primates (Blaize et al., 2020; Dizeux et al., 2019), as well as in neonate human brains (Demene et al., 2017; Imbault et al., 2017; Soloukey et al., 2019). In particular, Bruno-Félix Osmanski et al. reported the intrinsic connectivity mapping of 0.1 Hz spontaneous fUS signal fluctuation in rat brains (Baranger et al., 2021; Osmanski et al., 2014). Also, FUS has been applied to map dynamic functional connectivity changes in neonate human brains (Baranger et al., 2021). The fOA (Liao et al., 2012; Yao et al., 2016) can be used to detect the strong light absorption by hemoglobin in blood to map hemodynamic changes (Gottschalk et al., 2017). Similar to intrinsic optical imaging, specific spectral excitation enables the measurements of resting-state fluctuations in oxygen-hemoglobin (HbO), deoxygen-hemoglobin (Hb), and total hemoglobin (HbT) (Li et al., 2010; Mc Larney et al., 2020; Yao et al., 2013). Similar to FUS, resting-state functional connectivity has also been identified by the high-resolution optoacoustic tomography (Nasiriavanaki et al., 2014). Meanwhile, Gottschalk et al. also reported that fOA can be used to map genetically encoded calcium fluorescent indicators, enabling the direct measurement of neuronal activity. It makes fOA a promising method to bridge the direct neuronal activity measurement and hemodynamic-based resting-state functional connectivity. In summary, both methods enrich the multimodal imaging scheme of brain functional connectivity mapping with high spatiotemporal resolution.

Conclusion

In this chapter, we introduced state-of-the-art multimodal methodologies (mainly applied in animal studies, e.g., electrophysiological recording, 2-PM, fiber photometry, optogenetics, functional ultrasound, and optoacoustics), some of which can be directly combined with rs-fMRI, to map neuronal activity or hemodynamic responses, respectively. The goal of this cross-scale rs-fMRI mapping scheme is to target specific cellular or vascular components and verify their contribution to the systemic-level rs-fMRI signal fluctuation. Unique NGV dynamic features, e.g., vasomotion, or intrinsic astrocytic Ca^{2+} transients, can be directly involved in the formation or regulation of rs-fMRI signal fluctuation. Advanced fMRI methods, e.g., the single-vessel or line-scanning fMRI, can be used to elucidate distinct vascular dynamic contributions to the rs-fMRI signal fluctuation, as well as the circuit-specific regulation of laminar rs-fMRI signal fluctuation. Meanwhile, optogenetics, hemogenetics, and chemogenetics provide powerful tools to dissect neuronal cell-type and circuit contributions to the rs-fMRI signal fluctuation, of which the neuromodulation can be further specified by the fiber-photometry-based simultaneous recording of fluorescent signal changes from genetically encoded biosensors to a variety of neuromodulators. Here, we only outlined the list of multimodal techniques and their potential application to decipher detailed NGV events underlying rs-fMRI signal fluctuation. There is no doubt that more intriguing basic and translational results will be reported based on these cross-scale and multimodal neuroimaging platforms. Also, in contrast to fMRI, both fOA and fUS can provide high spatiotemporal resolution neuroimaging schemes to elucidate the low-frequency hemodynamic signal fluctuation across the cerebral vasculature, which will further enrich our understanding of the brain's dynamic changes during normal cognition and in the presence of brain disorders or cognitive impairment.

References

Albers, F., Schmid, F., Wachsmuth, L., Faber, C., 2018a. Line scanning fMRI reveals earlier onset of optogenetically evoked BOLD response in rat somatosensory cortex as compared to sensory stimulation. Neuroimage 164, 144–154.

Albers, F., Wachsmuth, L., Van Alst, T.M., Faber, C., 2018b. Multimodal functional neuroimaging by simultaneous BOLD fMRI and Fiber-optic calcium recordings and optogenetic control. Mol. Imaging Biol. 20, 171–182.

Alexander, G.M., Rogan, S.C., Abbas, A.I., Armbruster, B.N., Pei, Y., Allen, J.A., Nonneman, R.J., Hartmann, J., Moy, S.S., Nicolelis, M.A., McNamara, J.O., Roth, B.L., 2009. Remote control of neuronal activity in transgenic mice expressing evolved G protein-coupled receptors. Neuron 63, 27–39.

Allen, E.A., Damaraju, E., Plis, S.M., Erhardt, E.B., Eichele, T., Calhoun, V.D., 2014. Tracking whole-brain connectivity dynamics in the resting state. Cereb. Cortex 24, 663–676.

Andrews-Hanna, J.R., Snyder, A.Z., Vincent, J.L., Lustig, C., Head, D., Raichle, M.E., Buckner, R.L., 2007. Disruption of large-scale brain systems in advanced aging. Neuron 56, 924–935.

Anenberg, E., Chan, A.W., Xie, Y., Ledue, J.M., Murphy, T.H., 2015. Optogenetic stimulation of GABA neurons can decrease local neuronal activity while increasing cortical blood flow. J. Cereb. Blood Flow Metab. 35, 1579–1586.

Arciero, J.C., Secomb, T.W., 2012. Spontaneous oscillations in a model for active control of microvessel diameters. Math. Med. Biol. 29, 163–180.

Armbruster, B.N., Li, X., Pausch, M.H., Herlitze, S., Roth, B.L., 2007. Evolving the lock to fit the key to create a family of G protein-coupled receptors potently activated by an inert ligand. Proc. Natl. Acad. Sci. U. S. A. 104, 5163–5168.

Badea, A., Ali-Sharief, A.A., Johnson, G.A., 2007. Morphometric analysis of the C57BL/6J mouse brain. Neuroimage 37, 683–693.

Baranger, J., Demene, C., Frerot, A., Faure, F., Delanoe, C., Serroune, H., Houdouin, A., Mairesse, J., Biran, V., Baud, O., Tanter, M., 2021. Bedside functional monitoring of the dynamic brain connectivity in human neonates. Nat. Commun. 12, 1080.

Belliveau, J.W., Kennedy Jr., D.N., McKinstry, R.C., Buchbinder, B.R., Weisskoff, R.M., Cohen, M.S., Vevea, J.M., Brady, T.J., Rosen, B.R., 1991. Functional mapping of the human visual cortex by magnetic resonance imaging. Science 254, 716–719.

Berman, A.J.L., Grissom, W.A., Witzel, T., Nasr, S., Park, D.J., Setsompop, K., Polimeni, J.R., 2021. Ultra-high spatial resolution BOLD fMRI in humans using combined segmented-accelerated VFA-FLEET with a recursive RF pulse design. Magn. Reson. Med. 85, 120–139.

Biswal, B., Yetkin, F.Z., Haughton, V.M., Hyde, J.S., 1995. Functional connectivity in the motor cortex of resting human brain using echo-planar MRI. Magn. Reson. Med. 34, 537–541.

Biswal, B., Hudetz, A.G., Yetkin, F.Z., Haughton, V.M., Hyde, J.S., 1997. Hypercapnia reversibly suppresses low-frequency fluctuations in the human motor cortex during rest using echo-planar MRI. J. Cereb. Blood Flow Metab. 17, 301–308.

Blaize, K., Arcizet, F., Gesnik, M., Ahnine, H., Ferrari, U., Deffieux, T., Pouget, P., Chavane, F., Fink, M., Sahel, J.A., Tanter, M., Picaud, S., 2020. Functional ultrasound imaging of deep visual cortex in awake nonhuman primates. Proc. Natl. Acad. Sci. U. S. A. 117, 14453–14463.

Boido, D., Rungta, R.L., Osmanski, B.F., Roche, M., Tsurugizawa, T., Le Bihan, D., Ciobanu, L., Charpak, S., 2019. Mesoscopic and microscopic imaging of sensory responses in the same animal. Nat. Commun. 10, 1110.

Borden, P.M., Zhang, P., Shivange, A.V., Marvin, J.S., Cichon, J., Dan, C., Podgorski, K., Figueiredo, A., Novak, O., Tanimoto, M., Shigetomi, E., Lobas, M.A., Kim, H., Zhu, P.K., Zhang, Y., Zheng, W.S., Fan, C., Wang, G., Xiang, B., Gan, L., Zhang, G.-X., Guo, K., Lin, L., Cai, Y., Yee, A.G., Aggarwal, A., Ford, C.P., Rees, D.C., Dietrich, D., Khakh, B.S., Dittman, J.S., Gan, W.-B., Koyama, M., Jayaraman, V., Cheer, J.F., Lester, H.A., Zhu, J.J., Looger, L.L., 2020. A fast genetically encoded fluorescent sensor for faithful in vivo acetylcholine detection in mice, fish, worms and flies. bioRxiv.

Bourane, S., Duan, B., Koch, S.C., Dalet, A., Britz, O., Garcia-Campmany, L., Kim, E., Cheng, L., Ghosh, A., Ma, Q., Goulding, M., 2015. Gate control of mechanical itch by a subpopulation of spinal cord interneurons. Science 350, 550–554.

Boyden, E.S., Zhang, F., Bamberg, E., Nagel, G., Deisseroth, K., 2005. Millisecond-timescale, genetically targeted optical control of neural activity. Nat. Neurosci. 8, 1263–1268.

Cabral, J., Kringelbach, M.L., Deco, G., 2014. Exploring the network dynamics underlying brain activity during rest. Prog. Neurobiol. 114, 102–131.

Cardoso, M.M., Sirotin, Y.B., Lima, B., Glushenkova, E., Das, A., 2012. The neuroimaging signal is a linear sum of neurally distinct stimulus- and task-related components. Nat. Neurosci. 15, 1298–1306.

Carter, M.E., Soden, M.E., Zweifel, L.S., Palmiter, R.D., 2013. Genetic identification of a neural circuit that suppresses appetite. Nature 503, 111–114.

Chaigneau, E., Tiret, P., Lecoq, J., Ducros, M., Knopfel, T., Charpak, S., 2007. The relationship between blood flow and neuronal activity in the rodent olfactory bulb. J. Neurosci. 27, 6452–6460.

Chang, C., Leopold, D.A., Scholvinck, M.L., Mandelkow, H., Picchioni, D., Liu, X., Ye, F.Q., Turchi, J.N., Duyn, J.H., 2016. Tracking brain arousal fluctuations with FMRI. Proc. Natl. Acad. Sci. U. S. A. 113, 4518–4523.

Chen, T.W., Wardill, T.J., Sun, Y., Pulver, S.R., Renninger, S.L., Baohan, A., Schreiter, E.R., Kerr, R.A., Orger, M.B., Jayaraman, V., Looger, L.L., Svoboda, K., Kim, D.S., 2013. Ultrasensitive fluorescent proteins for imaging neuronal activity. Nature 499, 295–300.

Chen, X., Sobczak, F., Chen, Y., Jiang, Y., Qian, C., Lu, Z., Ayata, C., Logothetis, N.K., Yu, X., 2019. Mapping optogenetically-driven single-vessel fMRI with concurrent neuronal calcium recordings in the rat hippocampus. Nat. Commun. 10, 5239.

Chen, Y., Sobczak, F., Pais-Roldan, P., Schwarz, C., Koretsky, A.P., Yu, X., 2020. Mapping the brain-wide network effects by optogenetic activation of the corpus callosum. Cereb. Cortex 30, 5885–5898.

Chen, X., Jiang, Y., Choi, S., Pohmann, R., Scheffler, K., Kleinfeld, D., Yu, X., 2021. Assessment of single-vessel cerebral blood velocity by phase contrast fMRI. PLoS Biol. 19, e3000923.

Chen, Y., Wang, Q., Choi, S., Zeng, H., Takahashi, K., Qian, C., Yu, X., 2022. Focal fMRI signal enhancement with implantable inductively coupled detectors. Neuroimage 247, 118793.

Choi, S., Zeng, H., Chen, Y., Sobczak, F., Qian, C., Yu, X., 2022. Laminar-specific functional connectivity mapping with multi-slice line-scanning fMRI. Cereb. Cortex, bhab497.

Chuang, K.H., Van Gelderen, P., Merkle, H., Bodurka, J., Ikonomidou, V.N., Koretsky, A.P., Duyn, J.H., Talagala, S.L., 2008. Mapping resting-state functional connectivity using perfusion MRI. Neuroimage 40, 1595–1605.

Cordes, D., Haughton, V.M., Arfanakis, K., Carew, J.D., Turski, P.A., Moritz, C.H., Quigley, M.A., Meyerand, M.E., 2001. Frequencies contributing to functional connectivity in the cerebral cortex in "resting-state" data. Am. J. Neuroradiol. 22, 1326–1333.

Corsi-Cabrera, M., Figueredo-Rodríguez, P., Del Río-Portilla, Y., Sánchez-Romero, J., Galán, L., Bosch-Bayard, J., 2012. Enhanced frontoparietal synchronized activation during the wake-sleep transition in patients with primary insomnia. Sleep 35, 501–511.

De Luca, M., Beckmann, C.F., De Stefano, N., Matthews, P.M., Smith, S.M., 2006. fMRI resting state networks define distinct modes of long-distance interactions in the human brain. Neuroimage 29, 1359–1367.

Deco, G., Jirsa, V.K., Mcintosh, A.R., 2011. Emerging concepts for the dynamical organization of resting-state activity in the brain. Nat. Rev. Neurosci. 12, 43–56.

Deffieux, T., Demene, C., Pernot, M., Tanter, M., 2018. Functional ultrasound neuroimaging: a review of the preclinical and clinical state of the art. Curr. Opin. Neurobiol. 50, 128–135.

Demene, C., Baranger, J., Bernal, M., Delanoe, C., Auvin, S., Biran, V., Alison, M., Mairesse, J., Harribaud, E., Pernot, M., Tanter, M., Baud, O., 2017. Functional ultrasound imaging of brain activity in human newborns. Sci. Transl. Med. 9, p.eaah6756.

Denis, R.G., Joly-Amado, A., Webber, E., Langlet, F., Schaeffer, M., Padilla, S.L., Cansell, C., Dehouck, B., Castel, J., Delbes, A.S., Martinez, S., Lacombe, A., Rouch, C., Kassis, N., Fehrentz, J.A., Martinez, J., Verdie, P., Hnasko, T.S., Palmiter, R.D., Krashes, M.J., Guler, A.D., Magnan, C., Luquet, S., 2015. Palatability can drive feeding independent of Agrp neurons. Cell Metab. 22, 646–657.

DeSilva, J.M., Traniello, J.F.A., Claxton, A.G., Fannin, L.D., 2021. When and why did human brains decrease in size? A new change-point analysis and insights from brain evolution in ants. Front. Ecol. Evol. 9.

Detre, J.A., Wang, J., 2002. Technical aspects and utility of fMRI using BOLD and ASL. Clin. Neurophysiol. 113, 621–634.

Detre, J.A., Leigh, J.S., Williams, D.S., Koretsky, A.P., 1992. Perfusion imaging. Magn. Reson. Med. 23, 37–45.

Devor, A., Tian, P., Nishimura, N., Teng, I.C., Hillman, E.M., Narayanan, S.N., Ulbert, I., Boas, D.A., Kleinfeld, D., Dale, A.M., 2007. Suppressed neuronal activity and concurrent arteriolar vasoconstriction may explain negative blood oxygenation level-dependent signal. J. Neurosci. 27, 4452–4459.

Di Martino, A., Yan, C.G., Li, Q., Denio, E., Castellanos, F.X., Alaerts, K., Anderson, J.S., Assaf, M., Bookheimer, S.Y., Dapretto, M., Deen, B., Delmonte, S., Dinstein, I., ERTL-Wagner, B., Fair, D.A., Gallagher, L., Kennedy, D.P., Keown, C.L., Keysers, C., Lainhart, J.E., Lord, C., Luna, B., Menon, V., Minshew, N.J., Monk, C.S., Mueller, S., Muller, R.A., Nebel, M.B., Nigg, J.T., O'Hearn, K., Pelphrey, K.A., Peltier, S.J., Rudie, J.D., Sunaert, S., Thioux, M., Tyszka, J.M., Uddin, L.Q., Verhoeven, J.S., Wenderoth, N., Wiggins, J.L., Mostofsky, S.H., Milham, M.P., 2014. The autism brain imaging data exchange: towards a large-scale evaluation of the intrinsic brain architecture in autism. Mol. Psychiatry 19, 659–667.

Diez-Silva, M., Dao, M., Han, J., Lim, C.-T., Suresh, S., 2010. Shape and biomechanical characteristics of human red blood cells in health and disease. MRS Bull. 35, 382–388.

Dizeux, A., Gesnik, M., Ahnine, H., Blaize, K., Arcizet, F., Picaud, S., Sahel, J.A., Deffieux, T., Pouget, P., Tanter, M., 2019. Functional ultrasound imaging of the brain reveals propagation of task-related brain activity in behaving primates. Nat. Commun. 10, 1400.

Drew, P.J., Mateo, C., Turner, K.L., Yu, X., Kleinfeld, D., 2020. Ultra-slow oscillations in fMRI and resting-state connectivity: neuronal and vascular contributions and technical confounds. Neuron 107, 782–804.

Duyn, J., 2011. Spontaneous fMRI activity during resting wakefulness and sleep. Prog. Brain Res. 193, 295–305.

Duyn, J.H., Ozbay, P.S., Chang, C., Picchioni, D., 2020. Physiological changes in sleep that affect fMRI inference. Curr. Opin. Behav. Sci. 33, 42–50.

Farrell, M.S., Pei, Y., Wan, Y., Yadav, P.N., Daigle, T.L., Urban, D.J., Lee, H.M., Sciaky, N., Simmons, A., Nonneman, R.J., Huang, X.P., Hufeisen, S.J., Guettier, J.M., Moy, S.S., Wess, J., Caron, M.G., Calakos, N., Roth, B.L., 2013. A Galphas DREADD mouse for selective modulation of cAMP production in striatopallidal neurons. Neuropsychopharmacology 38, 854–862.

Feinberg, D.A., Vu, A.T., Beckett, A., 2018. Pushing the limits of ultra-high resolution human brain imaging with SMS-EPI demonstrated for columnar level fMRI. Neuroimage 164, 155–163.

Feng, J., Zhang, C., Lischinsky, J.E., Jing, M., Zhou, J., Wang, H., Zhang, Y., Dong, A., Wu, Z., Wu, H., Chen, W., Zhang, P., Zou, J., Hires, S.A., Zhu, J.J., Cui, G., Lin, D., Du, J., Li, Y., 2019. A genetically encoded fluorescent sensor for rapid and specific in vivo detection of norepinephrine. Neuron 102. 745–761e8.

Ferguson, S.M., Eskenazi, D., Ishikawa, M., Wanat, M.J., Phillips, P.E., Dong, Y., Roth, B.L., Neumaier, J.F., 2011. Transient neuronal inhibition reveals opposing roles of indirect and direct pathways in sensitization. Nat. Neurosci. 14, 22–24.

Ferguson, S.M., Phillips, P.E., Roth, B.L., Wess, J., Neumaier, J.F., 2013. Direct-pathway striatal neurons regulate the retention of decision-making strategies. J. Neurosci. 33, 11668–11676.

Fischl, B., Dale, A.M., 2000. Measuring the thickness of the human cerebral cortex from magnetic resonance images. Proc. Natl. Acad. Sci. 97, 11050–11055.

Fox, M.D., Raichle, M.E., 2007. Spontaneous fluctuations in brain activity observed with functional magnetic resonance imaging. Nat. Rev. Neurosci. 8, 700–711.

Fukuda, M., Poplawsky, A.J., Kim, S.-G., 2021. Time-dependent spatial specificity of high-resolution FMRI: insights into mesoscopic neurovascular coupling. Philos. Trans. R. Soc. Lond. B Biol. Sci. 376, 20190623.

Fukunaga, M., Horovitz, S.G., Van Gelderen, P., De Zwart, J.A., Jansma, J.M., Ikonomidou, V.N., Chu, R., Deckers, R.H., Leopold, D.A., Duyn, J.H., 2006. Large-amplitude, spatially correlated fluctuations in BOLD fMRI signals during extended rest and early sleep stages. Magn. Reson. Imaging 24, 979–992.

Ghosh, K.K., Burns, L.D., Cocker, E.D., Nimmerjahn, A., Ziv, Y., Gamal, A.E., Schnitzer, M.J., 2011. Miniaturized integration of a fluorescence microscope. Nat. Methods 8, 871–878.

Ghosh, S., Li, N., Schwalm, M., Bartelle, B.B., Xie, T., Daher, J.I., Singh, U.D., Xie, K., Dinapoli, N., Evans, N.B., Chung, K., Jasanoff, A., 2022. Functional dissection of neural circuitry using a genetic reporter for fMRI. Nat. Neurosci. 25, 390–398.

Goense, J.B.M., Logothetis, N.K., 2008. Neurophysiology of the BOLD fMRI signal in awake monkeys. Curr. Biol. 18, 631–640.

Gong, Y.Y., Huang, C., Li, J.Z., Grewe, B.F., Zhang, Y.P., Eismann, S., Schnitzer, M.J., 2015. High-speed recording of neural spikes in awake mice and flies with a fluorescent voltage sensor. Science 350, 1361–1366.

Gottschalk, S., Fehm, T.F., Dean-Ben, X.L., Tsytsarev, V., Razansky, D., 2017. Correlation between volumetric oxygenation responses and electrophysiology identifies deep thalamocortical activity during epileptic seizures. Neurophotonics 4, 011007.

Greicius, M.D., Krasnow, B., Reiss, A.L., Menon, V., 2003. Functional connectivity in the resting brain: a network analysis of the default mode hypothesis. Proc. Natl. Acad. Sci. U. S. A. 100, 253–258.

Guettier, J.M., Gautam, D., Scarselli, M., Ruiz De Azua, I., Li, J.H., Rosemond, E., Ma, X., Gonzalez, F.J., Armbruster, B.N., Lu, H., Roth, B.L., Wess, J., 2009. A chemical-genetic approach to study G protein regulation of beta cell function in vivo. Proc. Natl. Acad. Sci. U. S. A. 106, 19197–19202.

Gulati, S., Cao, V.Y., Otte, S., 2017. Multi-layer Cortical Ca^{2+} imaging in freely moving mice with prism probes and miniaturized fluorescence microscopy. J. Vis. Exp.

Hamel, E.J., Grewe, B.F., Parker, J.G., Schnitzer, M.J., 2015. Cellular level brain imaging in behaving mammals: an engineering approach. Neuron 86, 140–159.

Hampson, M., Driesen, N.R., Skudlarski, P., Gore, J.C., Constable, R.T., 2006. Brain connectivity related to working memory performance. J. Neurosci. 26, 13338–13343.

Harel, N., Lee, S.P., Nagaoka, T., Kim, D.S., Kim, S.G., 2002. Origin of negative blood oxygenation level-dependent FMRI signals. J. Cereb. Blood Flow Metab. 22, 908–917.

He, B.J., Raichle, M.E., 2009. The fMRI signal, slow cortical potential and consciousness. Trends Cogn. Sci. 13, 302–309.

He, Y., Wang, M., Chen, X., Pohmann, R., Polimeni, J.R., Scheffler, K., Rosen, B.R., Kleinfeld, D., Yu, X., 2018. Ultra-slow single-vessel BOLD and CBV-based fMRI spatiotemporal dynamics and their correlation with neuronal intracellular calcium signals. Neuron 97. 925–939e5.

Hirsch, S., Reichold, J., Schneider, M., Székely, G., Weber, B., 2012. Topology and hemodynamics of the cortical cerebrovascular system. J. Cereb. Blood Flow Metab. 32, 952–967.

Hodge, R.D., Bakken, T.E., Miller, J.A., Smith, K.A., Barkan, E.R., Graybuck, L.T., Close, J.L., Long, B., Johansen, N., Penn, O., Yao, Z., Eggermont, J., Hollt, T., Levi, B.P., Shehata, S.I., Aevermann, B., Beller, A., Bertagnolli, D., Brouner, K., Casper, T., Cobbs, C., Dalley, R., Dee, N., Ding, S.L., Ellenbogen, R.G., Fong, O., Garren, E., Goldy, J., Gwinn, R.P., Hirschstein, D., Keene, C.D., Keshk, M., Ko, A.L., Lathia, K., Mahfouz, A., Maltzer, Z., McGraw, M., Nguyen, T.N., Nyhus, J., Ojemann, J.G., Oldre, A., Parry, S., Reynolds, S., Rimorin, C., Shapovalova, N.V., Somasundaram, S., Szafer, A., Thomsen, E.R., Tieu, M., Quon, G., Scheuermann, R.H., Yuste, R., Sunkin, S.M., Lelieveldt, B., Feng, D., Ng, L., Bernard, A., Hawrylycz, M., Phillips, J.W., Tasic, B., Zeng, H., Jones, A.R., Koch, C., Lein, E.S., 2019. Conserved cell types with divergent features in human versus mouse cortex. Nature 573, 61–68.

Huber, L., Handwerker, D.A., Jangraw, D.C., Chen, G., Hall, A., Stüber, C., Gonzalez-Castillo, J., Ivanov, D., Marrett, S., Guidi, M., Goense, J., Poser, B.A., Bandettini, P.A., 2017. High-resolution CBV-fMRI allows mapping of laminar activity and connectivity of cortical input and output in human M1. Neuron 96. 1253–1263.e7.

Huber, L., Finn, E.S., Chai, Y., Goebel, R., Stirnberg, R., Stöcker, T., Marrett, S., Uludag, K., Kim, S.-G., Han, S., Bandettini, P.A., Poser, B.A., 2021. Layer-dependent functional connectivity methods. Prog. Neurobiol. 207, 101835.

Hyder, F., Rothman Douglas, L., 2010. Neuronal correlate of BOLD signal fluctuations at rest: err on the side of the baseline. Proc. Natl. Acad. Sci. 107, 10773–10774.

Imbault, M., Chauvet, D., Gennisson, J.L., Capelle, L., Tanter, M., 2017. Intraoperative functional ultrasound imaging of human brain activity. Sci. Rep. 7, 7304.

Jennings, J.H., Ung, R.L., Resendez, S.L., Stamatakis, A.M., Taylor, J.G., Huang, J., Veleta, K., Kantak, P.A., Aita, M., Shilling-Scrivo, K., Ramakrishnan, C., Deisseroth, K., Otte, S., Stuber, G.D., 2015. Visualizing hypothalamic network dynamics for appetitive and consummatory behaviors. Cell 160, 516–527.

Jiang, Y., Chen, X., Pais Roldán, P., Rosen, B.R., Yu, X., 2020. Deciphering the contribution of extracellular glutamate and intracellular calcium signaling to the BOLD fMRI signal. In: 2020 ISMRM & SMRT Virtual Conference & Exhibition.

Jing, M., Zhang, P., Wang, G., Feng, J., Mesik, L., Zeng, J., Jiang, H., Wang, S., Looby, J.C., Guagliardo, N.A., Langma, L.W., Lu, J., Zuo, Y., Talmage, D.A., Role, L.W., Barrett, P.Q., Zhang, L.I., Luo, M., Song, Y., Zhu, J.J., Li, Y., 2018. A genetically encoded fluorescent acetylcholine indicator for in vitro and in vivo studies. Nat. Biotechnol. 36, 726–737.

Jung Won, B., Jiang, H., Lee, S., Kim, S.-G., 2022. Dissection of brain-wide resting-state and functional somatosensory circuits by fMRI with optogenetic silencing. Proc. Natl. Acad. Sci. 119, e2113313119.

Keilholz, S.D., 2014. The neural basis of time-varying resting-state functional connectivity. Brain Connect. 4, 769–779.

Khanna, P.C., Poliakov, A.V., Ishak, G.E., Poliachik, S.L., Friedman, S.D., Saneto, R.P., Novotny Jr., E.J., Ojemann, J.G., Shaw, D.W., 2012. Preserved interhemispheric functional connectivity in a case of corpus callosum agenesis. Neuroradiology 54, 177–179.

Kim, T., Kim, S.G., 2011. Temporal dynamics and spatial specificity of arterial and venous blood volume changes during visual stimulation: implication for BOLD quantification. J. Cereb. Blood Flow Metab. 31, 1211–1222.

Kim, S.-G., Harel, N., Jin, T., Kim, T., Lee, P., Zhao, F., 2013. Cerebral blood volume MRI with intravascular superparamagnetic iron oxide nanoparticles. NMR Biomed. 26, 949–962.

Koenigsberger, M., Sauser, R., Beny, J.L., Meister, J.J., 2005. Role of the endothelium on arterial vasomotion. Biophys. J. 88, 3845–3854.

Kozorovitskiy, Y., Saunders, A., Johnson, C.A., Lowell, B.B., Sabatini, B.L., 2012. Recurrent network activity drives striatal synaptogenesis. Nature 485, 646–650.

Krashes, M.J., Koda, S., Ye, C., Rogan, S.C., Adams, A.C., Cusher, D.S., Maratos-Flier, E., Roth, B.L., Lowell, B.B., 2011. Rapid, reversible activation of AgRP neurons drives feeding behavior in mice. J. Clin. Invest. 121, 1424–1428.

Kugler, E.C., Greenwood, J., Macdonald, R.B., 2021. The "neuro-glial-vascular" unit: the Role of glia in neurovascular unit formation and dysfunction. Front. Cell Dev. Biol. 9, 2641.

Kwong, K.K., Belliveau, J.W., Chesler, D.A., Goldberg, I.E., Weisskoff, R.M., Poncelet, B.P., Kennedy, D.N., Hoppel, B.E., Cohen, M.S., Turner, R., et al., 1992. Dynamic magnetic resonance imaging of human brain activity during primary sensory stimulation. Proc. Natl. Acad. Sci. U. S. A. 89, 5675–5679.

Lachaux, J.-P., Fonlupt, P., Kahane, P., Minotti, L., Hoffmann, D., Bertrand, O., Baciu, M., 2007. Relationship between task-related gamma oscillations and BOLD signal: new insights from combined fMRI and intracranial EEG. Hum. Brain Mapp. 28, 1368–1375.

Lake, E.M.R., Ge, X., Shen, X., Herman, P., Hyder, F., Cardin, J.A., Higley, M.J., Scheinost, D., Papademetris, X., Crair, M.C., Constable, R.T., 2020. Simultaneous cortex-wide fluorescence Ca(2+) imaging and whole-brain fMRI. Nat. Methods 17, 1262–1271.

Lauritzen, M., 2001. Relationship of spikes, synaptic activity, and local changes of cerebral blood flow. J. Cereb. Blood Flow Metab. 21, 1367–1383.

Lauritzen, M., 2008. On the neural basis of FMRI signals. Clin. Neurophysiol. 119, 729–730.

Lee, J.H., Durand, R., Gradinaru, V., Zhang, F., Goshen, I., Kim, D.S., Fenno, L.E., Ramakrishnan, C., Deisseroth, K., 2010. Global and local fMRI signals driven by neurons defined optogenetically by type and wiring. Nature 465, 788–792.

Lev-Ram, V., Mehta, S.B., Kleinfeld, D., Tsien, R.Y., 2003. Reversing cerebellar long-term depression. Proc. Natl. Acad. Sci. U. S. A. 100, 15989–15993.

Li, C., Aguirre, A., Gamelin, J., Maurudis, A., Zhu, Q., Wang, L.V., 2010. Real-time photoacoustic tomography of cortical hemodynamics in small animals. J. Biomed. Opt. 15, 010509.

Li, J., Pan, L., Pembroke, W.G., Rexach, J.E., Godoy, M.I., Condro, M.C., Alvarado, A.G., Harteni, M., Chen, Y.-W., Stiles, L., Chen, A.Y., Wanner, I.B., Yang, X., Goldman, S.A., Geschwind, D.H., Kornblum, H.I., Zhang, Y., 2021. Conservation and divergence of vulnerability and responses to stressors between human and mouse astrocytes. Nat. Commun. 12, 3958.

Liang, Z., Ma, Y., Watson, G.D.R., Zhang, N., 2017. Simultaneous GCaMP6-based fiber photometry and fMRI in rats. J. Neurosci. Methods 289, 31–38.

Liao, L.D., Lin, C.T., Shih, Y.Y., Duong, T.Q., Lai, H.Y., Wang, P.H., Wu, R., Tsang, S., Chang, J.Y., Li, M.L., Chen, Y.Y., 2012. Transcranial imaging of functional cerebral hemodynamic changes in single blood vessels using in vivo photoacoustic microscopy. J. Cereb. Blood Flow Metab. 32, 938–951.

Liu, X., De Zwart, J.A., Scholvinck, M.L., Chang, C., Ye, F.Q., Leopold, D.A., Duyn, J.H., 2018. Subcortical evidence for a contribution of arousal to fMRI studies of brain activity. Nat. Commun. 9, 395.

Logothetis, N.K., 2008. What we can do and what we cannot do with fMRI. Nature 453, 869–878.

Logothetis, N.K., Eschenko, O., Murayama, Y., Augath, M., Steudel, T., Evrard, H.C., Besserve, M., Oeltermann, A., 2012. Hippocampal–cortical interaction during periods of subcortical silence. Nature 491 (7425), 547–553.

Logothetis, N.K., Pauls, J., Augath, M., Trinath, T., Oeltermann, A., 2001. Neurophysiological investigation of the basis of the fMRI signal. Nature 412, 150–157.

Lu, H., Golay, X., Pekar, J.J., Van Zijl, P.C., 2003. Functional magnetic resonance imaging based on changes in vascular space occupancy. Magn. Reson. Med. 50, 263–274.

Mace, E., Montaldo, G., Cohen, I., Baulac, M., Fink, M., Tanter, M., 2011. Functional ultrasound imaging of the brain. Nat. Methods 8, 662–664.

Magnuson, M., Majeed, W., Keilholz, S.D., 2010. Functional connectivity in blood oxygenation level-dependent and cerebral blood volume-weighted resting state functional magnetic resonance imaging in the rat brain. J. Magn. Reson. Imaging 32, 584–592.

Magnuson, M.E., Thompson, G.J., Pan, W.J., Keilholz, S.D., 2014. Effects of severing the corpus callosum on electrical and BOLD functional connectivity and spontaneous dynamic activity in the rat brain. Brain Connect. 4, 15–29.

Malinow, R., Tsien, R.W., 1990. Presynaptic enhancement shown by whole-cell recordings of long-term potentiation in hippocampal slices. Nature 346, 177–180.

Mandeville, J.B., 2012. IRON fMRI measurements of CBV and implications for BOLD signal. Neuroimage 62, 1000–1008.

Marchant, N.J., Whitaker, L.R., Bossert, J.M., Harvey, B.K., Hope, B.T., Kaganovsky, K., Adhikary, S., Prisinzano, T.E., Vardy, E., Roth, B.L., Shaham, Y., 2016. Behavioral and physiological effects of a novel kappa-opioid receptor-based DREADD in rats. Neuropsychopharmacology 41, 402–409.

Markicevic, M., Fulcher, B.D., Lewis, C., Helmchen, F., Rudin, M., Zerbi, V., Wenderoth, N., 2020. Cortical excitation: inhibition imbalance causes abnormal brain network dynamics as observed in neurodevelopmental disorders. Cereb. Cortex 30, 4922–4937.

Marvin, J.S., Borghuis, B.G., Tian, L., Cichon, J., Harnett, M.T., Akerboom, J., Gordus, A., Renninger, S.L., Chen, T.W., Bargmann, C.I., Orger, M.B., Schreiter, E.R., Demb, J.B., Gan, W.B., Hires, S.A., Looger, L.L., 2013. An optimized fluorescent probe for visualizing glutamate neurotransmission. Nat. Methods 10, 162–170.

Mateo, C., Knutsen, P.M., Tsai, P.S., Shih, A.Y., Kleinfeld, D., 2017. Entrainment of arteriole vasomotor fluctuations by neural activity is a basis of blood-oxygenation-level-dependent "resting-state" connectivity. Neuron 96. 936–948e3.

Mc Larney, B., Hutter, M.A., Degtyaruk, O., Dean-Ben, X.L., Razansky, D., 2020. Monitoring of stimulus evoked murine somatosensory cortex hemodynamic activity with volumetric multi-spectral optoacoustic tomography. Front. Neurosci. 14, 536.

Moon, C.H., Fukuda, M., Kim, S.-G., 2013. Spatiotemporal characteristics and vascular sources of neural-specific and -nonspecific fMRI signals at submillimeter columnar resolution. Neuroimage 64, 91–103.

Muller, A., Joseph, V., Slesinger, P.A., Kleinfeld, D., 2014. Cell-based reporters reveal in vivo dynamics of dopamine and norepinephrine release in murine cortex. Nat. Methods 11, 1245–1252.

Nagel, G., Szellas, T., Huhn, W., Kateriya, S., Adeishvili, N., Berthold, P., Ollig, D., Hegemann, P., Bamberg, E., 2003. Channelrhodopsin-2, a directly light-gated cation-selective membrane channel. Proc. Natl. Acad. Sci. U. S. A. 100, 13940–13945.

Nakamura, Y., Nakamura, Y., Pelosi, A., Djemai, B., Debacker, C., Herve, D., Girault, J.A., Tsurugizawa, T., 2020. fMRI detects bilateral brain network activation following unilateral chemogenetic activation of direct striatal projection neurons. Neuroimage 220, 117079.

Nasiriavanaki, M., Xia, J., Wan, H., Bauer, A.Q., Culver, J.P., Wang, L.V., 2014. High-resolution photoacoustic tomography of resting-state functional connectivity in the mouse brain. Proc. Natl. Acad. Sci. U. S. A. 111, 21–26.

Niessing, J., Ebisch, B., Schmidt, K.E., Niessing, M., Singer, W., Galuske, R.A., 2005. Hemodynamic signals correlate tightly with synchronized gamma oscillations. Science 309, 948–951.

Nir, Y., Fisch, L., Mukamel, R., Gelbard-Sagiv, H., Arieli, A., Fried, I., Malach, R., 2007. Coupling between neuronal firing rate, gamma LFP, and BOLD fMRI is related to Interneuronal correlations. Curr. Biol. 17, 1275–1285.

Nunes, D., Ianus, A., Shemesh, N., 2019. Layer-specific connectivity revealed by diffusion-weighted functional MRI in the rat thalamocortical pathway. Neuroimage 184, 646–657.

Obrig, H., Neufang, M., Wenzel, R., Kohl, M., Steinbrink, J., Einhaupl, K., Villringer, A., 2000. Spontaneous low frequency oscillations of cerebral hemodynamics and metabolism in human adults. Neuroimage 12, 623–639.

O'Connell, K.E., Mikkola, A.M., Stepanek, A.M., Vernet, A., Hall, C.D., Sun, C.C., Yildirim, E., Staropoli, J.F., Lee, J.T., Brown, D.E., 2015. Practical murine hematopathology: a comparative review and implications for research. Comp. Med. 65, 96–113.

Oliveira, Í.A.F., Cai, Y., Hofstetter, S., Siero, J.C.W., Van Der Zwaag, W., Dumoulin, S.O., 2022. Comparing BOLD and VASO-CBV population receptive field estimates in human visual cortex. Neuroimage 248, 118868.

O'Reilly, J.X., Croxson, P.L., Jbabdi, S., Sallet, J., Noonan, M.P., Mars, R.B., Browning, P.G., Wilson, C.R., Mitchell, A.S., Miller, K.L., Rushworth, M.F., Baxter, M.G., 2013. Causal effect of disconnection lesions on interhemispheric functional connectivity in rhesus monkeys. Proc. Natl. Acad. Sci. U. S. A. 110, 13982–13987.

Osmanski, B.F., Pezet, S., Ricobaraza, A., Lenkei, Z., Tanter, M., 2014. Functional ultrasound imaging of intrinsic connectivity in the living rat brain with high spatiotemporal resolution. Nat. Commun. 5, 5023.

Ouzounov, D.G., Wang, T., Wang, M., Feng, D.D., Horton, N.G., Cruz-Hernandez, J.C., Cheng, Y.T., Reimer, J., Tolias, A.S., Nishimura, N., Xu, C., 2017. In vivo three-photon imaging of activity of GCaMP6-labeled neurons deep in intact mouse brain. Nat. Methods 14, 388–390.

Owen, J.P., Li, Y.O., Yang, F.G., Shetty, C., Bukshpun, P., Vora, S., Wakahiro, M., Hinkley, L.B., Nagarajan, S.S., Sherr, E.H., Mukherjee, P., 2013. Resting-state networks and the functional connectome of the human brain in agenesis of the corpus callosum. Brain Connect. 3, 547–562.

Özbay, P.S., Chang, C., Picchioni, D., Mandelkow, H., Chappel-Farley, M.G., Van Gelderen, P., De Zwart, J.A., Duyn, J., 2019. Sympathetic activity contributes to the fMRI signal. Commun. Biol. 2, 421.

Pais-Roldan, P., Edlow, B.L., Jiang, Y., Stelzer, J., Zou, M., Yu, X., 2019. Multimodal assessment of recovery from coma in a rat model of diffuse brainstem tegmentum injury. Neuroimage 189, 615–630.

Pais-Roldan, P., Takahashi, K., Sobczak, F., Chen, Y., Zhao, X., Zeng, H., Jiang, Y., Yu, X., 2020. Indexing brain state-dependent pupil dynamics with simultaneous fMRI and optical fiber calcium recording. Proc. Natl. Acad. Sci. U. S. A. 117, 6875–6882.

Pais-Roldán, P., Mateo, C., Pan, W.-J., Acland, B., Kleinfeld, D., Snyder, L.H., Yu, X., Keilholz, S., 2021. Contribution of animal models toward understanding resting state functional connectivity. Neuroimage 245, 118630.

Palacios, E.M., Sala-Llonch, R., Junque, C., Roig, T., Tormos, J.M., Bargallo, N., Vendrell, P., 2013. Resting-state functional magnetic resonance imaging activity and connectivity and cognitive outcome in traumatic brain injury. JAMA Neurol. 70, 845–851.

Patriarchi, T., Cho, J.R., Merten, K., Howe, M.W., Marley, A., Xiong, W.H., Folk, R.W., Broussard, G.J., Liang, R., Jang, M.J., Zhong, H., Dombeck, D., Von Zastrow, M., Nimmerjahn, A., Gradinaru, V., Williams, J.T., Tian, L., 2018. Ultrafast neuronal imaging of dopamine dynamics with designed genetically encoded sensors. Science 360, eaat4422.

Poskanzer, K.E., Yuste, R., 2011. Astrocytic regulation of cortical UP states. Proc. Natl. Acad. Sci. U. S. A. 108, 18453–18458.

Poskanzer, K.E., Yuste, R., 2016. Astrocytes regulate cortical state switching in vivo. Proc. Natl. Acad. Sci. U. S. A. 113, E2675–E2684.

Pradier, B., Wachsmuth, L., Nagelmann, N., Segelcke, D., Kreitz, S., Hess, A., Pogatzki-Zahn, E.M., Faber, C., 2021. Combined resting state-fMRI and calcium recordings show stable brain states for task-induced fMRI in mice under combined ISO/MED anesthesia. Neuroimage 245, 118626.

Raichle, M.E., Macleod, A.M., Snyder, A.Z., Powers, W.J., Gusnard, D.A., Shulman, G.L., 2001. A default mode of brain function. Proc. Natl. Acad. Sci. U. S. A. 98, 676–682.

Rau, R., Kruizinga, P., Mastik, F., Belau, M., De Jong, N., Bosch, J.G., Scheffer, W., Maret, G., 2018. 3D functional ultrasound imaging of pigeons. Neuroimage 183, 469–477.

Ray, R.S., Corcoran, A.E., Brust, R.D., Kim, J.C., Richerson, G.B., Nattie, E., Dymecki, S.M., 2011. Impaired respiratory and body temperature control upon acute serotonergic neuron inhibition. Science 333, 637–642.

Rayshubskiy, A., Wojtasiewicz, T.J., Mikell, C.B., Bouchard, M.B., Timerman, D., Youngerman, B.E., McGovern, R.A., Otten, M.L., Canoll, P., Mckhann 2nd, G.M., Hillman, E.M., 2014. Direct, intraoperative observation of ~0.1 Hz hemodynamic oscillations in awake human cortex: implications for fMRI. Neuroimage 87, 323–331.

Rocchi, F., Canella, C., Noei, S., Gutierrez-Barragan, D., Coletta, L., Galbusera, A., Stuefer, A., Vassanelli, S., Pasqualetti, M., Iurilli, G., Panzeri, S., Gozzi, A., 2022. Increased fMRI connectivity upon chemogenetic inhibition of the mouse prefrontal cortex. Nat. Commun. 13, 1056.

Roelofs, T.J.M., Verharen, J.P.H., Van Tilborg, G.A.F., Boekhoudt, L., Van Der Toorn, A., De Jong, J.W., Luijendijk, M.C.M., Otte, W.M., Adan, R.A.H., Dijkhuizen, R.M., 2017. A novel approach to map induced activation of neuronal networks using chemogenetics and functional neuroimaging in rats: A proof-of-concept study on the mesocorticolimbic system. Neuroimage 156, 109–118.

Roth, B.L., 2016. DREADDs for neuroscientists. Neuron 89, 683–694.

Scheeringa, R., Fries, P., Petersson, K.-M., Oostenveld, R., Grothe, I., Norris, D.G., Hagoort, P., Bastiaansen, M.C.M., 2011. Neuronal dynamics underlying high- and low-frequency EEG oscillations contribute independently to the human BOLD signal. Neuron 69, 572–583.

Schölvinck, M.L., Maier, A., Ye, F.Q., Duyn, J.H., Leopold, D.A., 2010. Neural basis of global resting-state fMRI activity. Proc. Natl. Acad. Sci. U. S. A. 107, 10238–10243.

Schulz, K., Sydekum, E., Krueppel, R., Engelbrecht, C.J., Schlegel, F., Schroter, A., Rudin, M., Helmchen, F., 2012. Simultaneous BOLD fMRI and fiber-optic calcium recording in rat neocortex. Nat. Methods 9, 597–602.

Schwalm, M., Schmid, F., Wachsmuth, L., Backhaus, H., Kronfeld, A., Aedo Jury, F., Prouvot, P.H., Fois, C., Albers, F., Van Alst, T., Faber, C., Stroh, A., 2017. Cortex-wide BOLD fMRI activity reflects locally-recorded slow oscillation-associated calcium waves. Elife 6, e27602.

Sheline, Y.I., Raichle, M.E., 2013. Resting state functional connectivity in preclinical Alzheimer's disease. Biol. Psychiatry 74, 340–347.

Shmuel, A., Leopold, D.A., 2008. Neuronal correlates of spontaneous fluctuations in fMRI signals in monkey visual cortex: implications for functional connectivity at rest. Hum. Brain Mapp. 29, 751–761.

Smith, A.J., Blumenfeld, H., Behar, K.L., Rothman, D.L., Shulman, R.G., Hyder, F., 2002. Cerebral energetics and spiking frequency: the neurophysiological basis of fMRI. Proc. Natl. Acad. Sci. U. S. A. 99, 10765–10770.

Smith, S.M., Beckmann, C.F., Andersson, J., Auerbach, E.J., Bijsterbosch, J., Douaud, G., Duff, E., Feinberg, D.A., Griffanti, L., Harms, M.P., Kelly, M., Laumann, T., Miller, K.L., Moeller, S., Petersen, S., Power, J., Salimi-Khorshidi, G., Snyder, A.Z., Vu, A.T., Woolrich, M.W., Xu, J., Yacoub, E., Ugurbil, K., Van Essen, D.C., Glasser, M.F., Consortium, W.U.-M.H., 2013. Resting-state fMRI in the human connectome project. Neuroimage 80, 144–168.

Soloukey, S., Vincent, A., Satoer, D.D., Mastik, F., Smits, M., Dirven, C.M.F., Strydis, C., Bosch, J.G., Van Der Steen, A.F.W., De Zeeuw, C.I., Koekkoek, S.K.E., Kruizinga, P., 2019. Functional ultrasound (fUS) during awake brain surgery: the clinical potential of intra-operative functional and vascular brain mapping. Front. Neurosci. 13, 1384.

Sun, Y.W., Qin, L.D., Zhou, Y., Xu, Q., Qian, L.J., Tao, J., Xu, J.R., 2011. Abnormal functional connectivity in patients with vascular cognitive impairment, no dementia: a resting-state functional magnetic resonance imaging study. Behav. Brain Res. 223, 388–394.

Sun, F., Zeng, J., Jing, M., Zhou, J., Feng, J., Owen, S.F., Luo, Y., Li, F., Wang, H., Yamaguchi, T., Yong, Z., Gao, Y., Peng, W., Wang, L., Zhang, S., Du, J., Lin, D., Xu, M., Kreitzer, A.C., Cui, G., Li, Y., 2018. A genetically encoded fluorescent sensor enables rapid and specific detection of dopamine in flies, fish, and mice. Cell 174. 481–496e19.

Teissier, A., Chemiakine, A., Inbar, B., Bagchi, S., Ray, R.S., Palmiter, R.D., Dymecki, S.M., Moore, H., Ansorge, M.S., 2015. Activity of raphe serotonergic neurons controls emotional behaviors. Cell Rep. 13, 1965–1976.

Tian, L., Hires, S.A., Mao, T., Huber, D., Chiappe, M.E., Chalasani, S.H., Petreanu, L., Akerboom, J., McKinney, S.A., Schreiter, E.R., Bargmann, C.I., Jayaraman, V., Svoboda, K., Looger, L.L., 2009. Imaging neural activity in worms, flies and mice with improved GCaMP calcium indicators. Nat. Methods 6, 875–881.

Tong, Y., Hocke, L.M., Frederick, B.B., 2019. Low frequency systemic hemodynamic "noise" in resting state BOLD fMRI: characteristics, causes, implications, mitigation strategies, and applications. Front. Neurosci. 13, 787.

Tovar-Moll, F., Monteiro, M., Andrade, J., Bramati, I.E., Vianna-Barbosa, R., Marins, T., Rodrigues, E., Dantas, N., Behrens, T.E., De Oliveira-Souza, R., Moll, J., Lent, R., 2014. Structural and functional brain rewiring clarifies preserved interhemispheric transfer in humans born without the corpus callosum. Proc. Natl. Acad. Sci. U. S. A. 111, 7843–7848.

Turchi, J., Chang, C., Ye, F.Q., Russ, B.E., Yu, D.K., Cortes, C.R., Monosov, I.E., Duyn, J.H., Leopold, D.A., 2018. The basal forebrain regulates global resting-state fMRI fluctuations. Neuron 97. 940–952e4.

Tyszka, J.M., Kennedy, D.P., Adolphs, R., Paul, L.K., 2011. Intact bilateral resting-state networks in the absence of the corpus callosum. J. Neurosci. 31, 15154–15162.

Uddin, L.Q., Mooshagian, E., Zaidel, E., Scheres, A., Margulies, D.S., Kelly, A.M., Shehzad, Z., Adelstein, J.S., Castellanos, F.X., Biswal, B.B., Milham, M.P., 2008. Residual functional connectivity in the split-brain revealed with resting-state functional MRI. Neuroreport 19, 703–709.

Urban, D.J., Roth, B.L., 2015. DREADDs (designer receptors exclusively activated by designer drugs): chemogenetic tools with therapeutic utility. Annu. Rev. Pharmacol. Toxicol. 55, 399–417.

Vardy, E., Robinson, J.E., Li, C., Olsen, R.H.J., Diberto, J.F., Giguere, P.M., Sassano, F.M., Huang, X.P., Zhu, H., Urban, D.J., White, K.L., Rittiner, J.E., Crowley, N.A., Pleil, K.E., Mazzone, C.M., Mosier, P.D., Song, J., Kash, T.L., Malanga, C.J., Krashes, M.J., Roth, B.L., 2015. A new DREADD facilitates the multiplexed Chemogenetic interrogation of behavior. Neuron 86, 936–946.

Wang, M., He, Y., Sejnowski, T.J., Yu, X., 2018a. Brain-state dependent astrocytic ca(2+) signals are coupled to both positive and negative BOLD-fMRI signals. Proc. Natl. Acad. Sci. U. S. A. 115, E1647–E1656.

Wang, T., Ouzounov, D.G., Wu, C., Horton, N.G., Zhang, B., Wu, C.H., Zhang, Y., Schnitzer, M.J., Xu, C., 2018b. Three-photon imaging of mouse brain structure and function through the intact skull. Nat. Methods 15, 789–792.

Wang, R., Liu, N., Tao, Y.Y., Gong, X.Q., Zheng, J., Yang, C., Yang, L., Zhang, X.M., 2020. The application of rs-fMRI in vascular cognitive impairment. Front. Neurol. 11, 951.

Wearing, O.H., Ivy, C.M., Gutiérrez-Pinto, N., Velotta, J.P., Campbell-Staton, S.C., Natarajan, C., Cheviron, Z.A., Storz, J.F., Scott, G.R., 2021. The adaptive benefit of evolved increases in hemoglobin-O2 affinity is contingent on tissue O2 diffusing capacity in high-altitude deer mice. BMC Biol. 19, 128.

Webb, K.L., Dominelli, P.B., Baker, S.E., Klassen, S.A., Joyner, M.J., Senefeld, J.W., Wiggins, C.C., 2022. Influence of high Hemoglobin-oxygen affinity on humans during hypoxia. Front. Physiol. 12, 2528.

Williams, D.S., Detre, J.A., Leigh, J.S., Koretsky, A.P., 1992. Magnetic resonance imaging of perfusion using spin inversion of arterial water. Proc. Natl. Acad. Sci. U. S. A. 89, 212–216.

Wu, J., Liang, Y., Chen, S., Hsu, C.L., Chavarha, M., Evans, S.W., Shi, D., Lin, M.Z., Tsia, K.K., Ji, N., 2020. Kilohertz two-photon fluorescence microscopy imaging of neural activity in vivo. Nat. Methods 17, 287–290.

Xiong, H., Lacin, E., Ouyang, H., Naik, A., Xu, X., Xie, C., Youn, J., Kumar, K., Kern, T., Aisenberg, E., Kircher, D., Li, X., Zasadzinski, J.A., Mateo, C., Kleinfeld, D., Hrabetova, S., Slesinger, P.A., Qin, Z., 2021. Probing neuropeptide volume transmission in vivo by a novel all-optical approach. bioRxiv.

Yao, J., Xia, J., Maslov, K.I., Nasiriavanaki, M., Tsytsarev, V., Demchenko, A.V., Wang, L.V., 2013. Noninvasive photoacoustic computed tomography of mouse brain metabolism in vivo. Neuroimage 64, 257–266.

Yao, J., Xia, J., Wang, L.V., 2016. Multiscale functional and molecular photoacoustic tomography. Ultrason. Imaging 38, 44–62.

Yu, X., Glen, D., Wang, S., Dodd, S., Hirano, Y., Saad, Z., Reynolds, R., Silva, A.C., Koretsky, A.P., 2012. Direct imaging of macrovascular and microvascular contributions to BOLD fMRI in layers IV-V of the rat whisker-barrel cortex. Neuroimage 59, 1451–1460.

Yu, X., Qian, C., Chen, D.Y., Dodd, S.J., Koretsky, A.P., 2014. Deciphering laminar-specific neural inputs with line-scanning fMRI. Nat. Methods 11, 55–58.

Yu, X., He, Y., Wang, M., Merkle, H., Dodd, S.J., Silva, A.C., Koretsky, A.P., 2016. Sensory and optogenetically driven single-vessel fMRI. Nat. Methods 13, 337–340.

Zerbi, V., Floriou-Servou, A., Markicevic, M., Vermeiren, Y., Sturman, O., Privitera, M., Von Ziegler, L., Ferrari, K.D., Weber, B., De Deyn, P.P., Wenderoth, N., Bohacek, J., 2019. Rapid reconfiguration of the functional connectome after Chemogenetic locus coeruleus activation. Neuron 103. 702–718e5.

Zhang, L.I., Tao, H.W., Holt, C.E., Harris, W.A., Poo, M.-M., 1998. A critical window for cooperation and competition among developing retinotectal synapses. Nature 395, 37–44.

Zhang, F., Wang, L.P., Brauner, M., Liewald, J.F., Kay, K., Watzke, N., Wood, P.G., Bamberg, E., Nagel, G., Gottschalk, A., Deisseroth, K., 2007. Multimodal fast optical interrogation of neural circuitry. Nature 446, 633–639.

Zhang, K., Huang, D., Shah, N.J., 2018. Comparison of resting-state brain activation detected by BOLD, blood volume and blood flow. Front. Hum. Neurosci. 12, 443.

Zhang, T., Hernandez, O., Chrapkiewicz, R., Shai, A., Wagner, M.J., Zhang, Y., Wu, C.H., Li, J.Z., Inoue, M., Gong, Y., Ahanonu, B., Zeng, H., Bito, H., Schnitzer, M.J., 2019. Kilohertz two-photon brain imaging in awake mice. Nat. Methods 16, 1119–1122.

Zhu, H., Pleil, K.E., Urban, D.J., Moy, S.S., Kash, T.L., Roth, B.L., 2014. Chemogenetic inactivation of ventral hippocampal glutamatergic neurons disrupts consolidation of contextual fear memory. Neuropsychopharmacology 39, 1880–1892.

Zisis, E., Keller, D., Kanari, L., Arnaudon, A., Gevaert, M., Delemontex, T., Coste, B., Foni, A., Abdellah, M., Calì, C., Hess, K., Magistretti, P.J., Schürmann, F., Markram, H., 2021. Digital reconstruction of the neuro-glia-vascular architecture. Cereb. Cortex 31, 5686–5703.

10

Quality assurance: Best practices

Shasha Zhu, Aliza Ayaz, and Jonathan D. Power
Department of Psychiatry, Weill Cornell Medicine, New York, NY, United States

Introduction

This chapter focuses on the process of assessing fMRI data quality. Data quality is universally acknowledged as a crucial factor in fMRI studies. But what is data quality? What makes data high or low quality? How does an investigator judge the quality of a scan? These issues are fundamental to fMRI studies, because they determine how investigators evaluate the potential utility of an acquired data set, its eventual worth after data processing steps, and finally, the soundness of conclusions arising from analyses of those data. This chapter addresses questions like: What does fMRI data quality mean? How can it be quantified? How can it be visualized? Do data display particular properties that depend on particular qualities? How does an investigator reduce bad qualities while preserving good qualities in a data set? How can investigators distinguish among options for trying to improve data quality?

An overarching message of this chapter is that human attention and human vision are tremendously important parts of assessing fMRI data quality. Another overarching message is that much of the assessment process utilizes measures or patterns that are in (essentially) arbitrary units that depend on scan protocols or equipment. Thus, comparisons within a protocol are of value, but comparisons across protocols are often inappropriate or misguided. Additionally, quantitative effect sizes will vary by protocol and subject population and the kinds of problems that scans display. In order to encourage readers to pay attention to the qualitative patterns, which is the important skill to develop, and to explicitly discourage readers from latching onto thresholds or particular numbers, our figures entirely lack scales other than time. Within a particular figure, scales are held constant and visual comparisons across panels or scans are meaningful.

Advances in Resting-State Functional MRI. https://doi.org/10.1016/B978-0-323-91688-2.00003-5

Part 1: What is fMRI data quality?

Quality is multifaceted. In plain terms, high-quality data make it relatively easy to draw true and meaningful conclusions from fMRI data, whereas low-quality data make this process much more fragile and uncertain. Often, quality is distilled to a single number, which in practice is often related to how much head motion occurred in a scan. While head motion is no doubt one of the most important attributes of an fMRI scan, many other attributes are comparably influential in terms of data quality. We therefore urge readers to view quality not as a single property of a scan but as a collection of properties.

A thought experiment quickly illustrates our position. In an ideal resting state fMRI scan, a subject would lie perfectly motionless, watching a cross-hair or with their eyes closed, alert but quiet, and remain this way from the beginning to the end of a scan. Compare that scenario to (a) a person with allergies who sneezes several times, moving the head repeatedly; (b) a person who falls asleep; (c) a person who falls asleep and begins snoring, (d) a person who grows increasingly uncomfortable and restless throughout the scan; (e) someone who has a cold and keeps sniffling and swallowing throughout the scan; (f) a person with a cardiac arrhythmia that transiently disrupts cerebral blood flow without outward signs; (g) a person who perfectly complies with the scan but whose respiratory cycle gently moves the head throughout the scan; (h) a person who takes deep sighs several times in the scan, transiently changing cerebral blood flow; (i) a construction team turning on powerful generators close to the scanning center; (j) a malfunctioning element of a head coil. All these are recognized issues that occur in real scans, all of which degrade scan quality, and all with large effects on the scans. These examples illustrate that motion is one of the several major determinants of quality in a scan.

Accordingly, the assessment of data quality should be multifaceted and sensitive to as many major contributors to data quality as possible. From an external point of view, measurement of head motion, respiratory and cardiac information, and level of alertness are useful. From a data-driven point of view, space and time are the parameter spaces that are absolutely need to be checked, for quality problems manifest in one or both of these domains.

Part 2: How can data qualities be measured?

Because so many fMRI data quality issues can ultimately be traced to physiological or behavioral causes, we begin with the ways to measure physiological and behavioral effects during fMRI. An overriding message is that if care is not taken to obtain good external measures during acquisition, there will be a very limited basis for assessing corresponding fMRI data qualities later on. This section is the foundation of quality assessment.

Cardiac. Even in the stillest subject, the brain moves. Each heartbeat causes the brain to pulse within the skull, and in nearly all scanning situations, the only way to know when these pulsations occur is by recording the heartbeat. During scanning, by far, the most popular way to measure heartbeats is by using photoplethysmography (PPG), an optical technique that performs well in the magnetic environment. PPG equipment comes as part of virtually all scanners for the purpose of cardiac gating of scans, but it is used for different purposes in most fMRI research. Parenthetically, PPG is often referred to as pulse oximetry in the fMRI literature, but pulse oximetry is a particular kind of PPG application, and the PPG equipment that comes with most scanners does not actually perform or permit pulse oximetry measurement. Obtaining cardiac records permits two main kinds of quality assessment or intervention: one targeting pulsatile brain motion, and another targeting cardiac output and thus cerebral blood flow.

A nice PPG record is illustrated in Fig. 1A. PPG measures transmission of light by tissue, which changes as a function of the volume of blood in tissue over time. The peaks represent arrival of boluses of blood from ventricular contractions, and the time between peaks represents the time between heart beats. This record can be processed in many ways for application to fMRI data, such as deriving the phase of the cardiac cycle to index pulsation during acquisition of particular slices (Glover et al., 2000) or deriving the heart rate to index cerebral blood flow or autonomic tone (Chang et al., 2009; Shmueli et al., 2007). More sophisticated uses of PPG include secondary wave analyses to infer blood pressure or other properties (Attarpour et al., 2021), or deriving blood oxygen content, though these steps require specialized equipment and procedures.

Electrocardiography (ECG) is another option for measuring cardiac information, but this technique is substantially degraded by the magnetic field, and set-up time, subject comfort, and tear-down time are also considerations. If an investigator wishes to clearly understand cardiac phenomena, however, it is arguably worth the time and effort to acquire ECG due to the simple fact that it represents a source of cardiac information entirely independent from PPG: electrical rather than mechanical, with sensors on the trunk or limbs rather than on the fingers.

The reason that a second, independent source of cardiac information is so valuable is because physiology records, and PPG in particular, are often of poor quality or transiently corrupted. Having converging sources of cardiac information, or at a minimum one source working when the other transiently fails, greatly increases the

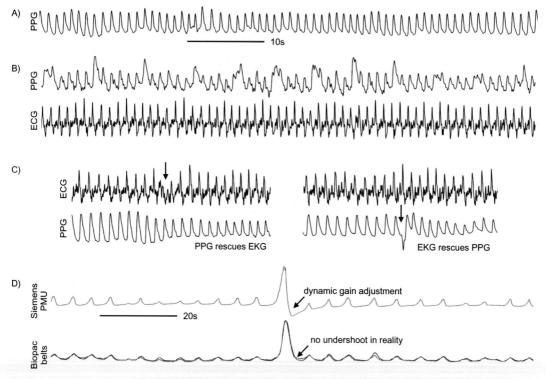

Fig. 1 Cardiopulmonary monitoring. (A) The photoplethysmography (PPG) trace from the Siemens Physiology Monitoring Unit (PMU) during fMRI. (B) A poorer PPG trace, with accompanying high-quality electrocardiography (ECG). (C) Examples of high-quality PPG or ECG transiently failing to provide clear cardiac information *(arrows)* and being rescued by the other modality. (D) Traces of breathing from the Siemens PMU and dual abdominal and chest respiratory belts from Biopac.

confidence an investigator can have in physiology data and in determining cardiovascular influences in fMRI. Fig. 1B shows an example of a poor PPG trace, which is rescued by a simultaneously acquired ECG trace. Fig. 1C shows examples of transient failures in good traces of each kind, each rescued by a redundant source of information.

Fully or transiently corrupted PPG records are common. In a manual curation of 900 young adult subjects from the Human Connectome Project data set, over half were judged to have unusable PPG records in one or more scans (Power, 2019). A large fraction of these failures were problems with setup itself and not with the subject, as evidenced by a strong tendency for quality to be either good or poor for all scans of a given day. Put differently, the scanner operator is the chief determinant of the quality of PPG records. A poorly situated PPG sensor will provide poor signals for an entire scan. This situation is clearly distinguishable from situations where a good signal becomes bad due to shifting the hand, the device falling off, etc. It is also distinguishable from the transient disruptions that often happen when subjects move, which are essentially unavoidable. Many scanners have on-bore visualization of PPG signals, and operators should attend to these waveforms when situating the subject, and double check them once the subject is placed within the bore.

Pulmonary. Even in the stillest subject, the brain *appears* to move. Each respiratory cycle causes instabilities in the magnetic field that cause the brain to appear to move primarily in the phase encode direction, a phenomenon known as pseudo-motion (Fair et al., 2019). Additionally, the head is truly moved by breathing in many people, especially when they take deep breaths like sighs and yawns (Power et al., 2019a). And, finally, breathing rate and depth governs the release of CO_2 from the bloodstream, and pCO_2 is the primary controlling variable for cerebral blood flow. Thus, there are many fMRI data qualities that can be assessed in terms of respiratory behavior, and a substantial body of literature treats this topic (Chang and Glover, 2009; Golestani and Chen, 2020).

All scanners come equipped with respiratory bellows: elastic belts that pin an air bladder against the subject's abdomen or chest. These bellows are provided with scanners for gating scans by breathing phase, but they are used for other purposes in fMRI research. A nice breathing record from a Siemens belt is illustrated in Fig. 1D. The peaks are inhalations, and from such traces, breathing rates and amplitudes can be derived. These traces are pseudo-quantitative, in the sense that they provide excellent rhythm information on breathing, and often a relative amplitude of breathing, but they do not provide absolute measures of breathing depth. These records are also liable to clip, meaning they hit floor and ceiling values, causing loss of amplitude information and degradation of rhythm timing.

More quantitative approaches require additional equipment and calibration with air flow, often paired with optical or mechanical monitoring of multiple body compartments. In Fig. 1D, records from calibrated abdominal and chest locations, acquired from a Biopac respiration belt (Biopac Systems, Inc., Goleta, CA), are also shown, acquired at the same time as the Siemens belt record. The subject takes a deep breath midway in this segment, and there is a difference in the records—an undershoot in the Siemens scanner belt that is not present in the calibrated Biopac belts. This difference reflects the fact that the scanner continuously adjusts the dynamic range on its belt in light of recent breathing behavior, a decision that reflects the priority of capturing breathing rhythm over breathing amplitude for scan gating. More elaborate approaches can measure end-tidal gases as part of respiratory monitoring, though this requires nonrebreather equipment, some discomfort or inconvenience to the subject, and set-up and tear-down time. Because scanner belts are available and take only a few seconds to put on and cause minimal discomfort, they are the dominant form of pulmonary monitoring in the field.

Despite the caveats above, a single pulmonary record using stock scanner respiratory belts is tremendously helpful for understanding fMRI data qualities, and these kinds of records have provided many major insights into breathing-related fMRI data qualities (Lynch et al., 2020b). Pulmonary records are far less likely to display transient corruption than PPG records, though it can happen if the belt slips during deep inhalations. Just as with PPG records, if a belt is not well placed as the subject is situated in the scanner, only a poor signal will result. For this reason, many scanners display the respiratory wave on the scanner bore. Operators should attend to these waveforms when situating the subject and double-check them once subjects are within the bore.

Motion. Head motion is the most famous data quality issue in fMRI, but, as was already seen, head motion occurs for many reasons, such as the breathing-related causes mentioned earlier. A host of other behaviors can cause head motion, such as sneezes, swallows, shifts to get comfortable, itchiness, or falling asleep. In a sense, head motion has attracted so much attention because it is an umbrella that captures and indexes many problematic attributes of a scan (Power et al., 2020). Regardless of the cause of the head motion, the problems it creates in fMRI data are often severe.

Motion can be measured externally with cameras or optical devices (Zaitsev et al., 2017), but most investigators measure motion from the fMRI data itself (Oakes et al., 2005). This data-driven process virtually always occurs after the scan has been transformed from the original series of k-space images into the more familiar series of three-dimensional (3D) brain volumes. The most common procedure is to treat the brain in each volume as a rigid body and solve for the affine transformations that align these brains to one another over time, a process usually called motion correction. Our preference is to perform this step prior to any data processing and specifically prior to any outlier removal or interpolation steps (e.g., de-spiking or slice time correction), because these steps inevitably smooth away motion from the data via temporal interpolation (Power et al., 2017a). From the translation and rotation estimates, derivatives can be created to describe the amount of motion from volume to volume. For example, the head motion measure Framewise Displacement (FD) (Power et al., 2011) is calculated as follows from the translation and rotation estimates: (1) convert rotations to arc displacement at 50 mm radius; (2) differentiate translation and rotation estimates in time by backward differences (from a volume to the prior volume); (3) take absolute values; (4) sum. For an 800-timepoint scan, this yields 799 values, and by convention, the first volume gets a value of 0 to restore a length

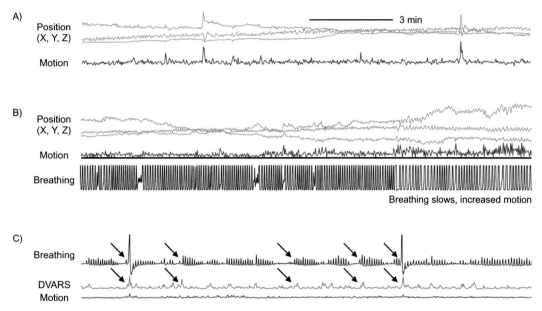

Fig. 2 Motion and DVARS. (A) Illustration of how head position traces *(gray)* are transformed into head motion traces *(red)*. This scan has two brief and large motions. (B) A scan in which breathing slowed considerably toward the end, concomitant with a large increase in motion at the respiratory rate. (C) Illustration of DVARS in a scan with very low motion, in which DVARS often reflects changes in breathing *(arrows)*.

of 800 to the quality vector. Fig. 2 illustrates several position traces in gray and their paired FD motion trace in red in two scans. The first scan shows several brief motions, and the second scan shows a worsening of respiratory motion toward the end of the scan. One other emerging form of motion estimation, mainly enabled by simultaneous multislice acquisitions, is to register slice stacks to each other for subrepetition-time (sub-TR) motion estimation (Teruel et al., 2018). This process yields more frequent motion estimates and ought to result in a better-preserved brain shape, but it is not yet widely in use and we do not develop this line of work further here.

A crucial point needs to be made about units of motion in fMRI. Motion is typically expressed in daily life as distance per unit time, such as meters per second or miles per hour. But in fMRI terms, motion is typically distance per repetition time (TR), a normalization specific to the scan protocol. If a brain moves 4 mm between sequential volumes, the fMRI "speed" will be 4 mm, no matter how much time separates the volumes. By contrast, the actual speed of the head may be 1 mm/s (if TR = 4000 ms), 8 mm/s (if TR = 500 ms), or 2 mm/s (if TR = 2000 ms). Common measures of head motion, like framewise displacement, make this TR basis explicit: the name itself indicates that motion is between frames of the scan (like a movie), not per unit time. The point is that investigators must keep in mind that motion is a biological, behavioral process with a characteristic set of speeds and timescales and that discussion of motion in fMRI is typically with respect to the protocol and not standard units of time. Two corollaries immediately follow: (1) fMRI motion values in protocols with different sampling rates are not directly comparable, and (2) various causes of motion at faster or slower rates may alias differentially depending on the sampling rate of the protocol. These points are so relevant to quality assessment that they are worth repeating: motion magnitudes from one protocol should not be viewed as automatically applicable or appropriate for another protocol, especially if the TR is substantially different (e.g., 4000 ms vs 1500 ms vs 600 ms). The motion measured will be quite different due to both aliasing characteristics and due to subdivision of motion in faster sampling rates (Gratton et al., 2020).

In an attempt to prevent certain kinds of physiological motion from being measured, or to emphasize other motions with characteristic timescales of a few seconds like deep breaths, investigators have proposed using frequency filters on motion estimates prior to FD calculation (Fair et al., 2019; Power et al., 2019a), dropping the respiratory-inflected phase encode directions in FD calculation (Power et al., 2019a), or adjusting the backward-difference gaps in FD calculation to target a given time interval irrespective of the actual repetition time (Power et al., 2019a). Such maneuvers are typically done to suppress respiratory motion and to draw out intermittent, large head motions that cause major signal artifacts. Parenthetically, all motions are not the same in their effects on fMRI data. In particular, pseudo-motion has very different consequences than real motion. Pseudo-motion causes a nearly pure within-plane translation in the phase encode direction, without spin history effects, and it is well worth the effort to separate away apparent but largely correctable pseudo-motion from real motion, which causes spin history effects and is much harder to correct. Further, slow and small real motions are generally much more correctable than large and fast motions.

DVARS. The final data quality measure discussed here is a measure of how rapidly and broadly the fMRI signal is changing from volume to volume, called DVARS (Smyser et al., 2010). It is calculated by differentiating the fMRI time series at each voxel in time by backward differences, then taking at each volume the root mean-squared value of those differentiated time series. DVARS is not directly comparable between protocols or even stages of data processing, as its magnitude depends on background rates of signal change across the brain, but normalization procedures have been proposed to enable such comparisons (Afyouni and Nichols, 2018). DVARS is closely related to head motion, since head motion produces so many of the rapid and aberrant signal changes in scans. But DVARS can also detect any other unusual changes in a scan, if sufficiently rapid or widespread. An example of DVARS is shown in Fig. 2, in a subject whose head is held very still by a customized head restraint (Lynch et al., 2021). In this scan, which is almost entirely free of motion artifact, DVARS detects transient abnormalities when breathing rates increase, probably due to transient changes in cerebral blood flow. As with FD above, DVARS can be aimed at specific (longer) timescales by calculating differences over multiple volumes rather than consecutive volumes. DVARS can also be targeted at particular spatial parts of an image, inside or outside of the brain, to examine motion in that compartment independently of the realignment estimates derived from the brain. One notable property of DVARS is that data processing steps affect its form. For example, a high value of DVARS that arises from motion might become a normal value after an ideal correction, or it can even become a trough value if censoring, de-spiking, or temporally targeted procedures like independent component analysis remove much or all of the variance at that point in time.

Part 3: How are data qualities visualized?

There is no substitute for seeing fMRI data. As a basic good practice to assure quality, when a scan is acquired (or downloaded), its 3D form should be visually checked by a careful human for completeness and appropriate appearance. This process can be batched, automated, and grouped into movies or series of images rather than interacting with the volume directly, enabling rapid assessment of hundreds of scans, but it should not be skipped. This process immediately discloses problems in reconstruction, dropout or ghosting artifacts, severe distortions, and problems in the brain coverage that usually affect the very top or bottom portions of the brain.

This process of 3D visual inspection should continue at crucial junctures of data processing. For example, if all images are registered to a common stereotaxic space, a slide show or movie format can confirm which images registered well or poorly. Fig. 3A shows the average fMRI signal of a well-registered scan after transformation to atlas space. Sequential examination of such images from different scans or subjects immediately discloses poorly registered scans and the variability in registration quality. Poor registration is not uncommon, especially in historical data sets with substantial distortion. Poor registration also occurs in subjects who are not the kinds of subjects used to design processing algorithms, such as those with anatomical variations, or subjects early or late in the lifespan. Field maps often assist with spatial registration of particular portions of images, but a poor field map can just as easily interfere with registration as assist it,

A)

Mean fMRI

GSCORR
(normal)

B)

GSCORR
(normal)

C)

GSCORR
(high motion)

D)

GSCORR
(coil failure)

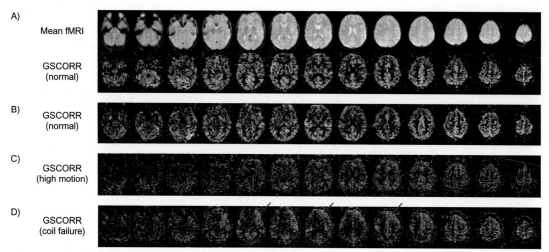

Fig. 3 Spatial examination. Data from several scans are displayed as slices, all after registration to a target atlas space. (A) The average fMRI signal over time shows a well-placed and well-formed T2* image. The GSCORR map displays the correlation of every voxel signal to the mean brain signal. This map is symmetric and well-localized to the gray matter, the typical, normal appearance of such maps. (B) Another scan with a normal GSCORR map. (C) A scan with a noisy GSCORR map; this scan has high amounts of head motion. (D) A scan during which a head coil element failed, leading to frontal asymmetry of the GSCORR map *(arrows)*.

and checking this step is therefore useful. Another juncture worth checking is the positioning of regions of interests or spatial masks on the fMRI data.

Another easy and informative 3D quality check is to calculate the average time series across all voxels and correlate every voxel's signal to that average signal, forming an image we call the global signal correlation map (GSCORR) (Power et al., 2017b). Barring anatomical abnormalities, a GSCORR map will be symmetric and smooth and well localized to the gray matter. Spatial abnormalities should trigger a hunt for a problem in the scan. Examples of typical maps are shown in Fig. 3A and B. Fig. 3C shows a noisy GSCORR map in a high motion subject, and Fig. 3D shows a scan during which a head coil element failed. In sum, we recommend that 3D checks of a scan's coverage and reconstruction, of its successful registration to a group space, and of its shared signal structure via GSCORR be basic, routine quality checks.

Our initial checks also include a two-dimensional (2D) check on the data, a format that we call a gray plot. This form is discussed in dedicated reviews (Power, 2017) and implementations in several software packages exist. The core of the format is to present fMRI as a grayscale raster plot with voxels or regions as rows and volumes as columns, so that the entire fMRI scan can be seen at once in a 2D plane. In temporal register with the fMRI data, other sources of information can be placed, like cardiac, pulmonary, motion, or DVARS traces. This placement allows the eye to instantly link known quality issues to the fMRI time series. This process too can be batched, automated, and grouped into slide shows or movies, enabling an investigator to examine rapidly hundreds or thousands of scans. The advantage of this format is that the eye is free and able to note relationships that would be missed by many prespecified models—even those that are well-reasoned and well-grounded.

Several gray plots are shown in Fig. 4. The first has essentially no motion and flat DVARS, both indicating excellent subject behavior. The breathing rate is even and normal. The fMRI data look like a gray wall of mildly fluctuating signals. The second has a large motion, with a corresponding large spike in DVARS, and a visible disruption in fMRI signals marked with the red arrow. The third has a large breath, also with a large motion, a large DVARS spike, and widespread changes in fMRI signal both at the time of motion (red arrow) and after (blue arrow). The fourth has a continuous waxing and waning breathing pattern, some motion, and fMRI signal changes time-locked to the breathing pattern. The final subject has a chaotic pattern of breathing, much head motion, and large, sharp signal changes in fMRI signals. For the most part, these kinds of properties are not evident in the 3D inspections above, nor are they cleanly delineated by summary statistics of things like motion or DVARS.

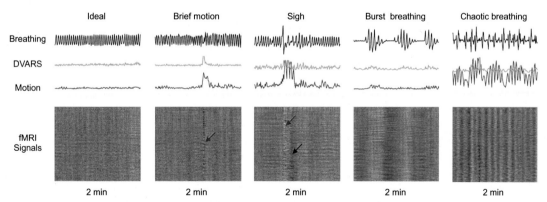

Fig. 4 *Gray* plots show fMRI scan qualities. Two-minute excerpts from several scans are shown, with breathing, DVARS, and motion (FD) traces plotted as in prior figures. The gray-scale heat maps show all gray matter voxels. The first scan shows no motion and only normal, tidal breathing, with little modulation of gray matter signals evident. The next scan has a brief motion seen in DVARS and motion traces, and a brief signal disruption at that moment *(red arrow)*. The next scan displays a sigh in the breathing trace, DVARS and motion elevations during and after the sigh, a brief, immediate signal disruption at the motion *(red arrow)*, and a long signal decrease after the sigh (*blue arrow* pointing to the *black* band). The burst breathing pattern shows periodic breathing modulations, periodic subtle DVARS and motion oscillations, and brain-wide fMRI signal modulation time-locked to the breathing. The chaotic breathing pattern is widely spaced, with much motion, and fMRI signal modulation time-locked to the breathing periodicity.

With regard to assessing scan qualities, the scans with desirable characteristics from all points of view are like the first scan. These will be a minority of scans in any data set, but they immediately show what the fMRI data can be like in the absence of problems. With time and repetition, patterns linking certain kinds of problems in the quality traces to fMRI signals become visually apparent. Gray plots can then serve as the launching pad for more formal and statistical descriptions of quality problems in fMRI data. Much of our work in data quality was prompted by features first noted in gray plots. In short, we find gray plots and the visual information they contain to be indispensable parts of our quality assessment process.

Part 4: Properties of fMRI data with quality problems

The qualities noted in individual scans translate into numeric scan properties in a variety of ways. Beyond the visual examination of scans discussed above, these computational properties can be used as checks or gauges of quality problems in data.

Head motion is prevalent and universally measured by investigators, and here, we point out its major properties. The properties of motion are (1) imaging artifacts are immediate upon motion, lasting a few volumes after the motion; (2) artifacts are spatially structured in ways that are dependent on the sequence and scanner. Consequences of these properties are several: (1) the influence of motion is mainly limited to the TRs during and just after motion; unless the head is very substantially shifted in the head coil, motion does not affect the scan beyond those portions of time, presuming the scan undergoes successful realignment; (2) artifacts caused by motion, being sporadic and affecting random portions of the scans, tend to cause signal covariance to rise between pairs of voxels that are spatially near and to drop between distant pairs of voxels. This is called distance-dependent motion artifact, and it is specifically and only found at times of head motion (Power et al., 2014). Thus, measuring this property during motion, or in a scan as a whole, is a useful gauge of how contaminated a scan is by motion artifact. Given a connectivity matrix of dimensions $R \times R \times N$, where Rs are regions of interest with known spatial distances between them and N is some set of scans, distance-dependent motion effects can be isolated in several ways: by comparing median split by motion (high-low difference), by correlating a motion vector (encoding a single summary measure of motion for each of the N scans) in the third dimension along the matrix, by contrasting matrix entries when including and excluding (scrubbing/censoring) high-motion time points, or by contrasting matrices of data before and after a motion-reducing intervention. As an example of the last point, if a custom-printed head case is made for a subject, providing greater restraint than typical possible, not only is an already-still subject made even more still but distance-dependent artifact is identified and eliminated, even within already very-low-motion data. Motion-targeting analyses are discussed in more detail in several articles (Lynch et al., 2021; Power et al., 2015, 2019b).

In consequence of these properties, from a functional connectivity point of view, a main signature of head motion is to cause nearby tissues to seem "connected" and distant tissues "disconnected." Another fMRI signature is to increase variance in sensorimotor representations of the face, since fMRI changes due to the neural activity associated with motion occur 4–8 s after a motion (Power et al., 2020). This variance is rarely targeted by de-noising or mentioned in discussions of functional connectivity, but it is readily found by general linear models of motion spikes.

Breathing is probably the most-measured behavior in the scanner after motion. The effects of breathing on fMRI are just as easily detected as those of head motion, but the spatial and temporal properties are quite different. As mentioned before, breathing modifies pCO_2 concentrations, altering blood flow and thus fMRI signal. All humans normally take intermittent deep breaths, and when that occurs, fMRI patterns like those shown in the third scan of Fig. 4 occur. At these times, brief activations in sensorimotor cortex are seen, followed by large global decreases in fMRI signals until about 15–20 s after the breath has finished. These look like vertical black stripes in gray plots. From a functional connectivity point of view, this is a time-locked, brain-wide modulation of signal, and it acts to increase correlations at essentially all parts of the brain. Other forms of breathing have different spatiotemporal profiles, but all share an aspect of global signal modulation and slow, sustained changes reflecting the gradual alteration of pCO_2 by changed breathing patterns. The fourth scan of Fig. 4 demonstrates a form of breathing that we have termed bursts, which have a characteristic timescale and signal lag structure, and which are biased to occur more frequently in males (Lynch et al., 2020b). Note that because respiration causes motion, motion analyses will capture respiratory properties. Put differently, the motion-targeting analyses above will reveal distance dependence, but usually also a global elevation in correlations regardless of distance, which is a consequence of the respiratory behaviors mentioned in this section. These kinds of variance can be, and have been, dissected apart using multiecho fMRI (Lynch et al., 2020a; Power et al., 2018).

Other canonical behavioral events exist in fMRI data, but for space reasons, we limit our discussion to the points already mentioned.

Part 5: Deciding how to suppress bad qualities to get sound conclusions

Readers will not find specific recommendations for processing steps in this section. For that, there exist many competing reviews of fMRI processing strategies and de-noising techniques (Ciric et al., 2018; Parkes et al., 2018). Rather, this section is about the process of making decisions in light of data quality in a given data set.

All fMRI data sets will contain a range of quality, and it is generally agreed that the best data will yield the soundest conclusions. To shore up the quality of work being done, it is customary within each lab to set criteria for data that are too bad for use. Often these criteria are single numbers, or conjunctions of single numbers, whether an absolute amount of motion, a fraction of the scan contaminated by motion, or similar kinds of criteria based on DVARS or other measures. From the prior discussion, it is clear that a single criterion cannot and should not apply broadly across all data sets and labs, for the quality measures themselves are in some ways custom to each data set and will reflect the behavior and physiology of particular populations.

How can investigators sensibly proceed in these conditions? Two broad considerations guide our thinking in which kinds of data to analyze, and which data to leave aside. First, which samples are not like the others due to quality problems? Second, does my data processing adequately shield my analyses and conclusions from certain kinds of quality problems? We discuss specific instances below to illustrate the principles in operation.

If we take volumes contaminated by motion as paradigmatic of the first consideration (which samples are not like the others), we find in study after study that censoring high-motion volumes reveals distance-dependent artifact specifically and only at times of motion, no matter what de-noising or signal processing strategies are employed (Ciric et al., 2018; Parkes et al., 2018; Power et al., 2017a, 2018). This is strong evidence that high-motion time points are not like the others and that none of our data correction techniques adequately remedied those volumes. To retain those volumes, whatever the reason, is to retain the properties of motion in analyses. In our view, this weakens the soundness of whatever claims arise from the data. The strictness of identifying high-motion volumes (i.e., the threshold) is the key consideration here. Typically distributions of motion over all volumes of a scan or group of scans show a unimodal distribution with a shoulder of higher values. Formal analyses designed to identify at what point motion clearly contributes distance-dependent artifact have twice settled on the shoulder region of values (Fair et al., 2019; Power et al., 2014). Not coincidentally, this location is where any person would visually mark a transition from one distribution to another, i.e., the "part that is not like the rest." Whether enough data remain in a particular scan once such data are censored is a decision made by each investigator. Two points are worth emphasizing: (1) eliminating phase encode pseudo-motion or otherwise reformulating FD to counter respiratory influences can greatly and helpfully alter the appearance of motion in a scan, especially in fast-TR protocols, and (2) if motion traces are not sensible and intelligible. There is little point in using them as the basis of censoring or quality decisions. This latter consideration is the entire basis of attempting to reformulate FD in data to target real problematic motions (Fair et al., 2019; Gratton et al., 2020; Power et al., 2019a).

This stance on data quality has a parallel across subjects and scans: when examining the quality composition of a group of scans, it is often the case that the bulk of the samples have some approximate quality value (like average FD, where lower numbers are generally better), and a smaller fraction have higher, and sometimes much higher, values. Outliers gain disproportionate influence in regression and correlation calculations and in many other kinds of analysis. An investigator must ask whether it is desirable that a minority of high motion scans, given that they probably have distinct fMRI properties resulting from motion, be given the chance to influence or perhaps dominate effects in the other scans. Or, whether it would be better simply to focus on the best collection of scans to reach the most useful kinds of conclusions. These considerations were illustrated with motion, but other quality measures could be substituted.

With regard to the second consideration (does my data processing fix certain problems), the rationale is straightforward. For example, if in a data set an investigator sees that sighs are common, meaning there are sensorimotor activations, and global respiratory fluctuations in gray plots, her question becomes whether de-noising strategies have negated or isolated these elements so that they are not factoring into analyses (presuming she is uninterested in these breathing events). General linear model analyses of the sighs before and after the de-noising, paired with gray plots before and after, will readily indicate whether these are problematic occurrences still that degrade data quality.

In the future, as the field's understanding of ongoing behavior and common patterns of nuisance fMRI signals evolves and improves, it seems that it could become customary to stop speaking of data quality as a unitary construct and instead to have a multidimensional empirical assessment that translates directly into spatiotemporal patterns in scans. Such a scale would presumably include elements of pseudo-motion, punctate motion, and continuous respiratory motion but also patterns of breathing, alterations in heart rate, subject alertness, and other factors we have not yet thought to quantify.

Conclusion

Data quality can be thought of on a continuum, but it is probably better to view it as a multidimensional construct. Investigators tend to prioritize measures of head motion when discussing quality, but head motion is an umbrella for several kinds of problems in scans, problems that confer a variety of spatiotemporal patterns to fMRI data. Above all, as an investigator seeks to understand data quality, we urge visual inspection and direct engagement with the data. This is the fastest, best, and most formative way to understand the widely varying qualities of fMRI data. We suggested several forms of 3D and 2D data representations that make apparent quality problems in data, and we linked these single-scan properties to analytic properties of different artifacts or data problems. Investigators who have grappled with data in this way will find themselves well-equipped to judge data qualities and to choose analysis strategies likely to generate sound conclusions.

Conflict of interest

The authors declare no conflicts of interest with respect to this report.

References

Afyouni, S., Nichols, T.E., 2018. Insight and inference for DVARS. NeuroImage 172, 291–312.

Attarpour, A., Ward, J., Chen, J.J., 2021. Vascular origins of low-frequency oscillations in the cerebrospinal fluid signal in resting-state fMRI: interpretation using photoplethysmography. Hum. Brain Mapp. 42, 2606–2622.

Chang, C., Glover, G.H., 2009. Relationship between respiration, end-tidal CO_2, and BOLD signals in resting-state fMRI. NeuroImage 47, 1381–1393.

Chang, C., Cunningham, J.P., Glover, G.H., 2009. Influence of heart rate on the BOLD signal: the cardiac response function. NeuroImage 44, 857–869.

Ciric, R., Rosen, A.F.G., Erus, G., Cieslak, M., Adebimpe, A., Cook, P.A., Bassett, D.S., Davatzikos, C., Wolf, D.H., Satterthwaite, T.D., 2018. Mitigating head motion artifact in functional connectivity MRI. Nat. Protoc. 13, 2801–2826.

Fair, D.A., Miranda-Dominguez, O., Snyder, A.Z., Perrone, A., Earl, E.A., Van, A.N., Koller, J.M., Feczko, E., Tisdall, M.D., van der Kouwe, A., Klein, R.L., Mirro, A.E., Hampton, J.M., Adeyemo, B., Laumann, T.O., Gratton, C., Greene, D.J., Schlaggar, B.L., Hagler Jr., D.J., Watts, R., Garavan, H., Barch, D.M., Nigg, J.T., Petersen, S.E., Dale, A.M., Feldstein-Ewing, S.W., Nagel, B.J., Dosenbach, N.U.F., 2019. Correction of respiratory artifacts in MRI head motion estimates. NeuroImage 208, 116400.

Glover, G.H., Li, T.Q., Ress, D., 2000. Image-based method for retrospective correction of physiological motion effects in fMRI: RETROICOR. Magn. Reson. Med. 44, 162–167.

Golestani, A.M., Chen, J.J., 2020. Controlling for the effect of arterial-CO_2 fluctuations in resting-state fMRI: comparing end-tidal CO_2 clamping and retroactive CO_2 correction. NeuroImage 216, 116874.

Gratton, C., Dworetsky, A., Coalson, R.S., Adeyemo, B., Laumann, T.O., Wig, G.S., Kong, T.S., Gratton, G., Fabiani, M., Barch, D.M., Tranel, D., Miranda-Dominguez, O., Fair, D.A.,

Dosenbach, N.U.F., Snyder, A.Z., Perlmutter, J.S., Petersen, S.E., Campbell, M.C., 2020. Removal of high frequency contamination from motion estimates in single-band fMRI saves data without biasing functional connectivity. NeuroImage 217, 116866.

Lynch, C.J., Power, J.D., Scult, M.A., Dubin, M., Gunning, F.M., Liston, C., 2020a. Rapid precision functional mapping of individuals using multi-echo fMRI. Cell Rep. 33, 108540.

Lynch, C.J., Silver, B.M., Dubin, M.J., Martin, A., Voss, H.U., Jones, R.M., Power, J.D., 2020b. Prevalent and sex-biased breathing patterns modify functional connectivity MRI in young adults. Nat. Commun. 11, 5290.

Lynch, C.J., Voss, H.U., Silver, B.M., Power, J.D., 2021. On measuring head motion and effects of head molds during fMRI. NeuroImage 225, 117494.

Oakes, T.R., Johnstone, T., Ores Walsh, K.S., Greischar, L.L., Alexander, A.L., Fox, A.S., Davidson, R.J., 2005. Comparison of fMRI motion correction software tools. NeuroImage 28, 529–543.

Parkes, L., Fulcher, B., Yucel, M., Fornito, A., 2018. An evaluation of the efficacy, reliability, and sensitivity of motion correction strategies for resting-state functional MRI. NeuroImage 171, 415–436.

Power, J.D., 2017. A simple but useful way to assess fMRI scan qualities. NeuroImage 154, 150–158.

Power, J.D., 2019. Temporal ICA has not properly separated global fMRI signals: a comment on Glasser et al. (2018). NeuroImage 197, 650–651.

Power, J.D., Cohen, A.L., Nelson, S.M., Wig, G.S., Barnes, K.A., Church, J.A., Vogel, A.C., Laumann, T.O., Miezin, F.M., Schlaggar, B.L., Petersen, S.E., 2011. Functional network organization of the human brain. Neuron 72, 665–678.

Power, J.D., Mitra, A., Laumann, T.O., Snyder, A.Z., Schlaggar, B.L., Petersen, S.E., 2014. Methods to detect, characterize, and remove motion artifact in resting state fMRI. NeuroImage 84, 320–341.

Power, J.D., Schlaggar, B.L., Petersen, S.E., 2015. Recent progress and outstanding issues in motion correction in resting state fMRI. NeuroImage 105, 536–551.

Power, J.D., Plitt, M., Kundu, P., Bandettini, P.A., Martin, A., 2017a. Temporal interpolation alters motion in fMRI scans: magnitudes and consequences for artifact detection. PLoS One 12, e0182939.

Power, J.D., Plitt, M., Laumann, T.O., Martin, A., 2017b. Sources and implications of whole-brain fMRI signals in humans. NeuroImage 146, 609–625.

Power, J.D., Plitt, M., Gotts, S.J., Kundu, P., Voon, V., Bandettini, P.A., Martin, A., 2018. Ridding fMRI data of motion-related influences: removal of signals with distinct spatial and physical bases in multiecho data. Proc. Natl. Acad. Sci. U. S. A. 115, E2105–E2114.

Power, J.D., Lynch, C.J., Silver, B.M., Dubin, M.J., Martin, A., Jones, R.M., 2019a. Distinctions among real and apparent respiratory motions in human fMRI data. NeuroImage 201, 116041.

Power, J.D., Silver, B.M., Silverman, M.R., Ajodan, E.L., Bos, D.J., Jones, R.M., 2019b. Customized head molds reduce motion during resting state fMRI scans. NeuroImage 189, 141–149.

Power, J.D., Lynch, C.J., Adeyemo, B., Petersen, S.E., 2020. A critical, event-related appraisal of denoising in resting-state fMRI studies. Cereb. Cortex 30, 5544–5559.

Shmueli, K., van Gelderen, P., de Zwart, J.A., Horovitz, S.G., Fukunaga, M., Jansma, J.M., Duyn, J.H., 2007. Low-frequency fluctuations in the cardiac rate as a source of variance in the resting-state fMRI BOLD signal. NeuroImage 38, 306–320.

Smyser, C.D., Inder, T.E., Shimony, J.S., Hill, J.E., Degnan, A.J., Snyder, A.Z., Neil, J.J., 2010. Longitudinal analysis of neural network development in preterm infants. Cereb. Cortex 20, 2852–2862.

Teruel, J.R., Kuperman, J.M., Dale, A.M., White, N.S., 2018. High temporal resolution motion estimation using a self-navigated simultaneous multi-slice echo planar imaging acquisition. J. Magn. Reson. Imaging.

Zaitsev, M., Akin, B., LeVan, P., Knowles, B.R., 2017. Prospective motion correction in functional MRI. NeuroImage 154, 33–42.

Advances in resting-state fMRI acquisition: Highly accelerated fMRI

Mark Chiew[a,b], Hsin-Ju Lee[a,b], and Fa-Hsuan Lin[a,b]
[a]Sunnybrook Research Institute, Toronto, ON, Canada, [b]Department of Medical Biophysics, University of Toronto, Toronto, ON, Canada

Introduction

With the rapid development of resting-state fMRI (rs-fMRI), there has been a latent but growing focus on the benefits of novel MRI acquisition techniques. Of particular note is the family of highly accelerated fMRI data acquisition methods, which hold great promise for enhancing the capability of rs-fMRI, providing faster sampling of whole-brain BOLD signals for increased functional sensitivity or improved physiological noise reduction. In this chapter, we first introduce signal and noise characteristics in fMRI and outline the benefits of the faster sampling for rs-fMRI. Then we will survey several methods for accelerated fMRI, including echo-planar imaging (EPI), echo-volumar imaging (EVI), simultaneous multislice (SMS)-EPI, inverse imaging (InI), and MR-encephalography (MR-EG). Finally, the chapter ends with some recommendations for fast fMRI acquisition protocols for rs-fMRI research.

Signal origins in resting-state fMRI

As described in earlier chapters, the workhorse of rs-fMRI is the blood-oxygenation-level-dependent (BOLD) contrast, which reflects changes in the transverse relaxation decay constant (T_2^*) of blood and surrounding tissue. Most commonly, fMRI signals are measured as magnitude fluctuations of T_2^*-weighted images, where T_2^* contains contributions from intrinsic tissue signal decay (T_2) as well as signal decay due to local magnetic field inhomogeneity (T_2'). These combine in such a way that $\dfrac{1}{T_2^*} = \dfrac{1}{T_2} + \dfrac{1}{T_2'}$, ensuring that T_2^* is never greater than T_2.

Advances in Resting-State Functional MRI. https://doi.org/10.1016/B978-0-323-91688-2.00002-3

255

As blood oxygenation levels change (in conjunction with other physiological factors such as blood volume and flow), so do the T_2^* values associated with voxels located near or downstream from regions of neuronal activation, which is reflected in the measured magnitude signal changes observed in the fMRI data.

However, the actual signals measured with T_2^*-weighted BOLD imaging depend on many additional factors as well, including

- transmit field strength (B_1^+: how much RF excitation is induced across space)
- receive sensitivities (B_1^-: how much weighting of the signal is due to receive-coil profiles)
- longitudinal relaxation time (T_1: reflecting the speed of signal recovery back to equilibrium)
- proton density (ρ: how many spins are present in the given voxel volume).

The precise contributions of these factors on image voxel values will depend on MRI pulse-sequence parameters, e.g., echo time (TE: the time between excitation and signal-contrast measurement, or equivalently the time between excitation and crossing the center of the so-called k-space, further discussed later in this chapter) and repetition time (TR: the time between successive excitations), which become important considerations when selecting highly accelerated imaging protocols. Choices made in image reconstruction also affect signal levels, e.g., different signal normalization (removal of bias fields) and head coil combination (transforming multichannel images to a single, composite image) methods. However, these additional factors are largely *static* during data acquisition and do not typically contribute to the *dynamic* component of the measured fMRI signal. The *dynamic, or time-varying*, component of the signal is typically associated only with T_2^* changes accompanying blood oxygenation fluctuations due to vascular effects (cerebral blood flow and cerebral blood volume) and metabolic effects ($CMRO_2$, or the cerebral metabolic rate of O_2). The vascular and metabolic impact on the measured signal in response to neuronal activity is often termed the hemodynamic response and incorporates all the mediating dynamic physiological factors that couple neuronal activation to fMRI signal changes. Net signal contributions can also be associated with *intravascular* and *extravascular* components (Uludağ et al., 2009), with each compartment exhibiting distinct characteristics. The net measured BOLD signal is a complex function of all these various physiological effects and is difficult to interpret quantitatively without quantitative modeling (e.g., qBOLD (He and Yablonskiy, 2006)) or calibrating for these other confounding factors (calibrated BOLD (Davis et al., 1998)).

Noise origins and signal-to-noise ratio in resting-state fMRI

In addition to the physiological signals of interest, other time-varying contributions to the measured image time series come from different sources of *noise*. The noise in fMRI data can be considered to come from two distinct sources: (1) signal-independent thermal noise and (2) signal-dependent physiological noise. While both sources of noise contribute to unwanted (i.e., nonfunctional) variance in the measured fMRI data, they arise from completely separate processes. Thermal noise is a purely additive noise component that depends only on acquisition parameters such as receiver bandwidth, number of measurements, and parallel imaging acceleration factors as well as physical factors such as tissue electrical properties and geometry. This noise is bivariate Gaussian-distributed in the complex data (real and imaginary components) and approximately Gaussian in high signal-to-noise ratio (SNR) magnitude data (and Rician-distributed in general). It may have spatially varying variance and intervoxel correlations (e.g., after parallel imaging reconstruction) but is completely characterized as a random process. In contrast, physiological noise is defined as fluctuations in the measured T_2^*-weighted signal that arise from dynamic physiological processes such as respiratory and cardiac cycles (Glover et al., 2000), changes in respiratory volume over time (RVT) (Birn et al., 2006), heart-rate variability (HRV) (Shmueli et al., 2007), and variations in CO_2 concentrations (Wise et al., 2004). While thermal noise is signal-independent and random, physiological noise processes differ in that they are signal-dependent and generally predictable using appropriate models. In addition to "physiological noise," other signal fluctuations not typically classified as noise can also corrupt fMRI data, and such noise sources include system instability due to scanner drift or gradient heating, and subject motion.

An important consideration for the quality of fMRI data is the image signal-to-noise ratio (SNR) and corresponding temporal SNR (tSNR), which quantify the amount of image signal strength relative to spatial or temporal noise standard deviations, respectively. Everything that contributes to SNR also contributes to tSNR, but tSNR has additional variance contributions from dynamic physiological noise processes and therefore is always equal to or less than SNR (Krüger and Glover, 2001). These metrics directly impact the statistical sensitivity of fMRI analyses and are particularly important in accelerated data acquisitions, which incur multiple penalties to SNR/tSNR compared to fully sampled acquisitions. These reductions in SNR/tSNR generally result from the following factors:

- signal magnitude reduction due to reduced equilibrium magnetization levels, assuming the TR is reduced and therefore less time is available for T_1 recovery of magnetization

- a $\sqrt{R}$ reduction in SNR/tSNR, where R is the in-plane acceleration factor (or reduction factor), which arises in effect due to having R fewer samples available for noise averaging during image reconstruction
- an additional "geometry-factor" (g-factor) penalty that reflects how much encoding power the receive-coil sensitivities had for the parallel imaging reconstruction.

Fig. 1 provides an example of how signal and noise change with acceleration factors, using a representative set of parameters. The degree to which SNR/tSNR is reduced depends on the specific TR reductions, acceleration factors, sampling patterns, and coil geometries used in a given acquisition, but SNR/tSNR is almost always reduced relative to equivalent, unaccelerated acquisitions.

Benefits of ultrafast sampling in rs-fMRI

The benefits of fast sampling in rs-fMRI have been explored for over a decade, and highly accelerated rs-fMRI acquisitions have already become de facto standards, owing to their use in large consortium studies such as the Human Connectome Project (Van Essen et al., 2013) and

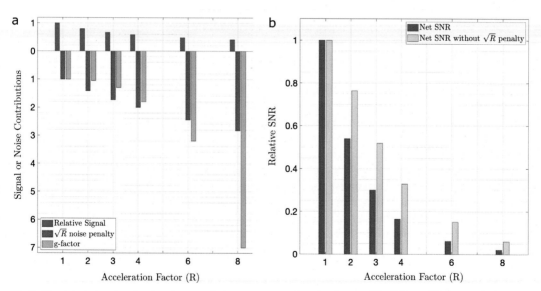

Fig. 1 Simulated signal and noise characteristics across acceleration factors (T_1=1500 ms, TR=3600 ms). On the left, relative signal (*blue*) decreases due to shortened T_1 recovery, resulting in approximately 40% of signal available at R = 8, compared to R = 1. Similarly, noise increases due to $\sqrt{R}$ (*red*) and g-factor (*orange*) penalties, highlighting the fact that g-factor is often the limiting factor at high acceleration. On the right, relative SNR (signal-to-noise ratio), with (*dark green*) and without (*light green*) $\sqrt{R}$ penalties, showing the rapid decrease in SNR as acceleration factor increases. The *light green* curve approximates SMS-EPI multiband SNR, which does not incur the $\sqrt{R}$ penalty.

the UK Biobank (Miller et al., 2016). At first glance, given the apparent "sluggish" nature of the BOLD hemodynamic response to neuronal activity, with its smooth and gradual signal peak 4–6 s after neuronal activation, it may not be obvious what benefits faster sampling can provide in rs-fMRI. However, there are several concrete advantages to faster sampling in rs-fMRI.

The most direct benefit comes from being able to probe neuronal oscillations or temporal dynamics of the functional signal at higher frequencies. While early reports of rs-fMRI found that dynamic functional information was predominantly located in low-frequency bands (0.01–0.1 Hz) (Damoiseaux et al., 2006), when faster sampling methods were employed, functional networks have been found at significantly higher frequencies. For example, two early studies, by Lee et al. (Lee et al., 2013) and Chen & Glover (Chen and Glover, 2015), reported evidence of resting-state networks up to 0.8 Hz (visual and sensorimotor networks) and 0.5 Hz (default mode and executive control networks). While much is still unknown about the precise nature of these high-frequency functional signals, sampling the fMRI signal at 0.8 Hz requires acquisition protocols with a minimum TR of 625 ms, making the use of highly accelerated sampling schemes for probing these high-frequency functional dynamics a necessity.

Another benefit of faster acquisitions comes from improved multimodal correspondence with modalities such as electroencephalography (EEG). Typically, EEG and fMRI probe functional dynamics at completely different timescales, with EEG sampling on the order of milliseconds and fMRI on the order of seconds. However, it has been reported that with MR-encephalography (MREG), which can decrease fMRI sampling times down an order of magnitude to ~100 ms, significantly higher sensitivity was found in the detection of spike-related BOLD responses in simultaneous EEG-fMRI of epilepsy (Jacobs et al., 2014). Task-based fMRI and simultaneous EEG of the visual system have also been found to detect driven oscillations of the fMRI signal up to 0.75 Hz (Lewis et al., 2016), providing further evidence that ultrafast fMRI can capture high-frequency signatures of functional information when coanalyzed with measurements from simultaneous EEG. While reports such as these do not directly involve resting-state analyses, they do provide supporting evidence that ultrafast fMRI is able to capture dynamics that are closer to electrophysiological timescales.

One very practical benefit of ultrafast acquisitions is better removal of physiological noise or unwanted signal. This benefit arises from the fact that physiological fluctuations, particularly those related to cardiac signals (~1 Hz), are often not adequately sampled with conventional acquisitions. Any acquisition TR greater than 0.5 s will result in aliasing of these signals to other frequencies within the sampling bandwidth and result in more difficulty separating and

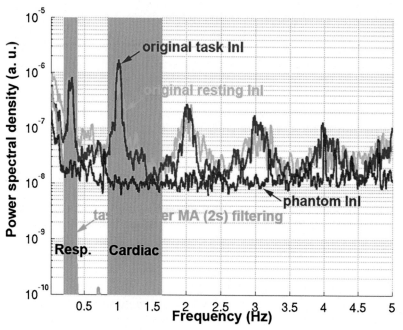

Fig. 2 Power spectral density plot featuring ultrafast sampling up to 5 Hz. In *blue* are task and rs-fMRI data acquired in a volunteer using InI, in *red* is the spectral density plot acquired in a phantom, and in *green* are the task InI data after filtering out physiological effects. Note the respiratory and cardiac peaks highlighted in *gray*, centered at 0.3 and 1.0 Hz, respectively, as well as the high-frequency cardiac harmonics. Conventional sampling would typically sample ≤0.5 Hz, resulting in aliasing of the cardiac physiological fluctuations. Figure reproduced with permission from Lin, F.-H., et al., 2012. Physiological noise reduction using volumetric functional magnetic resonance inverse imaging, Hum. Brain Mapp. 33(12), 2815–2830, https://doi.org/10.1002/hbm.21403.

removing these signals without impacting the functional signals of interest (see Fig. 2). In fact, using ultrafast sampling with inverse imaging (InI), physiological noise removal with straight-forward digital filtering was able to suppress physiological noise with performance better than expected with conventional methods such as RETROICOR (Glover et al., 2000), thus foregoing the need for external monitoring of respiratory or cardiac fluctuations (Lin et al., 2012). Furthermore, alternative methods for signal cleanup and artifact removal may also benefit from higher sampling rates. For example, the FIX method, which uses independent component analysis (ICA) to identify and remove unwanted signal components, demonstrates improved performance when used with simultaneous multislice accelerated acquisitions compared to conventional multislice imaging data, owing to the greater number of time points available for artifact classification (Griffanti et al., 2014).

Finally, faster acquisition comes with statistical benefits, by providing increased degrees of freedom for functional analyses. This statistical advantage was first reported in a study by Feinberg et al. (2010), demonstrating that the greater number of time points provided by simultaneous multislice imaging for a fixed acquisition duration (compared to a slower acquisition) resulted in improved z-scores for multiple-regression analyses, although the benefits did not persist for single-regression functional analyses. However, the statistical improvements are tempered by increased autocorrelation evident in highly accelerated fMRI data (Bollmann et al., 2018).

Overview of methods

This section will provide a description of several methods used for ultrafast fMRI. Echo-planar imaging (EPI) (Mansfield, 1977) is the backbone for the overwhelming majority of fMRI acquisition approaches and facilitates the measurement of a whole-brain volume in a few seconds by sequentially sampling 2-dimensional (2D) image slices to construct a 3-dimensional (3D) volume. Various extensions to conventional EPI have been developed that facilitate even faster imaging include echo-volumar imaging (EVI) (Song et al., 1994), which encodes an entire 3D volume with a single excitation (or small number of them), resulting in TRs of a few hundred milliseconds, and simultaneous multislice EPI (Moeller et al., 2010), which allows 2D multislice EPI to be acquired with multiple slices excited and sampled in parallel, achieving subsecond TRs.

Methods that facilitate even higher degrees of acceleration, and deviate more significantly from conventional EPI sampling schemes, are inverse imaging (InI) (Lin et al., 2006) and MR encephalography (MREG) (Hennig et al., 2007). These methods employ extreme undersampling to achieve whole-brain acquisitions at TRs of 100 ms or lower (capturing signal frequencies of 5 Hz or higher), relying on regularized image reconstruction techniques and the encoding power of multichannel receive arrays. Fig. 3 provides a schematic overview of these methods and their relative acquisition speeds.

The methods mentioned above target a broader rs-fMRI audience and are thus not an exhaustive representation of the published literature on accelerated fMRI. Beyond the scope of this book chapter, we note that several alternative approaches to ultrafast fMRI have been actively developed. These include EPI keyhole imaging (Zaitsev et al., 2001), which combines a static high-resolution reference image with a fast low-resolution "keyhole" acquisition for rapid sampling of contrast changes, compressed sensing approaches that take advantage of the sparse representation of fMRI signals (Jung et al., 2007), and low-rank methods which exploit spatiotemporal correlations for highly accelerated imaging (Chiew et al., 2015).

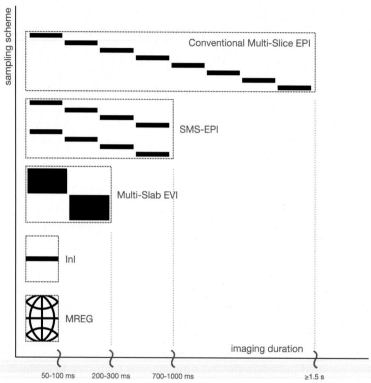

Fig. 3 Schematic diagram of the various rs-fMRI sampling schemes, in decreasing order of typical imaging durations for whole-brain sampling. In conventional multislice EPI, slices are acquired in sequential order resulting in relatively long acquisition times (≥1.5 s). In SMS-EPI, sets of slices are simultaneously acquired (multiband factor 2 shown here), significantly reducing imaging time (700–1000 ms). Multislab EVI covers the imaging volume in several slab acquisitions, for imaging times in the 200–300 ms range. Finally, InI and MREG acquire data with acquisition times of 100 ms or less with their highly accelerated imaging schemes.

Methods in detail

Echo-planar imaging (EPI)

Echo-planar imaging (EPI) was developed by Peter Mansfield (recipient of the 2003 Nobel Prize in Physiology or Medicine, alongside Paul Lauterbur, for the development of MRI) as a high-speed method to fill the so-called k-space (i.e., to acquire all of the spatial encoding needed to represent a 2D image) with a single gradient readout (Mansfield, 1977). The term k-space refers to a frequency-domain representation of the imaging data, and gradients in MRI are used to control the speed and direction of sampling in k-space. To obtain a 2D image, the corresponding 2D k-space needs to be sampled adequately before being transformed or reconstructed into a recognizable image. Mansfield described a process by which one gradient would rapidly switch polarity (causing the samples taken in k-space to be acquired moving left, then right, then left, etc.), while a second gradient had a small constant amplitude (causing the acquired sampling locations to slowly move forward). The net effect of this was zig-zag coverage of the rectangular slice of k-space, resulting in the acquisition of an entire image's worth of data after just a single excitation of the magnetization.

In modern use, EPI gradients typically employ slightly different gradient schemes to cover k-space in a rectangular raster. Typically, the direction of k-space moving along the sampled lines is termed the "frequency-encoding" or "readout" direction, and the direction of k-space perpendicular to the readout lines is termed the "phase-encoding" direction. It is more common in EPI to use small phase-encoding blips to move to different sampling lines at the end of each left or right raster, rather than a continuous motion, ensuring that the sampled lines are all parallel to one another. This Cartesian sampling (i.e., on a regular grid) process leads to many of the favorable properties of EPI, resulting in its near ubiquitous usage in fMRI data acquisition.

One of the drawbacks of conventional EPI is that opportunities for faster imaging are limited. The time needed to acquire a complete volume scales with the number of slices acquired, resulting in relatively long intervals before the same slice is acquired again. Several strategies can be used to minimize the total number of slices, such as using thicker slices, including a gap between slices, and changing the orientation of the imaging plane to minimize necessary slice coverage. However, these offer limited benefit and can result in anisotropic resolution (thicker slices) or missing information (slice gaps). In-plane acceleration, sometimes referred to by trade name as iPAT (Siemens), ASSET/ARC (GE), or SENSE (Philips), is a technique used to reduce the acquisition time of each slice by selectively undersampling k-space and using information from multichannel array head coils to facilitate image reconstruction. This does offer some benefit to reduced acquisition times, as well as benefits in terms of reduced off-resonance distortion effects and lower possible TEs, but is also limited. In-plane acceleration factors typically do not exceed factors of 3 or 4, using conventional hardware, and slice acquisition time overheads (excitation pulse duration, waiting for optimal TEs) are not decreased. Another strategy for reducing slice acquisition time is partial Fourier imaging (sometimes also referred to as half-Fourier, half-scan, or fractional-NEX), in which a fractional section of k-space is omitted entirely. Typically, these fractions range from 1/8 to 3/8 of k-space and are recovered using partial Fourier reconstruction algorithms. However, certain reconstruction schemes such as zero-filling k-space do result in loss of spatial resolution.

Pros

Key advantages of EPI include its robustness to many sources of artifact and image corruption. Off-resonance effects result in image distortion, which is possible to correct in image preprocessing with appropriate calibration data (either a field map or an opposite phase-encoding polarity EPI image). For alternative types of readouts, such as spiral imaging, off-resonance results in blurred images, which is significantly more difficult to correct. Similarly, chemical shift effects (such as with lipid signals from the scalp) result in easily detectable image offsets, rather than undetectable blurring. The rapid readout of EPI means that motion is typically captured as a snapshot, and individual images are not subject to motion-blurring, although slices and volumes can still be misregistered due to motion. Finally, its ease of use has led to EPI's ubiquitous availability on all scanners, meaning that virtually all clinical MRI systems offer EPI acquisition products.

Cons

The primary disadvantage of EPI (compared to the methods later in this chapter) is that it is still relatively slow, as it is restricted to acquiring only a single 2D slice per excitation and readout, resulting in time requirements of 1.5 s or more (i.e., a TR of 1.5 s or more) to acquire a whole-brain volume. Moreover, in addition to the distortion effects mentioned previously, EPI is also susceptible to other effects such as T_2^* blurring (loss of resolution due to T_2^* decay), signal dropout (loss of signal in regions with high field inhomogeneity), and peripheral nerve stimulation. Furthermore, EPI has limited flexibility for optimizing protocols with respect to TE, given that TE is constrained to be in the approximate center of the readout, which may be limiting at high ($\geq$7 T) and low ($\leq$1.5 T) field strengths, where optimal TEs are particularly short or long, respectively.

Echo-volumar imaging (EVI)

Echo-volumar imaging (EVI) generalizes the EPI sampling process from a 2D acquisition to a 3D acquisition. In contrast to 2D multislice imaging, 3D imaging encodes a single 3D k-space associated with the volumetric image, rather than a series of 2D k-spaces associated with each slice. While this distinction may seem subtle, it does have significant implications for SNR, as all else being equal, 3D imaging confers a $\sqrt{N_z}$ increase in SNR, where N_z is the number of image slices. EVI achieves 3D acquisition of k-space by using a sequence of EPI-like readouts to capture the T_2^*-weighted signal decay from a single excitation. Interestingly, this extension was originally suggested by Mansfield in 1977 at the same time EPI was proposed, but was not demonstrated experimentally until Song et al. (1994) showed the first images acquired with an EVI readout. Also note that multishot 3D EPI (Poser et al., 2010) is closely related to EVI in that it also encodes a 3D volumetric image, but does so using a purely planar readout (i.e., sampling just a single plane of k-space with an EPI readout) every excitation, and therefore does not have the same benefit to image speed as does EVI.

However, imaging with EVI can be challenging because while data are being sampled during the T_2^* signal decay, it can be difficult to sample all the necessary 3D k-space before the signal decays completely. Several enhancements to EVI have been developed to overcome this challenge. For example, interleaved EVI (van der Zwaag et al., 2006) addressed this problem by sampling one half of the 3D k-space in each excitation and combining the two samples to form a complete image. This ensures that sufficient signal is available for each readout, though being at the cost of additional scan time. Without interleaving, this study achieved a volume TR of 167 ms using 3×3×1.5 mm^3 voxels and a partial field-of-view (192×192×12 mm^3), while the interleaved approach with matched scan parameters results in a loss in temporal resolution by a factor of 2. Another approach, multislab EVI (Posse et al., 2012), partitioned the 3D image volume into thick slabs, and performs 3D EVI encoding on each slab prior to concatenating the slabs together to form the volume of interest. This also reduces the length of the EVI readout by reducing the size of the 3D k-space (due to the excited slab being smaller than the full acquired volume) but requires one excitation and readout for every slab. This multislab approach was able to achieve whole brain sampling at 4 mm isotropic resolution using 4 slabs with an overall temporal resolution of 286 ms.

To date, EVI has seen limited use in rs-fMRI, with one study using the multislab EVI method for high-speed rs-fMRI (Posse et al., 2013) (4 mm isotropic resolution, volume TR 276 ms), enabling ICA-based RSN connectivity analysis and the identification of signals related to cardiac pulsatility, the latter specifically enabled by the ultrafast sampling. Another study used the multislab EVI approach (4×4×6 mm^3 resolution, volume TR 136 ms) to examine high-frequency ($\geq$0.5 Hz) correlations in resting-state signals (Trapp et al., 2018), finding evidence of high-frequency correlations in default mode and auditory networks.

Pros

The primary advantage of EVI is the reduction in scan time by scanning more than a single plane of 3D k-space every excitation. This reduces overhead time costs associated with the RF excitation pulses and can facilitate very rapid acquisition.

Cons

There are several disadvantages to using EVI, owing largely to the long readout durations associated with the technique. As mentioned, long readout durations can suffer from loss of signal due to T_2^* decay, which not only reduces SNR but can also induces blurring in the image due to the filtering effects of the T_2^* decay envelope. This effect is made worse at higher field strengths (B_0), since T_2^* shortens as B_0 increases. Longer EVI readouts can also limit achievable TEs and suffer from signal dropout. As a consequence, EVI protocols are typically limited to modest resolutions or limited coverage in order to maintain image fidelity at these faster sampling rates. Multislab EVI strategies are also susceptible to artifacts at the boundaries between imaging slabs, appearing dark stripes across the image at slab interfaces.

Simultaneous multislice (SMS)-EPI

Simultaneous multislice EPI (SMS-EPI) has had rapid adoption following its development in the early 2010s. The approach is nearly identical to conventional multislice EPI, differing only in that instead of exciting and reading out signal from a single slice, multiple slices are excited using "multiband" pulses, and signals from all excited slices are sampled simultaneously. The number of simultaneously excited slices is often referred to as the "multiband (MB) factor." The acquired slices are then separated (or "unaliased") with reconstruction algorithms that use multichannel coil sensitivity information to distinguish slices from one another. The most common algorithm, slice GRAPPA (Setsompop et al., 2012), uses an approach similar to that for image reconstruction with parallel imaging. The success of SMS-EPI builds on work from Larkman et al. (Larkman et al., 2001), who introduced the concept of simultaneous multislice excitation and separation based on parallel imaging principles. Subsequently, Moeller et al. (Moeller et al., 2010) combined the SMS concept with EPI, with the first demonstration in fMRI using a multiband (MB) factor of 4 (i.e., four simultaneously acquired slices), for volume TRs of 1.25 s and 1.50 s in acquisitions with resolutions of $2{\times}2{\times}2\,\mathrm{mm}^3$ and $1{\times}1{\times}2\,\mathrm{mm}^3$, respectively.

In 2012, Setsompop et al. (Setsompop et al., 2012) developed a method for introducing relative shifts between the simultaneously excited slices in SMS-EPI (by way of small phase-encoding blips in the slice direction), making the unaliasing problem significantly easier (and therefore less prone to noise amplification). This extended work on CAIPIRINHA encoding for multislice imaging introduced by Breuer et al. (Breuer et al., 2005), hence the term "blipped-CAIPI." The blipped-CAIPI approach to SMS-EPI is the current standard, due to its advantages in SNR compared to nonblipped SMS-EPI, with SNR improvements up to 40% in high noise-amplification regions reported in the original study using a MB factor 3. Since this development, MB factors as high as 16 have been reported in the literature (Chen et al., 2015), resulting in a volume TR of 300 ms at 2.5×2.5×3 mm^3 spatial resolution. While the optimal balance of acceleration (MB factor) and SNR is still a topic of active inquiry (Risk et al., 2021), studies typically employ MB factors between 3 and 8.

It is not possible to highlight all the applications and discoveries facilitated by SMS-EPI, but one way to underscore the impact SMS-EPI has had on rs-FMRI is that it has been used to power the acquisition of rs-fMRI data in large cohort studies like the Human Connectome Project (HCP), the UK Biobank, and others. Both the HCP and UK Biobank projects use SMS-EPI with MB acceleration factor 8, resulting in subsecond volume TRs for whole-brain coverage at 2 mm (volume TR of 720 ms) and 2.4 mm (volume TR of 735 ms) resolutions, respectively. These consortium projects have produced data that have facilitated numerous advances in rs-fMRI, with selected examples including new brain parcellations (Glasser et al., 2016) and atlases (Fan et al., 2016), and discovery of novel associations between functional connectivity and lifestyle or health factors (Miller et al., 2016). As these large data sets have become important resources for rs-fMRI analysis, many individual studies are matching their imaging protocols to that of HCP or UK Biobank, resulting in SMS-EPI becoming a de-facto standard for acquisition of rs-fMRI data.

Pros

The key benefit of SMS-EPI is that it reduces volume acquisition times in proportion to the MB acceleration factor, allowing for subsecond whole-brain rs-fMRI. Furthermore, it achieves this without the $\sqrt{R}$ SNR penalty typically associated with parallel imaging acceleration, meaning that SMS-EPI data in theory have the same SNR as conventional 2D multislice data (when neglecting signal loss due to reduced T1 recovery from the shorter volume TR as well as the factor of acceleration-related SNR losses, also known as the *g-factor*). As SMS-EPI is based on EPI, all the same advantages (and disadvantages) of EPI acquisitions also apply to SMS-EPI.

Cons

While SMS-EPI is compatible with in-plane acceleration, the total acceleration factor (in-plane acceleration multiplied by MB acceleration) is limited by the geometry and encoding power of the multichannel receive array head coil. As the MB factor increases, *g*-factors can become significant sources of SNR degradation, particularly in central (subcortical) regions of the brain. However, with the blipped-CAIPI sampling scheme, *g*-factors are optimized for minimal SNR loss, and MB factors up to 8 are readily achievable. Another consideration of SMS-EPI is signal "leakage" between simultaneously acquired slices, possibly resulting in spatially misplaced signals. Strategies have been developed to minimize slice leakage, including split-slice GRAPPA (Cauley et al., 2014) (or "leak-block") reconstruction. As image reconstruction becomes increasingly complex, reconstruction time and necessary computing resources may be a concern, as well as increasing requirements for data storage (although this is common to all ultrafast sampling methods). Finally, multiband RF excitation pulses require more power than matched single-band excitations, which may be a concern for participant heating, although typically this is not a limiting factor in gradient-echo fMRI (in contrast to spin-echo fMRI, which requires high-power refocusing pulses).

Inverse imaging (InI)

Inverse imaging (InI) was introduced by Lin et al. in 2006 (Lin et al., 2006), based on the apparent similarities between the geometric configurations of high-channel count MRI head coils and those of whole-head magnetoencephalography (MEG) or EEG sensor systems. While MR imaging typically employs on a combination of encoding (k-space) using gradient fields and information from array coil sensitivities, MEG and EEG rely solely on the properties of their sensors for spatial localization, utilizing various source localization algorithms. However, with high-channel count MRI receive arrays, InI proposed the use of analogous source localization techniques generalized to MR imaging, reducing the need for time-intensive gradient encoding. In doing so, InI reduces the sampling of k-space to such an extent that the linear system is underdetermined (in contrast to the case of SMS-EPI for instance, which typically operates in an overdetermined regime), resulting in dramatically undersampled and ultrafast imaging.

To achieve this, InI (and its subsequent extensions) relies on two common themes: Cartesian sampling and linear constrained reconstruction. The original proof-of-concept InI approach reconstructed a 2D slice from a single sampled line of k-space using a 90-channel head coil and a minimum-norm constraint on the image reconstruction. The linear reconstruction allowed for analytical prediction of noise amplification (i.e., *g-factor*), facilitating direct estimation of dynamic statistical parametric maps. Subsequent extensions of InI employed alternative constraints inspired by MEG/EEG "beamformers" (Lin et al., 2008), or the use of multiple EPI-based projections at different angles (transverse, sagittal, and coronal) (Tsai et al., 2012), both able to achieve whole-brain coverage at 100 ms temporal resolution. Recent developments have even combined InI with SMS-EPI (SMS-InI) using blipped CAIPI EPI readouts for 100 ms whole-brain coverage at 5 mm effective resolution (Hsu et al., 2017).

The ultrafast sampling provided by the InI approach has been used for several clinical and resting-state applications. For example, SMS-InI has been used for improved localization of interictal spikes in patients with epilepsy, by sampling the brain volume for 100 ms every 2 s, to maintain a 95% duty cycle for the EEG, uncorrupted by MRI gradient artifacts. This resulted in a 54% reduction in EEG signal fluctuations compared to conventional simultaneous EEG-fMRI using multislice EPI, suggesting a decrease in residual gradient artifacts. Another example is the use of InI with additional phase constraints to explore resting state network frequency characteristics at 10 Hz (50 ms sampling rate) (Boyacioglu et al., 2013), which highlighted the fact that intersubject variation in network spectra was notably higher than internetwork variation, likely a reflection of individual hemodynamic response characteristics.

Pros

InI facilitates ultrafast sampling, with 50–100 ms volume TRs capable of providing 5–10 Hz sampling rates. As it does this using EPI-based readouts, implementation on scanners is also relatively straightforward. Furthermore, the use of linear constrained reconstructions is advantageous due to the ease of reconstruction and the ability to analytically predict noise characteristics, the latter of which is a particularly useful property for statistical inference.

Cons

One drawback of the InI approach is that because it relies on multichannel receive head coils to effectively replace spatial encoding with gradients, the fidelity of the acquired image is closely related to the number of channel and relatedly to the spatial encoding power of the available head coil. InI has been demonstrated on 32-channel coils, but more typically, 64-channel (or higher) coils are used. The consequence of the reduced gradient encoding in InI is reflected in reduced effective spatial resolution (compared to expected resolution given the nominal image matrix), and this effective blurring may be anisotropic (depending on the encoding scheme). Generally speaking, dimensions encoded with gradients will retain the nominal spatial resolution, whereas dimensions without gradient encoding will have a coil-dependent effective point-spread function.

MR-encephalography (MREG)

Similar to InI, MR-encephalography (MREG) is a technique that aims to reduce or eliminate gradient encoding, using only multichannel receive array information for spatial localization (Hennig et al., 2007).

However, unlike InI, the MREG method placed less emphasis on high-channel count arrays, focusing more on functional sensitivity rather than spatial specificity. Sometimes also referred to as the "one-voxel-one-coil" concept, this approach is a spiritual successor to early work on detecting brain activity using functional spectroscopy (Hennig et al., 1994), which also read out data without gradient encoding. In its original publication (Hennig et al., 2007), MREG was demonstrated on an 8-channel array coil, with no gradient encoding or 1D encoding for single-slice imaging at a TR of 50 ms. This initial demonstration also used a simple weighted combination of reference images for reconstruction, resulting in spatial resolutions reflecting the slowly varying spatial coil sensitivities.

In contrast to the direction taken by InI, subsequent development of the MREG technique focused on the development of (1) highly undersampled non-Cartesian trajectories (e.g., 3D rosettes (Zahneisen et al., 2011) or spherical stack of spirals (Assländer et al., 2013)) and (2) regularized nonlinear image reconstruction (e.g., nonlinear total-variation regularization). The stack of spirals MREG approach, for instance, was able to provide whole brain coverage at a 100-ms temporal resolution at 3 mm nominal isotropic resolution. The edge-preserving nature of total-variation constraints results in a significantly higher effective spatial resolution than the initial MREG method, at the cost of more complex reconstruction and reduced interpretability of resulting image noise.

A number of applications of MREG to rs-fMRI have been reported in the literature. Selected examples include a study by Lee et al. (Lee et al., 2013) in which the fast sampling of MREG was used to detect evidence of resting-state networks up to frequencies of 0.8 Hz (see Fig. 4). Additionally, a study by Akin et al. (Akin et al., 2017) demonstrated that compared to conventional EPI, MREG was able to detect a greater number of individual networks in subject-specific analyses and was also able to facilitate detection of networks with as little as 50 s of data. Furthermore, the ultrafast sampling characteristics of MREG have been used in studies examining the effects of sampling rate on an array of resting-state fMRI metrics (Huotari et al., 2019), such as seed-based connectivity, regional homogeneity, ICA maps, and dynamic connectivity metrics, which found that dynamic connectivity metrics (e.g., quasiperiodic patterns) benefit the most from fast sampling methods.

Pros

MREG provides ultrafast sampling, resulting in full brain coverage in ~100 ms. One advantage of MREG with 3D non-Cartesian encoding is that the resulting spatial resolution is more isotropic than approaches such as InI, as the undersampling is distributed evenly

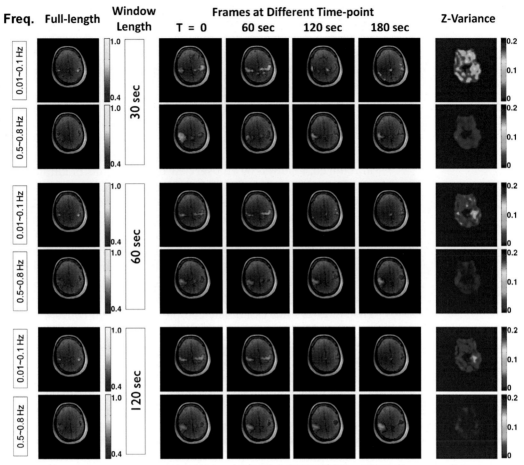

Fig. 4 Evidence of high-frequency resting-state sensorimotor networks (up to 0.8 Hz) measured using MREG (Lee et al., 2013). Results at two frequency bands are shown, 0.01–0.1 Hz and 0.5–0.8 Hz, highlighting the correspondence between the conventionally observed low-frequency rs-fMRI networks, and the high-frequency networks. Window length refers to the size of the sliding window used in the seed-based correlation analysis, and the z-variance refers to the temporal variance of the estimated correlation z-scores. Figure reproduced with permission from Lee, H.-L., Zahneisen, B., Hugger, T., LeVan, P., Hennig, J., 2013. Tracking dynamic resting-state networks at higher frequencies using MR-encephalography, NeuroImage 65, 216–222, https://doi.org/10.1016/j.neuroimage.2012.10.015.

across all dimensions. Furthermore, effective spatial resolution may further be enhanced by the nonlinear-reconstruction schemes.

Cons

Like in the case of EVI, signal decay and off-resonance during relatively long readouts in MREG can result in image blurring or other artifacts, and effective resolutions are typically larger than nominal.

Moreover, the sampling trajectories used in MREG are relatively complex, requiring custom gradient waveforms, and the corresponding image reconstructions can require more computational resources than conventional methods. Another potential downside of the nonlinear reconstruction is the unknown transformation of the resulting image noise characteristics.

Summary and conclusions

In this chapter, the factors defining the MRI signal and noise were introduced, followed by an overview of the benefits of ultrafast sampling in rs-fMRI, which have received rapidly growing recognition in rs-fMRI research. This chapter also provides relatively in-depth descriptions of dominant accelerated and highly accelerated rs-fMRI acquisition techniques, including EPI, EVI, SMS-EPI, InI, and MREG. All methods discussed have specific advantages and disadvantages, and there is no obvious "one-size-fits-all" solution to rs-fMRI data acquisition for diverse experimental contexts. However, several observations can be made about the current state of rs-fMRI data acquisition. First, SMS-EPI has essentially completely replaced conventional EPI for human functional neuroimaging, and there are very few reasons not to use SMS-EPI if it is supported and available on the MR scanner. SMS-EPI provides fast, subsecond whole brain sampling and should be adequate for most conventional, nondynamic connectivity studies. The best choice for the MB acceleration factor (i.e., how many slices to simultaneously acquire) is an actively debated question in the literature, with studies such as Risk et al. (Risk et al., 2021) recommending MB factors of 4 to balance acceleration with SNR, whereas HCP and Biobank protocols opted for a MB factor of 8, believing the benefit of faster sampling outweighs the SNR penalties. Other specific experimental considerations, such as SNR in subcortical regions (which typically suffers the greatest loss in SNR), may also factor into selection of MB factors.

For investigation of high-frequency rs-fMRI signals or dynamic connectivity, methods that provide faster sampling than SMS-EPI, namely, EVI (200–300 ms/volume) or InI/MREG (50–100 ms/volume), may be a more suitable choice. However, it should be noted that while SMS-EPI can achieve whole-brain sampling at 2.0–2.5 mm isotropic resolution, the effective resolution of whole images produced by EVI, InI, and MREG is significantly lower. Therefore, the choice to use ultrafast sampling typically comes at a cost of trading spatial resolution for temporal resolution, and it may be important to assess which is more important for any given study. Finally, it is important to note that using ultrafast sampling techniques may require special consideration during data analysis, particularly in confound modeling of autocorrelation structure, which is much more evident in ultrafast sampling regimes (Chen et al., 2019).

References

Akin, B., Lee, H.-L., Hennig, J., LeVan, P., 2017. Enhanced subject-specific resting-state network detection and extraction with fast fMRI. Hum. Brain Mapp. 38 (2), 817–830. https://doi.org/10.1002/hbm.23420.

Assländer, J., et al., 2013. Single shot whole brain imaging using spherical stack of spirals trajectories. NeuroImage 73, 59–70. https://doi.org/10.1016/j.neuroimage.2013.01.065.

Birn, R.M., Diamond, J.B., Smith, M.A., Bandettini, P.A., 2006. Separating respiratory-variation-related fluctuations from neuronal-activity-related fluctuations in fMRI. NeuroImage 31 (4), 1536–1548. https://doi.org/10.1016/j.neuroimage.2006.02.048.

Bollmann, S., Puckett, A.M., Cunnington, R., Barth, M., 2018. Serial correlations in single-subject fMRI with sub-second TR. NeuroImage 166, 152–166. https://doi.org/10.1016/j.neuroimage.2017.10.043.

Boyacioglu, R., Beckmann, C., Barth, M., 2013. An investigation of RSN frequency spectra using ultra-fast generalized inverse imaging. Front. Hum. Neurosci. 7, 156. https://doi.org/10.3389/fnhum.2013.00156.

Breuer, F.A., Blaimer, M., Heidemann, R.M., Mueller, M.F., Griswold, M.A., Jakob, P.M., 2005. Controlled aliasing in parallel imaging results in higher acceleration (CAIPIRINHA) for multi-slice imaging. Magn. Reson. Med. 53 (3), 684–691. https://doi.org/10.1002/mrm.20401.

Cauley, S.F., Polimeni, J.R., Bhat, H., Wald, L.L., Setsompop, K., 2014. Interslice leakage artifact reduction technique for simultaneous multislice acquisitions. Magn. Reson. Med. 72 (1), 93–102. https://doi.org/10.1002/mrm.24898.

Chen, J.E., Glover, G.H., 2015. BOLD fractional contribution to resting-state functional connectivity above 0.1Hz. NeuroImage 107, 207–218. https://doi.org/10.1016/j.neuroimage.2014.12.012.

Chen, J.E., Polimeni, J.R., Bollmann, S., Glover, G.H., 2019. On the analysis of rapidly sampled fMRI data. NeuroImage 188, 807–820. https://doi.org/10.1016/j.neuroimage.2019.02.008.

Chen, L., et al., 2015. Evaluation of highly accelerated simultaneous multi-slice EPI for fMRI. NeuroImage 104, 452–459. https://doi.org/10.1016/j.neuroimage.2014.10.027.

Chiew, M., Smith, S.M., Koopmans, P.J., Graedel, N.N., Blumensath, T., Miller, K.L., 2015. k-t FASTER: acceleration of functional MRI data acquisition using low rank constraints. Magn. Reson. Med. 74 (2), 353–364. https://doi.org/10.1002/mrm.25395.

Damoiseaux, J.S., et al., 2006. Consistent resting-state networks across healthy subjects. Proc. Natl. Acad. Sci. U. S. A. 103 (37), 13848–13853. https://doi.org/10.1073/pnas.0601417103.

Davis, T.L., Kwong, K.K., Weisskoff, R.M., Rosen, B.R., 1998. Calibrated functional MRI: mapping the dynamics of oxidative metabolism. Proc. Natl. Acad. Sci. U. S. A. 95 (4), 1834–1839.

Fan, L., et al., 2016. The human Brainnetome Atlas: a new brain atlas based on connectional architecture. Cereb. Cortex 26 (8), 3508–3526. https://doi.org/10.1093/cercor/bhw157.

Feinberg, D.A., et al., 2010. Multiplexed echo planar imaging for sub-second whole brain FMRI and fast diffusion imaging. PLoS One 5 (12), e15710. https://doi.org/10.1371/journal.pone.0015710.

Glasser, M.F., et al., 2016. A multi-modal parcellation of human cerebral cortex. Nature 536 (7615), 171–178. https://doi.org/10.1038/nature18933.

Glover, G.H., Li, T.Q., Ress, D., 2000. Image-based method for retrospective correction of physiological motion effects in fMRI: RETROICOR. Magn. Reson. Med. 44 (1), 162–167.

Griffanti, L., et al., 2014. ICA-based artefact removal and accelerated fMRI acquisition for improved resting state network imaging. NeuroImage 95, 232–247. https://doi.org/10.1016/j.neuroimage.2014.03.034.

He, X., Yablonskiy, D.A., 2006. Quantitative BOLD: mapping of human cerebral deoxygenated blood volume and oxygen extraction fraction: Default state. Magn. Reson. Med. 57 (1), 115–126. https://doi.org/10.1002/mrm.21108.

Hennig, J., Ernst, T., Speck, O., Deuschl, G., Feifel, E., 1994. Detection of brain activation using oxygenation sensitive functional spectroscopy. Magn. Reson. Med. 31 (1), 85–90.

Hennig, J., Zhong, K., Speck, O., 2007. MR-Encephalography: fast multi-channel monitoring of brain physiology with magnetic resonance. NeuroImage 34 (1), 212–219. https://doi.org/10.1016/j.neuroimage.2006.08.036.

Hsu, Y.-C., Chu, Y.-H., Tsai, S.-Y., Kuo, W.-J., Chang, C.-Y., Lin, F.-H., 2017. Simultaneous multi-slice inverse imaging of the human brain. Sci. Rep. 7 (1). https://doi.org/10.1038/s41598-017-16976-0. Art. no. 1.

Huotari, N., et al., 2019. Sampling rate effects on resting state fMRI metrics. Front. Neurosci. 13, 279. https://doi.org/10.3389/fnins.2019.00279.

Jacobs, J., et al., 2014. Fast fMRI provides high statistical power in the analysis of epileptic networks. NeuroImage 88, 282–294. https://doi.org/10.1016/j.neuroimage.2013.10.018.

Jung, H., Ye, J.C., Kim, E.Y., 2007. Improved k-t BLAST and k-t SENSE using FOCUSS. Phys. Med. Biol. 52 (11), 3201–3226. https://doi.org/10.1088/0031-9155/52/11/018.

Krüger, G., Glover, G.H., 2001. Physiological noise in oxygenation-sensitive magnetic resonance imaging. Magn. Reson. Med. 46 (4), 631–637.

Larkman, D.J., Hajnal, J.V., Herlihy, A.H., Coutts, G.A., Young, I.R., Ehnholm, G., 2001. Use of multicoil arrays for separation of signal from multiple slices simultaneously excited. J. Magn. Reson. Imag. 13 (2), 313–317.

Lee, H.-L., Zahneisen, B., Hugger, T., LeVan, P., Hennig, J., 2013. Tracking dynamic resting-state networks at higher frequencies using MR-encephalography. NeuroImage 65, 216–222. https://doi.org/10.1016/j.neuroimage.2012.10.015.

Lewis, L.D., Setsompop, K., Rosen, B.R., Polimeni, J.R., 2016. Fast fMRI can detect oscillatory neural activity in humans. Proc. Natl. Acad. Sci. 113 (43), E6679–E6685. https://doi.org/10.1073/pnas.1608117113.

Lin, F.-H., Wald, L.L., Ahlfors, S.P., Hämäläinen, M.S., Kwong, K.K., Belliveau, J.W., 2006. Dynamic magnetic resonance inverse imaging of human brain function. Magn. Reson. Med. 56 (4), 787–802. https://doi.org/10.1002/mrm.20997.

Lin, F.-H., Witzel, T., Zeffiro, T.A., Belliveau, J.W., 2008. Linear constraint minimum variance beamformer functional magnetic resonance inverse imaging. NeuroImage 43 (2), 297–311. https://doi.org/10.1016/j.neuroimage.2008.06.038.

Lin, F.-H., et al., 2012. Physiological noise reduction using volumetric functional magnetic resonance inverse imaging. Hum. Brain Mapp. 33 (12), 2815–2830. https://doi.org/10.1002/hbm.21403.

Mansfield, P., 1977. Multi-planar image-formation using NMR spin echoes. J. Phys. C-Solid State Phys. 10 (3), L55–L58.

Miller, K.L., et al., 2016. Multimodal population brain imaging in the UK Biobank prospective epidemiological study. Nat. Neurosci. 19 (11), 1523–1536. https://doi.org/10.1038/nn.4393.

Moeller, S., et al., 2010. Multiband multislice GE-EPI at 7 tesla, with 16-fold acceleration using partial parallel imaging with application to high spatial and temporal whole-brain fMRI. Magn. Reson. Med. 63 (5), 1144–1153. https://doi.org/10.1002/mrm.22361.

Poser, B.A., Koopmans, P.J., Witzel, T., Wald, L.L., Barth, M., 2010. Three dimensional echo-planar imaging at 7 Tesla. NeuroImage 51 (1), 261–266. https://doi.org/10.1016/j.neuroimage.2010.01.108.

Posse, S., et al., 2012. Enhancement of temporal resolution and BOLD sensitivity in real-time fMRI using multi-slab echo-volumar imaging. NeuroImage 61 (1), 115–130. https://doi.org/10.1016/j.neuroimage.2012.02.059.

Posse, S., et al., 2013. High-speed real-time resting-state fMRI using multi-slab echo-volumar imaging. Front. Hum. Neurosci. 7, 479. https://doi.org/10.3389/fnhum.2013.00479.

Risk, B.B., et al., 2021. Which multiband factor should you choose for your resting-state fMRI study? NeuroImage, 117965. https://doi.org/10.1016/j.neuroimage.2021.117965.

Setsompop, K., Gagoski, B.A., Polimeni, J.R., Witzel, T., Wedeen, V.J., Wald, L.L., 2012. Blipped-controlled aliasing in parallel imaging for simultaneous multislice echo planar imaging with reduced g-factor penalty. Magn. Reson. Med. 67 (5), 1210–1224. https://doi.org/10.1002/mrm.23097.

Shmueli, K., et al., 2007. Low-frequency fluctuations in the cardiac rate as a source of variance in the resting-state fMRI BOLD signal. NeuroImage 38 (2), 306–320. https://doi.org/10.1016/j.neuroimage.2007.07.037.

Song, A.W., Wong, E.C., Hyde, J.S., 1994. Echo-volume imaging. Magn. Reson. Med. 32 (5), 668–671. https://doi.org/10.1002/mrm.1910320518.

Trapp, C., Vakamudi, K., Posse, S., 2018. On the detection of high frequency correlations in resting state fMRI. NeuroImage 164, 202–213. https://doi.org/10.1016/j.neuroimage.2017.01.059.

Tsai, K.W.-K., Nummenmaa, A., Witzel, T., Chang, W.-T., Kuo, W.-J., Lin, F.-H., 2012. Multiprojection magnetic resonance inverse imaging of the human visuomotor system. NeuroImage 61 (1), 304–313. https://doi.org/10.1016/j.neuroimage.2012.01.115.

Uludağ, K., Müller-Bierl, B., Ugurbil, K., 2009. An integrative model for neuronal activity-induced signal changes for gradient and spin echo functional imaging. NeuroImage 48 (1), 150–165. https://doi.org/10.1016/j.neuroimage.2009.05.051.

Van Essen, D.C., et al., 2013. The WU-Minn Human Connectome Project: an overview. NeuroImage 80, 62–79. https://doi.org/10.1016/j.neuroimage.2013.05.041.

Wise, R.G., Ide, K., Poulin, M.J., Tracey, I., 2004. Resting fluctuations in arterial carbon dioxide induce significant low frequency variations in BOLD signal. NeuroImage 21 (4), 1652–1664. https://doi.org/10.1016/j.neuroimage.2003.11.025.

Zahneisen, B., et al., 2011. Three-dimensional MR-encephalography: fast volumetric brain imaging using rosette trajectories. Magn. Reson. Med. 65 (5), 1260–1268. https://doi.org/10.1002/mrm.22711.

Zaitsev, M., Zilles, K., Shah, N.J., 2001. Shared k-space echo planar imaging with keyhole. Magn. Reson. Med. 45 (1), 109–117. https://doi.org/10.1002/1522-2594(200101)45:1<109::AID-MRM1015>3.0.CO;2-X.

van der Zwaag, W., Francis, S., Bowtell, R., 2006. Improved echo volumar imaging (EVI) for functional MRI. Magn. Reson. Med. 56 (6), 1320–1327. https://doi.org/10.1002/mrm.21080.

12

Time-varying functional connectivity

Shella Keilholz

Biomedical Engineering, Emory University/Georgia Tech, Atlanta, GA, United States

Time-varying vs. time-averaged functional connectivity

In fMRI studies using cognitive tasks or sensory stimuli, researchers take advantage of the known timing of the chosen experimental paradigm to identify areas of the brain that are activated. Many separate incidents of the same stimulus are often combined during analysis, which increases the functional contrast for the chosen paradigm by suppressing the ongoing background fluctuations of intrinsic brain activity. These intrinsic fluctuations are not time-locked to the task or stimulus, and thus averaging across trials reduces their amplitude.

For resting state functional magnetic resonance imaging (rs-fMRI), the intrinsic fluctuations of brain activity are the feature of interest. In contrast to task-related fMRI experiments, the timing of expected changes in brain activity is unknown, forcing researchers to find other ways to describe the coordination of the spontaneous fluctuations across the brain. The earliest rs-fMRI analysis methods were based on statistical dependencies such as Pearson correlation that persisted for the entire length of the scan (Biswal et al., 1995). For example, if one is interested in the functional connectivity within the motor network, one might calculate the correlation between time courses from the left and right motor cortex during a ten-minute rs-fMRI scan, and obtain one correlation value for that scan. A more data-driven approach that does not require all areas of interest to be identified before analysis might involve taking the time course from a region of interest and displaying areas of significant correlation across the whole brain, in which case each voxel or region of the brain would have one correlation value for the entire scan. Another popular data-driven approach involves parcellating the brain using an atlas, and then calculating correlation pairwise between parcels. As an example, for the widely

Advances in Resting-State Functional MRI. https://doi.org/10.1016/B978-0-323-91688-2.00006-0

examined Human Connectome Project data (Van Essen et al., 2013), a scan that originally consists as thousands of voxels at 1200 time points might be parcellated into 360 brain areas (Glasser et al., 2016). Correlation can then be calculated pairwise between these areas, resulting in a single 360×360 matrix for the ten-minute scan. For all of these approaches, it is also common to concatenate subjects and/or multiple scans from the same subject.

Methods such as these that obtain a single metric of functional connectivity for an entire scan are often designated as "time-averaged" or, more loosely, "**static**" analysis methods. It is easy to see that time-averaged approaches discard substantial amounts of data. Moreover, they are relatively insensitive to changes that occur over the course of the scan. As subjects lie in the scanner, in the absence of a directed task, they may recall past events, make plans for the future, daydream—cognitive processes that can recapitulate tasks used for fMRI experiments, so that it might be more accurate to consider the subject as performing undirected tasks rather than "at rest" (Kucyi, 2018). Superimposed over this undirected cognition may be contributions from growing drowsiness or increasing discomfort from remaining immobile in the scanner (Fig. 1).

The desire to examine changes in the coordination of intrinsic activity that occur over the course of the scan motivated the development of new methods for the analysis of rs-fMRI. These methods obtain a time series of functional connectivity for each scan rather than a single measurement, and are referred to as "time-resolved,"

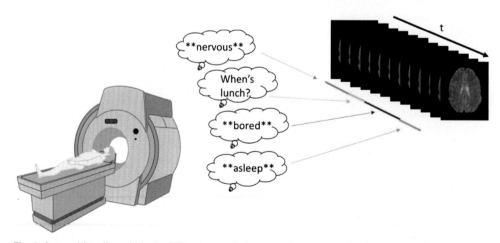

Fig. 1 As a subject lies within the MRI scanner during a resting state study, they may experience a variety of cognitive states, along with changes in emotion and variable arousal levels. Time-averaged analysis methods are relatively insensitive to these types of changes, which motivated the development of time-resolved methods.

"time-varying," or, more loosely, "**dynamic**" analysis methods. The "static" and "dynamic" terminology has more specific meaning within particular fields, which has led researchers to preferentially use other terms, although the term "dynamic" functional connectivity is still widely found in the literature.

Time-resolved analysis methods

Conceptually, the simplest approach to obtaining time-resolved information about functional connectivity is to apply the approaches that were developed for entire scans to segments of the scans instead. This is intuitively feasible, since rs-fMRI scans can range in length from a few minutes to 15 min or more, and one can readily imagine using a 15-min scan as five 3-min segments. In practice, the scan may be segmented into distinct windows or examined using a sliding window that provides a more continuous estimate of functional connectivity (Fig. 2).

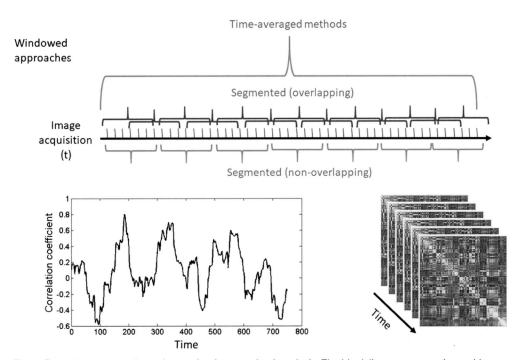

Fig. 2 Example segmentation schemes for time-resolved analysis. The black line represents time, with each hash representing the acquisition of an image. For time-averaged analysis, the entire time course is examined at once. For time-resolved analysis, the data are segmented into time windows, which may overlap to varying extents. The length of the window determines how many images will be used for each calculation. The result can be a time course of correlation between two areas of interest (bottom left) or, for parcellated data, a series of correlation matrices (bottom right).

Sliding window correlation (SWC) is one common approach to obtaining time-resolved information. Similar to methods that calculate correlation between seeds or parcels for the whole scan, correlation values can be calculated between regions of interest for successive windows. The result may be a time course of correlation values between two areas, or if the brain is parcellated, a matrix of correlation values for each time point (Allen et al., 2014; Chang and Glover, 2010; Keilholz et al., 2013). A similar approach can be taken with independent component analysis (ICA), another common technique for time-averaged analysis (Kiviniemi et al., 2011).

One of the major challenges for windowed approaches is determining the most appropriate length of the window. Longer windows are less sensitive to transient changes and asymptotically approach the values obtained from the whole scan; shorter windows are more sensitive to transient fluctuations but provide less accurate estimates of the correlation between signals and are prone to other artifacts (Hindriks et al., 2015; Leonardi and Van De Ville, 2015; Shakil et al., 2016). Some researchers have attempted to develop adaptive window techniques, where the window length can change over the course of the scan to better capture transient features of the data. However, these methods require knowledge of which features are likely to be important, while those features are still the topic of research and debate. Other researchers have taken more sophisticated approaches to identifying times of significant coherence, as in the seminal work by Chang et al. (Chang and Glover, 2010).

In order to avoid windows altogether, other analysis approaches consider each time point individually. (This can be considered the extreme end of windowed analysis, with each window consisting of a single time point.) Researchers have shown that individual, high-amplitude events in rs-fMRI account for much of the correlation that is observed (Caballero Gaudes et al., 2013; Petridou et al., 2013; Tagliazucchi et al., 2012a). This observation, and the fact that these individual events tend to co-occur in the same areas often over the course of the scan, led to the use of coactivation patterns (CAPs) for time-resolved analysis (Chen et al., 2015; Liu and Duyn, 2013). There are multiple variations on this approach, but in the original, when the BOLD signal passes a particular preset threshold in an area of interest, that time point is retained. The retained time points are then clustered based on spatial similarity. Other methods cluster the raw BOLD time series (Liu et al., 2013) or use the derivative to identify times of high change rather than high amplitude (Karahanoğlu and Van De Ville, 2015).

A different approach aims to capture the timing of common patterns of activity as well as their spatial distribution (Majeed et al., 2011; Yousefi et al., 2018; Zhang et al., 2021). This evolution of spatial activity over time is sometimes referred to as a "spatiotemporal trajectory." Research has shown that rs-fMRI can be described well by a fairly small number of these trajectories, and that they account for the structure of functional networks and functional connectivity gradients (Yousefi and Keilholz, 2021). These spatiotemporal trajectories are often obtained using a recursive pattern-finding algorithm that averages similar ~20 s segments of rs-fMRI time courses to create a template of the patter, also called a quasiperiodic pattern or QPP, and a time course that describes how strong it is at each time point of the scan (Majeed et al., 2011; Yousefi et al., 2018). Other methods have identified similar patterns in the latent variables obtained from a variational autoencoder (Zhang et al., 2021), and from complex PCA analysis of the Hilbert-transformed time courses (Bolt et al., 2022). The primary pattern in humans involves an alternation of high activity in the default-mode network and low activity in the task-positive network with its inverse pattern, along with propagation along the cortex as the pattern changes phases (Yousefi et al., 2018). Like CAPs, spatiotemporal trajectories are repeatedly expressed over the course of the scan and can be averaged to improve SNR; like sliding window correlation, they typically require the length of the pattern to be determined in advance. These trajectories can be considered a stereotyped sequence of individual CAPs states. Their relationship to SWC states is less clear, as the SWC windows typically average over at least one full cycle of the spatiotemporal pattern. Fig. 3 provides a conceptual overview of CAPs and QPPs or spatiotemporal trajectories.

The development of time-varying analysis techniques remains an active area of research, and a number of sophisticated methods have been proposed (e.g., Casorso et al., 2019; Faghiri et al., 2020; Iraji et al., 2019; Karahanoğlu and Van De Ville, 2015; Yaesoubi et al., 2018). The relatively simple approaches described here are widely used and serve as a foundation for understanding and interpreting new analysis methods.

Summarizing time-resolved analysis methods

Time-resolved analysis techniques are sometimes said to cause a "data explosion" (Hutchison et al., 2013; Keilholz et al., 2017; Preti et al., 2017). Both time-resolved matrices obtained with sliding window correlation and CAPs result in many measurements for a single scan (at the extreme, one spatial pattern per time point in the scan). This wealth of information can be summarized by clustering the matrices or CAPs into "brain states" (Allen et al., 2014; Liu and Duyn, 2013). Each time point is assigned to the state which it most closely resembles. The number of total states varies depending on the type of

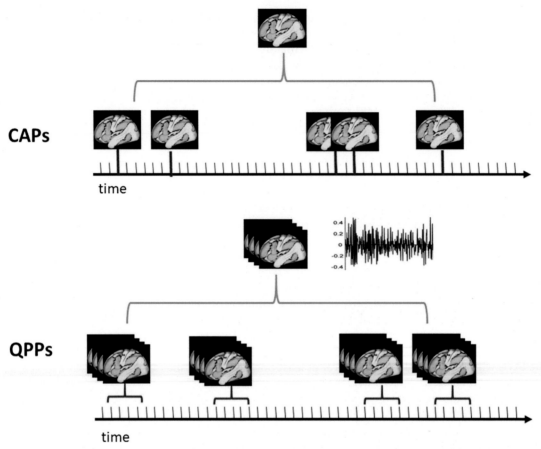

Fig. 3 Conceptual overview of CAPs and QPPs. For CAPs, images are retained at each time point when the signal in an area of interest passes a predetermined threshold. These images can be clustered based on spatial similarity and averaged to created activity maps (top image). Only one cluster is shown; another might be obtained from a different set of time points. For QPPs, a temporal series of images is selected based on random starting points. The spatial correlation between the original image series and the rest of the time course is calculated, and time points where the correlation is high are retained and averaged to create a template (top image series). The template is then used to calculate correlation with the time course, and the process is repeated until the template converges. A time course of correlation between the template and the scan is also obtained. Multiple seed points can be used to start the algorithm, and the resulting templates can be clustered to ensure that the one chosen is maximally representative of the data. For CAPs, similar to the correlation time course obtained with the QPP algorithm, the image at each time point can be assigned to the CAP that it most resembles.

analysis and the parameters chosen but common values are between 5 and 20 (Allen et al., 2014; Chen et al., 2016a; Karahanoğlu and Van De Ville, 2015; Liu and Duyn, 2013; Nomi et al., 2016). Using the time course assignment of states, one can then calculate metrics such as the dwell time, or average continuous duration of a state; fractional occupancy, or the number of time points assigned to a given state, or transition matrices, which describe the likelihood of transitioning from a given state to another given state. Alternatively, hidden Markov models (HMMs) model brain activity as sequential switching between brain states and can be used to describe the states and the transition probabilities simultaneously (Chen et al., 2016a; Vidaurre et al., 2017).

Another approach to summarizing the information obtained from time-resolved analysis comes from graph theory. A time-averaged matrix of correlation between areas can be represented as a graphical model, and characterized in terms of graph features such as centrality, path length, and efficiency (Bassett and Bullmore, 2006). A time-resolved series of correlation matrices can therefore be converted to a series of graphical models, and metrics that describe the network structure can be calculated for each time point. For example, the relative number of connections within a cluster of areas to the number of connections between different clusters can be used to describe the overall segregation or integration of the nodes of the brain over the course of the scan (Shine et al., 2016), giving a high-level view of the functional configuration (Bassett and Sporns, 2017).

Null models and statistical analysis

It was quickly recognized that time-varying analysis could exhibit variability that did not necessarily reflect variability in the underlying brain activity. Sliding window correlation is the most extensively studied of the time-resolved analysis techniques, and several groups have indicated its vulnerability to distortion and inaccuracy (Hindriks et al., 2015; Keilholz et al., 2013; Leonardi and Van De Ville, 2015; Shakil et al., 2016). Even for the high amplitude BOLD events that are used in CAPs analysis, Zhang et al. found a correlation of 0.44 with simultaneously measured local field potential power, and scatterplots make it clear that a very strong positive BOLD response can be associated with a negative change in local-field potential (LFP), and vice versa (Zhang et al., 2019). The average relationship between the BOLD signal and neural activity does not ensure a tight coupling at the level of individual events, a challenge for time-resolved analysis. On the other hand, sliding window correlation can provide better separation of patients from controls than time-averaged correlation (Chen et al., 2016b; de Vos et al., 2018; Rashid et al., 2014) and reflect changes in the underlying neural activity (Thompson et al., 2013b), suggesting that information about brain dynamics is preserved even if distorted. Gonzalez-Castillo et al. showed that the type of cognitive task could be better decoded from windowed correlation values than from activity patterns, suggesting that the information about coordinated (though possibly weak) activity across the brain is important for predictivity (Gonzalez-Castillo et al., 2015).

Statistical analysis for time-resolved methods has proven tricky. Null models must be carefully tailored to address the feature of interest. For example, null models created by pairing time courses that do not share temporal information can provide some insight into the amount of variability expected by chance, but they do not maintain the correlation distribution of the original data, which can affect variability. Moreover, simply because variability is within the limits expected by chance does not mean that it is meaningless or unrelated to brain function. For an excellent discussion of null models for time-resolved analysis, see Liégeois et al. (2021). As we learn more about how time-varying features of rs-fMRI are linked to neural activity, cognition, and behavior, new statistical approaches may become available. Until then, it is important to carefully consider appropriate null models in the design of experimental hypotheses, to ensure that the statistical tests are specific to the question of interest.

Sources of temporal variation in rs-fMRI

Generally speaking, the brain states that result from time-resolved analysis of rs-fMRI data typically involve the whole brain, persist for many time points, occur repeatedly over the course of the scan, and are clustered into a relatively small number of groups. These features are at odds with what we would expect from random cognitive changes over the course of the scan, raising the question of exactly what the observed variability means. While studies have shown that time-varying rs-fMRI reflects underlying neural activity and is linked to intra- and intersubject behavior, the relationship is far from being one-to-one (Bassett et al., 2011; Chang et al., 2013; Fong et al., 2019; Larabi et al., 2020; Pan et al., 2011, 2013; Sadaghiani et al., 2015; Spadone et al., 2015; Tagliazucchi et al., 2012b; Thompson et al., 2013a, 2013b, 2014), leaving room for contributions from other physiological processes.

Head motion is major source of unwanted variability for rs-fMRI. Small, abrupt head motions, associated with falling asleep for example, lead to short-lived but widespread changes across the brain that can easily influence time-resolved analysis (Power et al., 2017, 2020). Some aspects of this motion can be mitigated by scrubbing the affected time points. However, head motion has been shown to differ across patient populations (Huijbers et al., 2017) and over the developmental trajectory (Power et al., 2020), and remaining effects of motion can become the characteristic that distinguishes between groups, rather than the desired neural contrast. Global signal regression can reduce the contributions from physiological cycles and head motion, at the expense of possibly reducing neural contributions as well (Power et al., 2020).

As discussed in earlier chapters, physiological processes such as respiration and cardiac pulsation can introduce patterns of widespread correlation into rs-fMRI signals. The depth and rate of respiration, along with the rate of cardiac pulsation, vary over time in humans even during quiet rest, and unwanted contributions from these physiological cycles are particularly troublesome for time-varying analysis, which is sensitive to exactly the types of transient changes that occur. Respiration is especially problematic. Participants often take the occasional deep breath while lying still in the scanner, which has systemic effects of blood pressure and arterial gases that manifest as structured changes across the brain. Moreover, these changes in respiration can be tied to head motion (Power et al., 2020). A substantial amount of variability can be attributed to the combination of respiratory events and motion (Power et al., 2017, 2020).

Even variability that is related to brain function can be a potential confound for some studies. It has become apparent that changes in arousal have widespread influences on brain activity and functional connectivity, and may be the dominant source of neural variability in the rs-fMRI signal. Most participants become drowsy and/or fall asleep at some point in the rs-fMRI scan, and time-varying analysis methods may detect this change in arousal as, for example, a brain state that becomes more and more common as the scan progresses (Allen et al., 2014; Chang et al., 2016; Tagliazucchi and Laufs, 2014). Similar to the situation for head motion, many patient groups also have sleep disturbances and altered arousal levels (Fang et al., 2019; Mantovani et al., 2018; Wennberg et al., 2017), and so these changes can dominate differences observed between them and healthy controls. On the other hand, the brain systems involved in arousal are of great interest precisely because of their widespread alteration in neurodegenerative and psychiatric disorders. For example, subjects with autism spectrum disorder (ASD) exhibit atypical pupillary responses that suggest alterations in locus coeruleus modulation (Granovetter et al., 2020). It may be possible to selectively extract information about arousal fluctuations from rs-fMRI (Chang et al., 2016; Falahpour et al., 2018), which could prove extremely valuable.

Time-varying rs-fMRI analysis for basic science

Resting-state fMRI, while far from perfect, is arguably the best available tool for studying the systems-level organization of brain activity. It can cover the whole brain with spatial resolution that is adequate to resolve functionally relevant divisions, and increasingly, even mesoscopic features such as cortical layers and columns, while maintaining a temporal resolution that is comparable in scale to changes in behavior and cognition. MRI is noninvasive, meaning that it can be used in healthy humans, but it is also easily translated to rodent models, where it can be combined with other, more invasive measurement methods that can help to interpret rs-fMRI.

Rodent studies in particular have helped to place time-resolved analysis of rs-fMRI on a firm foundation. The combinations of rs-fMRI with modalities such as electrophysiology that are direct measurements of neural activity have consistently demonstrated a neural basis for the rs-fMRI results. Despite the inherent inaccuracies of sliding window correlation, correlation between SWC time courses of BOLD and bandlimited power from LFPs are significant in multiple frequency bands (Thompson et al., 2013b). The high-amplitude events that drive CAPs are linked to broadband changes in LFP power (Zhang et al., 2020). QPPs closely resemble the time-lagged correlation of infraslow electrical activity with rs-fMRI, suggesting that they capture the large-scale patterns that coordinate activity across the brain (Pan et al., 2013; Thompson et al., 2014). Wide field optical imaging of calcium indicators has identified "motifs" of activity that occur at both fast and slow time scales and which are largely consistent with the functional patterns obtained with rs-fMRI (Chan et al., 2015; Ma et al., 2016; Mohajerani et al., 2013). A thorough review the neurophysiological processes that contribute to rs-fMRI can be found in Pais-Roldán et al. (2021).

An impressive array of tools that are available in rodent models has the potential to elucidate the processes that lead to coordination of neural activity across the brain and the resulting time-varying functional connectivity. Optogenetic and chemogenetic approaches allow specific manipulation of particular brain areas and/or cell types, and fMRI allows monitoring of the response throughout the brain. Optogenetic stimulation has the advantage of allowing patterned temporal input, an important feature since neuromodulatory nuclei such as the locus coeruleus often have different firing modes (tonic vs. phasic). These tools are providing insight into the mechanisms that coordinate time-varying, brain-wide activity and improving the interpretation of rs-fMRI (Decot et al., 2017; Grandjean et al., 2019; Shah et al., 2016; Turchi et al., 2018; Zerbi et al., 2019).

Wide-field imaging of optical indicators has recently been combined with fMRI (Lake et al., 2020). As calcium and voltage indicators can be genetically targeted, this approach has the potential to elucidate the complementary roles of different types of cells in coordinating macroscale activity. Advances in large-scale neural recording are providing new insight into how single unit activity is related to coordinated activity across a population of cells, which may help in understanding how spontaneous activity fluctuations constrain individual neural response to incoming stimuli. Moreover, the complementary techniques available in rodents have the potential to bridge the gap between well-studied neural circuits, mesoscopic features such as columns or layers, and the macroscopic organization of activity across the brain.

With time-resolved rs-fMRI solidly grounded in animal models using established neuroscience techniques, the door has been opened for applications to outstanding questions in cognitive neuroscience and psychology in humans. For example, it has become apparent that many of the large-scale changes in brain activity over the course of a scan are linked to arousal levels, and a number of studies are beginning to explore how arousal and sympathetic activity manifest in brain-wide activity (Chang et al., 2016; Goodale et al., 2021; Li et al., 2021; Özbay et al., 2019; Raut et al., 2021; Teng et al., 2019). Other studies examine the overall organization of the network activity of the brain, to identify principles of brain organization related to task performance. For example, states in which the brain is more tightly integrated facilitate performance on a cognitive task (Shine et al., 2016). The better our understanding of the processes reflected in time-resolved rs-fMRI, the more interpretable information can be obtained with this technique.

As human MRI scanners move to fields of 7 Tesla (7T) and higher, it becomes feasible to image with enough spatial resolution to resolve features such as cortical layers, further bridging the gap across scales. In many brain areas, inputs arrive primarily at a particular layer (e.g., Layer 4 for thalamocortical input) and outputs to local or distant connections arise from other layers. This implies that if the spatial resolution is high enough to separate layers of the cortex, inputs can be separated from outputs, obtaining a directionality of influence that is not possible with standard rs-fMRI analysis (Sharoh et al., 2019; Stephan et al., 2019). Laminar imaging combined with time-varying analysis has therefore the potential to uncover the directional dynamics of the brain. The challenge is to maintain the temporal resolution needed for time-resolved analysis while obtaining the submillimeter spatial resolution required to resolve cortical layers.

Time-varying analysis for clinical research

In clinical studies, researchers often search for patterns of activity that distinguish patient groups from healthy controls, with the goal of identifying biomarkers that can be used to diagnose disorders, evaluate progression, and assess response to treatment. Resting state fMRI is particularly popular for clinical studies, as it requires minimal cooperation from the patient and is not confounded by lack of task performance, as can be the case for fMRI. Time-averaged analysis has had substantial success in identifying differences in patients that appear to be relevant to their specific disorder, but the predictive value of an individual scan remains low.

Time-resolved analysis has in many cases proven sensitive to changes associated with neurological or psychiatric disorders. For example, depression has been linked to longer dwell times in states where the DMN is active, and activity in a particular posterior portion of the DMN was linked to intrusive thoughts (Piguet et al., 2021). Time-varying functional connectivity has also identified changes that are unique to schizophrenia, bipolar disorder, and ADHD (Zhu et al., 2020), or to major depression and bipolar disorder (Chen et al., 2020), despite the difficulty that sometimes arises during differential diagnosis. Another study found that time-varying functional connectivity was associated with cognitive performance in patients with Alzheimer's disease (AD) and vascular dementia, and that some dynamic features were distinct to the individual dementia types (Fu et al., 2019). Children with attention-deficit hyperactivity disorder (ADHD) exhibit more variable state transitions than healthy controls, and differences in dwell times between sexes were observed (Scofield et al., 2019). Dynamic analysis may also help to separate highly variable disorders into subgroups. For example, two subtypes of ASD were detected with time-averaged and time-resolved functional connectivity (Easson et al., 2017). Classification of brain disorders is an active area of research, as reviewed in Du et al. (2018).

Despite the successes of time-resolved rs-fMRI, the differences between patients and controls remain in many cases are difficult to interpret. Many neurodegenerative disorders, for example, affect both the vasculature and neural activity. AD patients exhibit increased variability in the BOLD signal that arises from cardiovascular brain pulsations linked to abnormal cerebral perfusion (Tuovinen et al., 2020). In schizophrenia, hemodynamic responses are altered compared to healthy controls (Hanlon et al., 2015). From rs-fMRI alone, it is impossible to tell whether differences arise from neural activity, hemodynamics, or both.

Patient groups often differ from controls in other factors that can affect time-resolved rs-fMRI, such as their levels of motion and arousal, as described earlier. In some cases, these alterations might be vital to understanding the disorder. While there is little sense in using MRI as an expensive motion detector, time-varying rs-fMRI could play an important role in elucidating normal and pathological dynamics in the arousal system. The development of a template that can be used to estimate fluctuations in arousal over the course of a scan (Chang et al., 2016; Falahpour et al., 2018) is one important step toward this goal. Other large-scale phenomenon, such as the global signal and QPPs, may also prove sensitive to changes in arousal (Tal et al., 2013; Wong et al., 2013). For example, subjects with ADHD exhibit reduced strength of QPPs, consistent with reduced neuromodulatory input (Abbas et al., 2019).

One of the most powerful aspects of MRI is that its noninvasive nature allows it to be used in patient populations and rodent models of the disease of interest. This enables a progressive cycle of translation, where one can first identify differences in time-varying functional connectivity between patients and controls, then move to an appropriate animal model and determine whether it recapitulates the differences observed in humans. If not, researchers might develop improved models; if the same differences are observed, it is then possible to investigate potential sources of the differences using the wide array of tools available for rodents. Successful identification of a mechanism that accounts for the differences might then lead to treatment strategies in humans. An overview of this type of translational neuroimaging research is given in (in press, https://arxiv.org/abs/2111.05912).

Future directions

At this time, there is no single recommended approach to time-resolved analysis of rs-fMRI data. Leaving aside the challenges involved in choosing an analytical approach and the appropriate parameters (thresholds, window lengths, parcellations, etc.), larger questions about the experimental approach still linger. For example, when comparing a patient group to healthy controls, should brain states be defined based on concatenated data for both groups, or individually for each group? The first approach may lead to some states that are ill-represented in one group or the other, while the second may give states that are difficult to compare across groups. Some of these issues are present for time-averaged analysis (for example, the choice of parcellation), but can be exacerbated when using time-resolved techniques because of the increased variability. Null models and statistical analysis remain challenging, and interpretation of results can be difficult. Some groups have turned to deep learning to identify differences across groups in a data-driven manner (e.g., Parmar et al., 2020), but the results are heavily dependent on the model used and remain difficult to interpret.

It seems probable that our best path forward is to better understand the types of variability that are present in the rs-fMRI (as described in earlier chapters), so that we can design analysis methods that are best suited to extract the information of interest. Much has been done in terms of characterizing physiological processes such as respiration and motion, but the separation of neurophysiological processes is in its infancy. The promising work on extracting a signature of arousal level is encouraging, and the tools available in rodents make validation of such an approach feasible. One hopes that in the future, we will be able to selectively extract signatures of particular processes that are more sensitive and specific, which will transform basic and clinical research.

References

Abbas, A., Bassil, Y., Keilholz, S., 2019. Quasi-periodic patterns of brain activity in individuals with attention-deficit/hyperactivity disorder. NeuroImage Clin. 21, 101653. https://doi.org/10.1016/j.nicl.2019.101653.

Allen, E.A., Damaraju, E., Plis, S.M., Erhardt, E.B., Eichele, T., Calhoun, V.D., 2014. Tracking whole-brain connectivity dynamics in the resting state. Cereb. Cortex 24, 663–676. https://doi.org/10.1093/cercor/bhs352.

Bassett, D.S., Bullmore, E., 2006. Small-world brain networks. Neuroscientist 12, 512–523.

Bassett, D.S., Sporns, O., 2017. Network neuroscience. Nat. Neurosci. 20, 353–364. https://doi.org/10.1038/nn.4502.

Bassett, D.S., Wymbs, N.F., Porter, M.A., Mucha, P.J., Carlson, J.M., Grafton, S.T., 2011. Dynamic reconfiguration of human brain networks during learning. Proc. Natl. Acad. Sci. U. S. A. 108, 7641–7646. https://doi.org/10.1073/pnas.1018985108.

Biswal, B., Yetkin, F.Z., Haughton, V.M., Hyde, J.S., 1995. Functional connectivity in the motor cortex of resting human brain using echo-planar MRI. Magn. Reson. Med. 34, 537–541.

Bolt, T., Nomi, J.S., Bzdok, D., Salas, J.A., Chang, C., Thomas Yeo, B.T., Uddin, L.Q., Keilholz, S.D., 2022. A parsimonious description of global functional brain organization in three spatiotemporal patterns. Nat. Neurosci. 25 (8), 1093–1103. https://doi.org/10.1038/s41593-022-01118-1.

Caballero Gaudes, C., Petridou, N., Francis, S.T., Dryden, I.L., Gowland, P.A., 2013. Paradigm free mapping with sparse regression automatically detects single-trial functional magnetic resonance imaging blood oxygenation level dependent responses. Hum. Brain Mapp. 34, 501–518. https://doi.org/10.1002/hbm.21452.

Casorso, J., Kong, X., Chi, W., Van De Ville, D., Yeo, B.T.T., Liégeois, R., 2019. Dynamic mode decomposition of resting-state and task fMRI. Neuroimage 194, 42–54. https://doi.org/10.1016/J.NEUROIMAGE.2019.03.019.

Chan, A.W., Mohajerani, M.H., LeDue, J.M., Wang, Y.T., Murphy, T.H., 2015. Mesoscale infraslow spontaneous membrane potential fluctuations recapitulate high-frequency activity cortical motifs. Nat. Commun. 6, 7738. https://doi.org/10.1038/ncomms8738.

Chang, C., Glover, G.H.H., 2010. Time-frequency dynamics of resting-state brain connectivity measured with fMRI. Neuroimage 50, 81–98. https://doi.org/10.1016/j.neuroimage.2009.12.011.

Chang, C., Liu, Z., Chen, M.C.C., Liu, X., Duyn, J.H.H., 2013. EEG correlates of time-varying BOLD functional connectivity. Neuroimage 72, 227–236. https://doi.org/10.1016/j.neuroimage.2013.01.049.

Chang, C., Leopold, D.A., Schölvinck, M.L., Mandelkow, H., Picchioni, D., Liu, X., Ye, F.Q., Turchi, J.N., Duyn, J.H., 2016. Tracking brain arousal fluctuations with fMRI. Proc. Natl. Acad. Sci. U. S. A. 113, 4518–4523. https://doi.org/10.1073/pnas.1520613113.

Chen, J.E., Chang, C., Greicius, M.D., Glover, G.H., 2015. Introducing co-activation pattern metrics to quantify spontaneous brain network dynamics. Neuroimage. https://doi.org/10.1016/j.neuroimage.2015.01.057.

Chen, S., Langley, J., Chen, X., Hu, X., 2016a. Spatiotemporal modeling of brain dynamics using resting-state functional magnetic resonance imaging with Gaussian hidden Markov model. Brain Connect. 6, 326–334. https://doi.org/10.1089/brain.2015.0398.

Chen, X., Zhang, H., Gao, Y., Wee, C.-Y.Y., Li, G., Shen, D., Alzheimer's Disease Neuroimaging Initiative, 2016b. High-order resting-state functional connectivity network for MCI classification. Hum. Brain Mapp. 37, 3282–3296. https://doi.org/10.1002/hbm.23240.

Chen, G., Chen, P., Gong, J., Jia, Y., Zhong, S., Chen, F., Wang, J., Luo, Z., Qi, Z., Huang, L., Wang, Y., 2020. Shared and specific patterns of dynamic functional connectivity variability of striato-cortical circuitry in unmedicated bipolar and major depressive disorders. Psychol. Med. 1-10. https://doi.org/10.1017/S0033291720002378.

de Vos, F., Koini, M., Schouten, T.M., Seiler, S., van der Grond, J., Lechner, A., Schmidt, R., de Rooij, M., Rombouts, S.A.R.B., 2018. A comprehensive analysis of resting state fMRI measures to classify individual patients with Alzheimer's disease. Neuroimage 167, 62–72. https://doi.org/10.1016/J.NEUROIMAGE.2017.11.025.

Decot, H.K., Namboodiri, V.M.K., Gao, W., McHenry, J.A., Jennings, J.H., Lee, S.H., Kantak, P.A., Jill Kao, Y.C., Das, M., Witten, I.B., Deisseroth, K., Shih, Y.Y.I., Stuber, G.D., 2017. Coordination of brain-wide activity dynamics by dopaminergic neurons. Neuropsychopharmacology 42, 615–627. https://doi.org/10.1038/npp.2016.151.

Du, Y., Fu, Z., Calhoun, V.D., 2018. Classification and prediction of brain disorders using functional connectivity: promising but challenging. Front. Neurosci. 12, 525. https://doi.org/10.3389/fnins.2018.00525.

Easson, A.K., Fatima, Z., McIntosh, A.R., 2017. Defining subtypes of autism spectrum disorder using static and dynamic functional connectivity., https://doi.org/10.1101/198093.

Faghiri, A., Iraji, A., Damaraju, E., Belger, A., Ford, J., Mathalon, D., Mcewen, S., Mueller, B., Pearlson, G., Preda, A., Turner, J., Vaidya, J.G., Van Erp, T., Calhoun, V.D., 2020. Weighted average of shared trajectory: A new estimator for dynamic functional connectivity efficiently estimates both rapid and slow changes over time. J. Neurosci. Methods 334, 108600. https://doi.org/10.1016/j.jneumeth.2020.108600.

Falahpour, M., Chang, C., Wong, C.W., Liu, T.T., 2018. Template-based prediction of vigilance fluctuations in resting-state fMRI. Neuroimage 174, 317–327. https://doi.org/10.1016/j.neuroimage.2018.03.012.

Fang, H., Tu, S., Sheng, J., Shao, A., 2019. Depression in sleep disturbance: A review on a bidirectional relationship, mechanisms and treatment. J. Cell. Mol. Med. 23, 2324–2332. https://doi.org/10.1111/JCMM.14170.

Fong, A.H.C., Yoo, K., Rosenberg, M.D., Zhang, S., Li, C.-S.R., Scheinost, D., Constable, R.T., Chun, M.M., 2019. Dynamic functional connectivity during task performance and rest predicts individual differences in attention across studies. Neuroimage 188, 14–25. https://doi.org/10.1016/J.NEUROIMAGE.2018.11.057.

Fu, Z., Caprihan, A., Chen, J., Du, Y., Adair, J.C., Sui, J., Rosenberg, G.A., Calhoun, V.D., 2019. Altered static and dynamic functional network connectivity in Alzheimer's disease and subcortical ischemic vascular disease: shared and specific brain connectivity abnormalities. Hum. Brain Mapp. 40. https://doi.org/10.1002/hbm.24591.

Glasser, M.F., Coalson, T.S., Robinson, E.C., Hacker, C.D., Harwell, J., Yacoub, E., Ugurbil, K., Andersson, J., Beckmann, C.F., Jenkinson, M., Smith, S.M., Van Essen, D.C., 2016. A multi-modal parcellation of human cerebral cortex. Nature 536, 171–178. https://doi.org/10.1038/nature18933.

Gonzalez-Castillo, J., Hoy, C.W., Handwerker, D.A., Robinson, M.E., Buchanan, L.C., Saad, Z.S., Bandettini, P.A., 2015. Tracking ongoing cognition in individuals using brief, whole-brain functional connectivity patterns. Proc. Natl. Acad. Sci. U. S. A. https://doi.org/10.1073/pnas.1501242112.

Goodale, S.E., Ahmed, N., Zhao, C., de Zwart, J.A., Özbay, P.S., Picchioni, D., Duyn, J., Englot, D.J., Morgan, V.L., Chang, C., 2021. FMRI-based detection of alertness predicts behavioral response variability. Elife 10. https://doi.org/10.7554/eLife.62376.

Grandjean, J., Corcoba, A., Kahn, M.C., Upton, A.L., Deneris, E.S., Seifritz, E., Helmchen, F., Mann, E.O., Rudin, M., Saab, B.J., 2019. A brain-wide functional map of the serotonergic responses to acute stress and fluoxetine. Nat. Commun. 10. https://doi.org/10.1038/s41467-018-08256-w.

Granovetter, M.C., Burlingham, C.S., Blauch, N.M., Minshew, N.J., Heeger, D.J., Behrmann, M., 2020. Uncharacteristic task-evoked pupillary responses implicate atypical locus coeruleus activity in autism. J. Neurosci. 40. https://doi.org/10.1523/jneurosci.2680-19.2020.

Hanlon, F.M., Shaff, N.A., Dodd, A.B., Ling, J.M., Bustillo, J.R., Abbott, C.C., Stromberg, S.F., Abrams, S., Lin, D.S., Mayer, A.R., 2015. Hemodynamic response function abnormalities in schizophrenia during a multisensory detection task. Hum. Brain Mapp. https://doi.org/10.1002/hbm.23063.

Hindriks, R., Adhikari, M.H., Murayama, Y., Ganzetti, M., Mantini, D., Logothetis, N.K., Deco, G., 2015. Can sliding-window correlations reveal dynamic functional connectivity in resting-state fMRI? Neuroimage 127, 242–256. https://doi.org/10.1016/j.neuroimage.2015.11.055.

Huijbers, W., Van Dijk, K.R.A., Boenniger, M.M., Stirnberg, R., Breteler, M.M.B., 2017. Less head motion during MRI under task than resting-state conditions. Neuroimage 147, 111–120. https://doi.org/10.1016/J.NEUROIMAGE.2016.12.002.

Hutchison, R.M.M., Womelsdorf, T., Allen, E.A.E.A., Bandettini, P.A.P.A., Calhoun, V.D.V.D., Corbetta, M., Della Penna, S., Duyn, J.H.J.H., Glover, G.H.G.H., Gonzalez-Castillo, J., Handwerker, D.A.D.A., Keilholz, S., Kiviniemi, V., Leopold, D.A.D.A., de Pasquale, F., Sporns, O., Walter, M., Chang, C., 2013. Dynamic functional connectivity: promise, issues, and interpretations. Neuroimage 80, 360–378. https://doi.org/10.1016/j.neuroimage.2013.05.079.

Iraji, A., Deramus, T.P., Lewis, N., Yaesoubi, M., Stephen, J.M., Erhardt, E., Belger, A., Ford, J.M., McEwen, S., Mathalon, D.H., Mueller, B.A., Pearlson, G.D., Potkin, S.G., Preda, A., Turner, J.A., Vaidya, J.G., Erp, T.G.M., Calhoun, V.D., 2019. The spatial chronnectome reveals a dynamic interplay between functional segregation and integration. Hum. Brain Mapp. 40. https://doi.org/10.1002/hbm.24580.

Karahanoğlu, F.I., Van De Ville, D., 2015. Transient brain activity disentangles fMRI resting-state dynamics in terms of spatially and temporally overlapping networks. Nat. Commun. 6, 7751. https://doi.org/10.1038/ncomms8751.

Keilholz, S.D., Magnuson, M.E., Pan, W.J., Willis, M., Thompson, G.J., 2013. Dynamic properties of functional connectivity in the rodent. Brain Connect. 3, 31–40. https://doi.org/10.1089/brain.2012.0115.

Keilholz, S., Caballero-Gaudes, C., Bandettini, P., Deco, G., Calhoun, V., 2017. Time-resolved resting-state functional magnetic resonance imaging analysis: current status, challenges, and new directions. Brain Connect. 7, 465–481. https://doi.org/10.1089/brain.2017.0543.

Kiviniemi, V., Vire, T., Remes, J., Elseoud, A.A., Starck, T., Tervonen, O., Nikkinen, J., 2011. A sliding time-window ICA reveals spatial variability of the default mode network in time. Brain Connect. 1, 339–347. https://doi.org/10.1089/brain.2011.0036.

Kucyi, A., 2018. Just a thought: How mind-wandering is represented in dynamic brain connectivity. Neuroimage. https://doi.org/10.1016/j.neuroimage.2017.07.001. Academic Press.

Lake, E.M.R., Ge, X., Shen, X., Herman, P., Hyder, F., Cardin, J.A., Higley, M.J., Scheinost, D., Papademetris, X., Crair, M.C., Constable, R.T., 2020. Simultaneous cortex-wide fluorescence Ca2+ imaging and whole-brain fMRI. Nat. Methods 17, 1262–1271. https://doi.org/10.1038/s41592-020-00984-6.

Larabi, D.I., Renken, R.J., Cabral, J., Marsman, J.B.C., Aleman, A., Ćurčić-Blake, B., 2020. Trait self-reflectiveness relates to time-varying dynamics of resting state functional connectivity and underlying structural connectomes: Role of the default mode network. Neuroimage 219, 116896. https://doi.org/10.1016/j.neuroimage.2020.116896.

Leonardi, N., Van De Ville, D., 2015. On spurious and real fluctuations of dynamic functional connectivity during rest. Neuroimage 104, 430–436. https://doi.org/10.1016/j.neuroimage.2014.09.007.

Li, R., Ryu, J.H., Vincent, P., Springer, M., Kluger, D., Levinsohn, E.A., Chen, Y., Chen, H., Blumenfeld, H., 2021. The pulse: transient fMRI signal increases in subcortical arousal systems during transitions in attention. Neuroimage 232, 117873. https://doi.org/10.1016/j.neuroimage.2021.117873.

Liégeois, R., Yeo, B.T.T., Van De Ville, D., 2021. Interpreting null models of resting-state functional MRI dynamics: not throwing the model out with the hypothesis. Neuroimage. https://doi.org/10.1016/j.neuroimage.2021.118518.

Liu, X., Duyn, J.H.H., 2013. Time-varying functional network information extracted from brief instances of spontaneous brain activity. Proc. Natl. Acad. Sci. U. S. A. 110, 4392–4397. https://doi.org/10.1073/pnas.1216856110.

Liu, X., Chang, C., Duyn, J.H., 2013. Decomposition of spontaneous brain activity into distinct fMRI co-activation patterns. Front. Syst. Neurosci. 7. https://doi.org/10.3389/FNSYS.2013.00101/ABSTRACT.

Ma, Y., Shaik, M.A., Kozberg, M.G., Kim, S.H., Portes, J.P., Timerman, D., Hillman, E.M.C.C., 2016. Resting-state hemodynamics are spatiotemporally coupled to synchronized and symmetric neural activity in excitatory neurons. Proc. Natl. Acad. Sci. U. S. A. 113. https://doi.org/10.1073/pnas.1525369113.

Majeed, W., Magnuson, M., Hasenkamp, W., Schwarb, H., Schumacher, E.H.E.H.H., Barsalou, L., Keilholz, S.D.D.S.D., 2011. Spatiotemporal dynamics of low frequency BOLD fluctuations in rats and humans. Neuroimage 54, 1140–1150. https://doi.org/10.1016/j.neuroimage.2010.08.030.

Mantovani, S., Smith, S.S., Gordon, R., O'Sullivan, J.D., 2018. An overview of sleep and circadian dysfunction in Parkinson's disease. J. Sleep Res. 27. https://doi.org/10.1111/JSR.12673.

Mohajerani, M.H., Chan, A.W., Mohsenvand, M., Ledue, J., Liu, R., McVea, D.A., Boyd, J.D., Wang, Y.T., Reimers, M., Murphy, T.H., 2013. Spontaneous cortical activity alternates between motifs defined by regional axonal projections. Nat. Neurosci. 16, 1426–1435. https://doi.org/10.1038/nn.3499.

Nomi, J.S., Vij, S.G., Dajani, D.R., Steimke, R., Damaraju, E., Rachakonda, S., Calhoun, V.D., Uddin, L.Q., 2016. Chronnectomic patterns and neural flexibility underlie executive function. Neuroimage. https://doi.org/10.1016/j.neuroimage.2016.10.026.

Özbay, P.S., Chang, C., Picchioni, D., Mandelkow, H., Chappel-Farley, M.G., van Gelderen, P., de Zwart, J.A., Duyn, J., 2019. Sympathetic activity contributes to the fMRI signal. Commun. Biol. 2, 421. https://doi.org/10.1038/s42003-019-0659-0.

Pais-Roldán, P., Mateo, C., Pan, W.-J., Acland, B., Kleinfeld, D., Snyder, L.H., Yu, X., Keilholz, S., 2021. Contribution of animal models toward understanding resting state functional connectivity. Neuroimage. https://doi.org/10.1016/J.NEUROIMAGE.2021.118630.

Pan, W.-J., Thompson, G., Magnuson, M., Majeed, W., Jaeger, D., Keilholz, S., 2011. Broadband local field potentials correlate with spontaneous fluctuations in functional magnetic resonance imaging signals in the rat somatosensory cortex under isoflurane anesthesia. Brain Connect. 1. https://doi.org/10.1089/brain.2011.0014.

Pan, W.-J.J.W.-J., Thompson, G.J.G.J., Magnuson, M.E.M.E., Jaeger, D., Keilholz, S., 2013. Infraslow LFP correlates to resting-state fMRI BOLD signals. Neuroimage 74, 288–297. https://doi.org/10.1016/j.neuroimage.2013.02.035.

Parmar, H., Nutter, B., Long, R., Antani, S., Mitra, S., 2020. Spatiotemporal feature extraction and classification of Alzheimer's disease using deep learning 3D-CNN for fMRI data., https://doi.org/10.1117/1.JMI.7.5.056001.

Petridou, N., Gaudes, C.C., Dryden, I.L., Francis, S.T., Gowland, P.A., 2013. Periods of rest in fMRI contain individual spontaneous events which are related to slowly fluctuating spontaneous activity. Hum. Brain Mapp. 34, 1319–1329. https://doi.org/10.1002/hbm.21513.

Piguet, C., Karahanoğlu, F.I., Saccaro, L.F., Van De Ville, D., Vuilleumier, P., 2021. Mood disorders disrupt the functional dynamics, not spatial organization of brain resting state networks. NeuroImage Clin. 32, 102833. https://doi.org/10.1016/j.nicl.2021.102833.

Power, J.D., Plitt, M., Laumann, T.O., Martin, A., 2017. Sources and implications of whole-brain fMRI signals in humans. Neuroimage 146, 609–625. https://doi.org/10.1016/j.neuroimage.2016.09.038.

Power, J.D., Lynch, C.J., Adeyemo, B., Petersen, S.E., 2020. A critical, event-related appraisal of denoising in resting-state fMRI studies. Cereb. Cortex 30, 5544–5559. https://doi.org/10.1093/cercor/bhaa139.

Preti, M.G., Bolton, T.A., Van De Ville, D., 2017. The dynamic functional connectome: State-of-the-art and perspectives. Neuroimage 160, 41–54.

Rashid, B., Damaraju, E., Pearlson, G.D., Calhoun, V.D., 2014. Dynamic connectivity states estimated from resting fMRI Identify differences among Schizophrenia, bipolar disorder, and healthy control subjects. Front. Hum. Neurosci. 8, 897. https://doi.org/10.3389/fnhum.2014.00897.

Raut, R.V., Snyder, A.Z., Mitra, A., Yellin, D., Fujii, N., Malach, R., Raichle, M.E., 2021. Global waves synchronize the brain's functional systems with fluctuating arousal. Sci. Adv. 7, eabf2709. https://doi.org/10.1126/sciadv.abf2709.

Sadaghiani, S., Poline, J.-B., Kleinschmidt, A., D'Esposito, M., 2015. Ongoing dynamics in large-scale functional connectivity predict perception. Proc. Natl. Acad. Sci. U. S. A. 112, 8463–8468. https://doi.org/10.1073/pnas.1420687112.

Scofield, J.E., Johnson, J.D., Wood, P.K., Geary, D.C., 2019. Latent resting-state network dynamics in boys and girls with attention-deficit/hyperactivity disorder. PLoS One 14, e0218891. https://doi.org/10.1371/journal.pone.0218891.

Shah, D., Blockx, I., Keliris, G.A., Kara, F., Jonckers, E., Verhoye, M., Van der Linden, A., 2016. Cholinergic and serotonergic modulations differentially affect large-scale functional networks in the mouse brain. Brain Struct. Funct. 221, 3067–3079. https://doi.org/10.1007/s00429-015-1087-7.

Shakil, S., Lee, C.H.C.-H., Keilholz, S.D.S.D., 2016. Evaluation of sliding window correlation performance for characterizing dynamic functional connectivity and brain states. Neuroimage 133, 111–128. https://doi.org/10.1016/j.neuroimage.2016.02.074.

Sharoh, D., van Mourik, T., Bains, L.J., Segaert, K., Weber, K., Hagoort, P., Norris, D.G., 2019. Laminar specific fMRI reveals directed interactions in distributed networks during language processing. Proc. Natl. Acad. Sci. U. S. A. 116, 21185–21190. https://doi.org/10.1073/pnas.1907858116.

Shine, J.M., Bissett, P.G., Bell, P.T., Koyejo, O., Balsters, J.H., Gorgolewski, K.J., Moodie, C.A., Poldrack, R.A., 2016. The dynamics of functional brain networks: integrated network states during cognitive task performance. Neuron 92, 544–554. https://doi.org/10.1016/j.neuron.2016.09.018.

Spadone, S., Della Penna, S., Sestieri, C., Betti, V., Tosoni, A., Perrucci, M.G., Romani, G.L., Corbetta, M., 2015. Dynamic reorganization of human resting-state networks during visuospatial attention. Proc. Natl. Acad. Sci. U. S. A. 112, 8112–8117. https://doi.org/10.1073/pnas.1415439112.

Stephan, K.E., Petzschner, F.H., Kasper, L., Bayer, J., Wellstein, K.V., Stefanics, G., Pruessmann, K.P., Heinzle, J., 2019. Laminar fMRI and computational theories of brain function. Neuroimage 197, 699–706. https://doi.org/10.1016/J.NEUROIMAGE.2017.11.001.

Tagliazucchi, E., Laufs, H., 2014. Decoding wakefulness levels from typical fMRI resting-state data reveals reliable drifts between wakefulness and sleep. Neuron 82, 695–708. https://doi.org/10.1016/j.neuron.2014.03.020.

Tagliazucchi, E., Balenzuela, P., Fraiman, D., Chialvo, D.R., 2012a. Criticality in large-scale brain fmri dynamics unveiled by a novel point process analysis. Front. Physiol. 3, 15. https://doi.org/10.3389/fphys.2012.00015.

Tagliazucchi, E., von Wegner, F., Morzelewski, A., Brodbeck, V., Laufs, H., 2012b. Dynamic BOLD functional connectivity in humans and its electrophysiological correlates. Front. Hum. Neurosci. 6, 339. https://doi.org/10.3389/fnhum.2012.00339.

Tal, O., Diwakar, M., Wong, C.-W., Olafsson, V., Lee, R., Huang, M.-X., Liu, T.T., 2013. Caffeine-induced global reductions in resting-state BOLD connectivity reflect widespread decreases in MEG connectivity. Front. Hum. Neurosci. 7, 63. https://doi.org/10.3389/fnhum.2013.00063.

Teng, J., Ong, J.L., Patanaik, A., Tandi, J., Zhou, J.H., Chee, M.W.L., Lim, J., 2019. Vigilance declines following sleep deprivation are associated with two previously identified dynamic connectivity states. Neuroimage. https://doi.org/10.1016/J.NEUROIMAGE.2019.07.004.

Thompson, G.J.G.J., Magnuson, M.E.M.E., Merritt, M.D.M.D., Schwarb, H., Pan, W.-J.W.J., Mckinley, A., Tripp, L.D.L.D., Schumacher, E.H.E.H., Keilholz, S.D.S.D., 2013a. Short-time windows of correlation between large-scale functional brain networks predict vigilance intraindividually and interindividually. Hum. Brain Mapp. 34, 3280–3298. https://doi.org/10.1002/hbm.22140.

Thompson, G.J.G.J., Merritt, M.D.M.D., Pan, W.-J.W.J., Magnuson, M.E.M.E., Grooms, J.K.J.K., Jaeger, D., Keilholz, S.D.S.D., 2013b. Neural correlates of time-varying functional connectivity in the rat. Neuroimage 83, 826–836. https://doi.org/10.1016/j.neuroimage.2013.07.036.

Thompson, G.J.G.J., Pan, W.-J.W.J., Magnuson, M.E.M.E., Jaeger, D., Keilholz, S.D.S.D., 2014. Quasi-periodic patterns (QPP): Large-scale dynamics in resting state fMRI that correlate with local infraslow electrical activity. Neuroimage 84, 1018–1031. https://doi.org/10.1016/j.neuroimage.2013.09.029.

Tuovinen, T., Kananen, J., Rajna, Z., Lieslehto, J., Korhonen, V., Rytty, R., Mattila, H., Huotari, N., Raitamaa, L., Helakari, H., Elseoud, A.A., Krüger, J., Levan, P., Tervonen, O., Hennig, J., Remes, A.M., Nedergaard, M., Kiviniemi, V., 2020. The variability of functional MRI brain signal increases in Alzheimer's disease at cardiorespiratory frequencies. Sci. Rep. https://doi.org/10.1038/s41598-020-77984-1.

Turchi, J., Chang, C., Ye, F.Q., Russ, B.E., Yu, D.K., Cortes, C.R., Monosov, I.E., Duyn, J.H., Leopold, D.A., 2018. The basal forebrain regulates global resting-state fMRI fluctuations. Neuron 97. https://doi.org/10.1016/j.neuron.2018.01.032. 940–952.e4.

Van Essen, D.C., Smith, S.M., Barch, D.M., Behrens, T.E.J., Yacoub, E., Ugurbil, K., 2013. The WU-Minn human connectome project: an overview. Neuroimage 80, 62–79. https://doi.org/10.1016/j.neuroimage.2013.05.041.

Vidaurre, D., Smith, S.M., Woolrich, M.W., 2017. Brain network dynamics are hierarchically organized in time. Proc. Natl. Acad. Sci. U. S. A. 114, 12827–12832. https://doi.org/10.1073/pnas.1705120114.

Wennberg, A.M.V., Wu, M.N., Rosenberg, P.B., Spira, A.P., 2017. Sleep disturbance, cognitive decline, and dementia: a review. Semin. Neurol. 37, 395–406. https://doi.org/10.1055/S-0037-1604351.

Wong, C.W., Olafsson, V., Tal, O., Liu, T.T., 2013. The amplitude of the resting-state fMRI global signal is related to EEG vigilance measures. Neuroimage 83, 983–990. https://doi.org/10.1016/j.neuroimage.2013.07.057.

Yaesoubi, M., Adali, T., Calhoun, V.D., Adalı, T., Calhoun, V.D., 2018. A window-less approach for capturing time-varying connectivity in fMRI data reveals the presence of states with variable rates of change. Hum. Brain Mapp. https://doi.org/10.1002/hbm.23939.

Yousefi, B., Keilholz, S., 2021. Propagating patterns of intrinsic activity along macroscale gradients coordinate functional connections across the whole brain. Neuroimage 231, 117827. https://doi.org/10.1016/j.neuroimage.2021.117827.

Yousefi, B., Shin, J., Schumacher, E.H., Keilholz, S.D., 2018. Quasi-periodic patterns of intrinsic brain activity in individuals and their relationship to global signal. Neuroimage 167, 297–308. https://doi.org/10.1016/j.neuroimage.2017.11.043.

Zerbi, V., Floriou-Servou, A., Markicevic, M., Vermeiren, Y., Sturman, O., Privitera, M., von Ziegler, L., Ferrari, K.D., Weber, B., de Deyn, P.P., Wenderoth, N., Bohacek, J., 2019. Rapid reconfiguration of the functional connectome after chemogenetic locus coeruleus activation. bioRxiv. https://doi.org/10.1101/527457.

Zhang, X., Pan, W.J., Keilholz, S., 2019. The relationship between local field potentials and the blood-oxygenation-level dependent MRI signal can be non-linear. Front. Neurosci. 13. https://doi.org/10.3389/fnins.2019.01126.

Zhang, X., Pan, W.J., Keilholz, S.D., 2020. The relationship between BOLD and neural activity arises from temporally sparse events. Neuroimage 207, 116390. https://doi.org/10.1016/j.neuroimage.2019.116390.

Zhang, X., Maltbie, E.A., Keilholz, S.D., 2021. Spatiotemporal trajectories in resting-state FMRI revealed by convolutional variational autoencoder. Neuroimage 244, 118588. https://doi.org/10.1016/J.NEUROIMAGE.2021.118588.

Zhu, J., Zhang, S., Cai, H., Wang, C., Yu, Y., 2020. Common and distinct functional stability abnormalities across three major psychiatric disorders. NeuroImage Clin. 27, 102352. https://doi.org/10.1016/j.nicl.2020.102352.

13

Individual differences

Eyal Bergmann and Itamar Kahn

Department of Neuroscience, Mortimer B. Zuckerman Mind Brain Behavior Institute, Columbia University, New York, NY, United States

Introduction

Traditionally, human fMRI research has focused on group-level brain mapping in healthy participants or characterization of functional alterations in brain disorders. This approach, which assumes that organizational features are mostly shared between individual brains, has addressed many technical issues related to measurement stability and signal-to-noise ratio. Leveraging these advantages, group-level task-free (also termed resting-state and functional connectivity MRI [fcMRI]) and task-based fMRI studies have supported numerous insights into human brain organization and the involvement of different regions in sensation, action, and cognition in health and disease. Yet, given the richness of human experience, it is known that individuals are characterized by marked variation in personality and behavior, rooted in differences in brain function. In recent years, experimental and analytical advancements have set the stage for brain mapping at the level of the individual person, an approach termed "precision fMRI." This approach focuses on the differences between individuals, aiming to establish a link between variation in brain function and behavior and ultimately use this knowledge for personalized therapeutic interventions. In this chapter, we review different methods to characterize individual brains and discuss current challenges and putative solutions for application of these tools in basic research and clinical environments.

Quantitative metrics of individual differences in brain organization

Task-based and task-free fMRI can be used to characterize correlates of neural activity at the level of regions, large-scale networks and whole-brain organization, with transformation from lower to

Advances in Resting-State Functional MRI. https://doi.org/10.1016/B978-0-323-91688-2.00023-0

higher levels resulting in decreased variability and increased measurement stability (Noble et al., 2017). One of the challenges that emerged in the transition from group-level inferences to characterization of individual brains was the development of stable metrics which can quantify intersubject variability. This process required a novel experimental approach, namely multisession designs with dense sampling of individuals, and establishment of datasets with a large sample size. Together, these advancements set the stage for investigation of individual differences in the brain organization.

A prerequisite for studying functional brain organization in individuals is the ability to differentiate between true individual variability and noise sources (technical or biological). Such differentiation requires the acquisition of multiple resting-state sessions per subject. With this design, the variability between sessions from the same subject may represent day-to-day noise, while the variability between subjects captures both individual differences and noise. Then, accounting for the noise estimated from the intrasubject variability allows for isolation of individual differences. The first study that took this approach was published in 2013. By scanning 23 subjects over 5 fcMRI sessions, Mueller et al. (2013) were able to quantify within-subject variance and then regress it out from between-subject variance to isolate individual variability. By comparing seed-based functional connectivity maps of 1284 regions, the authors estimated the individual variability of different areas in the human cortex. While the overall individual variability was high (~0.6), marked differences were observed between different cortical networks. Strikingly, heteromodal association networks demonstrated higher individual-variability compared to unimodal sensory networks (Fig. 1A). Interestingly, this heterogenous distribution of individual variability in the human cortex was nonrandom, demonstrating significant correlation with evolutionary cortical expansion, sulcal depth, and degree of long-range connectivity. Moreover, increased variability in heteromodal association cortices was also observed in an independent sample using parcel-level metrics (Mira-Dominguez et al., 2014). These findings laid the foundations for further investigation of individual differences and the developmental mechanisms that underlie them.

After establishing the ability of fcMRI to measure individual variability in brain organization, the next step was to examine it in a larger cohort to better characterize the uniqueness of individual brains. In 2015, Finn et al. (2015) examined 126 subjects from the Human Connectome Project (HCP) to test whether individual functional connectivity profiles can act as an identifying fingerprint. The analysis was based on data acquired over two imaging sessions, where each session included a similar fcMRI scan and two different task-based fMRI scans. The researchers built a connectome using 268 predefined regions of

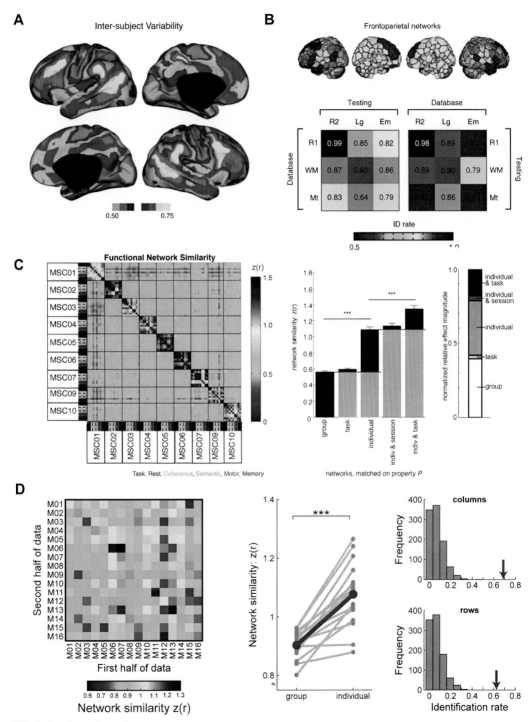

FIG. 1 See figure legend on next page.

Fig. 1, Cont'd Whole-brain metrics of individual variation in human and mouse. (A) Quantification of between-subjects variability across 23 subjects after correction for underlying within-subject variability. Values below the global mean are shown in *cool colors* while values above the global mean are shown in *warm colors*. (B) Fingerprinting analysis based on nodes from the combined frontoparietal networks (*top*) are highlighted in *color-coded matrices* (*bottom*) to compare identification (ID) rates across task and rest conditions. *Em*, emotion task; *Lg*, language task; *Mt*, motor task; *R1*, Rest1; *R2*, Rest2; *WM*, working-memory task. (C) A similarity matrix where every cell represents the similarity between two connectomes (*left*). The matrix is organized first by individual (marked by *solid black lines*), then by task *(colors along axes)*, and finally by sessions (indicated as *dashed lines*; here each split-half "session" represents data from five sessions). Similarity of human connectomes matched for different factors (*right*). The relative effects of each factor are highlighted in the black portion of each bar. Data are represented as mean ± SEM, ***P(FDR) < .001. Relative magnitude of these effects (*right*) calculated as a proportion of total effects. (D) Similarity of mouse connectomes (*left*) was used to compare group and individual similarities (*middle*) demonstrating higher similarity between connectomes from the same mouse ($n=16$ mice, ***$P<.001$). Fingerprint analysis in mice (*right*) showing above chance-level identification. (A) Reproduced with permission from Mueller, S., Wang, D., Fox, M.D., Yeo, B.T.T., Sepulcre, J., Sabuncu, M.R., Shafee, R., Lu, J., Liu, H., 2013. Individual variability in functional connectivity architecture of the human brain. Neuron 77, 586–595. https://doi.org/10.1016/j.neuron.2012.12.028. (B) From Finn, E.S., Shen, X., Scheinost, D., Rosenberg, M.D., Huang, J., Chun, M.M., Papademetris, X., Constable, R.T., 2015. Functional connectome fingerprinting: identifying individuals using patterns of brain connectivity. Nat. Neurosci. 18, 1664–1671. (C) Reproduced with permission from Gratton, C., Laumann, T.O., Nielsen, A.N., Greene, D.J., Gordon, E.M., Gilmore, A.W., Nelson, S.M., Coalson, R.S., Snyder, A.Z., Schlaggar, B.L., et al., 2018. Functional brain networks are dominated by stable group and individual factors, not cognitive or daily variation. Neuron 98, 439–452.e5. https://doi.org/10.1016/j.neuron.2018.03.035. (D) From Bergmann, E., Gofman, X., Kavushansky, A., Kahn, I., 2020. Individual variability in functional connectivity architecture of the mouse brain. Commun. Biol. 3. https://doi.org/10.1038/s42003-020-01472-5.

interest (ROIs) and calculated functional connectivity between each pair of ROIs. Then, they examined whether a subject's connectivity matrix, built using the first fcMRI session, can be used to identify the subject from a pool of connectomes built using data from the second fcMRI session. Strikingly, identification rates were remarkable, approaching 100%. Examining the amount of data per session that was required for successful identification, the researchers found that the measurement of more than approximately 6 min resulted in a stable measurement. Examination of identification rates based on connectivity profiles of specific networks revealed that all networks yielded identification rates higher than the chance level. Yet, in line with the observation of Mueller et al. (2013), superior identification rates were observed for frontoparietal networks (Fig. 1B). Moreover, the connectivity profile within the frontoparietal network supported slightly reduced but still successful identification even when comparing fcMRI to one or two different tasks. Further examination revealed that connections between and within association networks contribute more to identification, while connections within sensory networks are more consistent within subjects and across the groups. Beside confirming the heterogenous distribution of individual variability across the cortex, Finn et al. showed that similarity between connectivity matrices of the same subject is not only higher than their average similarity to the group but also higher than the similarity to all other subjects. This stringent criterion emphasizes the high degree of individual variation in human brain organization.

The works reviewed in the earlier section demonstrate unique organization of individual human brains and show some evidence for state-dependent effects (task vs rest). However, the temporal stability of brain organization and its relation to different cognitive states is not fully addressed. To tackle these issues, a new experimental design was needed, namely dense sampling of resting-state and task-based fMRI in individuals. In late 2015, the MyConnectome project was published by Poldrack et al. (2015), which included 100 resting-state sessions and multiomic data collected from a single subject over 18 months. Their analysis of this dataset sheds light on the contributions of gene expression, metabolic profile, psychological factors, and physical health to within-subject variance in functional connectivity, explaining some of the biological noise reported by Mueller et al. While the dense sampling approach taken in the MyConnectome project was novel and transformative, the sample size of one did not allow estimation of between-subject variances. In 2017, the Midnight Scan Club (MSC) dataset was published (Gordon et al., 2017), which included 10 individuals with several hours of fMRI data. Such a design is well positioned to study individual differences and to compare brain organization between task and resting conditions. Indeed, in 2018, Gratton et al. (2018) used this unique dataset to assess the contribution of individual variation, cognitive state, and time-to-brain organization. The authors developed a metric termed "network similarity," which represents the correlation between all edges in two connectomes. Examining the network similarity between functional connectomes built using data from different subjects, tasks, and sessions, Gratton et al. found differential weights of these factors (Fig. 1C). First, they noted that connectivity matrices from different subjects exhibited high baseline similarity independent of the task performed during the scan (including resting state). Next, they discovered that connectivity matrices from the same subject dramatically improved network similarity beyond this baseline, and in a session-independent manner. Finally, functional connectivity matrices constructed using data from the same subject and same task further increased network similarity values, indicating an interaction between individual organization and task-specific dynamics. Importantly, Mueller et al. and Finn et al., Gratton et al. found higher group similarity in sensory networks and higher individual similarity in association networks. Collectively, these findings indicate that functional brain organization is dominated by stable individual factors, and not cognitive content.

The ultimate goal of the studies described so far was to characterize individual human brain organization to understand brain-behavior relations in health and brain disorders. Yet, work in humans cannot effectively elucidate the biological principles that underlie individual variation in functional connectivity, such mechanistic work requires the development of an animal model. While brain organization in nonhuman primates is closest to humans, ethical and technical issues limit the amount of data and the number of subjects that can be scanned. On the other hand, rodents, and mice in particular, have brains that are less homologous to humans, but allow for data acquisition in large samples and utilization of genetic and molecular manipulations. Yet, a major challenge in mouse fMRI is the use of anesthesia, which limits the number of acquired sessions per animal and the translation to humans which is scanned in the awake state. To overcome this challenge, a platform for acquisition of fMRI data in passively awake head-fixed mice was developed (Bergmann et al., 2016), demonstrating reliable estimation of the mouse connectome across sessions (Melozzi et al., 2019) and a cross-species homology in brain networks and organization in health (Bergmann et al., 2016) and disease (Shofty et al., 2019). After group-level characterization was established, Bergmann et al. collected multisession data in 19 mice to characterize individual variability (Bergmann et al., 2020). Replicating the analyses from Gratton et al. (Fig. 1C), high network similarity was found across the group, though intraindividual similarity was significantly higher. Then, the authors tested whether the mouse connectome can act as a fingerprint in a group of mice that share the same genetic background (Fig. 1D) and found that while the identification rates are lower than the ones observed by Finn et al., they are still much higher than chance level. Taken together, these analyses establish the mouse as an adequate model for individual variation in brain organization, allowing for future investigations of the effects of genetic and environmental manipulations.

Precision functional mapping of individual brains

Before the era of precision fMRI, group-level fMRI studies characterized the organization of the human brain based on average connectivity profiles, yielding different network and regional parcellations (Yeo et al., 2011; Power et al., 2014; Shen et al., 2013; Glasser et al., 2016; Gordon et al., 2016). Such parcellations were used for construction of connectomes and extraction of global metrics of individual differences as described in the earlier section. However, this approach does not account for topographical differences in region and network organization across individuals. Examination of such features requires a whole-brain data-driven approach that is independent of assumptions regarding brain topography. The gist of such an approach was incorporated in the original work of Mueller et al., which was completely data-driven and provided a spatial output that estimates individual variation in connectivity across the human cortex. However, it only described the variation in functional connectivity across the cortex without exploring regional and network organization at the level of the individual person.

The first characterization of individual topographical brain organization was published in 2015 by Laumann et al. (2015). Using data from MyConnectome project, the authors were able to parcellate an individual brain in a reproducible and internally valid manner, demonstrating correspondence to individual patterns of task-evoked responses. Importantly, they show striking differences between group-level regional connectivity and unique individual patterns (Fig. 2A). While the study of Laumann et al. was transformative, providing evidence for divergent connectivity profiles between an individual and a large group of subjects did not compare brain organization between individuals. Such a comparison was done by Wang et al. (2015) using the longitudinal data from the study of Mueller et al., demonstrating that individual parcellation can capture intersubject variability. Yet, the relatively limited amount of data per subject did not support detailed examination of individual differences. A more detailed characterization of individual differences was performed by Gordon et al. in 2017 using the MSC dataset (Gordon et al., 2017), validating the observations reported by Laumann et al. in a group of 10 different individuals. Examining the amount of data needed for an accurate characterization of brain organization, Gordon et al. found that 30 and 90 min of motion censored data are needed for stable connectivity matrix and areal parcellation, respectively. This observation suggests that characterization of individual brain topography currently requires large amounts of data, far beyond standard fMRI study designs. However, rapid precision mapping may be supported by advanced fMRI sequences such as multiecho fMRI, which enables experimentally and clinically tractable scan times (Lynch et al., 2020). Finally, the study of Gordon et al. focused on individual differences in functional organization of the cerebral cortex. Several studies have now leveraged the MSC dataset to characterize individual differences in functional organization of other brain structures, including the cerebellum (Marek et al., 2018; Xue et al., 2021), striatum (Gordon et al., 2022), thalamus (Greene et al., 2019), and hippocampus (Zheng et al., 2021). Collectively, these studies demonstrate the feasibility of precision functional mapping of individual brain organization.

The aforementioned studies demonstrate the power of precision functional mapping in uncovering individual differences in brain organization. Yet, the MSC dataset, on which several of those studies were based, includes only 10 subjects from the Washington University community aged 24–34 years with above average IQ (range: 117–138) and years of education (17.5–22). The relatively homogenous sample in

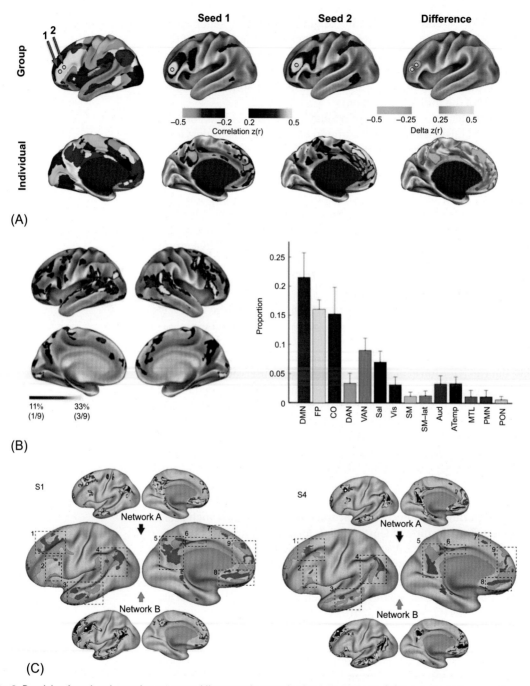

Fig. 2 Precision functional mapping uncovers idiosyncratic organization in individuals. (A) Example of idiosyncratic patterns of functional connectivity in an individual. Two nearby regions of interest *(white spheres)* in the lateral frontal cortex have the same system identity in the group (fronto-parietal) but different system identities in the individual (cingulo-opercular and fronto-parietal). Above, correlation maps from these two regions have very similar

the MSC dataset raises questions regarding the generalizability of the identified individual differences to more diverse populations. To address these questions, Seitzman et al. (2019) compared individual differences between the MSC subjects and 384 subjects from HCP. They defined individual deviations from group functional network organization as network variants and showed that they exist in all tested individuals. Examining the spatial distribution of network variants across the cortex, they found a nonhomogenous pattern with increased prevalence in high-order association networks (Fig. 2B), consistent with the connectome-based studies of individual differences. In addition, they showed that both the MSC and HCP cohort can be clustered to subgroups based on patterns of network variants, and that these subgroups are characterized by small but significant behavioral differences. This observation suggests distribution of network variants may reflect trait-like, functionally relevant individual differences.

In a recent review, Gordon and Nelson further discuss three types of individual variation in functional organization of brain networks (Gordon and Nelson, 2021). The first one is variation in connectivity strength, which represents differences in the magnitude of correlation between two distal regions. A second type is spatial variation, meaning that the same region is displaced or varies in size across individuals.

Fig. 2, Cont'd patterns in the group, with the largest differences occurring locally. Below, the same two regions demonstrate starkly different correlation patterns in the individual, with large regions of cortex showing large differences in correlation. (B) The overlap of network variant locations across individuals is displayed, with *brighter colors* indicating increasing levels of overlap for the MSC dataset (*Left*). Network variants occur commonly in lateral frontal cortex and near the temporo-occipito-parietal junction, and are rarely found in primary sensorimotor areas, the insula, superior parietal lobule, or posterior cingulate cortex. In addition to occurring in characteristic locations, network variants were also typically associated with a characteristic set of networks (*Right*). The mean proportion of variant functional network assignments to 14 canonical networks across individuals. *ATemp*, anterior temporal; *Aud*, auditory; *CO*, cingulo-opercular; *DAN*, dorsal attention; *DMN*, default mode network; *FP*, frontoparietal; *MTL*, medial temporal; *PMN*, medial parietal; *PON*, parieto-occipital; *Sal*, salience; *SM*, dorsal somatomotor; *SM-lat*, ventral somatomotor; *VAN*, ventral attention; *Vis*, visual. (C) fcMRI analysis in individuals reveals two dissociated networks near the canonical default network, Network A and Network B, which are shown for two subjects (S1 and S4) in a schematic form on the same cortical surface representation. The *dashed boxes* highlight nine cortical zones where neighboring representations of the two networks were found, including: (1) dorsolateral PFC, (2) inferior PFC, (3) lateral temporal cortex, (4) inferior parietal lobule extending into the temporoparietal junction, (5) posteromedial cortex, (6) midcingulate cortex, (7) dorsomedial PFC, (8) ventromedial PFC, and (9) anteromedial PFC. Some zones, including the dorsal region along the PFC (labeled 7), are subtle, but consistent, in all subjects, suggesting that there exist small, closely juxtaposed components of the two dissociated networks. (A) Modified with permission from Laumann, T.O., Gordon, E.M., Adeyemo, B., Snyder, A.Z., Joo, S.J., Chen, M.-Y., Gilmore, A.W., McDermott, K.B., Nelson, S.M., Dosenbach, N.U.F., et al., 2015. Functional system and areal organization of a highly sampled individual human brain. Neuron 87, 657–670. https://doi.org/10.1016/j.neuron.2015.06.037. (B) Modified from Seitzman, B.A., Gratton, C., Laumann, T.O., Gordon, E.M., Adeyemo, B., Dworetsky, A., Kraus, B.T., Gilmore, A.W., Berg, J.J., Ortega, M., et al., 2019. Trait-like variants in human functional brain networks. Proc. Natl. Acad. Sci. U. S. A. 116, 22851–22861. https://doi.org/10.1073/PNAS.1902932116/SUPPL_FILE/PNAS.1902932116.SAPP.PDF. (C) From Braga, R.M., Buckner, R.L., 2017. Parallel interdigitated distributed networks within the individual estimated by intrinsic functional connectivity. Neuron 95, 457. https://doi.org/10.1016/J.NEURON.2017.06.038.

A third type is topological variation, which means that the number of nodes per network varies across individuals such that a specific node might not exist in part of the population. Critically, these types of variation may confound the estimation of each other. For example, when using common a parcellation to estimate an individual connectome, we might find an edge that is dramatically different from an average group-level connectome, but the type of variation that contributes to this difference cannot be easily determined. Nevertheless, mapping brain networks in individuals has shown to support fine-grained characterization of network organization that could not be studied using group-averaged data. A prominent example is the organization of the Default Network, which was shown to be comprised of two different subnetworks (Braga et al., 2019; Braga and Buckner, 2017) that demonstrate marked spatial variation across subjects (Fig. 2C). Critically, this finding, which was based on resting-state functional connectivity, has been shown to be behaviorally relevant with the two subnetworks demonstrating preferential activation in different tasks (DiNicola et al., 2020). This observation is in line with previous reports on close agreement between resting-state functional connectivity and task-evoked response both in large cohorts (Tavor et al., 2016) and highly sampled individuals (Gordon et al., 2017; Laumann et al., 2015). Collectively, these studies demonstrate that studying functional brain organization in individuals enables analyses that cannot be achieved using traditional group-average approaches to advance our understanding of brain function in health and disease.

Using fcMRI to study brain-behavior relationship

A main motivation for studying brain organization at the level of the individual is to identify sources for behavioral differences across humans. Understanding how individual variation in brain organization contributes to behavioral or clinical measures is one of the holy grails in modern neuroscience. Historically, brain-behavior relationships were estimated using correlation between MRI-derived measures and cognitive traits or clinical parameters. However, this approach is prone to overfitting and as a result suffers from a generalizability problem (Vul et al., 2009). Nowadays, the gold standard approach is to identify meaningful brain features that can predict behavior using cross-validation, which improves replicability and establishes predictive power. Several algorithms support such predictions, including connectome-based predictive modeling (CPM; Fig. 3A) (Shen et al., 2017), support-vector regression (Smola and Schölkopf, 2004), and canonical correlation analysis (Wang et al., 2020). Yet, even

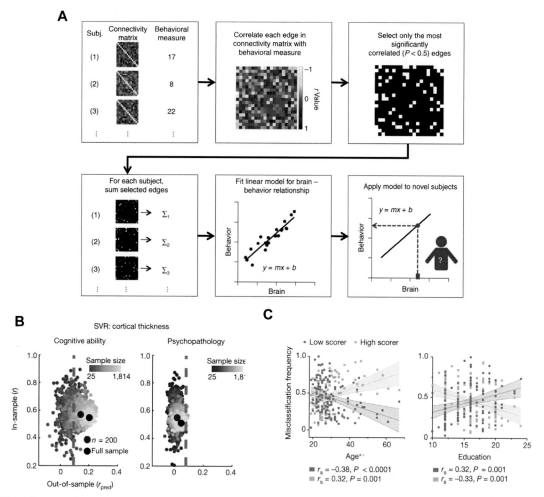

Fig. 3 Current challenges in brain-wide association studies (BWAS). (A) schematic of connectome-based predictive modeling (CPM). For each edge in the connectome, correlation between functional connectivity and a behavioral measure is calculated. Then, the brain-behavior correlation matrix is statistically thresholded. For each subject, connectivity in supra-threshold edges is summed and used as input for a linear regression model. The model is then applied on new subjects and predicted and observed behavioral values are compared. (B) Comparison between in-sample and out-of-sample predictions of cognitive ability and psychopathology using a multivariate model (SVR) as a functional of sample size, demonstrating that cross-sectional BWAS require large sample size. (C) Example of sociodemographic covariates related to predictive model failure in participant that defy sample stereotypes. The graphs show that subjects with underrepresented characteristics are prone to phenotype misclassification. (A) Reproduced with permission from Shen, X., Finn, E.S., Scheinost, D., Rosenberg, M.D., Chun, M.M., Papademetris, X., Constable, R.T., 2017. Using connectome-based predictive modeling to predict individual behavior from brain connectivity. Nat. Protoc. 12, 506–518. https://doi.org/10.1038/nprot.2016.178. (B) From Marek, S., Tervo-Clemmens, B., Calabro, F.J., Montez, D.F., Kay, B.P., Hatoum, A.S., Donohue, M.R., Foran, W., Miller, R.L., Hendrickson, T.J., et al., 2022. Reproducible brain-wide association studies require thousands of individuals. Nature 603. https://doi.org/10.1038/s41586-022-04492-9. (C) From Greene, A.S., Shen, X., Noble, S., Horien, C., Hahn, C.A., Arora, J., Tokoglu, F., Spann, M.N., Carrión, C.I., Barron, D.S., et al., 2022. Brain–phenotype models fail for individuals who defy sample stereotypes. Nature 609 (7925), 109–118. https://doi.org/10.1038/s41586-022-05118-w.

with proper validation, brain-wide association studies (BWAS) are still characterized by limited reproducibility. While part of this problem results from the flexibility in data analysis, software errors, and lack of direct replication (Poldrack et al., 2017; Botvinik-Nezer et al., 2020), a seminal study by Marek et al. has recently showed that a critical factor that contributes to the lack of reproducibility is sample size (Marek et al., 2022). This study demonstrated that reliable brain-behavior associations in cross-sectional studies require thousands of individuals (Fig. 3B), and that even in optimal cases (functional over structural data, cognitive tests over questionnaires, multivariate over univariate models) replicability is moderate. As most neuroimaging studies rely on sample sizes in the range of tens to hundreds, this observation motivates the need for a paradigm shift in the field.

Another important aspect that may limit current studies of brain-behavior relationships in cross-sectional data is the effect of nonstereotypical profiles of sociodemographic and clinical covariates, as well as biased phenotypic measures. Li et al. (2022) compared the performance of predictive models between White Americans and African Americans, demonstrating that the prediction of cognitive abilities and psychopathologies built using data dominated by White Americans fails to generalize to African Americans. Interestingly, while training the model using data from African Americans improved its performance in this population, it also resulted in better predictions in White Americans. These findings highlight the importance of diverse representation in neuroimaging studies to minimize bias in predictive modeling and emphasize the need for investigation of the relations between demographic variables and brain-behavior relations. Further examination of this issue was done by Greene et al. (2022), who showed that nonrandom model failure in multiple phenotypic measures is related to sociodemographic variables such as age, education, and race (Fig. 3C), demonstrating high misclassification frequency in participants who defy the stereotypic sociodemographic profile. This observation, which might explain the limited predictive value reported by Marek et al., suggests important directions to the field, namely data collection from more diverse and underrepresented populations and careful examination of sociodemographic confounds. Accounting for these aspects in study designs should improve generalizability and it will yield meaningful predictions in health and disease.

Given the challenges raised above for studies that aim to link brain and behavior, a recent perspective (Gratton et al., 2022) suggests two paths that the field can take. The first path is to create large consortiums that can collect behavioral and neuroimaging data from thousands of subjects to examine cross-sectional brain-behavior relationships that can inform about population tendency and guide social policy. Yet, as Marek et al. show using the ABCD Study, UK Biobank, and the Human Connectome Project, effect sizes of brain-behavior relationship are small and can explain very little variance of individual behavior. The second path is to increase the measured effect size. This can be achieved by adopting new study designs that improve signal-to-noise ratio such that they allow to characterize brain-behavior relationships in individuals and guide clinical care. A putative approach to improve the measured signal is to increase that variation in both brain and behavioral measures using subject selection criteria—for example, by studying clinical populations for whom better performance of predictive models has been found (Sui et al., 2020). Other options for such optimization include utilization of individual-specific areal parcellations (Kong et al., 2021), incorporation of causal manipulations, and development of theory-driven measures. Alternatively, noise reduction can be pursued in both brain and behavioral measures using extended data acquisition or rigorous control conditions, respectively. Critically, large consortium studies and boutique experiments in small but well-defined populations can be integrated using metamatching (He et al., 2022). This algorithm exploits the fact that different traits within defined behavioral domains are predicted by similar network features (Chen et al., 2022). By translating predictive models for one phenotype in a large group to models that predict related but more specific phenotypes in a small cohort, this approach can dramatically improve prediction accuracy.

A paradigm-shift in experimental design that can improve signal-to-noise ratio is incorporation of within-subject designs. These experiments aim to examine longitudinal variation in a behavioral measure and relate it to variation in brain function while controlling for cross-subject heterogeneity. Such studies were not common until recently, yet they have now been shown to support characterization of individual brain-behavior relationships such as predicting within-subject variation in attentional state following pharmacological manipulation (Rosenberg et al., 2020), utilizing natural stimuli to improve behavioral prediction (Finn and Bandettini, 2021), and characterizing dynamic reorganization of the brain following arm immobilization (Newbold et al., 2020) or along the menstrual cycle (Pritschet et al., 2020). These examples illustrate how focused and well-designed fMRI experiments can overcome issues of statistical power and generate behaviorally relevant information that can both advance our understanding of brain function and guide future clinical decision.

Functional connectivity MRI in clinical populations

While the diagnostic value of functional connectivity measurement in brain disorders is debatable, its ability to guide neuromodulation treatments may support a more accessible path for integration in clinical settings. Recent advances in the field of precision fMRI are highly relevant for developing and improving neuromodulatory treatments. Further, from a neuroscience perspective, treatment of patient populations may also allow for causal modulation of brain dynamics that cannot be performed in healthy volunteers. Hence, these treatments, as well as clinical data from patients with brain lesions, can contribute to basic understanding of brain networks underlying behavior and clinical symptoms.

In the previous section, we described the replication crisis in cognitive neuroscience and the complexity of characterization of cross-sectional brain-behavior associations. These works, as well as more traditional neuroimaging studies in clinical populations, start from a behavioral phenotype and aim to link it to specific features in the brain. However, in recent years, an alternative approach was established, in which cohorts of patients with brain lesions were built, and the connectivity profile of the lesion site was compared to a behavioral variable. This approach, also known as lesion network mapping (LNM), has shown promising results in linking behavioral variables to brain networks (Boes et al., 2015; Fox, 2018; Boes, 2021). The rationale behind this method is that anatomical lesions in different regions that belong to the same large-scale network will manifest in similar symptomology. Therefore, instead of looking at spatial overlap between anatomical lesions that produce similar symptoms, this technique uses fcMRI data from healthy subjects to examine the overlap between their connectivity profiles (Fig. 4A). This approach has uncovered brain networks underlying depression (Padmanabhan et al., 2019), hallucinations (Kim et al., 2021), addiction (Joutsa et al., 2022) and poststroke cognitive and motor outcomes (Bowren et al., 2022). These promising results suggest that LNM is a powerful tool for linking brain pathology to its clinical manifestation as well as for identifying new targets for neuromodulatory interventions. (Poldrack et al., 2017)

In parallel to the rise of LNM, a complementary approach for mapping clinical symptoms to brain network using neuromodulatory treatments was developed. Two of the more common neuromodulatory techniques approved in clinical populations are transcranial magnetic stimulation (TMS) for major depressive disorder (MDD) and deep brain stimulation (DBS) for Parkinson's Disease and obsessive-compulsive disorder. The approach is based on modulation of neural activity in a specific neuroanatomical location connected to a specific brain network. As described before, functional neuroanatomy varies across individuals, and thus brain coordinates defined at the group level do not necessarily target the same functional regions in the individual patient. As fcMRI is an excellent technique for identification of network variants, it is sensitive for these spatial differences, allowing guidance of the stimulation site.

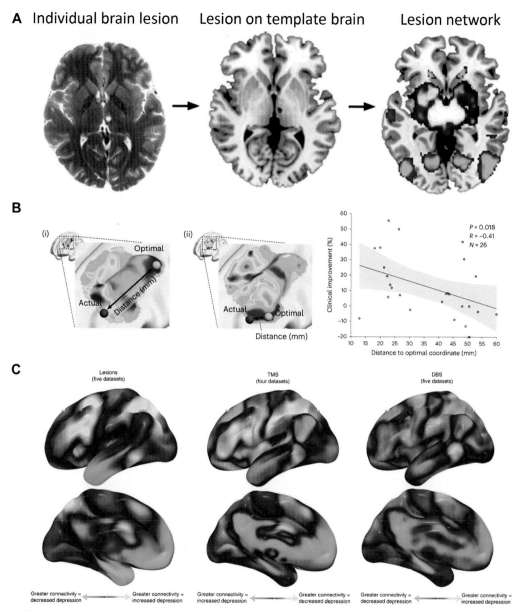

Fig. 4 Clinical samples provide support causal examination of brain-behavior associations. (A) Approach to lesion network mapping. A seed region is defined in the lesioned area and used to calculate functional-connectivity maps in a large dataset of healthy participants. The behavioral impact of different lesions is compared to their connectivity profiles to identify brain-behavior associations. (B) Calculation of optimized DLPFC TMS site for major depression based on negative correlation to the subgenual anterior cingulate cortex. Distance between the actual and optimal stimulation site was related to clinical outcome, demonstrating that treatment efficacy decreased with distance. (C) Comparison between depression circuit maps extracted using lesion, TMS and DBS data, demonstrating striking similarity. The color scale on TMS circuit maps is inverted to facilitate visual comparison with lesion and DBS circuit maps. (A) Reproduced with permission from Boes, A.D., 2021. Lesion network mapping: where do we go from here? Brain 144. https://doi.org/10.1093/brain/awaa350. (B) Reproduced with permission from Cash, R.F.H., Müller, V.I., Fitzgerald, P.B., Eickhoff, S.B., Zalesky, A., 2023. Altered brain activity in unipolar depression unveiled using connectomics. Nat. Mental Health 1, 174–185. (C) Reproduced with permission from Siddiqi, S.H., Schaper, F.L.W.V.J., Horn, A., Hsu, J., Padmanabhan, J.L., Brodtmann, A., Cash, R.F.H., Corbetta, M., Choi, K.S., Dougherty, D.D., et al., 2021. Brain stimulation and brain lesions converge on common causal circuits in neuropsychiatric disease. Nat. Hum. Behav. 5. https://doi.org/10.1038/s41562-021-01161-1.

A prominent example is fcMRI-guided TMS for MDD. TMS-based neuromodulation of the left dorsolateral prefrontal cortex (DLPFC) was approved by the Food and Drug Administration in 2008. However, at early stages, the treatment demonstrated large variability in clinical response. In 2012, Fox et al. (2012) examined whether the differences in response rates are associated with the targeted stimulation site within the DLPFC. Using fcMRI data from healthy subjects and MDD patients, they showed that stimulation sites with strong negative correlation ("anticorrelation") to the subgenual anterior cingulate cortex (sgACC) are characterized by better responses to TMS treatment. While this observation shed light on the neurophysiological mechanisms underlying TMS treatment in MDD and guided target selection at the population level, its relevance to the individual patient was limited. However, it laid the foundation for other studies to explore whether the stimulation site can be personalized using fcMRI data. It took several years, but more recent work has now demonstrated the feasibility of such a personalized approach, showing that connectivity between DLPFC stimulation site and sgACC is correlated with the treatment response in individuals (Cash et al., 2021a) (Fig. 4B) and can be used for identification of personalized optimal stimulation site (Cash et al., 2021b, 2023). Nowadays, fcMRI-based personalization is part of advanced TMS protocols that are showing promising clinical results (Cole et al., 2020, 2022). In parallel, similar findings were observed for DBS treatment for Parkinson's Disease (Horn et al., 2017; Younce et al., 2021) and for intraoperative direct cortical stimulation (Shofty et al., 2022). As neuromodulation is expected to be expanded to other neuropsychiatric disorders, the ability to target a desired network using fcMRI is a critical step for effective treatment (Lynch et al., 2022).

Brain mapping studies that use lesions and stimulation techniques represent a more casual approach to study brain-behavior correlation (Siddiqi et al., 2022). Since both methods emerged in recent years and have identified putative networks underlying the same pathologies (i.e., depression), a fundamental question in the field is to what extent they overlap. A recent study by Siddiqi et al. (2021) examined this by comparing brain lesion and brain stimulation analysis of depression. The results show remarkable spatial overlap (Fig. 4C), suggesting that these approaches are indeed consistent and complementary. The resulting cross-modal convergence of findings from both approaches supports stronger causal inference. Altogether, these techniques demonstrate how the field of precision fMRI can be transformed from basic research to everyday clinical work and vice versa.

Summary

In this chapter, we described the emergence of precision fMRI approaches within the human neuroimaging community. We reviewed the technical and conceptual advancements that support transformation from group-level to person-specific inferences using resting-state fMRI. We first presented connectome-based metrics for estimation of individual variation such as fingerprinting and network similarities. Next, we described methods for precision functional mapping, which characterize spatial differences in brain organization and detect network variants. Then, we discussed the challenges in linking individual variation in brain organization to behavioral differences, emphasizing the limitations of brain-wide association studies and the importance of participants' diversity. To address these limitations, we reviewed some alternative approaches, including consortium studies and within-subject experimental designs, and discussed how they complement each other. Finally, we showed how clinical populations can benefit from precision fcMRI in identification and personalization of neuromodulatory targets, and on the other hand how these populations can inform cognitive neuroscientists on the behavioral relevance of large-scale networks. Collectively, the literature presented here demonstrates how studying individual brains using fMRI has impacted human brain research within a 10-year period by supporting numerous discoveries on brain organization. Critically, while what was achieved so far is already transformative, and it is only the beginning, as the potential and utility of this approach is tremendous from both basic science and clinical perspectives.

References

Bergmann, E., Zur, G., Bershadsky, G., Kahn, I., 2016. The organization of mouse and human cortico-hippocampal networks estimated by intrinsic functional connectivity. Cereb. Cortex 26, 4497–4512. https://doi.org/10.1093/cercor/bhw327.

Bergmann, E., Gofman, X., Kavushansky, A., Kahn, I., 2020. Individual variability in functional connectivity architecture of the mouse brain. Commun. Biol. 3. https://doi.org/10.1038/s42003-020-01472-5.

Boes, A.D., 2021. Lesion network mapping: where do we go from here? Brain 144. https://doi.org/10.1093/brain/awaa350.

Boes, A.D., Prasad, S., Liu, H., Liu, Q., Pascual-Leone, A., Caviness, V.S., Fox, M.D., 2015. Network localization of neurological symptoms from focal brain lesions. Brain 138. https://doi.org/10.1093/brain/awv228.

Botvinik-Nezer, R., Holzmeister, F., Camerer, C.F., Dreber, A., Huber, J., Johannesson, M., Kirchler, M., Iwanir, R., Mumford, J.A., Adcock, R.A., et al., 2020. Variability in the analysis of a single neuroimaging dataset by many teams. Nature 582. https://doi.org/10.1038/s41586-020-2314-9.

Bowren, M., Bruss, J., Manzel, K., Edwards, D., Liu, C., Corbetta, M., Tranel, D., Boes, A.D., 2022. Post-stroke outcomes predicted from multivariate lesion-behaviour and lesion network mapping. Brain 145. https://doi.org/10.1093/brain/awac010.

Braga, R.M., Buckner, R.L., 2017. Parallel interdigitated distributed networks within the individual estimated by intrinsic functional connectivity. Neuron 95, 457. https://doi.org/10.1016/J.NEURON.2017.06.038.

Braga, R.M., van Dijk, K.R.A., Polimeni, J.R., Eldaief, M.C., Buckner, R.L., 2019. Parallel distributed networks resolved at high resolution reveal close juxtaposition of distinct regions. J. Neurophysiol. 121, 1513–1534. https://doi.org/10.1152/JN.00808.2018/ASSET/IMAGES/LARGE/Z9K0041950020014.JPEG.

Cash, R.F.H., Weigand, A., Zalesky, A., Siddiqi, S.H., Downar, J., Fitzgerald, P.B., Fox, M.D., 2021a. Using brain imaging to improve spatial targeting of transcranial magnetic stimulation for depression. Biol. Psychiatry 90. https://doi.org/10.1016/j.biopsych.2020.05.033.

Cash, R.F.H., Cocchi, L., Lv, J., Fitzgerald, P.B., Zalesky, A., 2021b. Functional magnetic resonance imaging-guided personalization of transcranial magnetic stimulation treatment for depression. JAMA Psychiatry 78. https://doi.org/10.1001/jamapsychiatry.2020.3794.

Cash, R.F.H., Müller, V.I., Fitzgerald, P.B., Eickhoff, S.B., Zalesky, A., 2023. Altered brain activity in unipolar depression unveiled using connectomics. Nat. Mental Health 1, 174–185.

Chen, J., Tam, A., Kebets, V., Orban, C., Ooi, L.Q.R., Asplund, C.L., Marek, S., Dosenbach, N.U.F., Eickhoff, S.B., Bzdok, D., 2022. Shared and unique brain network features predict cognitive, personality, and mental health scores in the ABCD study. Nat. Commun. 13, 2217.

Cole, E.J., Stimpson, K.H., Bentzley, B.S., Gulser, M., Cherian, K., Tischler, C., Nejad, R., Pankow, H., Choi, E., Aaron, H., et al., 2020. Stanford accelerated intelligent neuromodulation therapy for treatment-resistant depression. Am. J. Psychiatr. 177. https://doi.org/10.1176/appi.ajp.2019.19070720.

Cole, E.J., Phillips, A.L., Bentzley, B.S., Stimpson, K.H., Nejad, R., Barmak, F., Veerapal, C., Khan, N., Cherian, K., Felber, E., et al., 2022. Stanford neuromodulation therapy (SNT): a double-blind randomized controlled trial. Am. J. Psychiatry 179. https://doi.org/10.1176/appi.ajp.2021.20101429.

DiNicola, L.M., Braga, R.M., Buckner, R.L., 2020. Parallel distributed networks dissociate episodic and social functions within the individual. J. Neurophysiol. 123, 1144–1179. https://doi.org/10.1152/JN.00529.2019.

Finn, E.S., Bandettini, P.A., 2021. Movie-watching outperforms rest for functional connectivity-based prediction of behavior. NeuroImage 235, 117963. https://doi.org/10.1016/J.NEUROIMAGE.2021.117963.

Finn, E.S., Shen, X., Scheinost, D., Rosenberg, M.D., Huang, J., Chun, M.M., Papademetris, X., Constable, R.T., 2015. Functional connectome fingerprinting: identifying individuals using patterns of brain connectivity. Nat. Neurosci. 18, 1664–1671.

Fox, M.D., 2018. Mapping symptoms to brain networks with the human connectome. N. Engl. J. Med. 379. https://doi.org/10.1056/nejmra1706158.

Fox, M.D., Buckner, R.L., White, M.P., Greicius, M.D., Pascual-Leone, A., 2012. Efficacy of transcranial magnetic stimulation targets for depression is related to intrinsic functional connectivity with the subgenual cingulate. Biol. Psychiatry 72, 595–603. https://doi.org/10.1016/j.biopsych.2012.04.028.

Glasser, M.F., Coalson, T.S., Robinson, E.C., Hacker, C.D., Harwell, J., Yacoub, E., Ugurbil, K., Andersson, J., Beckmann, C.F., Jenkinson, M., et al., 2016. A multi-modal parcellation of human cerebral cortex. Nature 536, 171–178. https://doi.org/10.1038/nature18933.

Gordon, E.M., Nelson, S.M., 2021. Three types of individual variation in brain networks revealed by single-subject functional connectivity analyses. Curr. Opin. Behav. Sci. 40, 79–86. https://doi.org/10.1016/J.COBEHA.2021.02.014.

Gordon, E.M., Laumann, T.O., Adeyemo, B., Huckins, J.F., Kelley, W.M., Petersen, S.E., 2016. Generation and evaluation of a cortical area parcellation from resting-state correlations. Cereb. Cortex 26, 288–303. https://doi.org/10.1093/cercor/bhu239.

Gordon, E.M., Laumann, T.O., Gilmore, A.W., Newbold, D.J., Greene, D.J., Berg, J.J., Ortega, M., Hoyt-Drazen, C., Gratton, C., Sun, H., et al., 2017. Precision functional mapping of individual human brains. Neuron 95, 791–807.e7. https://doi.org/10.1016/j.neuron.2017.07.011.

Gordon, E.M., Laumann, T.O., Marek, S., Newbold, D.J., Hampton, J.M., Seider, N.A., Montez, D.F., Nielsen, A.M., Van, A.N., Zheng, A., et al., 2022. Individualized functional subnetworks connect human striatum and frontal cortex. Cereb. Cortex 32, 2868–2884. https://doi.org/10.1093/CERCOR/BHAB387.

Gratton, C., Laumann, T.O., Nielsen, A.N., Greene, D.J., Gordon, E.M., Gilmore, A.W., Nelson, S.M., Coalson, R.S., Snyder, A.Z., Schlaggar, B.L., et al., 2018. Functional brain networks are dominated by stable group and individual factors, not cognitive or daily variation. Neuron 98, 439–452.e5. https://doi.org/10.1016/j.neuron.2018.03.035.

Gratton, C., Nelson, S.M., Gordon, E.M., 2022. Brain-behavior correlations: two paths toward reliability. Neuron 110, 1446–1449. https://doi.org/10.1016/J.NEURON.2022.04.018.

Greene, D.J., Marek, S., Gordon, E.M., Siegel, J.S., Gratton, C., Laumann, T.O., Gilmore, A.W., Berg, J.J., Nguyen, A.L., Dierker, D., et al., 2019. Integrative and network-specific connectivity of the basal ganglia and thalamus defined in individuals. Neuron. https://doi.org/10.1016/j.neuron.2019.11.012.

Greene, A.S., Shen, X., Noble, S., Horien, C., Hahn, C.A., Arora, J., Tokoglu, F., Spann, M.N., Carrión, C.I., Barron, D.S., et al., 2022. Brain–phenotype models fail for individuals who defy sample stereotypes. Nature 609 (7925), 109–118. https://doi.org/10.1038/s41586-022-05118-w.

He, T., An, L., Chen, P., Chen, J., Feng, J., Bzdok, D., Holmes, A.J., Eickhoff, S.B., Yeo, B.T.T., 2022. Meta-matching as a simple framework to translate phenotypic predictive models from big to small data. Nat. Neurosci. 25. https://doi.org/10.1038/s41593-022-01059-9.

Horn, A., Reich, M., Vorwerk, J., Li, N., Wenzel, G., Fang, Q., Schmitz-Hübsch, T., Nickl, R., Kupsch, A., Volkmann, J., et al., 2017. Connectivity predicts deep brain stimulation outcome in Parkinson disease. Ann. Neurol. 82. https://doi.org/10.1002/ana.24974.

Joutsa, J., Moussawi, K., Siddiqi, S.H., Abdolahi, A., Drew, W., Cohen, A.L., Ross, T.J., Deshpande, H.U., Wang, H.Z., Bruss, J., 2022. Brain lesions disrupting addiction map to a common human brain circuit. Nat. Med. 28, 1249–1255.

Kim, N.Y., Hsu, J., Talmasov, D., Joutsa, J., Soussand, L., Wu, O., Rost, N.S., Morenas-Rodríguez, E., Martí-Fàbregas, J., Pascual-Leone, A., et al., 2021. Lesions causing hallucinations localize to one common brain network. Mol. Psychiatry 26. https://doi.org/10.1038/s41380-019-0565-3.

Kong, R., Yang, Q., Gordon, E., Xue, A., Yan, X., Orban, C., Zuo, X.N., Spreng, N., Ge, T., Holmes, A., et al., 2021. Individual-specific areal-level parcellations improve functional connectivity prediction of behavior. Cereb. Cortex 31. https://doi.org/10.1093/cercor/bhab101.

Laumann, T.O., Gordon, E.M., Adeyemo, B., Snyder, A.Z., Joo, S.J., Chen, M.-Y., Gilmore, A.W., McDermott, K.B., Nelson, S.M., Dosenbach, N.U.F., et al., 2015. Functional system and areal organization of a highly sampled individual human brain. Neuron 87, 657–670. https://doi.org/10.1016/j.neuron.2015.06.037.

Li, J., Bzdok, D., Chen, J., Tam, A., Ooi, L.Q.R., Holmes, A.J., Ge, T., Patil, K.R., Jabbi, M., Eickhoff, S.B., et al., 2022. Cross-ethnicity/race generalization failure of behavioral prediction from resting-state functional connectivity. Sci. Adv. 8. https://doi.org/10.1126/sciadv.abj1812.

Lynch, C.J., Power, J.D., Scult, M.A., Dubin, M., Gunning, F.M., Liston, C., 2020. Rapid precision functional mapping of individuals using multi-echo fMRI. Cell Rep. 33, 108540. https://doi.org/10.1016/J.CELREP.2020.108540.

Lynch, C.J., Elbau, I.G., Ng, T.H., Wolk, D., Zhu, S., Ayaz, A., Power, J.D., Zebley, B., Gunning, F.M., Liston, C., 2022. Automated optimization of TMS coil placement for personalized functional network engagement. Neuron 110, 3263–3277.

Marek, S., Siegel, J.S., Gordon, E.M., Raut, R.v., Gratton, C., Newbold, D.J., Ortega, M., Laumann, T.O., Adeyemo, B., Miller, D.B., et al., 2018. Spatial and temporal organization of the individual human cerebellum. Neuron 100, 977–993.e7. https://doi.org/10.1016/J.NEURON.2018.10.010.

Marek, S., Tervo-Clemmens, B., Calabro, F.J., Montez, D.F., Kay, B.P., Hatoum, A.S., Donohue, M.R., Foran, W., Miller, R.L., Hendrickson, T.J., et al., 2022. Reproducible brain-wide association studies require thousands of individuals. Nature 603. https://doi.org/10.1038/s41586-022-04492-9.

Melozzi, F., Bergmann, E., Harris, J.A., Kahn, I., Jirsa, V., Bernard, C., 2019. Individual structural features constrain the mouse functional connectome. Proc. Natl. Acad. Sci. U. S. A. 116, 26961–26969. https://doi.org/10.1073/pnas.1906694116.

Mira-Dominguez, O., Mills, B.D., Carpenter, S.D., Grant, K.A., Kroenke, C.D., Nigg, J.T., Fair, D.A., 2014. Connectotyping: model based fingerprinting of the functional connectome. PLoS One 9. https://doi.org/10.1371/journal.pone.0111048.

Mueller, S., Wang, D., Fox, M.D., Yeo, B.T.T., Sepulcre, J., Sabuncu, M.R., Shafee, R., Lu, J., Liu, H., 2013. Individual variability in functional connectivity architecture of the human brain. Neuron 77, 586–595. https://doi.org/10.1016/j.neuron.2012.12.028.

Newbold, D.J., Laumann, T.O., Hoyt, C.R., Hampton, J.M., Montez, D.F., Raut, R.v., Ortega, M., Mitra, A., Nielsen, A.N., Miller, D.B., et al., 2020. Plasticity and spontaneous activity pulses in disused human brain circuits. Neuron 107, 580–589.e6. https://doi.org/10.1016/J.NEURON.2020.05.007.

Noble, S., Spann, M.N., Tokoglu, F., Shen, X., Todd Constable, R., Scheinost, D., 2017. Influences on the test–retest reliability of functional connectivity MRI and its relationship with behavioral utility. Cereb. Cortex 27, 5415–5429. https://doi.org/10.1093/cercor/bhx230.

Padmanabhan, J.L., Cooke, D., Joutsa, J., Siddiqi, S.H., Ferguson, M., Darby, R.R., Soussand, L., Horn, A., Kim, N.Y., Voss, J.L., et al., 2019. A human depression circuit derived from focal brain lesions. Biol. Psychiatry 86. https://doi.org/10.1016/j.biopsych.2019.07.023.

Poldrack, R.A., Laumann, T.O., Koyejo, O., Gregory, B., Hover, A., Chen, M.Y., Gorgolewski, K.J., Luci, J., Joo, S.J., Boyd, R.L., et al., 2015. Long-term neural and physiological phenotyping of a single human. Nat. Commun. 6. https://doi.org/10.1038/ncomms9885.

Poldrack, R.A., Baker, C.I., Durnez, J., Gorgolewski, K.J., Matthews, P.M., Munafò, M.R., Nichols, T.E., Poline, J.B., Vul, E., Yarkoni, T., 2017. Scanning the horizon: towards transparent and reproducible neuroimaging research. Nat. Rev. Neurosci. 18. https://doi.org/10.1038/nrn.2016.167.

Power, J.D., Schlaggar, B.L., Petersen, S.E., 2014. Studying brain organization via spontaneous fMRI signal. Neuron 84, 681–696. https://doi.org/10.1016/j.neuron.2014.09.007.

Pritchet, L., Santander, T., Taylor, C.M., Layher, E., Yu, S., Miller, M.B., Grafton, S.T., Jacobs, E.G., 2020. Functional reorganization of brain networks across the human menstrual cycle. NeuroImage 220, 117091. https://doi.org/10.1016/J.NEUROIMAGE.2020.117091.

Rosenberg, M.D., Scheinost, D., Greene, A.S., Avery, E.W., Kwon, Y.H., Finn, E.S., Ramani, R., Qiu, M., Todd Constable, R., Chun, M.M., 2020. Functional connectivity predicts changes in attention observed across minutes, days, and months. Proc. Natl. Acad. Sci. U. S. A. 117, 3797–3807. https://doi.org/10.1073/PNAS.1912226117.

Seitzman, B.A., Gratton, C., Laumann, T.O., Gordon, E.M., Adeyemo, B., Dworetsky, A., Kraus, B.T., Gilmore, A.W., Berg, J.J., Ortega, M., et al., 2019. Trait-like variants in human functional brain networks. Proc. Natl. Acad. Sci. U. S. A. 116, 22851–22861. https://doi.org/10.1073/PNAS.1902932116/SUPPL_FILE/PNAS.1902932116. SAPP.PDF.

Shen, X., Tokoglu, F., Papademetris, X., Constable, R.T., 2013. Groupwise whole-brain parcellation from resting-state fMRI data for network node identification. NeuroImage 82, 403–415. https://doi.org/10.1016/j.neuroimage.2013.05.081.

Shen, X., Finn, E.S., Scheinost, D., Rosenberg, M.D., Chun, M.M., Papademetris, X., Constable, R.T., 2017. Using connectome-based predictive modeling to predict individual behavior from brain connectivity. Nat. Protoc. 12, 506–518. https://doi.org/10.1038/nprot.2016.178.

Shofty, B., Bergmann, E., Zur, G., Asleh, J., Bosak, N., Kavushansky, A., Castellanos, F.X., Ben-Sira, L., Packer, R.J., Vezina, G.L., et al., 2019. Autism-associated Nf1 deficiency disrupts corticocortical and corticostriatal functional connectivity in human and mouse. Neurobiol. Dis. 130. https://doi.org/10.1016/j.nbd.2019.104479.

Shofty, B., Gonen, T., Bergmann, E., Mayseless, N., Korn, A., Shamay-Tsoory, S., Grossman, R., Jalon, I., Kahn, I., Ram, Z., 2022. The default network is causally linked to creative thinking. Mol. Psychiatry 27. https://doi.org/10.1038/s41380-021-01403-8.

Siddiqi, S.H., Schaper, F.L.W.V.J., Horn, A., Hsu, J., Padmanabhan, J.L., Brodtmann, A., Cash, R.F.H., Corbetta, M., Choi, K.S., Dougherty, D.D., et al., 2021. Brain stimulation and brain lesions converge on common causal circuits in neuropsychiatric disease. Nat. Hum. Behav. 5. https://doi.org/10.1038/s41562-021-01161-1.

Siddiqi, S.H., Kording, K.P., Parvizi, J., Fox, M.D., 2022. Causal mapping of human brain function. Nat. Rev. Neurosci. 23, 361–375.

Smola, A.J., Schölkopf, B., 2004. A tutorial on support vector regression. Stat. Comput. 14. https://doi.org/10.1023/B:STCO.0000035301.49549.88.

Sui, J., Jiang, R., Bustillo, J., Calhoun, V., 2020. Neuroimaging-based individualized prediction of cognition and behavior for mental disorders and health: methods and promises. Biol. Psychiatry 88. https://doi.org/10.1016/j.biopsych.2020.02.016.

Tavor, I., Parker Jones, O., Mars, R.B., Smith, S.M., Behrens, T.E., Jbabdi, S., 2016. Task-free MRI predicts individual differences in brain activity during task performance. Science 352, 216–220. https://doi.org/10.1126/science.aad8127.

Vul, E., Harris, C., Winkielman, P., Pashler, H., 2009. Puzzlingly high correlations in fMRI studies of emotion, personality, and social cognition1. Perspect. Psychol. Sci. 4. https://doi.org/10.1111/j.1745-6924.2009.01125.x.

Wang, D., Buckner, R.L., Fox, M.D., Holt, D.J., Holmes, A.J., Stoecklein, S., Langs, G., Pan, R., Qian, T., Li, K., et al., 2015. Parcellating cortical functional networks in individuals. Nat. Neurosci. 18 (12), 1853–1860. https://doi.org/10.1038/nn.4164.

Wang, H.T., Smallwood, J., Mourao-Miranda, J., Xia, C.H., Satterthwaite, T.D., Bassett, D.S., Bzdok, D., 2020. Finding the needle in a high-dimensional haystack: canonical correlation analysis for neuroscientists. NeuroImage 216. https://doi.org/10.1016/j.neuroimage.2020.116745.

Xue, A., Kong, R., Yang, Q., Eldaief, M.C., Angeli, P.A., DiNicola, L.M., Braga, R.M., Buckner, R.L., Thomas Yeo, B.T., 2021. The detailed organization of the human cerebellum estimated by intrinsic functional connectivity within the individual. J. Neurophysiol. 125. https://doi.org/10.1152/jn.00561.2020.

Yeo, B.T., Krienen, F.M., Sepulcre, J., Sabuncu, M.R., Lashkari, D., Hollinshead, M., Roffman, J.L., Smoller, J.W., Zollei, L., Polimeni, J.R., et al., 2011. The organization of the human cerebral cortex estimated by intrinsic functional connectivity. J. Neurophysiol. 106, 1125–1165. https://doi.org/10.1152/jn.00338.2011.

Younce, J.R., Campbell, M.C., Hershey, T., Tanenbaum, A.B., Milchenko, M., Ushe, M., Karimi, M., Tabbal, S.D., Kim, A.E., Snyder, A.Z., et al., 2021. Resting-state functional connectivity predicts STN DBS clinical response. Mov. Disord. 36. https://doi.org/10.1002/mds.28376.

Zheng, A., Montez, D.F., Marek, S., Gilmore, A.W., Newbold, D.J., Laumann, T.O., Kay, B.P., Seider, N.A., Van, A.N., Hampton, J.M., et al., 2021. Parallel hippocampal-parietal circuits for self- and goal-oriented processing. Proc. Natl. Acad. Sci. U. S. A. 118. https://doi.org/10.1073/PNAS.2101743118/SUPPL_FILE/PNAS.2101743118.SAPP.PDF. e2101743118.

14

Resting-state fMRI and cerebrovascular reactivity

Peiying Liu[a] and Molly Bright[b,c]

[a]Department of Diagnostic Radiology & Nuclear Medicine, University of Maryland School of Medicine, Baltimore, MD, United States, [b]Department of Physical Therapy and Human Movement Sciences, Feinberg School of Medicine, Northwestern University, Chicago, IL, United States, [c]Department of Biomedical Engineering, McCormick School of Engineering, Northwestern University, Evanston, IL, United States

Introduction

Functional MRI (fMRI) is an important tool to study neural function and networks as it is noninvasive and sensitive to dynamic changes. It has been widely used in clinical research studies (Dickerson and Sperling, 2008; Smith et al., 2013; Calautti and Baron, 2003; Littlejohns et al., 2020). However, an important issue in the interpretation of fMRI data, both in task-evoked and in resting-state functional connectivity-based fMRI, is that the blood-oxygenation-level-dependent (BOLD) signal used in fMRI is based on hemodynamic responses secondary to neurometabolic activations, thus is modulated by vascular function. Several studies have demonstrated that alterations in the vascular function could confound the task-evoked fMRI findings (Bandettini and Wong, 1997; Liu et al., 2013; Handwerker et al., 2007, 2012; Kannurpatti et al., 2011; Riecker et al., 2003; Para et al., 2017). For instance, with aging, cerebral vessels become rigid and lose the ability to dilate upon stimulation, which can have a large effect on the quantification of brain activity using fMRI (Handwerker et al., 2007; Kannurpatti et al., 2011; Riecker et al., 2003). Recent studies have also suggested that vascular effect should also be accounted for in resting-state functional connectivity fMRI (fcMRI) (Murphy et al., 2013; Jahanian et al., 2018; Chang et al., 2008).

The ability of the cerebral blood vessels to dilate or constrict in response to stimulus is called cerebrovascular reactivity (CVR). CVR represents the dynamic property of the cerebral vessels, which is complementary to the steady-state vascular parameters such as cerebral

Advances in Resting-State Functional MRI. https://doi.org/10.1016/B978-0-323-91688-2.00008-4

blood flow (CBF) and cerebral blood volume (CBV). The knowledge of CVR is important for proper interpretation of BOLD fMRI signals.

In the meanwhile, CVR, by itself, has demonstrated clinical utility in a large number of diseases and conditions, including arterial stenosis (Mandell et al., 2008; Gupta et al., 2012; Mikulis et al., 2005; Donahue et al., 2013; De Vis et al., 2015a), stroke (Geranmayeh et al., 2015; Krainik et al., 2005), small vessel disease (Greenberg, 2006; Marstrand et al., 2002), brain tumors (Zaca et al., 2014; Pillai and Zaca, 2011; Fierstra et al., 2016), traumatic brain injury (Chan et al., 2015; Kenney et al., 2016), substance abuse (Han et al., 2008), and normal aging (Lu et al., 2011; Gauthier et al., 2013; De Vis et al., 2015b). Conventional CVR mapping techniques require breathing maneuvers or drug challenges, which could be cumbersome in clinical settings and impractical in some patient cohorts. Recently, a few CVR mapping techniques using resting-state fMRI data have been developed, which may largely broaden the clinical applications of CVR as a sensitive imaging marker of vascular function.

This chapter aims to describe current evidence and literature on how CVR influences resting-state fMRI measurements, and how CVR can be extracted from resting-state fMRI data and utilized in clinical patients.

How CVR influences rs-fMRI measurements?

It is potentially very beneficial that rs-fMRI data contains CVR information; one simple MRI acquisition can therefore provide information about cerebrovascular health that theoretically complements the neural connectivity information typically extracted. However, it is important to consider the negative or challenging ramifications of this relationship: systemic physiologic fluctuations that evoke CVR responses may simultaneously mask, bias, or alter true neural fluctuations in rs-fMRI data. In this section, we briefly review the role of arterial blood gas fluctuations as a confound in rs-fMRI, as well as the relationship between CVR metrics and functional connectivity estimates in typical and pathological circumstances. Finally, we discuss recent observations that CVR may demonstrate network-like properties, and the implications for the rs-fMRI field.

Mitigating the confounding effects of CO_2 fluctuations

Carbon dioxide is a robust and endogenous vasoactive agent, which can be readily modulated during BOLD fMRI using inhaled gas challenges (Lu et al., 2014; Spano et al., 2013; Tancredi et al., 2014;

Wise et al., 2007; Yezhuvath et al., 2009) or changes in ventilation (Geranmayeh et al., 2015; Krainik et al., 2005; Zaca et al., 2014; Chan et al., 2015; Bright and Murphy, 2013; Magon et al., 2009; Tancredi and Hoge, 2013; de Boorder et al., 2004; Bright et al., 2009) to evoke a CVR response. During these experiments, arterial CO_2 is typically estimated noninvasively through measurements of the end-tidal partial pressure of CO_2 ($P_{ET}CO_2$) via nasal cannula or face mask. These recordings can provide quantitative measures of the amplitude of the CO_2 change associated with the experimental paradigm, allowing CVR to be quantified in units of %BOLD/mmHg $P_{ET}CO_2$. However, CO_2 levels also fluctuate in the absence of any intentional experimental paradigm, typically in the range of 0–0.05 Hz (Murphy et al., 2013). In 2004, Wise et al. investigated whether these spontaneous low-frequency $PaCO_2$ fluctuations affected the 3T BOLD-weighted fMRI signal using correlation analysis (Wise et al., 2004). They observed CVR effects of approximately 0.1%BOLD/mmHg in gray matter during such spontaneous low-frequency physiologic fluctuations (with a smaller effect, 0.05%BOLD/mmHg, observed in white matter). This relationship was also heterogenous across gray matter regions, suggesting that the responsiveness of local vasculature to CO_2 may be spatially variable. With spontaneous variability of $\sim$ 2–3 mmHg $P_{ET}CO_2$ in typical rs-fMRI experiments, the associated CVR effects could be large enough to influence our characterization of intrinsic neural fluctuations. Subsequent work demonstrated that up to 15% of rs-fMRI signal variance could be explained by $P_{ET}CO_2$ fluctuations, although this effect shows substantial inter- and intra-subject variability (Golestani et al., 2015).

The immediate implication of this observation is that $P_{ET}CO_2$ became another important "nuisance regressor" for denoising rs-fMRI timeseries. Similar to the method original used by Wise and colleagues, expired CO_2 levels can be continuously recorded during rs-fMRI scanning, and $P_{ET}CO_2$ values can be extracted, convolved with a hemodynamic response function or similar smoothing kernel, and included in the general linear model design matrix as a regressor to model and remove the associated CVR effects (Murphy et al., 2013). Note that although the canonical hemodynamic response function is typically used, subject- and region-specific response functions may be more accurate (Golestani et al., 2015). Frequently, the $P_{ET}CO_2$ recording is temporally shifted to maximally correlate with the fMRI data, accounting for offset between the end-tidal measurements and the vasoactive effects on local brain vasculature. However, there is expected to be variation in this temporal offset, due to vascular transit delays and dynamics of the local vasodilatory response, and accounting for this varying delay at the voxel level is important for accurate modeling of CVR effects (Bright and Murphy, 2013; Moia et al., 2020). There are challenges with how these voxelwise delays are assessed, particularly in rs-fMRI data: there exist numerous sources of low-frequency fluctuations in BOLD rs-fMRI, and cross-correlation or similar approaches to optimize a temporal shift of the $P_{ET}CO_2$ regressor to the rs-fMRI timeseries can inflate statistics through spurious associations (Bright et al., 2017). Statistical inference on this relationship should account for the multiple temporal shifts considered in this analysis (e.g., Sidak correction or similar). It may also be helpful to evaluate these temporal shift estimates in test-retest data from the same subject and session, or reject voxels where the temporal shift was observed at the minimum or maximum values, suggesting an optimal solution was not identified (Bright et al., 2017; Stickland et al., 2021). Although challenging to achieve certainty, these types of methodological checks will improve confidence that $P_{ET}CO_2$ regression is not overfitting and removing neural fluctuations in an overlapping low-frequency domain. The addition of breath-hold task fMRI data, where CVR effects are more pronounced, may improve the robustness of this delay characterization, using either a separate scan (Golestani et al., 2015) or appending breath-hold tasks to the rs-fMRI paradigm (Stickland et al., 2021).

Removal of physiologic effects in rs-fMRI analysis is now an established necessity. CVR effects may be modeled using $P_{ET}CO_2$ recordings; however, many research groups and clinical studies use measures of ventilation (via respiratory belt worn around the diaphragm) as a more readily available surrogate (Caballero-Gaudes and Reynolds, 2017). Although many data-driven approaches exist for removing other types of confounds, like head motion and arterial pulsations, CVR effects are challenging to differentiate from the BOLD-weighted neural signals of interest due to their common hemodynamic mechanisms. Nuisance regression is therefore still the primary means of capturing and removing CVR effects, along with other low-frequency BOLD-weighted respiratory and cardiac phenomenon. Golestani et al. evaluated the effect of such corrections on the quality of rs-fMRI measures in healthy individuals (Golestani et al., 2017). They demonstrated that correction for $P_{ET}CO_2$ (CVR) effects produced the most consistently beneficial effects on the reproducibility of rs-fMRI measures of the physiologic denoising models tested. Furthermore, $P_{ET}CO_2$ correction enhanced the separability of functional connectivity networks, demonstrating that global CVR phenomena artificially enhanced internetwork similarity.

An alternative strategy to nuisance regression is to reduce variability in $P_{ET}CO_2$ during rs-fMRI, and thus reduce the associated CVR confound in the BOLD timeseries. Although this may be possible using paced breathing strategies, this would involve active task performance by the participant, deviating from true "resting-state" behavior. Advanced computer-controlled gas blending and delivery systems can be used to "clamp" arterial blood gases at specific target levels, and thus minimize variability in CO_2 during an rs-fMRI scan without the participant's knowledge or active participation. $P_{ET}CO_2$ clamping has been shown to reduce overall seed-based functional connectivity (reducing global physiologic correlations) and reduce intersubject variability in functional connectivity (Golestani and Chen, 2020).

Although numerous strategies exist to capture and remove CVR effects from rs-fMRI data, it is important to acknowledge that there may be residual effects of $P_{ET}CO_2$ fluctuations that persist and propagate in subsequent analysis of intrinsic neural fluctuations. Although the literature is mixed as to whether these effects are large enough to be problematic in most studies (Bright et al., 2019), $P_{ET}CO_2$ is known to have some degree of impact on neural activity. For example, using magnetoencephalography, low-frequency spontaneous fluctuations in $P_{ET}CO_2$ were shown to influence resting neural activity in multiple-frequency bands (Driver et al., 2016). The denoising strategies discussed above target the removal of the CVR effect, and may not adequately capture concurrent neural effects in the data. $P_{ET}CO_2$ fluctuations may also have long-lasting impact on brain activity and dynamics: studying whole-brain dynamic functional connectivity during intermittent inhaled CO_2 challenges, Lewis and colleagues observed that healthy participants exhibited altered connectivity for at least 1 min after a CO_2 challenge finished (Lewis et al., 2020). It is not clear how such "after-effects" may impact rs-fMRI data where smaller, spontaneous fluctuations in $P_{ET}CO_2$ are present.

CVR-informed rs-fMRI measures

CVR amplitude has frequently been considered as a tool for normalizing BOLD fMRI activations to account for variable hemodynamics and more accurately capture neuronal differences between brain regions, individual subjects, cohorts, and scan sessions (Bandettini and Wong, 1997; Liu et al., 2013; Handwerker et al., 2007, 2012; Kannurpatti et al., 2011; Riecker et al., 2003; Para et al., 2017). Particularly in subjects with vascular pathology, fMRI activation maps can be extremely challenging to interpret without careful consideration of how altered CVR may influence the coupling between neural activity, the hemodynamic response, and the measured BOLD-weighted signal (Para et al., 2017). However, it is less clear how CVR amplitude can be used to correct for functional connectivity estimates in rs-fMRI, which are typically measured through correlation statistics that are not as directly impacted by the amplitude of fluctuations.

CVR timing, however, is increasingly highlighted as a key bias to functional connectivity estimates. For example, Jahanian and colleagues demonstrated that spatially varying hemodynamic delays in Moyamoya disease result in inaccurate and incomplete characterization of functional connectivity through standard independent component analysis (ICA) or seed-correlation approaches (Jahanian et al., 2018). The authors suggested that hundreds of published rs-fMRI studies in patients may be biased because of uncorrected hemodynamic delay effects, and discuss correction strategies.

Chang et al. proposed a correction based on CVR timing in healthy individuals, where hemodynamic delays can still vary by several seconds throughout the cortex (Chang et al., 2008). In their approach, a breathing task was used to evoke a larger whole-brain CVR response, and the relative voxelwise timing differences in this response were measured and used to correct (e.g., shift) rs-fMRI timeseries, resulting in significant connectivity changes within the default-mode network. However, the timing of BOLD fMRI signals associated with neurovascular coupling (e.g., in intrinsic functional connectivity networks) and CVR phenomena likely reflect distinct physiological mechanisms. Furthermore, typical hemodynamic delays reported in the literature vary substantially; pathology may enhance intra-subject variation, breathing tasks may alter the range of delays observed, and parameters of the acquisition and analysis could further influence how delays are estimated (Gong et al., 2023). Further work is needed to determine how best to use CVR timing to improve the accuracy of rs-fcMRI analyses.

Network-specific CVR effects

Finally, there is growing evidence that CVR may influence BOLD timeseries in a network-specific manner. In 2008, Birn et al. observed that ICA of rs-fMRI data could frequently resolve two spatially similar "default-mode" networks, where one network reflected respiration-related changes (Birn et al., 2008). In fact, automatic methods for selecting the default-mode network frequently selected the component thought to be more associated with respiration than neural activity. CVR may therefore exhibit spatial patterns that can strongly resemble canonical *neuronal* brain networks.

This concept was more directly probed using inhaled CO_2 to evoke CVR responses during orthogonal working memory and visual stimuli. In this experiment, pairs of brain networks were identified that were spatially similar to each of the canonical default-mode, task-positive, and visual networks (Bright et al., 2020). Here, one network of the pair exhibited a timeseries reflecting the neural stimuli, and the other network exhibited a timeseries reflecting the CVR phenomenon. This observation supports the idea of network-specific vascular regulation, which may provide efficient vascular support for brain regions that typically function coherently. The relationship between spatially similar network pairs (one appearing more neural and one more vascular in their temporal patterns) was also observed in a dataset without CO_2 inhalation (i.e., where spontaneous fluctuations in $P_{ET}CO_2$

were present). Furthermore, similar network features have been observed in data simulated purely from cardiorespiratory signals in rs-fMRI (Chen et al., 2020), further highlighting the complex relationship of CVR physiology and network features of the human brain.

This relationship may be an exciting avenue for studying neurovascular pathology. For example, cardiovascular risk factors are associated with reduced CVR in the default-mode network but not in visual network regions (Haight et al., 2015), suggesting that pathological vascular changes in specific networks may contribute to network-specific dysfunction in numerous diseases. However, the network-specific regulation of the cerebrovasculature may also be the next important challenge for isolating true neural fluctuations in rs-fMRI.

CVR mapping using resting-state fMRI

Since CVR is a dynamic property of the cerebral blood vessels, CVR mapping is different from other perfusion imaging techniques in that a physiological challenge is usually required during the MRI scan to introduce vasodilation or constrictions. As mentioned earlier, CO_2 gas inhalation is a vasoactive challenge that is typically used in CVR mapping (Lu et al., 2014; Spano et al., 2013; Tancredi et al., 2014; Wise et al., 2007; Yezhuvath et al., 2009). CO_2 gas inhalation increases the blood concentration of CO_2, which, as a potent vasodilator, dilates blood vessels and increases cerebral perfusion (Brian Jr., 1998). CVR can then be quantified by the BOLD signal increases associated with CO_2 inhalation. This CVR mapping method has been successfully applied in a number of research studies in healthy and chronic disease patients (Liu et al., 2013; Mandell et al., 2008; Mikulis et al., 2005; Donahue et al., 2013, 2014; Han et al., 2008; Lu et al., 2011; Thomas et al., 2013). However, the inherent need for gas inhalation and the associated apparatus setup require additional time and expertise for handling and monitoring, which may limit the applications of this technique. This is especially the case when examining acute patients (e.g., acute stoke, acute traumatic brain injury). Other approaches to manipulate blood CO_2 concentration include breath-holding (Geranmayeh et al., 2015; Zaca et al., 2014; Chan et al., 2015; Bright and Murphy, 2013; Magon et al., 2009; Tancredi and Hoge, 2013; de Boorder et al., 2004) and hyperventilation (Krainik et al., 2005; Bright et al., 2009). For both breath-hold and hyperventilation tasks, the requirement of subject's cooperation in performing the breathing tasks also makes it difficult for patients with acute or severe conditions. Subject motion is another potential concern, although methodological advances in the differentiation and mitigation of motion artifacts using multiecho fMRI techniques have shown promise in this area (Moia et al., 2021).

In addition, since breath-hold and hyperventilation could be considered motor tasks, the concomitant motor activation may affect the CVR estimation across the task-activated brain regions.

Recently, resting-state BOLD fMRI data has been shown to have the potential to provide an estimation of CVR maps in addition to the conventional functional connectivity mappings. Instead of manipulating the blood CO_2 concentration by hypercapnia breathing or breath-holding, resting-state CVR mapping approaches utilize spontaneous fluctuations in the breathing pattern, and thus the blood CO_2 level, to extract CVR information from the resting-state data. There are two types of CVR mapping approaches using rs-fMRI data, which are based on the temporal pattern of spontaneous fluctuations in breathing and the degree of signal variation during the scan, respectively.

Temporal pattern-based rs-CVR mapping approaches

Spontaneous fluctuations in breathing pattern during the resting-state fMRI scans, which can be quantified by $P_{ET}CO_2$ and respiration volume per time (RVT), have been shown to be a significant contributor of low-frequency fluctuations in BOLD signal (Wise et al., 2004; Birn et al., 2006; Chang and Glover, 2009), and considered a nuisance signal, which is often discarded in preprocessing of fcMRI analysis (Murphy et al., 2013). A few attempts have been made to use $P_{ET}CO_2$ time course during the resting-state fMRI scan as a regressor in the general linear model (GLM) for CVR mapping (Golestani et al., 2015, 2016; Lipp et al., 2015), where $P_{ET}CO_2$ time course was the independent variable and voxelwise BOLD time course was the dependent variable. Note that this information is also natural byproduct of using $P_{ET}CO_2$ as a nuisance regressor when modeling rs-fMRI timeseries. However, the resulting CVR map only showed a fair reproducibility (intraclass correlation coefficient, ICC = 0.42), which is much lower than that obtained using CO_2 inhalation or breath-holding methods (Golestani et al., 2015; Liu et al., 2021a). One possible reason is that $P_{ET}CO_2$ measured at the end of each breath may not be sensitive and fast enough to capture the actual changes of blood CO_2 concentration during resting state.

More recently, other surrogates of the $P_{ET}CO_2$ time course, which are derived from BOLD signal fluctuations, have been used for resting-state CVR mapping. Although these techniques do not produce *quantitative* CVR maps (%BOLD/mmHg $P_{ET}CO_2$), they may provide more robust access to *relative* CVR maps when $P_{ET}CO_2$ recordings inadequately capture the effects of interest. Liu et al. first demonstrated the proof-of-principle to use the whole-brain averaged gray matter time course as a regressor for CVR mapping in healthy subjects and patients with Moyamoya disease in 2015 (Liu et al., 2015). Fig. 1 shows

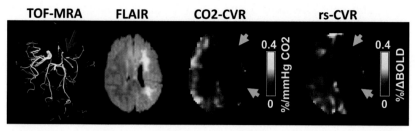

Fig. 1 Example of resting-state CVR map from a patient with Moyamoya disease. Time-of-flight MRA, FLAIR, and CVR map obtained from a CO_2 inhalation scan from the same patient are also shown. *Red arrow* indicates the stenosis of the middle cerebral artery. *Green arrows* point to the deficit regions in the CVR maps.

the resulted CVR maps obtained from patients with Moyamoya disease using both resting-state fMRI and the CO_2 inhalation method. This approach was further optimized by applying a temporal filter of 0.02–0.04 Hz to the BOLD data to improve the correlation between the regressor signal and $P_{ET}CO_2$ time course (Liu et al., 2017), and by applying a temporal filter of 0–0.1164 Hz to the BOLD data to improve the spatial correlation between resting-state CVR maps and the conventional CO_2-CVR maps (Liu et al., 2021b). In both studies, the optimization was conducted in healthy subjects and validated with CO_2 inhalation CVR mapping method in patients with Moyamoya disease (Liu et al., 2017, 2021b). In 2017, Jahanian et al. proposed to use the respiratory variation (RV) timeseries, defined as the standard deviation of the respiration waveform across a 6-s sliding window, as regressors in the GLM analysis for CVR mapping (Jahanian et al., 2017). In healthy elderly subjects, the resulting CVR maps showed a strong correlation with those obtained from the breath-holding method (Jahanian et al., 2017). These approaches can provide relative CVR maps when $P_{ET}CO_2$ is not acquired during the resting-state scan, and therefore, can be applied retrospectively to all resting-state fMRI data that have been collected for functional connectivity assessments.

Despite the efforts to improve and optimize the data analysis of resting-state CVR mapping, in subjects with very small fluctuations in their spontaneous breathing pattern, the CVR results could be noisy or unreliable. In a study of 170 healthy subjects and 50 Moyamoya patients with both resting-state and CO_2 inhalation CVR maps obtained, it was found that 96% of healthy controls and 88% of participants with Moyamoya disease had a fair spatial correlation ($r \geq 0.4$) between the resting-state and CO_2-CVR maps (Liu et al., 2021b). To further improve the sensitivity of resting-state CVR mapping and improve the success rate, additional brief periods of guided-breathing could be added to the resting-state scans to increase the variations of the spontaneous breathing (Bright et al., 2009; Stickland et al., 2021; Liu et al., 2020).

The regression-based resting-state CVR analysis is often performed with the assumption that the "lung-to-brain" delay time is the same for every voxel and every region in the brain. However, it has been demonstrated that white matter manifests a substantially longer delay in response to CO_2 inhalation (by 19 s), compared to the gray matter, even in healthy subjects (Thomas et al., 2014). Even within gray matter voxels, the hemodynamic response delays to breath-holding can still vary by several seconds throughout the cortex (Chang et al., 2008). As mentioned in "How CVR influences rs-fMRI measurements?" section, this CVR timing information, usually referred to as bolus arrival time (BAT), could be a key bias to functional connectivity estimates. Therefore, algorithms have been developed to estimate voxelwise or regional BAP using resting-state fMRI data. The most intuitive approach is to shifting the $P_{ET}CO_2$ time course relative to the BOLD time course of each voxel or each brain region and determining the time shift that provides the best correlation, using either maximum CC or minimal residual (Stickland et al., 2021). When $P_{ET}CO_2$ time course is not obtained, reference time course from superior sagittal sinus (Christen et al., 2015; Tong et al., 2017) or whole-brain gray matter (Amemiya et al., 2014) could also be used for BAT mapping. The resulted BAT maps are found to be consistent with the time to maximum of the residue function (Tmax) maps obtained using gadolinium-based dynamic susceptibility contrast (DSC) MRI (Christen et al., 2015; Amemiya et al., 2014). A recursive tracking approach using the "Regressor Interpolation at Progressive Time Delays (RIPTiDe)" procedure with whole-brain-averaged BOLD time course as the initial regressor has also been reported and shown to improve global noise removal from resting-state functional connectivity maps (Erdoğan et al., 2016).

Degree of variation-based rs-CVR mapping approaches

There have also been studies using the resting-state fluctuation of amplitude (RSFA) and the amplitude of low-frequency fluctuation (ALFF) (Lipp et al., 2015; Kannurpatti et al., 2014; Kannurpatti and Biswal, 2008; Tsvetanov et al., 2015; Di et al., 2013) to approximate relative CVR maps. The resting-state fluctuation amplitude (RSFA) was originally defined as the temporal standard deviation of the BOLD time course (Kannurpatti and Biswal, 2008). A normalized version of RSFA, i.e., coefficient of variance (CV) of the BOLD time course, which is defined as the temporal standard deviation divided by the temporal mean (Liu et al., 2013; Jahanian et al., 2014), has also been used to estimate CVR. RSFA and CV have been reported to have a significant correlation with CO_2-CVR or breath-holding CVR (Liu et al., 2013; Lipp et al., 2015; Kannurpatti et al., 2014; Kannurpatti and Biswal, 2008).

The amplitude of low-frequency fluctuations (ALFF), computed as the square root of the power spectrum across a specific low-frequency band of the resting-state BOLD signal, was also shown to be correlated with CO_2-CVR in healthy subjects (Golestani et al., 2016). A recent study showed that ALFF has a moderate relation with CO_2-induced BOLD signal change (R^2 of 0.6), but this relation decreased in carotid occlusion patients (R^2 of 0.33) (De Vis et al., 2018). It was also found that fractional ALFF (fALFF), calculated as ALFF normalized by the power of the whole spectrum, has a weaker relationship with CO_2-induced BOLD signal change than ALFF (De Vis et al., 2018), suggesting that fALFF contains less vascular contribution than ALFF. Both ALFF and fALFF have been employed as a scaling factor of task-based fMRI analysis to reduce nonneural-related intersubject variability (Kalcher et al., 2013).

A physiological noise metric, calculated by subtracting the variance in signal attributed to thermal noise from the total BOLD signal variance, was also reported to be correlated with age and whiter matter lesion load in patients with small vessel disease (Makedonov et al., 2013). In general, without considering the temporal feature of the BOLD signal fluctuations, these degree of variation-based approaches are thought to be more prone to non-CVR physiological noise sources.

Clinical application of resting-state CVR

Although a recent advance, resting-state CVR mapping has been applied in a number of diseases with vascular conditions and has showed great promise in clinical studies. Using the filtered whole-brain resting-state BOLD signal as a regressor, in a study of 20 stroke patients, resting-state CVR in stroke lesion regions was found to be significantly lower than that in control regions ($p = 0.0002$), and was also lower in the perilesional regions in a graded manner (Taneja et al., 2019). In a subgroup of 6 patients with follow-up resting-state scans, a strong correlation was observed between lesion CVR ($r^2 = 0.91$, $p = 0.006$) as well as between control CVR ($r^2 = 0.79$, $p = 0.036$) measured at two time points (Taneja et al., 2019). In Moyamoya patients whose middle cerebral artery (MCA) flow territories are affected by the intracranial stenosis, using the least-affected cerebellum gray matter BOLD signal as regressor, resting-state CVR in the MCA territory was lower without revascularization surgery than that after surgery ($P = .006$), which was consistent with the CO_2 inhalation CVR data ($P = .003$) (Liu et al., 2021b). In another study of 38 glaucoma patients and 21 healthy subjects, relative CVR in visual cortex derived from resting-state fMRI showed a decreasing trend with glaucoma severity ($p < 0.05$) and was coupled with visual-evoked response and functional connectivity ($p < 0.001$) (Chan et al., 2021).

The BAT maps obtained from resting-state fMRI also have important clinical values. Amemiya et al. demonstrated that the resting-state BAP maps can delineate the extent and degree of perfusion delay in 5 patients with chronic hypoperfusion without neurologic impairment and 6 patients with acute stroke (Amemiya et al., 2014). Christen et al. showed similar results in 10 Moyamoya patients and demonstrated that accounting for these delays may affect the results of functional connectivity maps in these patients (Christen et al., 2015).

The degree of variation-based approaches has also been applied in clinical cohorts. Jahanian et al. demonstrated that CV of the BOLD signals was significantly higher in hypertensive elderly subjects with chronic kidney disease than that in young healthy controls (Jahanian et al., 2014). Two studies have reported abnormal ALFF in patients with leukoaraiosis (Cheng et al., 2017; Wang et al., 2019), and the ALFF in left precuneus was also found to be positively correlated with impairment in executive functions (Wang et al., 2019). However, in a study of 15 patients with internal carotid artery occlusive disease and 15 matched healthy controls, ALFF and fALFF did not differ between patients and controls, even though CO_2 inhalation-induced BOLD signal showed significant group difference ($p < 0.005$) (De Vis et al., 2018). This suggests that more work is needed before these alternative rs-fMRI CVR mapping techniques can be used to inform clinical practice. At present, CVR maps based on temporal patterns appear more robust in clinical studies: a recent study in 25 patients with glioma compared the approach using filtered whole-brain BOLD time course as regressor and RSFA and found that the regressor approach outperformed the RSFA in this clinical cohort by showing higher spatial correlation with the standard breath-holding CVR maps (Yeh et al., 2022).

Conclusion

In this chapter, we reviewed current literature of resting-state fMRI and CVR in the context of how CVR could influence resting-state fMRI measurements and how CVR can be extracted from resting-state fMRI data and utilized in clinical patients. CVR may play an important role in the interpretation of resting-state fMRI data by better differentiating vascular from neural contributions in resting-state BOLD signal. CVR and surrogates of CVR can be derived from resting-state BOLD data in addition to the standard functional connectivity analysis. CVR derived from resting-state fMRI may be a promising biomarker in understanding functional connectivity and in the diagnosis and treatment monitoring of cerebrovascular diseases and conditions.

References

Amemiya, S., et al., 2014. Cerebral hemodynamic impairment: assessment with resting-state functional MR imaging. Radiology 270 (2), 548–555.

Bandettini, P.A., Wong, E.C., 1997. A hypercapnia-based normalization method for improved spatial localization of human brain activation with fMRI. NMR Biomed. 10 (4–5), 197–203.

Birn, R.M., et al., 2006. Separating respiratory-variation-related fluctuations from neuronal-activity-related fluctuations in fMRI. NeuroImage 31 (4), 1536–1548.

Birn, R.M., Murphy, K., Bandettini, P.A., 2008. The effect of respiration variations on independent component analysis results of resting state functional connectivity. Hum. Brain Mapp. 29 (7), 740–750.

Brian Jr., J.E., 1998. Carbon dioxide and the cerebral circulation. Anesthesiology 88 (5), 1365–1386.

Bright, M.G., Murphy, K., 2013. Reliable quantification of BOLD fMRI cerebrovascular reactivity despite poor breath-hold performance. NeuroImage 83, 559–568.

Bright, M.G., et al., 2009. Characterization of regional heterogeneity in cerebrovascular reactivity dynamics using novel hypocapnia task and BOLD fMRI. NeuroImage 48 (1), 166–175.

Bright, M.G., Tench, C.R., Murphy, K., 2017. Potential pitfalls when denoising resting state fMRI data using nuisance regression. NeuroImage 154, 159–168.

Bright, M.G., et al., 2019. Multiparametric measurement of cerebral physiology using calibrated fMRI. NeuroImage 187, 128–144.

Bright, M.G., et al., 2020. Vascular physiology drives functional brain networks. NeuroImage 217, 116007.

Caballero-Gaudes, C., Reynolds, R.C., 2017. Methods for cleaning the BOLD fMRI signal. NeuroImage 154, 128–149.

Calautti, C., Baron, J.-C., 2003. Functional neuroimaging studies of motor recovery after stroke in adults. Stroke 34 (6), 1553–1566.

Chan, S.T., et al., 2015. A case study of magnetic resonance imaging of cerebrovascular reactivity: a powerful imaging marker for mild traumatic brain injury. Brain Inj. 29 (3), 403–407.

Chan, R.W., et al., 2021. Relationships between cerebrovascular reactivity, visual-evoked functional activity, and resting-state functional connectivity in the visual cortex and basal forebrain in glaucoma. In: 2021 43rd Annual International Conference of the IEEE Engineering in Medicine & Biology Society (EMBC).

Chang, C., Glover, G.H., 2009. Relationship between respiration, end-tidal CO2, and BOLD signals in resting-state fMRI. NeuroImage 47 (4), 1381–1393.

Chang, C., Thomason, M.E., Glover, G.H., 2008. Mapping and correction of vascular hemodynamic latency in the BOLD signal. NeuroImage 43 (1), 90–102.

Chen, J.E., et al., 2020. Resting-state "physiological networks". NeuroImage 213, 116707.

Cheng, R., et al., 2017. Abnormal amplitude of low-frequency fluctuations and functional connectivity of resting-state functional magnetic resonance imaging in patients with leukoaraiosis. Brain Behav. 7 (6), e00714.

Christen, T., et al., 2015. Noncontrast mapping of arterial delay and functional connectivity using resting-state functional MRI: a study in Moyamoya patients. J. Magn. Reson. Imaging 41 (2), 424–430.

de Boorder, M.J., Hendrikse, J., van der Grond, J., 2004. Phase-contrast magnetic resonance imaging measurements of cerebral autoregulation with a breath-hold challenge: a feasibility study. Stroke 35 (6), 1350–1354.

De Vis, J.B., et al., 2015a. Calibrated MRI to evaluate cerebral hemodynamics in patients with an internal carotid artery occlusion. J. Cereb. Blood Flow Metab. 35 (6), 1015–1023.

De Vis, J.B., et al., 2015b. Age-related changes in brain hemodynamics; a calibrated MRI study. Hum. Brain Mapp. 36 (10), 3973–3987.

De Vis, J.B., et al., 2018. Effect sizes of BOLD CVR, resting-state signal fluctuations and time delay measures for the assessment of hemodynamic impairment in carotid occlusion patients. NeuroImage 179, 530–539.

Di, X., et al., 2013. Calibrating BOLD fMRI activations with neurovascular and anatomical constraints. Cereb. Cortex 23 (2), 255–263.

Dickerson, B.C., Sperling, R.A., 2008. Functional abnormalities of the medial temporal lobe memory system in mild cognitive impairment and Alzheimer's disease: insights from functional MRI studies. Neuropsychologia 46 (6), 1624–1635.

Donahue, M.J., et al., 2013. Relationships between hypercarbic reactivity, cerebral blood flow, and arterial circulation times in patients with moyamoya disease. J. Magn. Reson. Imaging 38 (5), 1129–1139. https://doi.org/10.1002/jmri.24070 (in press).

Donahue, M.J., et al., 2014. Routine clinical evaluation of cerebrovascular reserve capacity using carbogen in patients with intracranial stenosis. Stroke 45 (8), 2335–2341.

Driver, I.D., et al., 2016. Arterial CO_2 fluctuations modulate neuronal rhythmicity: implications for MEG and fMRI studies of resting-state networks. J. Neurosci. 36 (33), 8541–8550.

Erdoğan, S.B., et al., 2016. Correcting for blood arrival time in global mean regression enhances functional connectivity analysis of resting state fMRI-BOLD signals. Front. Hum. Neurosci. 10.

Fierstra, J., et al., 2016. Altered intraoperative cerebrovascular reactivity in brain areas of high-grade glioma recurrence. Magn. Reson. Imaging 34 (6), 803–808.

Gauthier, C.J., et al., 2013. Age dependence of hemodynamic response characteristics in human functional magnetic resonance imaging. Neurobiol. Aging 34 (5), 1469–1485.

Geranmayeh, F., et al., 2015. Measuring vascular reactivity with breath-holds after stroke: a method to aid interpretation of group-level BOLD signal changes in longitudinal fMRI studies. Hum. Brain Mapp. 36 (5), 1755–1771.

Golestani, A.M., Chen, J.J., 2020. Controlling for the effect of arterial-CO2 fluctuations in resting-state fMRI: comparing end-tidal CO2 clamping and retroactive CO2 correction. NeuroImage 216, 116874.

Golestani, A.M., et al., 2015. Mapping the end-tidal CO2 response function in the resting-state BOLD fMRI signal: spatial specificity, test-retest reliability and effect of fMRI sampling rate. NeuroImage 104, 266–277.

Golestani, A.M., Wei, L.L., Chen, J.J., 2016. Quantitative mapping of cerebrovascular reactivity using resting-state BOLD fMRI: validation in healthy adults. NeuroImage 138, 147–163.

Golestani, A.M., et al., 2017. The effect of low-frequency physiological correction on the reproducibility and specificity of resting-state fMRI metrics: functional connectivity, ALFF, and ReHo. Front. Neurosci. 11.

Gong, J., et al., 2023. Hemodynamic timing in resting-state and breathing-task BOLD fMRI. NeuroImage 274, 120120.

Greenberg, S.M., 2006. Small vessels, big problems. N. Engl. J. Med. 354 (14), 1451–1453.

Gupta, A., et al., 2012. Cerebrovascular reserve and stroke risk in patients with carotid stenosis or occlusion: a systematic review and meta-analysis. Stroke 43 (11), 2884–2891.

Haight, T.J., et al., 2015. Vascular risk factors, cerebrovascular reactivity, and the default-mode brain network. NeuroImage 115, 7–16.

Han, J.S., et al., 2008. BOLD-MRI cerebrovascular reactivity findings in cocaine-induced cerebral vasculitis. Nat. Clin. Pract. Neurol. 4 (11), 628–632.

Handwerker, D.A., et al., 2007. Reducing vascular variability of fMRI data across aging populations using a breathholding task. Hum. Brain Mapp. 28 (9), 846–859.

Handwerker, D.A., et al., 2012. The continuing challenge of understanding and modeling hemodynamic variation in fMRI. NeuroImage 62 (2), 1017–1023.

Jahanian, H., et al., 2014. Spontaneous BOLD signal fluctuations in young healthy subjects and elderly patients with chronic kidney disease. PLoS One 9 (3), e92539.

Jahanian, H., et al., 2017. Measuring vascular reactivity with resting-state blood oxygenation level-dependent (BOLD) signal fluctuations: a potential alternative to the breath-holding challenge? J. Cereb. Blood Flow Metab. 37 (7), 2526–2538.

Jahanian, H., et al., 2018. Erroneous resting-state fMRI connectivity maps due to prolonged arterial arrival time and how to fix them. Brain Connect. 8 (6), 362–370.

Kalcher, K., et al., 2013. RESCALE: voxel-specific task-fMRI scaling using resting state fluctuation amplitude. NeuroImage 70, 80–88.

Kannurpatti, S.S., Biswal, B.B., 2008. Detection and scaling of task-induced fMRI-BOLD response using resting state fluctuations. NeuroImage 40 (4), 1567–1574.

Kannurpatti, S.S., et al., 2011. Increasing measurement accuracy of age-related BOLD signal change: minimizing vascular contributions by resting-state-fluctuation-of-amplitude scaling. Hum. Brain Mapp. 32 (7), 1125–1140.

Kannurpatti, S.S., et al., 2014. Assessment of unconstrained cerebrovascular reactivity marker for large age-range FMRI studies. PLoS One 9 (2), e88751.

Kenney, K., et al., 2016. Cerebral vascular injury in traumatic brain injury. Exp. Neurol. 275 (Pt 3), 353–366.

Krainik, A., et al., 2005. Regional impairment of cerebrovascular reactivity and BOLD signal in adults after stroke. Stroke 36 (6), 1146–1152.

Lewis, N., et al., 2020. Static and dynamic functional connectivity analysis of cerebrovascular reactivity: an fMRI study. Brain Behav. 10 (6), e01516.

Lipp, I., et al., 2015. Agreement and repeatability of vascular reactivity estimates based on a breath-hold task and a resting state scan. NeuroImage 113, 387–396.

Littlejohns, T.J., et al., 2020. The UK Biobank imaging enhancement of 100,000 participants: rationale, data collection, management and future directions. Nat. Commun. 11 (1), 2624.

Liu, P., et al., 2013. A comparison of physiologic modulators of fMRI signals. Hum. Brain Mapp. 34 (9), 2078–2088.

Liu, P., et al., 2015. Can resting state fMRI be used to map cerebrovascular reactivity? In: Proceedings of the International Society for Magnetic Resonance in Medicine.

Liu, P., et al., 2017. Cerebrovascular reactivity mapping without gas challenges. NeuroImage 146, 320–326.

Liu, P., et al., 2020. Cerebrovascular reactivity mapping using intermittent breath modulation. NeuroImage 215, 116787.

Liu, P., et al., 2021a. Multi-vendor and multisite evaluation of cerebrovascular reactivity mapping using hypercapnia challenge. NeuroImage 245, 118754.

Liu, P., et al., 2021b. Cerebrovascular reactivity mapping using resting-state BOLD functional MRI in healthy adults and patients with Moyamoya disease. Radiology 299 (2), 419–425.

Lu, H., et al., 2011. Alterations in cerebral metabolic rate and blood supply across the adult lifespan. Cereb. Cortex 21 (6), 1426–1434.

Lu, H., et al., 2014. MRI mapping of cerebrovascular reactivity via gas inhalation challenges. J. Vis. Exp. 94.

Magon, S., et al., 2009. Reproducibility of BOLD signal change induced by breath holding. NeuroImage 45 (3), 702–712.

Makedonov, I., Black, S.E., Macintosh, B.J., 2013. BOLD fMRI in the white matter as a marker of aging and small vessel disease. PLoS One 8 (7), e67652.

Mandell, D.M., et al., 2008. Mapping cerebrovascular reactivity using blood oxygen level-dependent MRI in patients with arterial steno-occlusive disease: comparison with arterial spin labeling MRI. Stroke 39 (7), 2021–2028.

Marstrand, J.R., et al., 2002. Cerebral perfusion and cerebrovascular reactivity are reduced in white matter hyperintensities. Stroke 33 (4), 972–976.

Mikulis, D.J., et al., 2005. Preoperative and postoperative mapping of cerebrovascular reactivity in moyamoya disease by using blood oxygen level-dependent magnetic resonance imaging. J. Neurosurg. 103 (2), 347–355.

Moia, S., et al., 2020. Voxelwise optimization of hemodynamic lags to improve regional CVR estimates in breath-hold fMRI. In: 2020 42nd Annual International Conference of the IEEE Engineering in Medicine & Biology Society (EMBC).

Moia, S., et al., 2021. ICA-based denoising strategies in breath-hold induced cerebrovascular reactivity mapping with multi echo BOLD fMRI. NeuroImage 233, 117914.

Murphy, K., Birn, R.M., Bandettini, P.A., 2013. Resting-state fMRI confounds and cleanup. NeuroImage 80, 349–359.

Para, A.E., et al., 2017. Invalidation of fMRI experiments secondary to neurovascular uncoupling in patients with cerebrovascular disease. J. Magn. Reson. Imaging 46 (5), 1448–1455.

Pillai, J.J., Zaca, D., 2011. Clinical utility of cerebrovascular reactivity mapping in patients with low grade gliomas. World J. Clin. Oncol. 2 (12), 397–403.

Riecker, A., et al., 2003. Relation between regional functional MRI activation and vascular reactivity to carbon dioxide during normal aging. J. Cereb. Blood Flow Metab. 23 (5), 565–573.

Smith, S.M., et al., 2013. Resting-state fMRI in the human connectome project. NeuroImage 80, 144–168.

Spano, V.R., et al., 2013. CO2 blood oxygen level-dependent MR mapping of cerebrovascular reserve in a clinical population: safety, tolerability, and technical feasibility. Radiology 266 (2), 592–598.

Stickland, R.C., et al., 2021. A practical modification to a resting state fMRI protocol for improved characterization of cerebrovascular function. NeuroImage 239, 118306.

Tancredi, F.B., Hoge, R.D., 2013. Comparison of cerebral vascular reactivity measures obtained using breath-holding and CO2 inhalation. J. Cereb. Blood Flow Metab. 33 (7), 1066–1074.

Tancredi, F.B., Lajoie, I., Hoge, R.D., 2014. A simple breathing circuit allowing precise control of inspiratory gases for experimental respiratory manipulations. BMC Res. Notes 7, 235.

Taneja, K., et al., 2019. Evaluation of cerebrovascular reserve in patients with cerebrovascular diseases using resting-state MRI: a feasibility study. Magn. Reson. Imaging 59, 46–52.

Thomas, B., et al., 2013. Assessment of cerebrovascular reactivity using real-time BOLD fMRI in children with moyamoya disease: a pilot study. Childs Nerv. Syst. 29 (3), 457–463.

Thomas, B.P., et al., 2014. Cerebrovascular reactivity in the brain white matter: magnitude, temporal characteristics, and age effects. J. Cereb. Blood Flow Metab. 34 (2), 242–247.

Tong, Y., et al., 2017. Perfusion information extracted from resting state functional magnetic resonance imaging. J. Cereb. Blood Flow Metab. 37 (2), 564–576.

Tsvetanov, K.A., et al., 2015. The effect of ageing on fMRI: correction for the confounding effects of vascular reactivity evaluated by joint fMRI and MEG in 335 adults. Hum. Brain Mapp. 36 (6), 2248–2269.

Wang, J., et al., 2019. Low-frequency fluctuations amplitude signals exhibit abnormalities of intrinsic brain activities and reflect cognitive impairment in leukoaraiosis patients. Med. Sci. Monit. 25, 5219–5228.

Wise, R.G., et al., 2004. Resting fluctuations in arterial carbon dioxide induce significant low frequency variations in BOLD signal. NeuroImage 21 (4), 1652–1664.

Wise, R.G., et al., 2007. Dynamic forcing of end-tidal carbon dioxide and oxygen applied to functional magnetic resonance imaging. J. Cereb. Blood Flow Metab. 27 (8), 1521–1532.

Yeh, M.Y., et al., 2022. Cerebrovascular reactivity mapping using resting-state functional MRI in patients with gliomas. J. Magn. Reson. Imaging.

Yezhuvath, U.S., et al., 2009. On the assessment of cerebrovascular reactivity using hypercapnia BOLD MRI. NMR Biomed. 22 (7), 779–786.

Zaca, D., et al., 2014. Cerebrovascular reactivity mapping in patients with low grade gliomas undergoing presurgical sensorimotor mapping with BOLD fMRI. J. Magn. Reson. Imaging 40 (2), 383–390.

15

Clinical applications of resting-state fMRI

Rui Duarte Armindo[a] and Greg Zaharchuk[b]

[a]Department of Neuroradiology, Hospital Beatriz Ângelo, Loures—Lisbon, Portugal, [b]Department of Radiology, Stanford University, Stanford, CA, United States

Introduction

The field of neuroimaging has evolved significantly over the last 40 years, in large part due to the introduction of new MR techniques to the neuroscientist's armamentarium. One elusive objective of advanced neuroimaging has been to visualize brain function and understand pathological mechanisms. The development of functional MRI (fMRI) has long been believed to be a promising way to solve these questions. Traditional functional MRI typically involves "task-based" methods, in which the subject is presented a stimulus or specific instructions that elicit brain responses in specific language or sensorimotor areas, for example (Raichle, 2000; Spitzer et al., 1995). The alternative approach of focusing on synchronous activations in different brain regions and consequent modulations in the BOLD signal proposed by resting-state functional MRI (rs-fMRI) promises significant practical advantages. First and foremost, the need for patient compliance is significantly reduced, becoming comparable to that of routine MR imaging (Sair et al., 2017). This opens the possibility of performing fMRI studies in virtually any population, even in clinical settings, particularly in patients who are aphasic, paretic, or sedated (Lee et al., 2016). Additional studies have shown the stability of connectivity during light sleep (Horovitz et al., 2008; Larson-Prior et al., 2009) and for some functional connections even under conscious sedation (Greicius et al., 2008; Wang et al., 2021). The less demanding requirements on patient collaboration also allow the recruitment of more research subjects/patients, leading to more representative samples for group analyses such as comparing brain-network connectivity differences between patients with specific conditions and healthy controls without task-related confounders (Fox, 2010).

Advances in Resting-State Functional MRI. https://doi.org/10.1016/B978-0-323-91688-2.00014-X

Another classically described advantage of rs-fMRI over task-based methods is that by accounting for more than 50% of the total brain BOLD variance, it can facilitate image acquisition with significant signal-to-noise improvements (Fox et al., 2007), given the ability to interpret most of these spontaneous fluctuations and separate them from actual (nonneural) noise.

As discussed in earlier chapters, it was the observation of the residual low-frequency BOLD fMRI signal that led to the demonstration that these spontaneous fluctuations correlated within areas of the somatomotor system (Biswal et al., 1995). Later studies showed that a variety of networks—sets of spatially remote brain areas with correlated intrinsic activity—can be reproducibly mapped with multiple technical approaches. Importantly, multiple neural processes and networks can be studied from a single rs-fMRI data set, giving more information to the researcher/clinician without increasing the study duration.

Previous chapters focused on these resting-state networks (RSNs), but Table 1 in this chapter summarizes those most important for clinical interpretation of rs-fMRI studies. This chapter will focus on the current state of clinical use of rs-fMRI in advanced imaging clinical centers and look at specific applications in which this technique has been shown useful, including presurgical sensorimotor and language mapping, neurodegenerative diseases, neuropsychiatric conditions, and epilepsy.

Current use of rs-fMRI in clinical practice

In 2019, the members of the American Society of Functional Neuroradiology (ASFNR) were invited to participate in an electronic survey concerning their use of rs-fMRI in clinical and research settings (O'Connor and Zeffiro, 2019). The information collected gave readers a good glimpse of the state of clinical applicability of this technique then, even if mostly limited to academic centers in the United States. Among the reported results, it was clear that while most of the respondents had adequate equipment to conduct rs-fMRI, only 40% were planning on using it for clinical purposes the following year. The main concerns identified by the survey were related to the analysis, interpretation, and reproducibility of rs-fMRI data.

To determine how these views have evolved over the last three years, as well as to expand the sample to a broader and more international group of neuroradiologists, we prepared an adapted 13-item electronic survey for members of the largest continental neuroradiology societies, namely the American Society of Neuroradiology (ASNR), the European Society of Neuroradiology (ESNR), and the Asian-Oceanian Society of Neuroradiology and Head & Neck Radiology (AOSNHNR).

Table 1 Main features of commonly defined resting-state networks (RSNs).

Designation	Location	Functions
Default mode network (DMN)	Posterior cingulate, precuneus, medial prefrontal, and lateral temporal cortices; inferior parietal lobule	*Task-negative* network. Cognitive and social processing
Dorsal attention network	Intraparietal sulcus, frontal eye field	Goal-directed behavior. Spatial attention, orientation
Ventral attention network	Temporo-parietal junction and ventral frontal cortices	Detection of environmental stimuli
Fronto-parietal control network	Dorsal prefrontal cortex, inferior parietal lobule	Working memory, control processes for goal-directed behavior
Cingulo-opercular network/salience network	Operculum, insula, anterior cingulate cortex	Vigilance, arousal, and maintained executive control
Somatosensory network	Perirolandic cortex (Brodmann areas 1–4), supplementary motor cortex	Connects primary and supplementary motor and sensory areas
Language network	Superior temporal (Wernicke-like), inferior frontal (Broca-like), adjoining prefrontal, and temporo-parietal cortices	Homotopic connectivity of areas specifically associated with language
Visual network	Occipital cortex	Vision
Auditory network	Superior temporal gyrus, Heschl (transverse temporal) gyrus, insula	Hearing

The survey was additionally distributed through the authors' social media channels and via the American Society of Functional Neuroradiology (ASFNR) and the Glioma MR imaging COST Action (GliMR 2.0) newsletters. The survey responders included 68 individuals, of whom 31% identified as neuroradiology division chief/head of department, 44% as attending/staff radiologists, and 9% as MR researchers, with the rest being mostly junior clinical staff. There was good global representation, with 60% of respondents being based at academic institutions outside the United States.

Responses for five questions regarding general thoughts on the current state of rs-fMRI were collected using a five-point Likert scale and are presented in Fig. 1. Questions concerning the current use of the technique at the respondent's institution are presented as answers to *Yes/No* questions (Fig. 2). In summary, the results of this survey reflect a more diverse population than that of the previously referred ASFNR survey. Only 50% agreed or strongly agreed that rs-fMRI is easy to acquire (vs 82% in the previous study), but the percentage of respondents

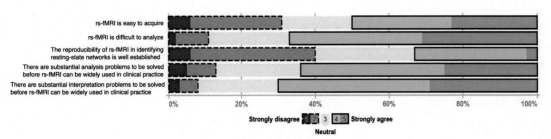

Fig. 1 Survey responses regarding the use of rs-fMRI in 2021–2022, using a five-point likert scale: 1, strongly disagree; 2, disagree; 3, neutral; 4, agree; 5, strongly agree. Interpretation and analysis continue to be major concerns of clinicians when considering use of rs-fMRI. The sample size is 68.

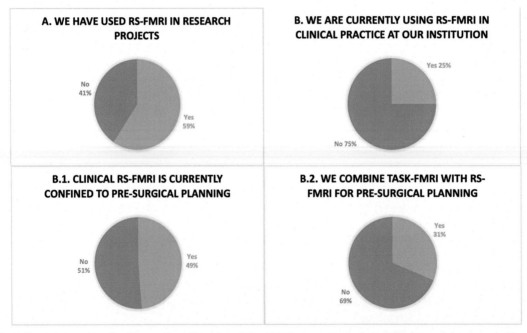

Fig. 2 Responses to the *Yes/No* questions. Questions B.1 and B.2 are applied only to 24% of respondents currently using rs-fMRI in clinical practice (the sample size is 68 responders for A and B and 17 responders for B.1 and B.2).

that reported using rs-fMRI in past research was similar at 59%. Twenty-five percent are currently using rs-fMRI in clinical settings, and this represents a broader use than previously reported, with less than half (49%) confining the clinical applicability of the technique to presurgical planning. Only 11% of the respondents reported performing presurgical planning using rs-fMRI exclusively. Epilepsy, clinical psychology, and perfusion evaluation were mentioned as additional applications currently being used at the respondents' institutions.

There are continued concerns regarding rs-fMRI analysis, with 67% agreeing or strongly agreeing that rs-fMRI is difficult to analyze. Seventy percent of the respondents expressed additional concerns with interpretation issues that should be solved before rs-fMRI is widely used in clinical practice. The reproducibility of rs-fMRI in the identification of resting-state networks is still considered an obstacle by around 41% of the respondents. Most of these challenges/barriers will be briefly addressed in this chapter.

The results of this survey show that there is a clear interest in the field of neuroradiology globally to use rs-fMRI in clinical settings, expanding it from the traditional presurgical planning to other emerging applications, but with continued concerns regarding some of its technical aspects that are still delaying the adoption of the technique in a significant number of institutions. The authors would like to thank all the individuals and organizations that contributed to the distribution of the survey and to the survey respondents for providing the information requested.

Challenges/barriers

The main issues that have traditionally been associated with the reluctance to routinely include rs-fMRI in the clinical evaluation of neurological patients can be separated into (1) those common to all fMRI studies and (2) those specific to rs-fMRI. Most clinical applications require reliable within-individual results, ideally with diagnostic or prognostic information applicable to the subject being studied. The development of fMRI as a clinical tool has been hindered over the years by high levels of reported individual variability (Matthews et al., 2006), inconsistent results across studies (Fox, 2010), and lack of statistical power due to small sample and effect sizes (Turner et al., 2018). These are common problems for most applications of resting-state and task-based fMRI. Potentially related to this is a problem with current knowledge of the theoretical basis of fMRI; the BOLD signal itself is a neurophysiological effect that is still incompletely understood, especially in the context of diverse sets of diseases. Previous studies have noted how neurovascular coupling can be affected in cases of traumatic and hypoxic brain injury (Arngrim et al., 2019; Stevens et al., 2012), and the relationship between BOLD signal and infiltrative lesions, associated mass effect, and edema is still mostly unclear (Roland et al., 2017). In contrast to clinically established modalities such as diffusion-weighted or T1-weighted imaging, fMRI's interpretation requires increased caution, as the BOLD signal represents a fundamentally indirect measurement of neuronal activity. Caution should be taken to exclude other possible mechanisms affecting this physiological response, which are a source of variability in fMRI studies. To mention just a few, localized and/or disease-specific changes in the hemodynamic response function related to underlying differences in cortical thickness, blood volume, and arterial arrival times can all modulate the BOLD signal. For example, how much of the decreased functional connectivity of the default mode network that is observed in Alzheimer's dementia (Greicius et al., 2004) is related to reductions in blood flow, delayed blood flow, and/or to reduced cortical thickness? After controlling for these factors, what becomes of the functional-connectivity decrease? Assessing the role of all of these factors in routine clinical imaging will likely be challenging even once fMRI acquisition methods and analysis strategies eventually become standardized (Specht, 2020).

Specific to rs-fMRI, the most significant barrier to widespread clinical implementation is the lack of availability of standardized software tools for data processing and the high degree of expertise currently needed to perform the analyses (Lee et al., 2016; Leuthardt et al., 2018). While the focus of this chapter is not to detail the individual aspects of each rs-fMRI analysis method, there are some considerations with implications on the clinical interpretation of rs-fMRI. Traditional seed-based analysis relies on *a priori* selection and extraction of the BOLD time course from a region of interest and correlation of this extracted signal with the time course from the other brain voxels (Fox and Raichle, 2007). This approach is heavily influenced by our previous knowledge of functional networks, making it vulnerable to potential bias, and usually limited to one functional system per analysis (Khosla et al., 2019). Additionally, in patients with infarcts or infiltrative lesions, there may be cortical reorganization such that expected seed locations for networks may become unreliable.

Independent component analysis (ICA) is a promising data-driven alternative to ROI-driven seed-based correlation. ICA-based methods use statistically driven blind-source separation to determine mean time-series for several independent networks and then correlate them with the time course of each voxel. The number of independent components to identify is usually the only necessary input, with the analysis then extracting all detectable networks within the subject with high reliability (Zuo et al., 2010). The interpretation of ICA results can be complicated by synchronous activity of nonneural origin, such as cardiac or respiratory cycle-induced artifacts. Also, the selection of the number of independent components could lead to erroneous grouping or subdivision of networks into modules without clinical significance (Lv et al., 2018). It has also been suggested that the number of anatomical correlations may be significantly higher than determined by temporal correlation-based methods due to the relatively slow temporal resolution of fMRI (Kroll et al., 2017).

Understanding and accepting that different analysis methods have different strengths and potential to bias, with varying levels of reliability (Holiga et al., 2018), should lead to caution in the clinical interpretation of rs-fMRI results and efforts to standardize time-series extraction and optimal connectivity metrics. A recent review has shown that there is still a significant degree of variability in the current reporting practices of rs-fMRI acquisition and preprocessing parameters, adding to the complexity of comparing results across different studies (Waheed et al., 2016). Published reports of the use of rs-fMRI in real-world clinical scenarios have nonetheless shown promising results, with study failure rates significantly lower than for task-based fMRI (Leuthardt et al., 2018). Moreover, reproducibility studies show maintained functional-connectivity differences between patient groups despite different approaches in processing and data analysis (Griffanti et al., 2016). The following section focuses on proposed applications of rs-fMRI to clinical neuroimaging, with emphasis on those that have been more extensively validated and considered to have more potential to be used as reliable biomarkers.

rs-fMRI for presurgical planning

The primary application of fMRI in clinical practice is currently preoperative brain mapping for neurosurgical planning (Håberg et al., 2004). This is driven by demonstrations that maximal tumor resection with sparing of "eloquent cortex" in the vicinity of the lesion can lead to improved outcomes and reduced morbidity (Keles et al., 2006; McGirt et al., 2009). This has been confirmed using gold-standard data from intraoperative cortical stimulation (Duffau, 2011; Sanai et al., 2008) and comparison with preoperative task-based fMRI protocols (Petrella et al., 2006). There is some evidence that the use of task-based fMRI for this purpose reduces patient morbidity after surgery (Luna et al., 2021). The advent of rs-fMRI, with the previously mentioned advantages over task-based methods in terms of widening the candidate patient populations to less collaborative, younger, or sedated individuals, was seen as promising for this type of clinical study.

Additionally, while the definition of eloquent cortex is still under debate (Kahn et al., 2017), there is growing interest in mapping higher functions like visuospatial cognition, nonverbal semantic processing, sociocognitive, and executive functions for presurgical planning (Prat-Acín et al., 2021; Vilasboas et al., 2017). The lack of appropriate tasks to evaluate these functions makes it difficult to access their integrity intraoperatively or with task-based fMRI, but some have been shown to correlate with RSNs that can be localized with rs-fMRI (Tordjman et al., 2021).

The earliest studies showing the applicability of rs-fMRI to the preoperative setting focused on localizing the motor system (Kokkonen et al., 2009; Shimony et al., 2009; Zhang et al., 2009). Case reports and small proof-of-concept case series, comparative studies between rs-fMRI, task-fMRI, and intraoperative cortical stimulation showed similar results with good test-retest reliability (Liu et al., 2009; Mannfolk et al., 2011; Rosazza et al., 2014). Another comparative study looking at hand motor-area localization with the three techniques reported slightly higher sensitivity and specificity for rs-fMRI than task-fMRI when compared with direct cortical stimulation (Qiu et al., 2014).

Mapping language function with preoperative MRI comes with the additional challenge of determining language lateralization, which is more problematic with rs-fMRI due to the homotopic connectivity characteristic of RSNs (Park et al., 2020). Task-based fMRI studies have shown similar results to the intracarotid amobarbital procedure ("Wada" test), the gold standard for lateralization evaluation, suggesting that it may be a safe and noninvasive alternative method in most subjects (Binder, 2011; Janecek et al., 2013). Different methods have been developed to compute language lateralization from functional connectivity rs-fMRI data, based on evidence that the major language areas have greater intrahemispheric connectivity in the language dominant side and more interhemispheric connectivity in the contralateral side (Rolinski et al., 2020). Multiple studies compared the results of rs-fMRI laterality determination with task-fMRI and, less frequently, the Wada test. High concordance between these techniques seems obtainable (DeSalvo et al., 2016; Doucet et al., 2015), even in the pediatric population (Desai et al., 2019). Some evidence exists that rs-fMRI can be used to "salvage" cases in which task-based fMRI is inconclusive, as reported by a recent retrospective study that rs-fMRI analysis was beneficial in over 80% of cases in which task-based studies were deemed inadequate (36% of all the cases included) (Kumar et al., 2020). Alternative methods may be needed in cases with lesion-related locoregional functional connectivity disruption (Teghipco et al., 2016).

To localize language areas irrespective of hemispheric dominance, additional advantages of rs-fMRI over task-based fMRI have been suggested, such as the ability to more consistently map Wernicke-like areas, which are sometimes difficult to activate with expressive language tasks. Another advantage is the absence of data contamination due to speech motor and attention functions that are often present with task-based fMRI (Park et al., 2020). Language impairment itself is a frequent cause of task-based fMRI and/or direct cortical stimulation failure, a problem that rs-fMRI can circumvent in a significant number of cases (Kumar et al., 2020).

Around one-third of our survey's respondents stated that clinical rs-fMRI is currently limited to these applications, and most of the published institutional reports of clinical experience with rs-fMRI focus on its integration in neurosurgical planning protocols. In a single-center report of 18 months of experience with rs-fMRI, Leuthardt et al. (2018) reported the results and failure rates of rs-fMRI in 191 patients, giving examples of comparable localization with task-based fMRI, successful rs-fMRI mapping in cases of task-based fMRI failure (including language localization in aphasic patients), and successful rs-fMRI in sedated pediatric patients. In a separate report, the same institution (Roland et al., 2017) shared their experience with presurgical rs-fMRI in 20 pediatric patients, 30% of which cases were performed under sedation, with successful localization of the somatosensory and motor networks in all but two patients. Another possible application of rs-fMRI for surgical planning is as part of an intraoperative MR monitoring protocol, a possibility that has been shown feasible but would benefit from faster processing methods to enhance clinical relevance within the surgical time frame (Roder et al., 2016). It is important to note that no randomized clinical trial on the use of rs-fMRI for surgical planning or patient outcomes has been performed.

Other applications

Additional applications are still more frequently limited to the research setting, but promising results in recent literature warrant a closer look at some of the possibilities for rs-fMRI to contribute to the characterization of diverse clinical populations. In general, two approaches can be identified in the clinical rs-fMRI literature: the assessment of group-level differences and the application of rs-fMRI at the individual level. The latter is necessary to turn rs-fMRI into a clinically useful tool, capable of measuring treatment effects and/or of dividing large heterogenous disease states into smaller groups.

It is currently possible to find literature evaluating differences in specific RSNs between groups of healthy controls and patients in almost all major neurological or psychiatric disease states (Fox, 2010). We will therefore emphasize conditions associated with more consistent results and more advanced analyses, enabling rs-fMRI to become the aforementioned type of clinical tool.

Epilepsy

The application of rs-fMRI to epilepsy is rapidly developing, probably due to the large number of intractable epilepsy patients undergoing presurgical MRI evaluations that include rs-fMRI. As previously discussed, information from RSN mapping can help surgeons to evaluate the need for intraoperative cortical stimulation or intraoperative imaging, and to avoid disrupting eloquent cortex (Lee et al., 2016). But apart from these applications, common to other rs-fMRI studies in the preoperative setting, the use of rs-fMRI to map epileptic foci or networks has been consistently demonstrated, with significantly higher spatial resolution than electroencephalography can provide. First, group-level analyses showed that individuals with temporal lobe epilepsy tend to have a pattern of disconnection of the medial temporal lobe from the rest of the DMN (Pittau et al., 2012; Zhang et al., 2010). Second, promising case series reported the possibility of detecting within-patient focal connectivity increases related with electroencephalography-defined epileptogenic areas (Stufflebeam et al., 2011; Tang et al., 2021). Future developments may lead to the understanding of how epilepsy-associated brain damage affects presurgical networks and how these networks might become altered by surgery (Barnett et al., 2017).

Alzheimer's disease (AD)

The use of rs-fMRI in the context of AD diagnosis and evaluation has long been thought of as the next potential clinical application of this technique. Early rs-fMRI studies in AD patients identified deficient DMN activity when compared with healthy elderly adults, particularly in the hippocampi (Greicius et al., 2004; Li et al., 2002). This has been one of the most reproducible rs-fMRI abnormalities found to date, documentable across the different types of rs-fMRI analysis methods (Supekar et al., 2008). Longitudinal studies have shown that the decrease in functional connectivity progresses with clinical deterioration in both AD and mild cognitive impairment (MCI) patients, further adding to the clinical value of rs-fMRI as a possible biomarker (Bai et al., 2011; Damoiseaux et al., 2012). The next step would be to determine the possibility of identifying patients with MCI or AD from healthy controls, and multiple studies exploring that hypothesis report promising results. Koch et al. (2012) compared rs-fMRI data from 21 healthy subjects, 17 MCI patients, and 15 AD patients and found a diagnostic power of 64% for a single DMN region that could be increased up to 97% with a multivariate approach involving multiple brain regions. A more recent review of the diagnostic power of DMN connectivity in the detection of AD and MCI included the results of more than 30 papers with reported diagnostic accuracies for AD ranging from 63.9% to 97.2%, and concluded that there is great potential to use rs-fMRI evaluation of the DMN as a diagnostic marker for AD, especially when multivariate analysis methods are used (Ibrahim et al., 2021). However, it is important to realize that there are important structural and hemodynamic differences between normal subjects and those with cognitive decline, and whether the rs-fMRI findings are reflective of these differences or fundamental to the disease state remains an open question.

Other neurodegenerative conditions

Studies on the effect of other neurodegenerative diseases on RSN connectivity have mainly reported group-level differences, with altered correlations in different networks when compared with healthy individuals. There have been conflicting reports for rs-fMRI evaluation of patients with amyotrophic lateral sclerosis (ALS), with more studies being needed to clarify the functional underpinning of the disease process (Renga, 2022). Interesting results have been published regarding the behavioral variant of frontotemporal dementia (bvFTD), in which consistently reduced functional connectivity within the salience network has been noted (Filippi et al., 2013; Whitwell et al., 2011). The divergence in findings between FTD and AD, which can show increased salience network connectivity besides the typical DMN abnormalities, seems to allow the distinction of these two entities with more than 90% accuracy (Zhou et al., 2010). Attenuated connectivity in the more recently described and less frequently studied basal ganglia network (BGN) has been shown in early Parkinson's Disease (PD), with dopaminergic therapy resulting in connectivity increase (Szewczyk-Krolikowski et al., 2014); the between-group differences were immune to variations in rs-fMRI processing methods (Griffanti et al., 2016).

Psychiatric diseases

In general, clinical rs-fMRI in psychiatry deals with difficulties inherent to psychiatric patient populations, namely, more common movement during image acquisition, more frequent use of agents modifying the central nervous system, and more use of recreational drugs, confounding the functional imaging findings. Disease-related confounds like sleep alteration and elevated stress levels (Pearlson, 2017) can further obscure findings of functional mechanisms of disease. Additionally, there is a significant heterogeneity in the patient populations, and individual conditions can represent a disease spectrum of disease subtypes with sometimes ill-defined distinction from normality. Furthermore, different disease subtypes have different diagnostic criteria, making it more difficult to find significant differences even in group-level studies (Specht, 2020). A summary of previously reported rs-fMRI findings in neuropsychiatric disorders is presented in Table 2, with this literature reflecting these challenges. Analysis of the rs-fMRI data from the multicenter Bipolar and Schizophrenia Network for Intermediate Phenotypes (BSNIP) study showed

Table 2 Connectivity abnormalities in neuropsychiatric disorders as reported in the literature.

Disease	Findings	References
ADHD	Decreased: DMN, executive control, fronto-parietal, and ventral attention networks; increased: somatosensory and visual networks	Li et al. (2020)
Autism	Decreased connectivity in the dorsal posterior cingulate cortex and the medial paracentral lobule	Lau et al. (2019)
Bipolar disorder	Decreased corticolimbic connectivity. Normal FC when euthymic	Syan et al. (2018) and Vargas et al. (2013)
Depressive disorder	Variable DMN abnormalities	Zhu et al. (2012)
Generalized anxiety	Increased: amygdala, insula, putamen, thalamus, and posterior cingulate cortex; decreased: frontal and temporal cortex	Qiao et al. (2017)
OCD	Increased: corticostriatal connectivity	Harrison et al. (2009), Sakai et al. (2011), and Takagi et al. (2017)
PTSD	Variable with clinical severity and different patterns for specific types of trauma	Santarnecchi et al. (2019)
Schizophrenia	Variable correlations within DMN, generalized disconnection	Garrity et al. (2007) and Pearlson (2017)

significant overlap in altered resting-state connectivity in patients with schizophrenia, psychotic bipolar disorder, and their first-degree relatives (Tamminga et al., 2014). A more recent systematic review including cumulatively 897 bipolar disorder patients during interepisodic periods and 1030 controls found conflicting results in most studies investigating resting-state functional connectivity in this scenario. The authors suggested that a state of remission may stabilize the RSN abnormalities seen in acute mood episodes and that discrepancies can be attributed to variables associated with illness history and psychotropic medication use (Syan et al., 2018).

In the study of rs-fMRI in patients with generalized anxiety disorder, the authors reported significantly different functional brain connectivity patterns between patients and matched healthy controls: patients showed stronger functional connectivity in amygdala, insula, putamen, thalamus, and posterior cingulate cortex but weaker in frontal and temporal cortex compared with controls. These features led to the development of a discriminator with a classification accuracy of 87.5% (Qiao et al., 2017). Another recent study used publicly available rs-fMRI data from children and adolescents with attention-deficit/hyperactivity disorder (ADHD) and controls to identify functional connectivity discriminators of ADHD subjects and neurotypic children. The authors found an overall classification accuracy of 65.6% using rs-fMRI only, and of 79.6% when adding other personal attributes to the classifier. This analysis showed decreased activity in DMN and other executive related networks but increased activity in SMN and VN in ADHD relative to controls (Li et al., 2020). These results contribute to the belief that rs-fMRI is a promising way to study the neural underpinnings of the typically and atypically developing children, though more robust studies and meta-analyses are needed to confirm these findings (Lau et al., 2019).

Altered consciousness, coma, and brain death

Another potential application of rs-fMRI with significant clinical impact would be in aiding the determination of cognitive integrity in different states of consciousness. Consistent results have shown a progressive decrease in DMN connectivity when examining normal consciousness, locked-in states to minimally conscious patients and vegetative comatose states (Vanhaudenhuyse et al., 2010). The absence or severe impairment of long-distance brain connectivity in brain death (Boly et al., 2009) is another relevant finding that can potentially give rs-fMRI a role in outcome prediction and therapeutic management in disorders of consciousness (Boerwinkle et al., 2019; Wagner et al., 2020). Nonetheless, the potential for bias in rs-fMRI findings by baseline hemodynamics is probably quite significant, raising the question of whether rs-fMRI adds value beyond traditional assessments that evaluate the presence of intracranial blood flow, such as nuclear medicine methods or arterial-spin labeling MRI.

Stroke, perfusion, and vascular reactivity

The contribution of rs-fMRI to the evaluation of stroke patients can be divided in two important types of application: in acute-stroke management and in follow-up studies of stroke patients. The fact that rs-fMRI assesses the temporal coherence of the BOLD effect in different brain regions allows the extraction of brain perfusion metrics such as the mean transit time (MTT) (Tong and Frederick, 2010). Multiple studies have shown that rs-fMRI-based perfusion metrics correlate well with maps derived from contrast-based perfusion MRI for both stroke and Moyamoya disease (Amemiya et al., 2014; Christen et al., 2015; Lv et al., 2013). Notably, in rs-fMRI, correlations of the times series of individual voxels with either the global signal or large draining veins can be maximized by time-shifting the two curves relative to one another; the time shift at which this maximum is potentially observed is related to the arterial arrival time delay of the voxel. This arrival delay is closely related to the hemodynamic response function and has traditionally been measured using contrast-enhanced perfusion techniques. This time delay can indicate hypoperfusion and be useful in patient selection for endovascular therapy, for example. Thus, the use of a delayed rs-fMRI BOLD signal to identify severely hypoperfused tissue in acute ischemic stroke (penumbra or at-risk tissue) could make it an alternative to contrast-based perfusion methods (Khalil et al., 2017). This could be of value when there are contraindications to contrast media administration, with the additional advantage of giving the clinician more information than traditional perfusion imaging, including microvascular perfusion and oxygenation (Kroll et al., 2017). The biggest clinical challenge to these types of measurements is the increased head motion often seen in acute stroke patients and the limitations on acquisition times, though adequate maps for perfusion metrics can be obtained using scans around 3 min in duration (Christen et al., 2015).

Another important metric for stroke imaging is cerebrovascular reactivity (CVR), which is defined as the ability of blood vessels to enlarge in response to a vasodilatory challenge; it essentially reflects the vasculature's response to a "stress test." Interestingly, spontaneous BOLD fluctuations seem to reflect, in part, changes in the partial pressure of end-tidal CO_2, which have been shown to allow the determination of CVR maps from rs-fMRI data. While typical CVR maps require the addition of external factors such as a pharmacologic vasodilator (such as acetazolamide) or gas inhalation or breath-holding challenges, significant correlations between rs-fMRI amplitude of fluctuations and traditional methods have been demonstrated (Jahanian et al., 2017). In this study, it was shown that rs-fMRI measures of CVR could be obtained even in elderly patients who could not comply with the requirements for a breath-holding challenge. It was further shown that relative CVR could be mapped noninvasively with rs-fMRI even in patients with conditions associated with cerebral blood flow disturbance, such as Moyamoya disease (Liu et al., 2021). Finally, quantitative CVR mapping using nothing more than rs-fMRI has been demonstrated (Golestani et al., 2016; Prokopiou et al., 2019), although its clinical utility, either in the acute or the subacute stage, remains to be demonstrated.

Adding rs-fMRI to the follow-up imaging protocols of stroke patients could help explain symptoms that structural MR imaging does not, such as the observation of the relatively weak link between infarct size and functional outcomes. rs-fMRI may additionally provide prognostic information and help to understand the pathophysiology of stroke-related white matter tract injury (Corbetta, 2012; Corbetta et al., 2015; Rehme et al., 2011). Attention to persistent changes in the hemodynamic response function, particularly delayed perfusion, may be critical to properly interpret the presence, absence, and strength of network connections (Jahanian et al., 2018).

Summary

As described in this chapter, rs-fMRI is already being used for multiple clinical applications, though it is clearly still early days and there is much still to be understood about the best practices for this new and powerful technique. Studies on measurement reproducibility and patient outcomes should be prioritized to help us understand where rs-fMRI studies add value. There is little doubt that the number of clinical applications of rs-fMRI will continue to growly quickly with the development of new analysis techniques.

References

Amemiya, S., Kunimatsu, A., Saito, N., Ohtomo, K., 2014. Cerebral hemodynamic impairment: assessment with resting-state functional MR imaging. Radiology 270, 548–555. https://doi.org/10.1148/radiol.13130982.

Arngrim, N., Hougaard, A., Schytz, H.W., Vestergaard, M.B., Britze, J., Amin, F.M., Olsen, K.S., Larsson, H.B., Olesen, J., Ashina, M., 2019. Effect of hypoxia on BOLD fMRI response and total cerebral blood flow in migraine with aura patients. J. Cereb. Blood Flow Metab. 39, 680–689. https://doi.org/10.1177/0271678X17719430.

Bai, F., Watson, D.R., Shi, Y., Wang, Y., Yue, C., Teng, Y., Wu, D., Yuan, Y., Zhang, Z., 2011. Specifically progressive deficits of brain functional marker in amnestic type mild cognitive impairment. PloS One 6, e24271. https://doi.org/10.1371/journal.pone.0024271.

Barnett, A., Audrain, S., McAndrews, M.P., 2017. Applications of resting-state functional MR imaging to epilepsy. Neuroimaging Clin. N. Am. 27, 697–708. https://doi.org/10.1016/j.nic.2017.06.002.

Binder, J.R., 2011. Functional MRI is a valid noninvasive alternative to Wada testing. Epilepsy Behav. 20, 214–222. https://doi.org/10.1016/j.yebeh.2010.08.004.

Biswal, B., Zerrin Yetkin, F., Haughton, V.M., Hyde, J.S., 1995. Functional connectivity in the motor cortex of resting human brain using echo-planar MRI. Magn. Reson. Med. 34, 537–541. https://doi.org/10.1002/mrm.1910340409.

Boerwinkle, V.L., Torrisi, S.J., Foldes, S.T., Marku, I., Ranjan, M., Wilfong, A.A., Adelson, P.D., 2019. Resting-state fMRI in disorders of consciousness to facilitate early therapeutic intervention. Neurol Clin Pract 9, e33–e35. https://doi.org/10.1212/CPJ.0000000000000596.

Boly, M., Tshibanda, L., Vanhaudenhuyse, A., Noirhomme, Q., Schnakers, C., Ledoux, D., Boveroux, P., Garweg, C., Lambermont, B., Phillips, C., Luxen, A., Moonen, G., Bassetti, C., Maquet, P., Laureys, S., 2009. Functional connectivity in the default network during resting state is preserved in a vegetative but not in a brain dead patient. Hum. Brain Mapp. 30, 2393–2400. https://doi.org/10.1002/hbm.20672.

Christen, T., Jahanian, H., Ni, W.W., Qiu, D., Moseley, M.E., Zaharchuk, G., 2015. Noncontrast mapping of arterial delay and functional connectivity using resting-state functional MRI: a study in Moyamoya patients. J. Magn. Reson. Imaging 41, 424–430. https://doi.org/10.1002/jmri.24558.

Corbetta, M., 2012. Functional connectivity and neurological recovery. Dev. Psychobiol. 54, 239–253. https://doi.org/10.1002/dev.20507.

Corbetta, M., Ramsey, L., Callejas, A., Baldassarre, A., Hacker, C.D., Siegel, J.S., Astafiev, S.V., Rengachary, J., Zinn, K., Lang, C.E., Connor, L.T., Fucetola, R., Strube, M., Carter, A.R., Shulman, G.L., 2015. Common behavioral clusters and subcortical anatomy in stroke. Neuron 85, 927–941. https://doi.org/10.1016/j.neuron.2015.02.027.

Damoiseaux, J.S., Prater, K.E., Miller, B.L., Greicius, M.D., 2012. Functional connectivity tracks clinical deterioration in Alzheimer's disease. Neurobiol. Aging 33, 828.e19–828.e30. https://doi.org/10.1016/j.neurobiolaging.2011.06.024.

Desai, V.R., Vedantam, A., Lam, S.K., Mirea, L., Foldes, S.T., Curry, D.J., Adelson, P.D., Wilfong, A.A., Boerwinkle, V.L., 2019. Language lateralization with resting-state and task-based functional MRI in pediatric epilepsy. J. Neurosurg. Pediatr. 23, 171–177. https://doi.org/10.3171/2018.7.PEDS18162.

DeSalvo, M.N., Tanaka, N., Douw, L., Leveroni, C.L., Buchbinder, B.R., Greve, D.N., Stufflebeam, S.M., 2016. Resting-state functional MR imaging for determining language laterality in intractable epilepsy. Radiology 281, 264–269. https://doi.org/10.1148/radiol.2016141010.

Doucet, G.E., Pustina, D., Skidmore, C., Sharan, A., Sperling, M.R., Tracy, J.I., 2015. Resting-state functional connectivity predicts the strength of hemispheric lateralization for language processing in temporal lobe epilepsy and normals. Hum. Brain Mapp. 36, 288–303. https://doi.org/10.1002/hbm.22628.

Duffau, H., 2011. The necessity of preserving brain functions in glioma surgery: the crucial role of intraoperative awake mapping. World Neurosurg. 76, 525–527. https://doi.org/10.1016/j.wneu.2011.07.040.

Filippi, M., Agosta, F., Scola, E., Canu, E., Magnani, G., Marcone, A., Valsasina, P., Caso, F., Copetti, M., Comi, G., Cappa, S.F., Falini, A., 2013. Functional network connectivity in the behavioral variant of frontotemporal dementia. Cortex 49, 2389–2401. https://doi.org/10.1016/j.cortex.2012.09.017.

Fox, M.D., 2010. Clinical applications of resting state functional connectivity. Front Syst Neurosci. https://doi.org/10.3389/fnsys.2010.00019

Fox, M.D., Raichle, M.E., 2007. Spontaneous fluctuations in brain activity observed with functional magnetic resonance imaging. Nat. Rev. Neurosci. 8, 700–711. https://doi.org/10.1038/nrn2201.

Fox, M.D., Snyder, A.Z., Vincent, J.L., Raichle, M.E., 2007. Intrinsic fluctuations within cortical systems account for intertrial variability in human behavior. Neuron 56, 171–184. https://doi.org/10.1016/j.neuron.2007.08.023.

Garrity, A.G., Pearlson, G.D., McKiernan, K., Lloyd, D., Kiehl, K.A., Calhoun, V.D., 2007. Aberrant "default mode" functional connectivity in schizophrenia. Am. J. Psychiatry 164, 450–457. https://doi.org/10.1176/ajp.2007.164.3.450.

Golestani, A.M., Wei, L.L., Chen, J.J., 2016. Quantitative mapping of cerebrovascular reactivity using resting-state BOLD fMRI: validation in healthy adults. Neuroimage 138, 147–163. https://doi.org/10.1016/j.neuroimage.2016.05.025.

Greicius, M.D., Kiviniemi, V., Tervonen, O., Vainionpää, V., Alahuhta, S., Reiss, A.L., Menon, V., 2008. Persistent default-mode network connectivity during light sedation. Hum. Brain Mapp. 29, 839–847. https://doi.org/10.1002/hbm.20537.

Greicius, M.D., Srivastava, G., Reiss, A.L., Menon, V., 2004. Default-mode network activity distinguishes Alzheimer's disease from healthy aging: evidence from functional MRI. Proc. Natl. Acad. Sci. U.S.A. 101, 4637–4642. https://doi.org/https://doi.org/10.1073/pnas.0308627101.

Griffanti, L., Rolinski, M., Szewczyk-Krolikowski, K., Menke, R.A., Filippini, N., Zamboni, G., Jenkinson, M., Hu, M.T.M., Mackay, C.E., 2016. Challenges in the reproducibility of clinical studies with resting state fMRI: an example in early Parkinson's disease. Neuroimage 124, 704–713. https://doi.org/10.1016/j.neuroimage.2015.09.021.

Håberg, A., Kvistad, K.A., Unsgård, G., Haraldseth, O., 2004. Preoperative blood oxygen level-dependent functional magnetic resonance imaging in patients with primary brain tumors: clinical application and outcome. Neurosurgery 54, 902–915. https://doi.org/10.1227/01.NEU.0000114510.05922.F8.

Harrison, B.J., Soriano-Mas, C., Pujol, J., Ortiz, H., López-Solà, M., Hernández-Ribas, R., Deus, J., Alonso, P., Yücel, M., Pantelis, C., Menchon, J.M., Cardoner, N., 2009. Altered corticostriatal functional connectivity in obsessive-compulsive disorder. Arch. Gen. Psychiatry 66, 1189. https://doi.org/10.1001/archgenpsychiatry.2009.152.

Holiga, Š., Sambataro, F., Luzy, C., Greig, G., Sarkar, N., Renken, R.J., Marsman, J.-B.C., Schobel, S.A., Bertolino, A., Dukart, J., 2018. Test-retest reliability of task-based and resting-state blood oxygen level dependence and cerebral blood flow measures. PloS One 13, e0206583. https://doi.org/10.1371/journal.pone.0206583.

Horovitz, S.G., Fukunaga, M., de Zwart, J.A., van Gelderen, P., Fulton, S.C., Balkin, T.J., Duyn, J.H., 2008. Low frequency BOLD fluctuations during resting wakefulness and light sleep: a simultaneous EEG-fMRI study. Hum. Brain Mapp. 29, 671–682. https://doi.org/10.1002/hbm.20428.

Ibrahim, B., Suppiah, S., Ibrahim, N., Mohamad, M., Hassan, H.A., Nasser, N.S., Saripan, M.I., 2021. Diagnostic power of resting-state fMRI for detection of network connectivity in Alzheimer's disease and mild cognitive impairment: a systematic review. Hum. Brain Mapp. 42, 2941–2968. https://doi.org/10.1002/hbm.25369.

Jahanian, H., Christen, T., Moseley, M.E., Pajewski, N.M., Wright, C.B., Tamura, M.K., Zaharchuk, G., 2017. Measuring vascular reactivity with resting-state blood oxygenation level-dependent (BOLD) signal fluctuations: a potential alternative to the breath-holding challenge? J. Cereb. Blood Flow Metab. 37, 2526–2538. https://doi.org/10.1177/0271678X16670921.

Jahanian, H., Christen, T., Moseley, M.E., Zaharchuk, G., 2018. Erroneous resting-state fMRI connectivity maps due to prolonged arterial arrival time and how to fix them. Brain Connect. 8, 362–370. https://doi.org/10.1089/brain.2018.0610.

Janecek, J.K., Swanson, S.J., Sabsevitz, D.S., Hammeke, T.A., Raghavan, M., Rozman, E., Binder, J.R., 2013. Language lateralization by fMRI and Wada testing in 229 patients with epilepsy: rates and predictors of discordance. Epilepsia 54, 314–322. https://doi.org/10.1111/epi.12068.

Kahn, E., Lane, M., Sagher, O., 2017. Eloquent: history of a word's adoption into the neurosurgical lexicon. J. Neurosurg. 127, 1461–1466. https://doi.org/10.3171/2017.3.JNS17659.

Keles, G.E., Chang, E.F., Lamborn, K.R., Tihan, T., Chang, C.-J., Chang, S.M., Berger, M.S., 2006. Volumetric extent of resection and residual contrast enhancement on initial surgery as predictors of outcome in adult patients with hemispheric anaplastic astrocytoma. J. Neurosurg. 105, 34–40. https://doi.org/10.3171/jns.2006.105.1.34.

Khalil, A.A., Ostwaldt, A.-C., Nierhaus, T., Ganeshan, R., Audebert, H.J., Villringer, K., Villringer, A., Fiebach, J.B., 2017. Relationship between changes in the temporal dynamics of the blood-oxygen-level-dependent signal and hypoperfusion in acute ischemic stroke. Stroke 48, 925–931. https://doi.org/10.1161/STROKEAHA.116.015566.

Khosla, M., Jamison, K., Ngo, G.H., Kuceyeski, A., Sabuncu, M.R., 2019. Machine learning in resting-state fMRI analysis. Magn. Reson. Imaging 64, 101–121. https://doi.org/10.1016/j.mri.2019.05.031.

Koch, W., Teipel, S., Mueller, S., Benninghoff, J., Wagner, M., Bokde, A.L.W., Hampel, H., Coates, U., Reiser, M., Meindl, T., 2012. Diagnostic power of default mode network resting state fMRI in the detection of Alzheimer's disease. Neurobiol. Aging 33, 466–478. https://doi.org/10.1016/j.neurobiolaging.2010.04.013.

Kokkonen, S.-M., Nikkinen, J., Remes, J., Kantola, J., Starck, T., Haapea, M., Tuominen, J., Tervonen, O., Kiviniemi, V., 2009. Preoperative localization of the sensorimotor area using independent component analysis of resting-state fMRI. Magn. Reson. Imaging 27, 733–740. https://doi.org/10.1016/j.mri.2008.11.002.

Kroll, H., Zaharchuk, G., Christen, T., Heit, J.J., Iv, M., 2017. Resting-state BOLD MRI for perfusion and ischemia. Top. Magn. Reson. Imaging 26, 91–96. https://doi.org/10.1097/RMR.0000000000000119.

Kumar, V.A., Heiba, I.M., Prabhu, S.S., Chen, M.M., Colen, R.R., Young, A.L., Johnson, J.M., Hou, P., Noll, K., Ferguson, S.D., Rao, G., Lang, F.F., Schomer, D.F., Liu, H.-L., 2020. The role of resting-state functional MRI for clinical preoperative language mapping. Cancer Imaging 20, 47. https://doi.org/10.1186/s40644-020-00327-w.

Larson-Prior, L.J., Zempel, J.M., Nolan, T.S., Prior, F.W., Snyder, A.Z., Raichle, M.E., 2009. Cortical network functional connectivity in the descent to sleep. Proc. Natl. Acad. Sci. U.S.A. 106, 4489–4494. https://doi.org/10.1073/pnas.0900924106.

Lau, W.K.W., Leung, M.-K., Lau, B.W.M., 2019. Resting-state abnormalities in autism spectrum disorders: a meta-analysis. Sci. Rep. 9, 3892. https://doi.org/10.1038/s41598-019-40427-7.

Lee, M.H., Miller-Thomas, M.M., Benzinger, T.L., Marcus, D.S., Hacker, C.D., Leuthardt, E.C., Shimony, J.S., 2016. Clinical resting-state fMRI in the preoperative setting. Top. Magn. Reson. Imaging 25, 11–18. https://doi.org/10.1097/RMR.0000000000000075.

Leuthardt, E.C., Guzman, G., Bandt, S.K., Hacker, C., Vellimana, A.K., Limbrick, D., Milchenko, M., Lamontagne, P., Speidel, B., Roland, J., Miller-Thomas, M., Snyder, A.Z., Marcus, D., Shimony, J., Benzinger, T.L.S., 2018. Integration of resting state functional MRI into clinical practice—a large single institution experience. PloS One 13, e0198349. https://doi.org/10.1371/journal.pone.0198349.

Li, J., Joshi, A.A., Leahy, R.M., 2020. A network-based approach to study of Adhd using tensor decomposition of resting state Fmri data, in: 2020 IEEE 17th International Symposium on Biomedical Imaging (ISBI). IEEE, pp. 1–5. https://doi.org/10.1109/ISBI45749.2020.9098584.

Li, S.-J., Li, Z., Wu, G., Zhang, M.-J., Franczak, M., Antuono, P.G., 2002. Alzheimer disease: evaluation of a functional MR imaging index as a marker. Radiology 225, 253–259. https://doi.org/10.1148/radiol.2251011301.

Liu, H., Buckner, R.L., Talukdar, T., Tanaka, N., Madsen, J.R., Stufflebeam, S.M., 2009. Task-free presurgical mapping using functional magnetic resonance imaging intrinsic activity. J. Neurosurg. 111, 746–754. https://doi.org/10.3171/2008.10.JNS08846.

Liu, P., Liu, G., Pinho, M.C., Lin, Z., Thomas, B.P., Rundle, M., Park, D.C., Huang, J., Welch, B.G., Lu, H., 2021. Cerebrovascular reactivity mapping using resting-state BOLD functional MRI in healthy adults and patients with Moyamoya disease. Radiology 299, 419–425. https://doi.org/10.1148/radiol.2021203568.

Luna, L.P., Sherbaf, F.G., Sair, H.I., Mukherjee, D., Oliveira, I.B., Köhler, C.A., 2021. Can preoperative mapping with functional MRI reduce morbidity in brain tumor resection? A systematic review and meta-analysis of 68 observational studies. Radiology 300, 338–349. https://doi.org/10.1148/radiol.2021204723.

Lv, Y., Margulies, D.S., Cameron Craddock, R., Long, X., Winter, B., Gierhake, D., Endres, M., Villringer, K., Fiebach, J., Villringer, A., 2013. Identifying the perfusion deficit in acute stroke with resting-state functional magnetic resonance imaging. Ann. Neurol. 73, 136–140. https://doi.org/10.1002/ana.23763.

Lv, H., Wang, Z., Tong, E., Williams, L.M., Zaharchuk, G., Zeineh, M., Goldstein-Piekarski, A.N., Ball, T.M., Liao, C., Wintermark, M., 2018. Resting-state functional MRI: everything that nonexperts have always wanted to know. Am. J. Neuroradiol. https://doi.org/10.3174/ajnr.A5527.

Mannfolk, P., Nilsson, M., Hansson, H., Ståhlberg, F., Fransson, P., Weibull, A., Svensson, J., Wirestam, R., Olsrud, J., 2011. Can resting-state functional MRI serve as a complement to task-based mapping of sensorimotor function? A test-retest reliability study in healthy volunteers. J. Magn. Reson. Imaging 34, 511–517. https://doi.org/10.1002/jmri.22654.

Matthews, P.M., Honey, G.D., Bullmore, E.T., 2006. Applications of fMRI in translational medicine and clinical practice. Nat. Rev. Neurosci. 7, 732–744. https://doi.org/10.1038/nrn1929.

McGirt, M.J., Mukherjee, D., Chaichana, K.L., Than, K.D., Weingart, J.D., Quinones-Hinojosa, A., 2009. Association of surgically acquired motor and language deficits on overall survival after resection of glioblastoma multiforme. Neurosurgery 65, 463–470. https://doi.org/10.1227/01.NEU.0000349763.42238.E9.

O'Connor, E.E., Zeffiro, T.A., 2019. Why is clinical fMRI in a resting state? Front. Neurol. 10. https://doi.org/10.3389/fneur.2019.00420.

Park, K.Y., Lee, J.J., Dierker, D., Marple, L.M., Hacker, C.D., Roland, J.L., Marcus, D.S., Milchenko, M., Miller-Thomas, M.M., Benzinger, T.L., Shimony, J.S., Snyder, A.Z., Leuthardt, E.C., 2020. Mapping language function with task-based vs. resting-state functional MRI. PloS One 15, e0236423. https://doi.org/10.1371/journal.pone.0236423.

Pearlson, G.D., 2017. Applications of resting state functional MR imaging to neuropsychiatric diseases. Neuroimaging Clin. N. Am. 27, 709–723. https://doi.org/10.1016/j.nic.2017.06.005.

Petrella, J.R., Shah, L.M., Harris, K.M., Friedman, A.H., George, T.M., Sampson, J.H., Pekala, J.S., Voyvodic, J.T., 2006. Preoperative functional MR imaging localization of language and motor areas: effect on therapeutic decision making in patients with potentially resectable brain tumors. Radiology 240, 793–802. https://doi.org/10.1148/radiol.2403051153.

Pittau, F., Grova, C., Moeller, F., Dubeau, F., Gotman, J., 2012. Patterns of altered functional connectivity in mesial temporal lobe epilepsy. Epilepsia 53, 1013–1023. https://doi.org/10.1111/j.1528-1167.2012.03464.x.

Prat-Acín, R., Galeano-Senabre, I., López-Ruiz, P., Ayuso-Sacido, A., Espert-Tortajada, R., 2021. Intraoperative brain mapping of language, cognitive functions, and social cognition in awake surgery of low-grade gliomas located in the right non-dominant hemisphere. Clin. Neurol. Neurosurg. 200, 106363. https://doi.org/10.1016/j.clineuro.2020.106363.

Prokopiou, P.C., Pattinson, K.T.S., Wise, R.G., Mitsis, G.D., 2019. Modeling of dynamic cerebrovascular reactivity to spontaneous and externally induced CO_2 fluctuations in the human brain using BOLD-fMRI. Neuroimage 186, 533–548. https://doi.org/10.1016/j.neuroimage.2018.10.084.

Qiao, J., Li, A., Cao, C., Wang, Z., Sun, J., Xu, G., 2017. Aberrant functional network connectivity as a biomarker of generalized anxiety disorder. Front. Hum. Neurosci. 11. https://doi.org/10.3389/fnhum.2017.00626.

Qiu, T., Yan, C., Tang, W., Wu, J., Zhuang, D., Yao, C., Lu, J., Zhu, F., Mao, Y., Zhou, L., 2014. Localizing hand motor area using resting-state fMRI: validated with direct cortical stimulation. Acta Neurochir. 156, 2295–2302. https://doi.org/10.1007/s00701-014-2236-0.

Raichle, M.E., 2000. A brief history of human functional brain mapping. In: Brain Mapping: The Systems. Elsevier, pp. 33–75. https://doi.org/10.1016/B978-012692545-6/50004-0.

Rehme, A.K., Fink, G.R., von Cramon, D.Y., Grefkes, C., 2011. The role of the contralesional motor cortex for motor recovery in the early days after stroke assessed with longitudinal FMRI. Cereb. Cortex 21, 756–768. https://doi.org/10.1093/cercor/bhq140.

Renga, V., 2022. Brain connectivity and network analysis in amyotrophic lateral sclerosis. Neurol. Res. Int. 2022, 1–20. https://doi.org/10.1155/2022/1838682.

Roder, C., Charyasz-Leks, E., Breitkopf, M., Decker, K., Ernemann, U., Klose, U., Tatagiba, M., Bisdas, S., 2016. Resting-state functional MRI in an intraoperative MRI setting: proof of feasibility and correlation to clinical outcome of patients. J. Neurosurg. 125, 401–409. https://doi.org/10.3171/2015.7.JNS15617.

Roland, J.L., Griffin, N., Hacker, C.D., Vellimana, A.K., Akbari, S.H., Shimony, J.S., Smyth, M.D., Leuthardt, E.C., Limbrick, D.D., 2017. Resting-state functional magnetic resonance imaging for surgical planning in pediatric patients: a preliminary experience. J. Neurosurg. Pediatr. 20, 583–590. https://doi.org/10.3171/2017.6.PEDS1711.

Rolinski, R., You, X., Gonzalez-Castillo, J., Norato, G., Reynolds, R.C., Inati, S.K., Theodore, W.H., 2020. Language lateralization from task-based and resting state functional MRI in patients with epilepsy. Hum. Brain Mapp. 41, 3133–3146. https://doi.org/10.1002/hbm.25003.

Rosazza, C., Aquino, D., D'Incerti, L., Cordella, R., Andronache, A., Zacà, D., Bruzzone, M.G., Tringali, G., Minati, L., 2014. Preoperative mapping of the sensorimotor cortex: comparative assessment of task-based and resting-state fMRI. PLoS One 9, e98860. https://doi.org/10.1371/journal.pone.0098860.

Sair, H.I., Agarwal, S., Pillai, J.J., 2017. Application of resting state functional MR imaging to presurgical mapping. Neuroimaging Clin. N. Am. 27, 635–644. https://doi.org/10.1016/j.nic.2017.06.003.

Sakai, Y., Narumoto, J., Nishida, S., Nakamae, T., Yamada, K., Nishimura, T., Fukui, K., 2011. Corticostriatal functional connectivity in non-medicated patients with obsessive-compulsive disorder. Eur. Psychiatry 26, 463–469. https://doi.org/10.1016/j.eurpsy.2010.09.005.

Sanai, N., Mirzadeh, Z., Berger, M.S., 2008. Functional outcome after language mapping for glioma resection. N. Engl. J. Med. 358, 18–27. https://doi.org/10.1056/NEJMoa067819.

Santarnecchi, E., Bossini, L., Vatti, G., Fagiolini, A., la Porta, P., di Lorenzo, G., Siracusano, A., Rossi, S., Rossi, A., 2019. Psychological and brain connectivity changes following trauma-focused CBT and EMDR treatment in single-episode PTSD patients. Front. Psychol. 10. https://doi.org/10.3389/fpsyg.2019.00129.

Shimony, J.S., Zhang, D., Johnston, J.M., Fox, M.D., Roy, A., Leuthardt, E.C., 2009. Resting-state spontaneous fluctuations in brain activity. Acad. Radiol. 16, 578–583. https://doi.org/10.1016/j.acra.2009.02.001.

Specht, K., 2020. Current challenges in translational and clinical fMRI and future directions. Front. Psych. 10. https://doi.org/10.3389/fpsyt.2019.00924.

Spitzer, M., Kwong, K.K., Kennedy, W., Rosen, B.R., Belliveau, J.W., 1995. Category-specific brain activation in fMRI during picture naming. Neuroreport 6, 2109–2112. https://doi.org/10.1097/00001756-199511000-00003.

Stevens, M.C., Lovejoy, D., Kim, J., Oakes, H., Kureshi, I., Witt, S.T., 2012. Multiple resting state network functional connectivity abnormalities in mild traumatic brain injury. Brain Imaging Behav. 6, 293–318. https://doi.org/10.1007/s11682-012-9157-4.

Stufflebeam, S.M., Liu, H., Sepulcre, J., Tanaka, N., Buckner, R.L., Madsen, J.R., 2011. Localization of focal epileptic discharges using functional connectivity magnetic resonance imaging. J. Neurosurg. 114, 1693–1697. https://doi.org/10.3171/2011.1.JNS10482.

Supekar, K., Menon, V., Rubin, D., Musen, M., Greicius, M.D., 2008. Network analysis of intrinsic functional brain connectivity in Alzheimer's disease. PLoS Comput. Biol. 4, e1000100. https://doi.org/10.1371/journal.pcbi.1000100.

Syan, S.K., Smith, M., Frey, B.N., Remtulla, R., Kapczinski, F., Hall, G.B.C., Minuzzi, L., 2018. Resting-state functional connectivity in individuals with bipolar disorder during clinical remission: a systematic review. J. Psychiatry Neurosci. 43, 298–316. https://doi.org/10.1503/jpn.170175.

Szewczyk-Krolikowski, K., Menke, R.A.L., Rolinski, M., Duff, E., Salimi-Khorshidi, G., Filippini, N., Zamboni, G., Hu, M.T.M., Mackay, C.E., 2014. Functional connectivity in the basal ganglia network differentiates PD patients from controls. Neurology 83, 208–214. https://doi.org/10.1212/WNL.0000000000000592.

Takagi, Y., Sakai, Y., Lisi, G., Yahata, N., Abe, Y., Nishida, S., Nakamae, T., Morimoto, J., Kawato, M., Narumoto, J., Tanaka, S.C., 2017. A neural marker of obsessive-compulsive disorder from whole-brain functional connectivity. Sci. Rep. 7, 7538. https://doi.org/10.1038/s41598-017-07792-7.

Tamminga, C.A., Pearlson, G., Keshavan, M., Sweeney, J., Clementz, B., Thaker, G., 2014. Bipolar and schizophrenia network for intermediate phenotypes: outcomes across the psychosis continuum. Schizophr. Bull. 40, S131–S137. https://doi.org/10.1093/schbul/sbt179.

Tang, Y., Choi, J.Y., Alexopoulos, A., Murakami, H., Daifu-Kobayashi, M., Zhou, Q., Najm, I., Jones, S.E., Wang, Z.I., 2021. Individual localization value of resting-state fMRI in epilepsy presurgical evaluation: a combined study with stereo-EEG. Clin. Neurophysiol. 132, 3197–3206. https://doi.org/10.1016/j.clinph.2021.07.028.

Teghipco, A., Hussain, A., Tivarus, M.E., 2016. Disrupted functional connectivity affects resting state based language lateralization. Neuroimage Clin 12, 910–927. https://doi.org/10.1016/j.nicl.2016.10.015.

Tong, Y., Frederick, B.D., 2010. Time lag dependent multimodal processing of concurrent fMRI and near-infrared spectroscopy (NIRS) data suggests a global circulatory origin for low-frequency oscillation signals in human brain. Neuroimage 53, 553–564. https://doi.org/10.1016/j.neuroimage.2010.06.049.

Tordjman, M., Madelin, G., Gupta, P.K., Cordova, C., Kurz, S.C., Orringer, D., Golfinos, J., Kondziolka, D., Ge, Y., Wang, R.L., Lazar, M., Jain, R., 2021. Functional connectivity of the default mode, dorsal attention and fronto-parietal executive control networks in glial tumor patients. J. Neurooncol 152, 347–355. https://doi.org/10.1007/s11060-021-03706-w.

Turner, B.O., Paul, E.J., Miller, M.B., Barbey, A.K., 2018. Small sample sizes reduce the replicability of task-based fMRI studies. Commun Biol 1, 62. https://doi.org/10.1038/s42003-018-0073-z.

Vanhaudenhuyse, A., Noirhomme, Q., Tshibanda, L.J.-F., Bruno, M.-A., Boveroux, P., Schnakers, C., Soddu, A., Perlbarg, V., Ledoux, D., Brichant, J.-F., Moonen, G., Maquet, P., Greicius, M.D., Laureys, S., Boly, M., 2010. Default network connectivity reflects the level of consciousness in non-communicative brain-damaged patients. Brain 133, 161–171. https://doi.org/10.1093/brain/awp313.

Vargas, C., López-Jaramillo, C., Vieta, E., 2013. A systematic literature review of resting state network—functional MRI in bipolar disorder. J. Affect. Disord. 150, 727–735. https://doi.org/10.1016/j.jad.2013.05.083.

Vilasboas, T., Herbet, G., Duffau, H., 2017. Challenging the myth of right nondominant hemisphere: lessons from corticosubcortical stimulation mapping in awake surgery and surgical implications. World Neurosurg. 103, 449–456. https://doi.org/10.1016/j.wneu.2017.04.021.

Wagner, F., Hänggi, M., Weck, A., Pastore-Wapp, M., Wiest, R., Kiefer, C., 2020. Outcome prediction with resting-state functional connectivity after cardiac arrest. Sci. Rep. 10, 11695. https://doi.org/10.1038/s41598-020-68683-y.

Waheed, S.H., Mirbagheri, S., Agarwal, S., Kamali, A., Yahyavi-Firouz-Abadi, N., Chaudhry, A., DiGianvittorio, M., Gujar, S.K., Pillai, J.J., Sair, H.I., 2016. Reporting of resting-state functional magnetic resonance imaging preprocessing methodologies. Brain Connect. 6, 663–668. https://doi.org/10.1089/brain.2016.0446.

Wang, J., Xu, Y., Deshpande, G., Li, K., Sun, P., Liang, P., 2021. The effect of light sedation with midazolam on functional connectivity of the dorsal attention network. Brain Sci. 11, 1107. https://doi.org/10.3390/brainsci11081107.

Whitwell, J.L., Josephs, K.A., Avula, R., Tosakulwong, N., Weigand, S.D., Senjem, M.L., Vemuri, P., Jones, D.T., Gunter, J.L., Baker, M., Wszolek, Z.K., Knopman, D.S., Rademakers, R., Petersen, R.C., Boeve, B.F., Jack, C.R., 2011. Altered functional connectivity in asymptomatic MAPT subjects: a comparison to bvFTD. Neurology 77, 866–874. https://doi.org/10.1212/WNL.0b013e31822c61f2.

Zhang, D., Johnston, J.M., Fox, M.D., Leuthardt, E.C., Grubb, R.L., Chicoine, M.R., Smyth, M.D., Snyder, A.Z., Raichle, M.E., Shimony, J.S., 2009. Preoperative sensorimotor mapping in brain tumor patients using spontaneous fluctuations in neuronal activity imaged with functional magnetic resonance imaging. Operat. Neurosurg. 65. https://doi.org/10.1227/01.NEU.0000350868.95634.CA. ons226–ons236.

Zhang, Z., Lu, G., Zhong, Y., Tan, Q., Liao, W., Wang, Z., Wang, Z., Li, K., Chen, H., Liu, Y., 2010. Altered spontaneous neuronal activity of the default-mode network in mesial temporal lobe epilepsy. Brain Res. 1323, 152–160. https://doi.org/10.1016/j.brainres.2010.01.042.

Zhou, J., Greicius, M.D., Gennatas, E.D., Growdon, M.E., Jang, J.Y., Rabinovici, G.D., Kramer, J.H., Weiner, M., Miller, B.L., Seeley, W.W., 2010. Divergent network connectivity changes in behavioural variant frontotemporal dementia and Alzheimer's disease. Brain 133, 1352–1367. https://doi.org/10.1093/brain/awq075.

Zhu, X., Wang, X., Xiao, J., Liao, J., Zhong, M., Wang, W., Yao, S., 2012. Evidence of a dissociation pattern in resting-state default mode network connectivity in first-episode, treatment-naive major depression patients. Biol. Psychiatry 71, 611–617. https://doi.org/10.1016/j.biopsych.2011.10.035.

Zuo, X.-N., Kelly, C., Adelstein, J.S., Klein, D.F., Castellanos, F.X., Milham, M.P., 2010. Reliable intrinsic connectivity networks: test-retest evaluation using ICA and dual regression approach. Neuroimage 49, 2163–2177. https://doi.org/10.1016/j.neuroimage.2009.10.080.

Index

Note: Page numbers followed by *f* indicate figures, *t* indicate tables, and *b* indicate boxes.